# Handbook of Research on Applied Optimization Methodologies in Manufacturing Systems


Ömer Faruk Yılmaz
*Istanbul Technical University, Turkey & Yalova University, Turkey*

Süleyman Tüfekçí
*University of Florida, USA*



A volume in the Advances in Logistics, Operations, and Management Science (ALOMS) Book Series

IGI Global
DISSEMINATOR OF KNOWLEDGE

Published in the United States of America by
IGI Global
Business Science Reference (an imprint of IGI Global)
701 E. Chocolate Avenue
Hershey PA, USA 17033
Tel: 717-533-8845
Fax: 717-533-8661
E-mail: cust@igi-global.com
Web site: http://www.igi-global.com



Library of Congress Cataloging-in-Publication Data

Names: Yilmaz, Omer Faruk, 1989- editor. | Tufekci, Suleyman, editor.
Title: Handbook of research on applied optimization methodologies in manufacturing systems / Omer Faruk Yilmaz and Suleyman Tufekci, editors.
Description: Hershey, PA : Business Science Reference, [2018] | Includes bibliographical references.
Identifiers: LCCN 2017012416| ISBN 9781522529446 (hardcover) | ISBN 9781522529453 (ebook)
Subjects: LCSH: Manufacturing processes--Mathematical models--Handbooks, manuals, etc. | Mathematical optimization--Handbooks, manuals, etc. | Heuristic algorithms--Handbooks, manuals, etc.
Classification: LCC TS183 .H3595 2018 | DDC 670--dc23 LC record available at https://lccn.loc.gov/2017012416

This book is published in the IGI Global book series Advances in Logistics, Operations, and Management Science (ALOMS) (ISSN: 2327-350X; eISSN: 2327-3518)

British Cataloguing in Publication Data
A Cataloguing in Publication record for this book is available from the British Library.

# Advances in Logistics, Operations, and Management Science (ALOMS) Book Series

John Wang
Montclair State University, USA

ISSN:2327-350X
EISSN:2327-3518

## Mission

Operations research and management science continue to influence business processes, administration, and management information systems, particularly in covering the application methods for decision-making processes. New case studies and applications on management science, operations management, social sciences, and other behavioral sciences have been incorporated into business and organizations real-world objectives.

The **Advances in Logistics, Operations, and Management Science** (ALOMS) Book Series provides a collection of reference publications on the current trends, applications, theories, and practices in the management science field. Providing relevant and current research, this series and its individual publications would be useful for academics, researchers, scholars, and practitioners interested in improving decision making models and business functions.

## Coverage

- Marketing engineering
- Networks
- Risk management
- Services management
- Finance
- Operations management
- Information management
- Decision analysis and decision support
- Organizational behavior
- Production management

IGI Global is currently accepting manuscripts for publication within this series. To submit a proposal for a volume in this series, please contact our Acquisition Editors at Acquisitions@igi-global.com or visit: http://www.igi-global.com/publish/.

The Advances in Logistics, Operations, and Management Science (ALOMS) Book Series (ISSN 2327-350X) is published by IGI Global, 701 E. Chocolate Avenue, Hershey, PA 17033-1240, USA, www.igi-global.com. This series is composed of titles available for purchase individually; each title is edited to be contextually exclusive from any other title within the series. For pricing and ordering information please visit http://www.igi-global.com/book-series/advances-logistics-operations-management-science/37170. Postmaster: Send all address changes to above address. 

## Titles in this Series

*For a list of additional titles in this series, please visit: www.igi-global.com/book-series*

*Novel Six Sigma Approaches to Risk Assessment and Management*
Vojo Bubevski (Independent Researcher, UK)
Business Science Reference • copyright 2018 • 251pp • H/C (ISBN: 9781522527039) • US $200.00 (our price)

*Enterprise Resiliency in the Continuum of Change Emerging Research and Opportunities*
Raj Kumar Bhattarai (Tribhuvan University, Nepal)
Business Science Reference • copyright 2018 • 186pp • H/C (ISBN: 9781522526278) • US $150.00 (our price)

*Examining Cultural Influences on Leadership Styles and Learning From Chinese Approaches to Management Emerging Research and Opportunities*
Valerie Zhu (Xi'an University of Science and Technology, China)
Business Science Reference • copyright 2017 • 207pp • H/C (ISBN: 9781522522775) • US $125.00 (our price)

*Globalization and the Ethical Responsibilities of Multinational Corporations Emerging Research and Opportunities*
Tarnue Johnson (Argosy University - Chicago, USA)
Business Science Reference • copyright 2017 • 110pp • H/C (ISBN: 9781522525349) • US $125.00 (our price)

*Multi-Criteria Decision Making for the Management of Complex Systems*
Albert Voronin (National Aviation University of Ukraine, Ukraine)
Business Science Reference • copyright 2017 • 201pp • H/C (ISBN: 9781522525097) • US $175.00 (our price)

*Handbook of Research on Manufacturing Process Modeling and Optimization Strategies*
Raja Das (VIT University, India) and Mohan Pradhan (Maulana Azad National Institute of Technology, Bhopal, India)
Business Science Reference • copyright 2017 • 530pp • H/C (ISBN: 9781522524403) • US $285.00 (our price)

*Managerial Strategies and Green Solutions for Project Sustainability*
Gilman C.K. Tam (Independent Researcher, China)
Business Science Reference • copyright 2017 • 255pp • H/C (ISBN: 9781522523710) • US $180.00 (our price)

*Optimal Management Strategies in Small and Medium Enterprises*
Milan B. Vemić (Higher School of Academic Studies "DOSITEJ", Serbia)
Business Science Reference • copyright 2017 • 437pp • H/C (ISBN: 9781522519492) • US $225.00 (our price)

IGI Global
DISSEMINATOR OF KNOWLEDGE

701 East Chocolate Avenue, Hershey, PA 17033, USA
Tel: 717-533-8845 x100 • Fax: 717-533-8661
E-Mail: cust@igi-global.com • www.igi-global.com

# Editorial Advisory Board

# List of Contributors

# Table of Contents

# Detailed Table of Contents



**Section 1**
**Applications of Heuristic and Metaheuristic Algorithms in Manufacturing Systems**

*This section discusses several heuristic and metaheuristic algorithms to manage and optimize the problems in manufacturing systems and examines two main levels of decision systems in production: the middle term (tactical) and short term (operational). As each level encounters specific problems, appropriate approaches to deal with these are introduced and explained. These problems include the scheduling in flexible manufacturing systems, the project scheduling, the optimization of extrusion manufacturing process, the sequencing in assembly lines, the hub location and allocation, the clustering, the order release, the worker assignment, the batch scheduling, and the multi-objective scheduling in hybrid manufacturing systems.*



Flexible manufacturing systems have many advantages like adaptation to changes and reduction of lateness. But flexible machines are expensive. The scheduling is a central functionality in manufacturing systems. Optimizing the job routing through the system, while taking advantage from the flexibility of the machines, aims at improving the system's profitability. The introduction of the flexibility defines a variant of the scheduling problems known as flexible job shop scheduling. This variant is more difficult than the classical job shop since two sub-problems are to be solved the assignment and the routing. To guarantee the generation of efficient schedules in reasonable computation time, the metaheuristic approach is largely explored. Particularly, much research has addressed the resolution of the flexible job shop problem by genetic algorithms. This chapter presents the different adaptations of the genetic scheme to the flexible job shop problem. The solution encodings and the genetic operators are presented and illustrated by examples.

A car factory contains three main workshops; body shop, paint shop and assembly shop. Each of these three workshops has their set of constraints which have to be met in a production day by arranging the vehicles. The car sequencing problem is used to create a production sequence that meets these constraints. Car sequencing problem first handled in the literature by optimization of assembly constraints including ratio constraints. After that, color constraints are integrated to assembly constraints. At this chapter, the scenario in which high priority ratio constraints are primary, color constraints are secondary is tackled and a heuristic approach is proposed. For optimization of ratio constraints, an initial algorithm based on the greedy algorithm is used. The developed algorithm is coded and used on data set which is proposed by Renault at the ROADEF'2005 challenge. According to results, it is achieved the range of results which is achieved by ROADEF finalists.



The Hub Location-Allocation Problem is one of the most important topics in industrial engineering and operations research, which aims to find a form of distribution strategy for goods, services, and information. There are plenty of applications for hub location problem, such as Transportation Management, Urban Management, locating service centers, Instrumentation Engineering, design of sensor networks, Computer Engineering, design of computer networks, Communication Networks Design, Power Engineering, localization of repair centers, maintenance and monitoring power lines, and Design of Manufacturing Systems. In order to define the hub location problem, the present chapter offers two different metaheuristic algorithms, namely Particle Swarm Optimization or PSO and Differential Evolution. The presented algorithms, then, are applied to one of the hub location problems. Finally, the performances of the given algorithms are compared in term of benchmarking.



Clustering is an approach used in data mining to classify objects in parallel with similarities or separate according to dissimilarities. The aim of clustering is to decrease the amount of data by grouping similar data items together. There are different methods to cluster. One of the most popular techniques is K-means algorithm and widely used in literature to solve clustering problem is discussed. Although it is a simple and fast algorithm, there are two main drawbacks. One of them is that, in minimizing problems, solution may trap into local minimum point since objective function is not convex. Since the clustering is an NP-hard problem and to avoid converging to a local minimum point, several heuristic algorithms applied to clustering analysis. The heuristic approaches are a good way to reach solution in a short time. Five approaches are mentioned briefly in the chapter and given some directions for details. For an example, particle swarm optimization approach was used for clustering problem. In example, iris dataset including 3 clusters and 150 data was used.



There are three main problems that could impact the performance of a Hybrid Manufacturing System (HMS): (1) order release (OR), (2) batch scheduling and (3) worker assignment. This paper deals with these three main problems hierarchically for an HMS. Three different mathematical models are developed to describe the problems more clearly. A novel methodology is proposed to adopt a holistic approach to these problems and find an effective solution. Implementation of the proposed methodology permits integrating batch scheduling and worker timetabling. Feasible solutions in the best-known Pareto front are evaluated as alternative solutions. The goal is to select a preferred solution that satisfies worker constraints, creates effective worker teams in cells, minimizes the number of utility workers, and the average flow time. The study also presents several improvements, which are made following the application of the proposed methodology to a real company that produces expansion joints.



Problems encountered in real manufacturing environments are complex to solve optimally, and they are expected to fulfill multiple objectives. Such problems are called multi-objective optimization problems(MOPs) involving conflicting objectives. The use of multi-objective evolutionary algorithms (MOEAs) to find solutions for these problems has increased over the last decade. It has been shown that MOEAs are well-suited to search solutions for MOPs having multiple objectives. In this chapter, in addition to comprehensive information, two different MOEAs are implemented to solve a MOP for comparison purposes. One of these algorithms is the non-dominated sorting genetic algorithm (NSGA-II), the effectiveness of which has already been demonstrated in the literature for solving complex MOPs. The other algorithm is fast Pareto genetic algorithm (FastPGA), which has population regulation operator to adapt the population size. These two algorithms are used to solve a scheduling problem in a Hybrid Manufacturing System (HMS). Computational results indicate that FastPGA outperforms NSGA-II.

## Section 2
## Supply Chain and Inventory Management

*This section reveals the principles of Supply Chain and Inventory Management. The first chapter deals with the pricing, the lotsizing and the shipment for a two-echelon supply chain. Metaheuristic algorithms are used in the first chapter to solve the problem for two-echelon supply chain. The second chapter considers the key supply chain risks which could cause abnormalities and occur from rapid changes in customer demand, unpredictable price fluctuations, defect variations and delivery delays and provides the correction of these problems automatically. The third chapter focuses the components which help to constitute a supply chain strategy and classify the supply chain strategies described in the literature.*



A two-echelon supply chain is studied that involves a retailer who faces demand from two or more market segments and enable to set different prices and marketing expenditures and a supplier who desires to find optimal number of shipments through an integrated system. A new mixed-integer non-linear fractional programming (MINLFP) model is developed. In order to solve the resultant MINLFP model, the constrained non-linear programming model is reformulated as an unconstrained one using penalty terms. Two meta-heuristics, namely simulated annealing (SA) and imperialist competitive algorithm (ICA), are applied to solve the relaxed unconstrained model. Numerical results show that ICA can reach better solutions in comparison with SA. However, SA has the ability of providing more robust solutions which are converged to a good solution. The chapter concludes with superiority of SA.



Supply chain management paradigms are becoming increasingly common management perspectives all over the world due to violent global competition of trade organizations and rapid changes in technology. In recent years, thanks to the communication improvements, customers have become more conscious about purchasing goods or services. Furthermore, organizations have to be customer oriented and more flexible against the dynamism of supply chain environment which increases uncertainties in supply chain parameters. Although a considerable amount of risk factors appearing in supply chain operations, this study concentrates on detecting key supply chain risks which could cause abnormalities and occur from rapid changes in customer demand, unpredictable price fluctuations, defect variations and delivery delays and provides the correction of these problems automatically. Thus, a system dynamics model is established for determining risks. This combined approach would be helpful for integrated supply chain risk management.

Uncertainties and unpredictability in the market force companies to develop strategies which enable them to perform better than their competitors. Developing proper strategies for a supply chain is crucial. Strategies are affected by the nature of the firm's products or services, customer preferences, operations, process design of the firm, etc. Companies should form adaptive supply chain strategies which enable them to be resilient and flexible enough in the flow of materials, products, information, and money along the supply chain. There are many studies about supply chain management and supplier selection in the literature. However, the number of studies about the selection of the right supply chain strategy are very limited. This study presents the components which help to constitute a supply chain strategy and classify the supply chain strategies described in the literature. Lastly, it offers a strategy and criteria matrix which can be used as a road map for selecting the most appropriate supply chain strategy by firms.



*This section introduces most used techniques that increase the efficiency of manufacturing systems. These techniques include the mathematical models, the simulation methods, the optimal control theory, the lean manufacturing principles, and the multi-criteria decision making.*



This chapter initially presents a brief information about production systems. At these systems, different types of maintenance policies are developed to cope with wear out failures. Mainly used maintenance policies can be classified as corrective, preventive, and condition-based maintenance. In the corrective maintenance, repair or replacement is applied whenever components of the machine breakdown. In the preventive maintenance approach maintenance activities are applied to the critical components on a periodic basis. On the other hand, maintenance activities are applied whenever critical reliability level is reached or exceeded. These types of maintenance policies are modeled using mathematical modeling techniques such as linear programming, goal programming, dynamic programming, and simulation. A review of current literature about the mathematical models, the simulation-based optimization studies examining these maintenance policies are categorized and explained. Besides, the solution methodologies are discussed. Finally, the opportunities for future research are presented.

The considered problem (P) concerns the production of strains (also called jobs or batches), which are the used components in the final products that are bought by the consumers. (P) contains two components that have to be tackled sequentially: the inventory management problem (IMP) and the job scheduling problem (JSP). (IMP) is solved with a reorder-point policy, defined on the basis of critical demand coverage. To tackle (JSP), a descent local search (DLS) is used, based on swap moves. In other words, for a given job sequence, a series of modifications is performed on it in order to try to improve the solution, where each modification consists of exchanging the positions of two jobs. Because of random events (some jobs might be rejected if they do not meet predefined standards) and stochasticity (the duration of each job follows a normal distribution), simulation is required to evaluate any sequence of jobs that is a solution to (JSP). A simulation-optimization approach is therefore proposed to accurately tackle (JSP). This work is motivated by a real pharmaceutical company.



This chapter addresses some issues related to some Optimal Control Theory (OCT) problems (for example, impossible analytical solution because of an unsolvable integral, or punctual parameters that were unrealistic). It is proposed the use of OCT as a benchmarking tool to analyze inventory control systems to enhance parameters. In addition, the application of methods and heuristics in solving these problems is also described. These methods are discussed and applied in calculating the production and inventory functions using data of accounting variables of USA and Brazil companies, available in the Economatica software data base. Eventually, the results are compared and some recommendations about the advantages and disadvantages of each method are accomplished.



In a system though the analog circuits occupy very less space but they require far more design time than the digital circuits. This is due to the fact that the number of performance measures of an analog circuit is more than those for digital circuits. Predicting and improving the performance, robustness and overall cost of such systems is a major concern in the process of automation. In the automation process, optimization of performances subjected to a verity of environmental constraints is a central task. In this chapter, efficient analog circuit sizing techniques and their optimization are surveyed.

Lean has become a new mantra in today's manufacturing sector. In this millennium, companies are facing a challenge to be economically competitive in manufacturing. Many of them have realized that the old style of mass manufacturing is no longer successful. Hence, lean manufacturing is being embraced by the companies to simultaneously achieve a competitive edge and economic growth. Many studies have shown that lean organizations are capable of meeting customer's expectations consistently, at each step of the production systems. Lean manufacturing also implies efficient use of non-renewable resources in order to maintain a sustainable environment. To reach the full potential of an organization, lean must be embraced as a holistic business strategy. In this chapter, the history of lean innovation will be briefly discussed, followed by the principles of lean manufacturing and various tools in implementing lean practices. Examples of organizations that have experienced significant improvements once transformed to lean manufacturing will also be cited.



*Smart Factory concepts describe fully networked, autonomous factories and form an essential part of flexible, however still highly efficient manufacturing systems. From a business perspective, the term industry 4.0 stands for a new organizational step of controlling the entire value chain along the product life cycle. The requirements for the further development of existing manufacturing systems towards a Smart Factory are analyzed and studied in this section.*



Smart Factory concepts describe fully networked, autonomous factories and form an essential part of flexible, but still highly efficient production systems. The requirements for the further development of existing production environments towards a Smart Factory are multidimensional and vastly complex. Many companies therefore fail in the structured realization of a holistic Smart Factory concept. They either focus one dimension of the challenge or merely address the maximum penetration of powerful technologies. This chapter addresses this issue and describes a systematic development path towards a Smart Factory by means of a domain specific maturity model. Based on the analysis of existing maturity models, requirements are derived which must be considered when realizing a Smart Factory. In total, 20 design fields (e.g., degree of intelligence, communication protocols, human-machine-interface and IT security) and respective detail descriptions result from this research. They holistically structure the relevant fields of action to pursue a Smart Factory.

The new form of future generation machines and automated systems could be synchronized by IoT adaptation. By this way, a very large size data can be carefully stored in data repositories and have to be analyzed for extracting knowledge. Thus, optimization techniques are becoming invaluable tools for finding patterns from parallel distributed machines. On the other hand, statistical methods and optimization models could not be utilized efficiently due to excessive dimension of data. Additionally, data analytics should be applied and results should be gathered by using practical approaches especially for security, access control and fault detection issues. In this study, optimization techniques are evaluated in the perspective of big data analytics and both mathematical and statistical methods will be extensively analyzed for different versions of problem solving and decision making in Industry 4.0 era.

# Foreword

The manufacturing process and its wide supporting activities from planning to delivery are in need of faster, more robust and more accurate decisions and actions than ever, especially with the introduction of Industry 4.0. Concepts such as big data, deep learning, etc. are evolving and finding their applications in all industries. This evolution may remain at a decision-making level for traditional job shops. Intelligent algorithms aid decisions in inventory control, production scheduling and sequencing, worker/machine assignment, job routing, lot sizing, etc. As the traditional Integer Programming solving methods may require excessive time to finalize their results, problem-based heuristic and metaheuristic methods lend a crucial hand for finalizing a decision in a reasonable time. One of the most fundamental metaheuristics is Genetic Algorithm (GA) which aids optimization problems with a global search scheme. Particle Swarm Optimization (PSO) embodies a similar approach with a little more local search concentration. Since GA and PSO, numerous metaheuristics are offered to balance the global and the local search on the optimization problem space. Some latest metaheuristics are Flower Pollination Algorithm (FPA) and Bee-Based Algorithms (BBA), which are known to overcome disadvantages of these algorithms such as slow or early convergence, sticking to local optima. The application of these algorithms involve but are not limited to scheduling, obtaining optimal process parameters, lot sizing, depot locating problems. However, as the related literature points, there is no free lunch for decision making in manufacturing systems. Heuristics and metaheuristics provide a shorter route for an ultimate decision; yet, the optimality of this decision is never guaranteed. On one side these algorithms provide "more than good enough" solutions when exact solution methods fail to provide a solution, and on the other side, the search for the global optimum continues. However computationally costly, exact solution algorithms are fundamental for reaching the global optimum, especially in large-scale problems. Decomposing the larger problem into small and manageable sub-problems and solving each sub-problem using exact solution methods is also an approach for handling similar manufacturing issues. Furthermore, in the presence of qualitative data, Multi-Criterion Decision Making methods also aid in the planning phases of manufacturing. Furthermore, these approaches assume the demand and other parameters to be known beforehand with certainty. However, in case of not producing to order, the demand is typically unknown and mostly assumed to be stochastic. In such cases, simulation serves as a proper method for optimization.

The evolution in technology also may extend to the physical levels of the company such as smart factories, RFID, e-commerce, etc. The traditional conflict in manufacturing systems is that as product variety increases, the manufacturing speed decreases. Smart factories target a flexible and rapid and a safe "manufacturing ecosystem". The implementation of the state-of-the-art technology is crucial for this notion; hence, sensors, robots or cobots (robots with cognitive abilities) and wireless information transmission are issues that are currently emergent, unlocking a vast area for researchers. The very nature of this research area is a fountain of data for Big Data Analytics. Statistical and Machine Learning methods in planning and operations of smart factories.

The objectives of manufacturing-system based optimization problems differ from industry or the problem at hand. The most common objectives in manufacturing problems involve minimizing cost, maximizing profit, minimizing the flow time, maximizing quality. The recent addition to these objectives are green objectives such as minimizing waste. These objectives also depend on the planning horizon. For example, green objectives are mostly exploited in problems involving multiple periods or a rolling horizon. Moreover, most of these objectives are conflicting which leads to either exact or heuristic multi-objective solution methods.

In this book, a wide range of manufacturing problems and their recent advances are addressed with various methods. It starts with a very introductory application of parameter optimization in a cold extrusion process with an approach of a very recent metahuristic: Flower Pollination Algorithm. Manufacture of some products are especially more delicate; analog circuit board production serves as a good example for product-based manufacturing systems. A review of optimization method in analog circuit board production provides insights for such manufacturing systems. As data analytics become more and more important, an overview of PSO clustering is presented, which is especially beneficial in cellular manufacturing systems. Being probably the most fundamental managerial challenge of manufacturing, the scheduling problem is dealt with a number of approaches and in different environments. The conventional job shop scheduling problem is handled with a novel greedy heuristic. Metaheuristic approaches dominate the world of scheduling due to the large size of the real-world cases. Bee-Based metaheuristic algorithm approaches in literature are reviewed and a guideline is presented for readers of interest. Aside from traditional manufacturing conditions, flexible and hybrid manufacturing environments are also analyzed. Scheduling in a flexible manufacturing system is achieved with a GA that maximizes profitability. Scheduling in a hybrid manufacturing system is dealt with a multi-objective GA, that is the NSGA-II algorithm. Scheduling in a hybrid manufacturing system is also extended with a worker assignment and order release echelons with a holistic approach.

Inventory management and lot sizing problems are the main supporting decisions of a manufacturing environment. Due to the complicated nature of the mathematical formulation of the exact situations, inventory model problems may be solved directly or may require approximation methods. One chapter demonstrates how complex inventory management problems can benefit from Optimal Control Theory. One of the recent approaches of the last decade is the use of RFID systems in inventory control, and optimization of an RFID-controlled inventory system is achieved through Fuzzy Programming. Lot sizing is a subset of inventory management problems and determines the size of the production/purchase orders. In general, production orders involve two sides or in technical terms, echelons: the producer and the customer, which have different constraints or objectives. A Mixed-Integer Nonlinear and Fuzzy Programming model is built to analyze the trade-off between these two echelons. In the real-world applications, lot sizing and scheduling go hand in hand, as scheduling uses the resources obtained in the lot sizing phase. A pharmaceutical industry example is solved using simulation under stochastic conditions. Other supporting activities involve the retrieval and delivery of goods and products. Supplier selection problem is important for directly affecting the production process and evaluating suppliers may involve qualitative criteria. Hence, a multi-criteria approach is presented for selecting the best supplier. Hub allocation problem deals with both delivery and retrieval of goods. In this book, a PSO and a Differential Evolution approach are introduced and their results are compared using a real-world problem. For the most recent topics in manufacturing, Data Analytic techniques used in Industry 4.0. are reviewed and a holistic approach for smart factories is presented.

This book combines the traditional yet still ongoing manufacturing challenges with the ones of state-of-the-art. The reader can obtain a grasp on the traditional manufacturing systems and the reflections of these problems on the recent solution approaches, as well as on the recent advances and the future of these systems. It is also an insightful guide for those who pursue research on manufacturing from a managerial perspective.

*Ayça Altay*
*Rutgers University, USA*

# Preface

Manufacturing systems of today and the near future are going through significant changes both in planning and executing the production, receiving raw material input to the manufacturing system and delivering the final products to markets for the customers. Managing these complex manufacturing systems in harmony and coordination in an optimal way is becoming a core issue for management to continue to survive in a globally very competitive world. To make the issue more complex, an enormous amount of raw data is collected in today's manufacturing systems from end to end through various sources including data collected through electronic means, internet, sensors, etc. This very large data needs to be converted into useful information which can be fed into various optimization models used in manufacturing systems.

Applied Optimization in Manufacturing Systems is an ongoing field of research and development. The ultimate goal of all optimization models developed for manufacturing systems is to provide applicable solutions for various aspects of the manufacturing systems, including production planning, supply chain planning, quality control, just-in-time manufacturing, cost minimization, shorter manufacturing times, etc. Decisions such as when to order raw material, how much to order, which delivery channels to use, how to schedule production, how much to produce, how to deliver the final products to customers, etc., are some of the questions today's manufacturing managers are faced with every day.

The goal of Applied Optimization methodology is to provide useful and usable data driven optimization models for the problems of manufacturing mentioned above. Some of the important characteristics of these problems are nonlinearity of performance metrics, too many variables affecting the performance, too many restriction (or constraints) that makes decision process more complex, and some decision parameters can only be selected from a discrete set which itself can be extremely large such as integer choices for certain decision parameters. On top of all these complexities stochastic nature of some elements of the manufacturing process makes decision making a challenging task.

This book provides an end-to-end optimization models including optimization in supply chain, scheduling in manufacturing and assembly process, optimization of maintenance process, risk management in manufacturing and extensive heuristic models for solving manufacturing and supply chain related problems in manufacturing. The book provides up to date survey of the literature in heuristic models used in clustering problems in manufacturing, sequencing of cars in automobile assembly lines, multi-objective decision models in scheduling manufacturing process, and risk management models in e-commerce supply chains. The models provided in this book will be extremely useful to our intended audience, practitioners in the manufacturing ad supply chain area as well as researcher joining this vast field of optimization models in manufacturing systems.

# Preface

Design and development of mathematical/heuristic/simulation models for various aspects of decision making in manufacturing/supply chain systems is an ongoing commitment by the scientific community. The fast-changing technology and its adoption to manufacturing/supply chain processes makes some of the older developed mathematical/optimization/simulation models either insufficient, obsolete or incapable of solving the current problems emerging due to changing technologies and/or increased demand on number of decision variables, the wider range the variables may operate and the number of constraints that will be imposed on the problem. Application of these models to real problems in industry depends very much on the availability of an appropriate model. Given the existence of appropriate model(s) the second important aspect of applicability is the availability of the appropriate software model which the practitioners can easily obtain for implementation. In many real-life applications, the customized software development follows the building of the appropriate decision model for the problem under review. This book contains chapters in modeling manufacturing systems, design and application of optimization models for particular industries and up to date literature survey in some areas of manufacturing, scheduling and supply chain fields.

With this book, we intended to reach to a wide and complementary group of the population including researchers in the field of optimization, simulation, manufacturing management and practitioners who are looking for solutions to their daily decision problems in the areas of manufacturing, scheduling, supply chain management, project management, risk management and other decision problems emerging in manufacturing systems.

As has been shown in many publications the decision problems in manufacturing and supply chain field is complex in size, involving decision parameters, nonlinearity in relationships between these parameters and discrete nature of some of the decision variables makes the developed mathematical optimization models very difficult or impossible to solve optimally within a reasonable computer time. Most of these problems are categorized as NP-hard problems. Implying that it will not be possible to develop algorithms with polynomial complexity in the size of the problem which guarantees the optimal solution. Therefore, researchers have focused on developing heuristic solution algorithms to get optimal or near optimal solutions to these intractable optimization models with minimal computational burden. There has been a significant number of heuristic solution procedures for some of the manufacturing/supply chain related decision problems in the last two decades. Some of these heuristic algorithms get their inspiration from the nature. Among these nature-inspired heuristic algorithms we can name Flower Pollination, Particle Swarm, Genetic, and Ant Colony algorithms. Simulated Annealing, Tabu Search, Clustering and K-means are also widely used heuristics for solving many decision problem in manufacturing and supply chain problems.

Production scheduling and inventory management in job shops, assembly lines, cellular manufacturing systems and hybrid manufacturing systems has been one of the most popular areas for research. This is primarily due to the continuous and rapid change in manufacturing systems infrastructures and modern technologies making some of the older decision models obsolete. Thus, the need to develop new and more comprehensive and realistic models capable of solving today's decision problems in manufacturing and supply chain management.

## ORGANIZATION OF THE BOOK

This book is organized in 18 chapters. A brief summary of each chapter follows.

Chapter 1 discusses earlier approaches on flexible job shop scheduling. Optimizing the job routing through a flexible manufacturing system, while taking advantage from the flexibility of the machines, aims at improving the system's profitability. The introduction of the flexibility defines a variant of the scheduling problems known as flexible job shop scheduling. This added flexibility creates more difficult scheduling problem than the classical job shop. The two sub-problems, the assignment and routing problems, needs to be solved simultaneously. To guarantee the generation of efficient schedules in reasonable computation time, the metaheuristic approach is largely explored. Earlier research has addressed the resolution of the flexible job shop problem by genetic algorithms. This chapter presents the different adaptations of the genetic scheme to the flexible job shop problem. The solution encodings and the genetic operators are presented and illustrated by examples.

Chapter 2 discusses the application and evaluation of bee-based algorithms (BBAs) in scheduling. Most complex scheduling problems that needs to be tackled with in manufacturing environment can be classified as NP-hard problems. Finding the optimal solution for multi-variable scheduling problems with polynomial computation times is extremely hard. Scheduling problems of this nature can be solved up to certain problem size by using these traditional methodologies. In order to get optimal or near optimal solutions to larger size problems we need new intelligent optimization tools such as BBA. BBAs are inspired by the food foraging behavior of honey bees and capable of locating optimal or near optimal solutions efficiently. The experiments on some benchmark problems show that BBA outperforms other heuristic methods which are used solve scheduling problems in terms of the speed of optimization and accuracy of the results. This chapter first highlights the use of BBA and its variants for scheduling and provides classification of scheduling problems with BBA applications. Following this, a step by step example is provided for multi-mode project scheduling problem in order to show how a BBA algorithm is implemented.

Chapter 3 provides metaheuristic approaches for extrusion manufacturing process. It is very important to find the optimal values of the decision parameters for an extrusion process in order to prevent waste. In this work two metaheuristic algorithms, Flower Pollination Algorithm (FPA) and Particle Swarm Optimization (PSO) are used to replace the traditional trial and error method to optimize the cold forward extrusion process. It is shown in this work that the optimization performance of the FPA is comparable to the performance of the PSO.

Chapter 4 provides a heuristic approach to sequencing problem including assembly ratio and color constraints in a car manufacturing factory. A car manufacturing factory contains three main workshops; body shop, paint shop and assembly shop. Each of these three workshops has their own set of constraints which have to be met in a production day by sequencing the cars in each of these shops. Earlier literature on car sequencing problem focused on optimization of assembly constraints including ratio constraints. In later literature color constraints are integrated to assembly constraints. In this chapter ratio constraints are treated as primary, color constraints are treated as secondary in the proposed heuristic approach. For optimization of ratio constraints, an initial algorithm based on greedy algorithm is used. The developed algorithm is coded and tested on data set which is proposed by Renault at the ROADEF'2005 challenge. The results of the proposed algorithm are comparable to the achievements obtained by ROADEF finalists.

Chapter 5 presents the Hub Location Allocation Problem (HLAP) and proposed solution algorithms in the literature. The Hub Location Allocation Problem deals with finding a form of distribution strategy for goods, services and information. There are plenty of applications for HLAP, including transportation management, urban management, locating service centers, instrumentation engineering, computer engineering, design of computer networks, communication networks design, power engineering, localization of repair centers, maintenance and monitoring power lines, and design of manufacturing systems. This chapter presents two different metaheuristic algorithms, namely Particle Swarm Optimization or PSO and Differential Evolution used in solving the HLAP. The presented algorithms, then, are applied to one of the hub location problems and their performances are compared.

Chapter 6 reviews the heuristic approaches in clustering problems. Clustering is an approach used in data mining to classify objects in parallel with similarities or separate according to dissimilarities. The aim of clustering is to decrease the amount of data by grouping similar data items together. There are different methods to cluster. A commonly used partitioned clustering method K-means algorithm is discussed. Clustering problems fall into the class of NP-hard problems. Therefore, heuristic approaches are a good way to reach solution in a short time. Five approaches are mentioned briefly in the chapter and given some directions for details. A numerical example has been provided for the particle swarm optimization approach for a clustering problem. In the example, iris dataset including 3 clusters and 150 data was used.

Chapter 7 addresses problems facing Hybrid Manufacturing Systems (HMS). There are three major decision needs to be made that could impact the performance of an HMS: (1) order release (OR), (2) part/batch scheduling and (3) worker assignment. This chapter deals with the order release, the batch scheduling and the worker assignment problems hierarchically in the HMS. Three different mathematical model are developed to describe the problems more clearly. A novel methodology is proposed to adopt a holistic approach to these problems and find an effective solution. Implementation of the proposed methodology permits integrating batch scheduling and worker timetabling. Feasible solutions in the best-known Pareto front are evaluated as alternative solutions. The goal is to select a preferred solution that meets worker constraints, creates effective worker teams in cells, minimizes the number of utility workers, and minimizes average flow time. The study also presents improvements made, following the application of the proposed methodology, for a scheduling horizon of one shift at a real company that produces expansion joints.

Chapter 8 reviews the evolutionary algorithms for multi-objective scheduling in hybrid manufacturing systems. Problems faced in the real manufacturing environments are complex and computationally expensive, and they are generally composed of multiple objectives. This chapter provides information on the general structures and fields of use of multi-objective evolutionary algorithms (MOEAs). Following this survey two different MOEAs are also used for an expensive multi-objective optimization problem (MOP). Two algorithms, the NSGA-II, the FastPGA, which have shown their effectiveness in MOPs are used in Hybrid Manufacturing Systems (HMSs) to solve multi-objective product scheduling problem. Their effectiveness on solving problems in hybrid manufacturing systems as well. Computational results indicate that the FastPGA algorithm outperforms the NSGAII algorithm.

Chapter 9 addresses the differential return on investment optimization pricing, lot sizing and shipment considerations in two-echelon supply chain. This chapter addresses them simultaneously aiming to maximize return on inventory investment. A two-echelon supply chain modeled involves a retailer who faces demand from two or more market segments and enable to set different prices and marketing expenditures and a supplier who desires to find optimal number of shipments through an integrated system. A new mixed-integer non-linear fractional programming (MINLFP) model is developed to determine optimal ordering, shipping and differential pricing and marketing expenditure quantities simultaneously. In order to solve the resultant MINLFP model, the constrained non-linear programming model is reformulated as an unconstrained one using penalty terms. Then a suitable penalty function is used which incurs a positive penalty for infeasible points and no penalty for feasible points.

Chapter 10 presents a comprehensive risk management tool based on multi agents and system dynamics for traditional and e-commerce supply chain. Supply chain management has gained a significant importance in the last two decades all over the world due to violent global competition of trade organizations and rapid changes in technology. Dramatic losses due to inefficient supply chain management support, supply chain risk management applications are sought after by many organizations. Decision making tools such as ERP, CRM etc. are not sufficient enough for detecting risk and making adjustments according to these risks. This chapter presents a comprehensive risk management tool based on multi agents and system dynamics for traditional and e-commerce supply chain.

Chapter 11 analyzes an optimal strategy selection in a supply chain. Uncertainties and unpredictability in business environment forces companies to develop strategies to perform better than their competitors. Strategies are affected by the nature of the firm's products or services, customer preferences, operations and process design of the firm etc. Companies should form adaptive supply chain strategies for their customers which enables to be resilient and flexible enough in the flow of materials, products, information and money along the supply chain. There are many studies about supply chain management and supplier selection in the literature. However, the number of studies about selection the right strategy in a supply chain is very limited. This chapter identifies the components of a supply chain which help in building an optimal supply chain strategy and classifies the supply chain strategies described in the literature. Finally, the chapter offers a strategy and criteria matrix which can be applied as a road map for selecting the most appropriate supply chain strategy by firms.

Chapter 12 presents brief information about production systems and different types of maintenance policies for the production systems. A review of current literature about the mathematical models, the simulation-based optimization studies examining these maintenance policies are categorized and explained. Besides, the solution methodologies are discussed, the authors presented the opportunities for future research directions.

Chapter 13 provides a simulation-optimization approach for the production of components for a pharmaceutical company. The considered problem concerns the production of strains (also called jobs or batches), which are the used in manufacturing the final product. Two decision problems need to be tackled sequentially: One is the inventory management problem (IMP) and the second is the job scheduling problem (JSP). (IMP) is solved with a reorder-point policy, defined on the basis of critical demand coverage. A descent local search (DLS) is used for solving JSP, based on swap moves. Because of random events (some jobs might be rejected if they do not meet predefined standards) and stochasticity (the duration of each job follows a normal distribution), simulation is required to evaluate any sequence of jobs that is a solution to (JSP). A simulation-optimization approach is therefore proposed to accurately tackle (JSP). This work is motivated by a real pharmaceutical company.

Chapter 14 presents the use of optimal control theory as a benchmarking tool in production-Inventory Systems. This chapter addresses some difficulties related to Optimal Control Theory (OCT) problems (for example, impossible analytical solution because of an unsolvable integral). It also addresses the application of many methods and heuristics in solving these problems. Particularly application of these methods in production and inventory control problems are discussed implemented on some US and Brazilian companies whose data were available in the software data base Economatica. Finally the results are compared and recommendations about the advantages and disadvantages of each method are provided.

Chapter 15 reviews recent advances in analog integrated circuit optimization. Although the analog circuits occupy very small space in electronic components, they require far more design time than the digital circuits. Predicting and improving the performance, robustness and overall cost of such systems is a major concern in the process of automation. This chapter provides efficient analog circuit sizing techniques and their optimization are surveyed.

Chapter 16 discusses the history of lean innovation and describes the principles of lean manufacturing and various tools in implementing lean principles. Examples of organizations that have experienced significant improvements once transformed to lean manufacturing are also cited in the chapter.

Chapter 17 presents a maturity model to organize the multidimensionality of digitalization in industrial facilities. Smart factory concepts describe fully networked, autonomous factories and form an essential part of flexible, but still highly efficient production systems. The requirements for the transition of existing production environments towards a smart factory are multidimensional and vastly complex. Many companies therefore fail in the structured realization of a holistic smart factory concept. They either focus one dimension of the challenge or merely address the maximum penetration of powerful technologies. This chapter addresses these issues and describes a systematic development path towards a smart factory by means of a domain specific maturity model. In total, 20 design fields (e.g. degree of intelligence, communication protocols, human- machine-interface and it security) and respective detail descriptions result from this research. They holistically structure the relevant fields of action to pursue a smart factory.

Chapter 18 reviews the Data Analytics in Industry 4.0. The new form of future generation machines and automated systems could be synchronized by IoT adaptation. This adaptation creates a huge size of data. This data should be carefully stored in data repositories and being extracted to be analyzed for creating information or input to various decision models. Thus, optimization techniques are invaluable tools for finding patterns from parallel working and distributed machines. On the other hand, statistical methods and optimization models are not be able to work efficiently due to the dimension of data. Additionally, data analytics should be applied and results should be gathered by using practical approaches especially for security, access control and fault detection issues. In this chapter, optimization techniques are evaluated in the perspective of big data analytics and both mathematical and statistical methods are presented in detail for helping the decision making in Industry 4.0 era.

# Acknowledgment

The editors would like to acknowledge the help of all the people involved in this project and, more specifically, to the authors and reviewers that took part in the review process. Without their support, this book would not have become a reality.

First, the editors would like to thank each one of the authors for their contributions. Our sincere gratitude goes to the chapter's authors who contributed their time and expertise to this book.

Second, the editors wish to acknowledge the valuable contributions of the reviewers regarding the improvement of quality, coherence, and content presentation of chapters. Most of the authors also served as referees; we highly appreciate their double task.

*Ömer Faruk Yılmaz*
*Istanbul Technical University, Turkey & Yalova University, Turkey*

*Süleyman Tüfekçi*
*University of Florida, USA*

# Section 1

# Applications of Heuristic and Metaheuristic Algorithms in Manufacturing Systems

*This section discusses several heuristic and metaheuristic algorithms to manage and optimize the problems in manufacturing systems and examines two main levels of decision systems in production: the middle term (tactical) and short term (operational). As each level encounters specific problems, appropriate approaches to deal with these are introduced and explained. These problems include the scheduling in flexible manufacturing systems, the project scheduling, the optimization of extrusion manufacturing process, the sequencing in assembly lines, the hub location and allocation, the clustering, the order release, the worker assignment, the batch scheduling, and the multi-objective scheduling in hybrid manufacturing systems.*

# Chapter 1
# Scheduling in Flexible Manufacturing Systems:
## Genetic Algorithms Approach

**Fraj Naifar**
*Digital Research Center of Sfax, Tunisia*

**Mariem Gzara**
*University of Monastir, Tunisia*

**Taicir Loukil Moalla**
*Tabuk University, Saudi Arabia*

## ABSTRACT

*Flexible manufacturing systems have many advantages like adaptation to changes and reduction of lateness. But flexible machines are expensive. The scheduling is a central functionality in manufacturing systems. Optimizing the job routing through the system, while taking advantage from the flexibility of the machines, aims at improving the system's profitability. The introduction of the flexibility defines a variant of the scheduling problems known as flexible job shop scheduling. This variant is more difficult than the classical job shop since two sub-problems are to be solved the assignment and the routing. To guarantee the generation of efficient schedules in reasonable computation time, the metaheuristic approach is largely explored. Particularly, much research has addressed the resolution of the flexible job shop problem by genetic algorithms. This chapter presents the different adaptations of the genetic scheme to the flexible job shop problem. The solution encodings and the genetic operators are presented and illustrated by examples.*

## INTRODUCTION

Flexible manufacturing systems are characterized by multipurpose operations and flexible job routing. A multipurpose operation can be processed at least by one machine, with possibility of variable performances. In ordinary systems, the route of every job is fixed and every operation of a job is allocated to

DOI: 10.4018/978-1-5225-2944-6.ch001

a unique machine. In flexible manufacturing, an operation can be allocated to a suitable machine from a set of alternatives which are identical or similar in functionalities. One can distinguish two types of flexible manufacturing systems: totally flexibility where every operation can be processed by every machine and partially flexible systems where a pool of machines is associated to each operation (Figure 1).

A fabrication order can evolve in the system through different paths. It can visit the same machine more than once. The flexibility has demonstrated its efficiency for system performance by reducing latencies and work in progress. Flexible machines allow rapid adaptation to changs and re-routing flexibility in presence of breakdowns or bottlenecks. Flexible manufacturing systems have a high development cost since flexible machines are more expensive than the mono-purpose ones. The scheduling functionality is a central key of profitability in flexible manufacturing systems. The allocation in space (machines) and in time of parts processing have to be optimized to take advantage from machine flexibility.

The use of flexible machines defines a generalization of the scheduling problem known as the flexible job shop problem (FJSP). This problem has attracted many researches. Due to the combinatorial number of possible schedules and the strongly NP-hard nature of FJSP (Garey et al., 1976), exact methods that can guarantee solution optimality will not return results in a reasonable amount of time. Indeed, a significant attention has been paid in the literature to techniques of achieving efficient approximation algorithms which offer a good compromise between computation time and solutions qualities including genetic algorithms, tabu search, simulated annealing among other heuristics and metaheuristics.

Actually, much research literature addressed the genetic algorithm approach to solve the FJSP because of their ability to perform global search and provide good solutions in a short computation time. Otherwise, crossover and mutation operators in a genetic process can easily combine affectation schemes and adopt different sequencing strategies to conduct an effective parallel and sampled search to intelligently locate promising area in the solution space. The objective of this chapter is to present how the genetic resolution scheme was adapted to solve the flexible job shop problems. The solution encoding and the genetic operators, crossover and mutation, are described and illustrated with examples.

*Figure 1. Flexible manufacturing systems*

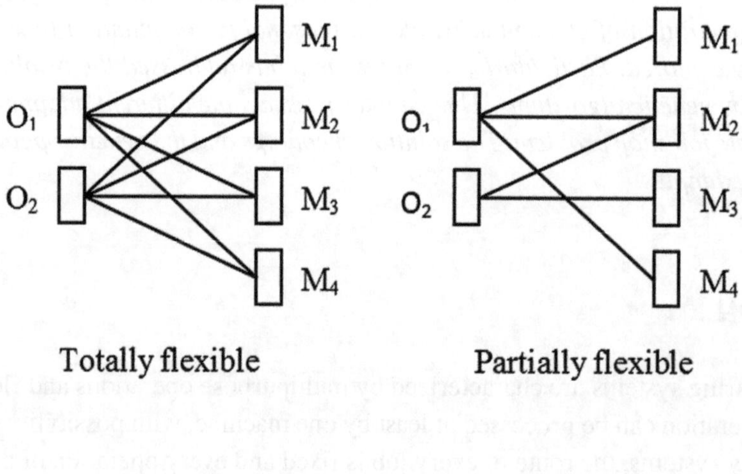

## FLEXIBLE JOB SHOP SCHEDULING

The FJSP can be described as follows. There are n independent jobs (no precedence constraints among operations of different jobs) $J = \{1, \ldots, j, \ldots, n\}$, each job j is composed of a predefined sequence of $n_j$ operations $O_{ij}$, where $O_{ij}$ denotes the $i^{th}$ operation of the job j. The order of operations of each job is predefined and cannot be modified. A set of K machines $M = \{M_k, 1 \leq k \leq m\}$ is available. At any time, each machine can process at most one operation. One job can visit at least one machine at a time. An operation can be allocated to a machine from a set of alternatives with variables performances to be processed without interruption. Hence, to each operation $O_{ij}$ is associated a pool of machines $m_{ij}$. Given a machine $m_k$ from $m_{ij}$ than the execution time of $O_{ij}$ on $m_k$ is $e_{ijk}$. A scheduling is a definition for each operation of a machine $m_k$ from its pool, a starting date $s_{ij}$ and a completion time $c_{ij}$ of $O_{ij}$ on that machine. The objective on the FJSP is to both determine an assignment and a sequence of the operations on the machines so that some criteria have to be optimized. The widely used criterion in the literature is the maximum completion time (makespan) to be minimized.

The following notations are used:

$j = 1 \ldots n$: a set of n jobs
$n_j$: number of operations of job j
$m_k$, $k = 1 \ldots K$: a set of K machines
$O_{ij}$, $i = 1 \ldots n_j$: ith operation of the job j
$m_{ij}$, $i = 1 \ldots n_j$, $j = 1 \ldots n$: pool of the operation $O_{ij}$
$e_{ijk}$: the execution time of the operation $O_{ij}$ on the machine $m_k$
$s_{ij}$: starting time of the execution of the operation $O_{ij}$
$c_{ij}$: completion time of the operation $O_{ij}$
$c_j = \max(c_{ij})_{1 \leq i \leq nj}$: completion time of the job j
$c_{max} = \max(c_j)_{1 \leq j \leq n}$: completion time of all jobs known as makespan
$\Gamma^+_{ij}$: set of direct successors of the operation $O_{ij}$
$\Gamma^-_{ij}$: set of direct predecessors of the operation $O_{ij}$
$r_{ij}$: the ready time of the operation $O_{ij}$
N: total number of operations of all jobs

The Table 1 gives an instance of the FJSP with 3 jobs and 3 machines. A feasible solution is presented in Table 2. The Figure 2 models the problem solution in Table 2 with a disjunctive graph (Roy & Sussmann, 1964). A feasible schedule of the FJSP can be mapped by a disjunctive graph $G = (V, C \cup D)$. In which, V is a set of all the nodes including dummy starting and terminating ones, each node corresponds to an operation in the FJSP; C means the set of all the conjunctive arcs which connect two adjacent operations within one job; D refers to a set of all the disjunctive arcs connecting two adjacent operations processed on the same machine and the arc directions show the precedence relations between operations. The weight of the node which is simply the processing time for each operation is labelled above its corresponding node. The makespan is the length of the longest path in the disjunctive graph. This path is said to be the critical path and a graph may admit more than one critical path.

*Table 1. Processing time table of an instance of FJSP*

| Job | Operation | M1 | M2 | M3 |
|---|---|---|---|---|
| J1 | O1 | 2 | - | 2 |
| | O2 | 2 | 2 | 2 |
| | O3 | - | - | 2 |
| J2 | O1 | - | 3 | 3 |
| | O2 | 3 | - | - |
| | O3 | 3 | 3 | 3 |
| J3 | O1 | 5 | 5 | 5 |
| | O2 | 5 | - | 5 |
| | O3 | - | 5 | - |

*Table 2. A possible solution to the instance given in Table 1*

| Job | Operation | M1 | M2 | M3 |
|---|---|---|---|---|
| J1 | O1 | (0,2) | - | ∞ |
| | O2 | (2,4) | ∞ | ∞ |
| | O3 | - | - | (10,12) |
| J2 | O1 | - | (0,3) | ∞ |
| | O2 | (4,7) | - | - |
| | O3 | (7,10) | ∞ | ∞ |
| J3 | O1 | ∞ | ∞ | (0,5) |
| | O2 | ∞ | - | (5,10) |
| | O3 | - | (10,15) | - |

*Figure 2. Illustration of disjunctive graph*

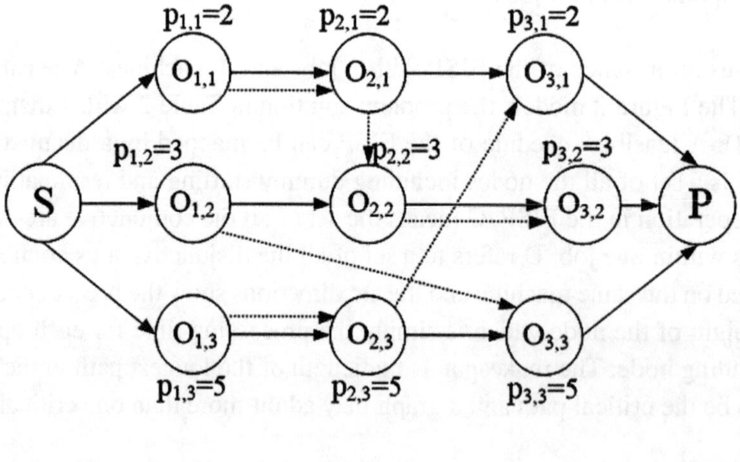

## BASIC GENETIC ALGORITHM

The genetic algorithm is a popular population based metaheuristic which has a wide area of application domains including both discrete and continuous optimisation. The genetic scheme takes inspiration from the Darwinian's theory of survival of the fittest. The main components of a genetic algorithm are the chromosome, the population, the fitness, the selection, the crossover and the mutation. The Genetic evolution starts from a population of initial solutions represented by chromosomes. The well adapted chromosomes survive and are combined by crossover and diversified by mutation. The selection operators determine at each generation the survivors from the parents and descendants. The process is repeated until some stopping condition is satisfied. The basic genetic algorithm (GAs) is outlined as below:

**Step 1:** Generate initial population of chromosomes, that is, feasible solutions for the problem.
**Step 2:** Evaluate the fitness of each chromosome in the population.
**Step 3:** Create a new population by reproduction

- **Step 3a:** Select parent's chromosomes from the current population.
- **Step 3b:** Combine the parents by crossover operator with a crossover probability to form new chromosome
- **Step 3c:** Mutate each new generated chromosome with a mutation probability
- **Step 3d:** Repeat a, b and c until desired number of children is reached

**Step 4:** Use the selection operator to form the population of survivors.
**Step 5:** If the termination condition is reached stop, and return the best solution, else go to step 2.

The genetic algorithms performance is largely influenced by the way the solution is encoded and the crossover and the mutation operators are designed.

## GENERATION OF THE INITIAL POPULATION

The initial population gives the starting points for the genetic evolution. Multiple strategies were used to generate the initial solutions such as:

- **Random Rule:** Every solution in the population is generated in random way, which enhances global searching capability of the algorithm but this strategy may slowdown the convergence of the algorithm.
- **Heuristic:** Several heuristics are utilized to produce solutions using different rules. Unfortunately, the evolving population could easily fall into local optimal solutions far from the best one.
- **Mixed population of Both Random and Good Solutions:** This combination is applied to maintain certain population diversity.
- **Duplication and Evolution:** Originally, the population contains a unique solution which is duplicated or modified by applying crossover and mutation repeatedly until generating all individuals.

Furthermore, the population size is considered as a critical genetic algorithm parameter. On the one hand, it is hard to a population within a reduced size to avoid being trapped in a locality. On the other hand, it is unclear how a large population can help in finding good solutions because of the dramatic increase of computation time. Only experiments can provide an idea about population size while taking into account inherent specifications of each problem. Indeed, a compromise has to be established between running time and solutions quality.

## SOLUTION ENCODING

The encodings transform a solution into a chromosome. Solution representations specific to the flexible job shop problem are whether indirect or direct. Indirect encodings gives somehow a solution can be constructed. Direct encodings generally use a scheduling algorithm computes the starting times and the completion times of the operations. The chromosome must lead to feasible scheduling and thus reparation mechanisms are employed when the representation doesn't guarantee cycle prevention and problem constraints.

- **List of Fabrication Order/Plan Encoding (Bagchi et al., 1991; Uckun et al., 1993):** This encoding is proposed for the generalized flexible job shop problem where an operation can be processed by every machine and the jobs have many fabrication plans. A chromosome is a list of job's operations where a plan is affected to each operation. This encoding is indirect since a scheduling generator assigns to every operation a machine and an execution time interval on the elected machine.
- **List of Fabrication Order/Plan/Resource Encoding (Bagchi et al., 1991; Uckun et al., 1993):** This encoding is specific for the generalized flexible job shop problem where a fabrication order has more than one manufacturing plan. A chromosome is a list of jobs where for each one of its operations the production plan and the machine in charge of its execution are fixed. In the example (Figure 3), operation 1 of the plan B is assigned to the machine $M_9$, etc.
- **Parallel Machine Encoding (Mesghouni et al., 2004):** This encoding extends the representation proposed in (Kobayashi et al., 1995) to the flexible job shop problem. Each chromosome is composed of k vectors each one corresponds to the sequence of one machine. Each cell gives three data: job number (j), operation number (i) and the starting execution time ($s_{ijk}$). A given coding is a feasible solution if neither cycle is detected otherwise the individual is repaired (Figure 4).

*Figure 3. Order/plan encoding*

| $Job_7/plan_B$ | $Job_8/plan_C$ | $Job_4/plan_A$ | … | $Job_7/plan_B$ |
|---|---|---|---|---|

(a) *List of fabrication order/plan encoding*

| $OpB_1$ $OpB_2$ $OpB_3$ $Job_3$- $M_9$ $M_7$ $M_5$ | $OpA_1$ $OpA_2$ $Job_8$- $M_4$ $M_2$ | … |
|---|---|---|

(b) *List of fabrication order/plan/resource encoding*

*Figure 4. Parallel machine encoding*

| | | | | |
|---|---|---|---|---|
| $M_1$ | $(O_{11}, 1, 0)$ | $(O_{21}, 1, 2)$ | $(O_{22}, 2, 4)$ | $(O_{32}, 2, 7)$ |
| $M_2$ | $(O_{12}, 2, 0)$ | $(O_{33}, 3, 10)$ | | |
| $M_3$ | $(O_{31}, 1, 10)$ | $(O_{13}, 3, 0)$ | $(O_{23}, 3, 5)$ | |

- **Parallel Job Encoding (Mesghouni et al., 2004):** This encoding extends the one presented (Yamada et al., 1992) for the job shop problem. A chromosome is composed of n vectors. Each vector is related to one job. A cell contains two data, the machine that will execute the operation and the starting execution time. The cell (i, j) correspond to the $i^{th}$ operation of the job j (Figure 5). This encoding always generates a feasible solution.
- **Operation/Rank/Machine Encoding (Ghedjati, 1994):** This encoding is similar to the one given in (Bagchi et al., 1991) and (Uckun et al., 1993). A chromosome is composed of three vectors of length N where N is the total number of the operations of all the jobs. The vector operation is numbered from 1 to N. The vector machine gives the machine's assignment. The vector rank gives the execution order of the operation on its machine. For example, in the Figure 6, the operation number 2 is performed on the machine $M_2$ and it occupies the first place in the sequence of $M_2$.
- **Operation/Scheduling Rule/Assignment Rule (Ghedjati, 1994):** This encoding is indirect. It codes a constructive heuristic that applies at each decision point a scheduling rule to choose an operation from the waiting queue and a heuristic that assigns a multipurpose operation to a machine from its pool. A chromosome is composed of three vectors of N cells. The vector operation is composed of N cells numbered from 1 to N. The vector scheduling rule gives the application order of the rules. The vector heuristic gives the order of application of the assignment rules. In the example of Figure 7, at the 2nd decision point, the scheduling rule number 1 is applied and the

*Figure 5. Parallel job encoding*

| | | | |
|---|---|---|---|
| $J_1$ | $(M_1, 0)$ | $(M_1, 2)$ | $(M_3, 10)$ |
| $J_2$ | $(M_2, 0)$ | $(M_1, 4)$ | $(M_3, 7)$ |
| $J_3$ | $(M_3, 0)$ | $(M_3, 5)$ | $(M_2, 10)$ |

*Figure 6. Operation/Rank/Machine encoding*

| Operation | 1 | 2 | 3 | 4 | 5 | 6 | 7 | 8 | 9 |
|---|---|---|---|---|---|---|---|---|---|
| Machine | $M_1$ | $M_1$ | $M_3$ | $M_2$ | $M_1$ | $M_1$ | $M_3$ | $M_3$ | $M_2$ |
| Rank | 1 | 2 | 3 | 1 | 3 | 4 | 1 | 2 | 2 |

*Figure 7. Operation/scheduling rule/assignment rule*

| Operation | 1 | 2 | 3 | 4 | 5 | 6 | 7 | 8 | 9 |
|---|---|---|---|---|---|---|---|---|---|
| Scheduling rule | 2 | 1 | 2 | 1 | 2 | 3 | 2 | 1 | 2 |
| Assignment rule | 2 | 4 | 3 | 2 | 2 | 1 | 5 | 2 | 3 |

assignment heuristic number 4. Particularly, (Ghedjati, 1994) has used four scheduling rules and seven assignment heuristics.

- **Operation-Machine Encoding (Kacem et al., 2001b):** A chromosome is an N by k matrix. The operations are on the lines and the machines are on the columns. The cell ((i, j),k) gives firstly, a binary value that specifies whether the operation $O_{ij}$ is assigned to the machine $M_k$ or not (Figure 8). After that, a scheduling algorithm is applied and the assignments are replaced by the couple ($s_{i,j}$, $c_{i,j}$) of the starting and the finishing execution time.
- **Operation List Encoding (Kacem et al., 2001a):** This encoding is composed of three vectors of N cells. Three information are coded which are the operation (i,j), the machine assignment $M_k$ and the scheduling by giving the starting and the finishing execution time ($s_{i,j}$, $c_{i,j}$) of the operation $O_{i,j}$ on the machine $M_k$ (Figure 9). This encoding gives the assignment of operations to machines in the first step then computes a feasible schedule by an algorithm that resolves conflicts by scheduling rules.
- **List of the Operation's Sequences Encoding (Kacem et al., 2001c)**: This encoding is a generalization of the one used by (Lee et al., 1998) for the one machine scheduling problem. A chromosome is a vector of N cells. It gives both the assignment of the operations to the machines and the execution order of the operations on the machines. A given cell contains the information (i, j, k) which means that the operation $O_{ij}$ is performed on the machine $M_k$ (Figure 10). This encoding was applied in (Zandieh et al., 2008).

*Figure 8. Operation-machine encoding*

| Job | Operation | M1 | M2 | M3 |
|---|---|---|---|---|
| J1 | O1 | 1 | 0 | 0 |
| | O2 | 1 | 0 | 0 |
| | O3 | 0 | 0 | 1 |
| J2 | O1 | 0 | 1 | 0 |
| | O2 | 1 | 0 | 0 |
| | O3 | 1 | 0 | 0 |
| J3 | O1 | 0 | 0 | 1 |
| | O2 | 0 | 0 | 1 |
| | O3 | 0 | 1 | 0 |

*Assignement table*

| Job | Operation | M1 | M2 | M3 |
|---|---|---|---|---|
| J1 | O1 | 0,2 | 0 | 0 |
| | O2 | 2,4 | 0 | 0 |
| | O3 | 0 | 0 | 10,12 |
| J2 | O1 | 0 | 0,3 | 0 |
| | O2 | 4,7 | 0 | 0 |
| | O3 | 7,10 | 0 | 0 |
| J3 | O1 | 0 | 0 | 0,5 |
| | O2 | 0 | 0 | 5,10 |
| | O3 | 0 | 10,15 | 0 |

*Scheduling table*

*Figure 9. Operation list encoding*

| $O_{i,j}$ | $M_k$ | $T_{i,j}$ $tf_{i,j}$ |
|---|---|---|
| $O_{1,1}$ | 1 | 0,2 |
| $O_{2,1}$ | 1 | 2,4 |
| $O_{3,1}$ | 3 | 10,12 |
| $O_{1,2}$ | 2 | 0,3 |
| $O_{2,2}$ | 1 | 4,7 |
| $O_{3,2}$ | 1 | 7,10 |
| $O_{1,3}$ | 3 | 0,5 |
| $O_{2,3}$ | 3 | 5,10 |
| $O_{3,3}$ | 2 | 10,15 |

- **Encoding of Tay and Wibowo, 2004:** This encoding is composed of three parts $D_1$, $D_2$ and $D_3$. It combines the parts $C_1$ and $C_2$ ($D_1$, $D_2$) from (Paredis, 1992) and the part $B_2$ ($D_3$) from (Ho & Tay, 2004). The vector $D_1$ gives the operations sequence where the cell h gives the job $f_{job}(O_h)$ of the operation $O_h$. The order of the operation within the same job is implicitly deduced from the index value. The vector $D_2$ gives the assignment of the operation $O_{ij}$ to the machine $f_{idx}\left(M_{O_{ij}}\right)$. In the vector $D_3$, the term $b_{ijik}$ is a binary value that indicates the presence of a precedence relation between the operations $O_{ij}$ and $O_{ik}$.
- **Machine Order With Bits Encoding (Paredis, 1992):** A chromosome is represented by two vectors $B_1$ and $B_2$ (Figure 11). The first vector $B_1$contains N cells corresponding to the N operations. The cell i indicates the number of the machine that will execute the $i^{th}$ operation. The second is a binary vector that specifies for each pair of operations which one of them is executed before the other. The binary value is equal to 0 if the first operation is performed before the second and to 1 otherwise. The length of $B_2$ is equal to the number of precedence conflicts.
- **Encoding of Ho & Tay, 2004:** The representation comprises two parts: the Operator order vector and a selection binary vector. The first one gives the order of the operations to be processed by specifying the job number which the operation belongs to. The number of the operation is deduced from the number of the occurrence of the job. The machine selection vector represents the assignment of machines to operations. It is a binary vector that associates to each operation a

*Figure 10. List of the operation's sequences encoding*

| $T_1$ | $T_2$ | $T_3$ | $T_4$ | $T_5$ | $T_6$ | $T_7$ | $T_8$ | $T_9$ |
|---|---|---|---|---|---|---|---|---|
| (1,1,1) | (2,1,1) | (3,1,3) | (1,2,2) | (2,2,1) | (3,2,1) | (1,3,3) | (2,3,3) | (3,3,2) |

*Figure 11. Machine order with bit encodings*

| $J_1$ | $J_1$ | $J_2$ | $J_1$ | $J_3$ | $J_3$ | $J_2$ | $J_2$ | $J_3$ |
|---|---|---|---|---|---|---|---|---|

| $O_{11}$ | $O_{21}$ | $O_{31}$ | $O_{12}$ | $O_{22}$ | $O_{32}$ | $O_{13}$ | $O_{23}$ | $O_{33}$ |
|---|---|---|---|---|---|---|---|---|
| $M_1$ | $M_2$ | $M_3$ | $M_2$ | $M_1$ | $M_1$ | $M_3$ | $M_3$ | $M_2$ |

*Figure 12. Encoding of Ho and Tay, 2004*

| $J_1$ | $J_1$ | $J_2$ | $J_1$ | $J_3$ | $J_3$ | $J_2$ | $J_2$ | $J_3$ |
|---|---|---|---|---|---|---|---|---|

<table>
<tr><td colspan="2">$O_{11}$</td><td colspan="3">$O_{21}$</td><td>$O_{31}$</td><td colspan="2">$O_{12}$</td><td>$O_{22}$</td><td colspan="3">$O_{32}$</td><td colspan="3">$O_{13}$</td><td colspan="2">$O_{13}$</td><td>$O_{13}$</td></tr>
<tr><td>$M_1$</td><td>$M_3$</td><td>$M_1$</td><td>$M_2$</td><td>$M_3$</td><td>$M_3$</td><td>$M_2$</td><td>$M_3$</td><td>$M_1$</td><td>$M_1$</td><td>$M_2$</td><td>$M_3$</td><td>$M_1$</td><td>$M_2$</td><td>$M_3$</td><td>$M_1$</td><td>$M_3$</td><td>$M_2$</td></tr>
<tr><td>1</td><td>0</td><td>1</td><td>0</td><td>0</td><td>1</td><td>0</td><td>1</td><td>1</td><td>0</td><td>0</td><td>1</td><td>0</td><td>0</td><td>1</td><td>1</td><td>0</td><td>1</td></tr>
</table>

pool of machines. When a machine is selected it receives 1 otherwise 0 (Figure 12). To obtain a solution, a scheduling algorithm is launched to compute operation's starting time. This encoding does not form cycles and prevent the formation of infeasible solutions but the quality depends on the scheduling algorithm.

- **Machine Order With Integers Encoding (Chen et al., 1999):** A chromosome is coded by two integer vector of length N. The first vector assigns a machine to each operation. The second vector gives the order of operations on each machine (Figure 13).
- **Multi-Stage Encoding (Zhang & Gen, 2005):** In this encoding, the operations are the stages and the machines are the states. A multipurpose machine has a number of possible states that equals the number of machines of its pool. An instance composed of 3 jobs and 4 machines is formulated as a problem of 8 stages and 4 states. A feasible schedule is obtained by connecting the nodes by dashed arcs one node from each stage.

*Figure 13. Machine order with integers encoding*

Parent$_1$

| Operation | 11 | 12 | 13 | 14 | 15 | 21 | 22 | 31 | 32 | 33 | 41 | 42 |
|---|---|---|---|---|---|---|---|---|---|---|---|---|
| Machine | 1 | 2 | 5 | 3 | 4 | 1 | 2 | 3 | 1 | 2 | 4 | 2 |
| Rang | 1 | 2 | 1 | 2 | 2 | 3 | 3 | 1 | 2 | 4 | 1 | 1 |

Parent$_2$

| Operation | 11 | 12 | 13 | 14 | 15 | 21 | 22 | 31 | 32 | 33 | 41 | 42 |
|---|---|---|---|---|---|---|---|---|---|---|---|---|
| Machine | 3 | 2 | 5 | 3 | 4 | 1 | 2 | 3 | 1 | 2 | 4 | 2 |
| Rang | 1 | 2 | 1 | 3 | 2 | 1 | 3 | 2 | 2 | 4 | 1 | 1 |

Child$_1$ with anomalies

| Operation | 11 | 12 | 13 | 14 | 15 | 21 | 22 | 31 | 32 | 33 | 41 | 42 |
|---|---|---|---|---|---|---|---|---|---|---|---|---|
| Machine | **1** | 2 | 5 | **3** | 4 | **1** | 2 | **3** | **1** | 2 | 4 | 2 |
| Rang | **1** | 2 | 1 | **3** | 2 | **1** | 3 | 2 | 2 | 4 | 1 | 1 |

Child$_1$ after rank correction

| Operation | 11 | 12 | 13 | 14 | 15 | 21 | 22 | 31 | 32 | 33 | 41 | 42 |
|---|---|---|---|---|---|---|---|---|---|---|---|---|
| Machine | **1** | 2 | 5 | **3** | 4 | **1** | 2 | **3** | **1** | 2 | 4 | 2 |
| Rang | **1** | 2 | 1 | **2** | 2 | **2** | 3 | **1** | **3** | 4 | 1 | 1 |

The Table 3 shows the space complexity of the chromosome and the time complexity of the scheduling algorithm that computes the starting execution times where N denotes the total number of job operations in the FJSP, c denotes the number of precedence constraints and d denotes the length of the string D. cycle detection in O(N+c) and cycle removal in O(c) time.

## CROSSOVER OPERATORS

The crossovers exchange genes between parents to form new chromosomes in different ways. The combination performs swap of operations assignment, of operations sequences or both of them. It is difficult

*Table 3. Flexible Job Shop Problem encodings*

| Chromosome Encodings | Chromosome Space Complexity | Scheduling Algorithm Complexity |
|---|---|---|
| Machine order with integer Chen et al. | 2N | O(N+c) |
| Encoding of Paredis | N+0.5N(N-1) | $O(N^2+c)$ |
| Encoding of Ho and Tay | 2N | O(N+c) |
| J.C. Tay and D. Wibowo | 2N+d | O(N+c+d) |
| Encoding of Ho & Tay, 2004) | N+N*K | O(N+c) |
| Machine order with integers encoding | 3N | O(N+c) |
| Multi-stage encoding | N*K | O(N+c) |

to generate feasible chromosomes and verification of the validity and eventually reparation procedure are launched after recombination.

- **Permutation of Two Machines Crossover:** Let $P_1$ and $P_2$ be two parents. Select randomly one machine $M_{k1}$ from the parent $P_1$ and a machine $M_{k2}$ from the parent $P_2$ than exchange the sequences between the two machines. So, the child $E_1$ (respectively $E_2$) receives the same sequences as in $P_1$ ($P_2$) except for the machine $M_{k1}$ ($M_{k2}$). It has the same sequence of the machine $M_{k2}$ ($M_{k1}$) in the parent $P_2$. This crossover was applied by (Kacem, 2003).
- **One Machine Permutation Crossover:** Let $P_1$ and $P_2$ be two parents. Select randomly one machine $M_k$. The child $E_1$ receives the sequence of the machine $M_k$ from the parent $P_2$ and the remaining one's from $P_1$. The child $E_2$ is constructed while inverting the roles of $P_1$ and $P_2$. This crossover was applied by [Mesghouni, K., et al., 2004] on the parallel machine encoding.
- **One Job Crossover:** Let $P_1$ and $P_2$ be two parents. Select randomly one job j. The child $E_1$ ($E_2$) receives the same assignment for all operations of the job j from parent $P_1$ ($P_2$). The assignments of the remaining jobs are swapped (Mesghouni et al., 2004; Kacem, 2003).
- **Uniform Crossover:** The uniform crossover constructs a descendent by copying the genes one by one with equal probabilities from the parent $P_1$ and the Parent $P_2$. The obtained child requires a correction treatment. This crossover is applied by (Ghedjati, 1994) for the operation/rank/machine crossover (Figure 14).
- **One Point Crossover:** Let $P_1$ and $P_2$ be two parents. Select a random cut point that divides the chromosome into two blocks. The child $E_1$ (respectively $E_2$) receives the block before the cut point from $P_1$ (respectively $P_2$) and that after the cut point from $P_2$ (respectively $P_1$). This crossover was

*Figure 14. Uniform crossover on the Operation/Rank/Machine*

Parent$_1$

| Operation | 11 | 12 | 13 | 14 | 15 | 21 | 22 | 31 | 32 | 33 | 41 | 42 |
|---|---|---|---|---|---|---|---|---|---|---|---|---|
| Machine | 1 | 2 | 5 | 3 | 4 | 1 | 2 | 3 | 1 | 2 | 4 | 2 |
| Rang | 1 | 2 | 1 | 2 | 2 | 3 | 3 | 1 | 2 | 4 | 1 | 1 |

Parent$_2$

| Operation | 11 | 12 | 13 | 14 | 15 | 21 | 22 | 31 | 32 | 33 | 41 | 42 |
|---|---|---|---|---|---|---|---|---|---|---|---|---|
| Machine | 3 | 2 | 5 | 3 | 4 | 1 | 2 | 3 | 1 | 2 | 4 | 2 |
| Rang | 1 | 2 | 1 | 3 | 2 | 1 | 3 | 2 | 2 | 4 | 1 | 1 |

Child$_1$ with anomalies

| Operation | 11 | 12 | 13 | 14 | 15 | 21 | 22 | 31 | 32 | 33 | 41 | 42 |
|---|---|---|---|---|---|---|---|---|---|---|---|---|
| Machine | **1** | 2 | 5 | **3** | 4 | **1** | 2 | **3** | **1** | 2 | 4 | 2 |
| Rang | **1** | 2 | 1 | **3** | 2 | **1** | 3 | 2 | 2 | 4 | 1 | 1 |

Child$_1$ after rank correction

| Operation | 11 | 12 | 13 | 14 | 15 | 21 | 22 | 31 | 32 | 33 | 41 | 42 |
|---|---|---|---|---|---|---|---|---|---|---|---|---|
| Machine | **1** | 2 | 5 | 3 | 4 | **1** | 2 | **3** | **1** | 2 | 4 | 2 |
| Rang | **1** | 2 | 1 | 2 | 2 | 2 | 3 | **1** | **3** | 4 | 1 | 1 |

adapted by (Ghedjati, 1994) for the operation/scheduling rule/assignment rule heuristic encoding (Figure 15).

- **Two-Point Crossover:** Let $P_1$ and $P_2$ be two parents. Select two cut points that divide the encoding into three blocks. The child $E_1$ (respectively $E_2$) receives the central block from $P_1$(respectively $P_2$) and the outer one's from $P_2$ (respectively $P_1$) (Figure 16). This crossover was adopted in [Kacem, I., 2003] for both encoding schemes operation/machine and operation list.
- **Assignment Crossover (Kacem, 2003):** Let $P_1$ and $P_2$ be two parents. Select randomly a set of operations. The first child $E_1$ receives the parent $P_1$ after modification of the assignment of the selected operations according to parent $P_2$. The second child is constructed in the same manner by changing in parent $P_2$ the assignment of the selected operations according to $P_1$ (Figure 17).
- **Selective Machine Sequences Crossover (SMSC):** Only one offspring is produced by The SMSC operator (Figure 18). This child inherits, for each machine, the sequence satisfying one of the following criteria: the minimum completion time, the least loaded machine or one randomly chosen machine amongst the two parents (Naifar et al., 2006).

## MUTATION OPERATORS

A mutation operator makes a slight modification on the solution encoding. A mutation operator has a diversification effect. It enriches the population with new genetic material. In the case of the FJSP, new operation's assignment and block of sequences are incorporated.

- **Permutation of Two Operations:** This mutation selects randomly one machine and two successive operations on that machine and permutes them (Ghedjati, 1994).

*Figure 15. One point crossover on the operation/scheduling rule/assignment rule heuristic encoding*

Parent$_1$

| Operation | 11 | 12 | 13 | 14 | 15 | 21 | 22 | 31 | 32 | 33 | 41 | 42 |
|---|---|---|---|---|---|---|---|---|---|---|---|---|
| Machine | 2 | 1 | 2 | 1 | 2 | 2 | 2 | 1 | 2 | 1 | 2 | 2 |
| Rang | 2 | 4 | 3 | 2 | 2 | 2 | 5 | 2 | 3 | 3 | 1 | 2 |

Parent$_2$

| Operation | 11 | 12 | 13 | 14 | 15 | 21 | 22 | 31 | 32 | 33 | 41 | 42 |
|---|---|---|---|---|---|---|---|---|---|---|---|---|
| Machine | 1 | 2 | 1 | 2 | 2 | 2 | 1 | 2 | 2 | 1 | 2 | 1 |
| Rang | 5 | 4 | 2 | 4 | 3 | 3 | 1 | 1 | 1 | 4 | 5 | 4 |

Child$_1$

| Operation | 11 | 12 | 13 | 14 | 15 | 21 | 22 | 31 | 32 | 33 | 41 | 42 |
|---|---|---|---|---|---|---|---|---|---|---|---|---|
| Machine | 2 | 1 | 2 | 1 | 2 | 2 | 2 | 1 | 2 | 1 | 2 | 1 |
| Rang | 2 | 4 | 3 | 2 | 2 | 2 | 5 | 2 | 1 | 4 | 5 | 4 |

*Figure 16. Two points crossover for the COM and CLO (cut points j = 1, j' = 2, i = 2, i' = 2)*

$P_1$

| Job | Operation | M1 | M2 | M3 | M4 |
|---|---|---|---|---|---|
| J1 | O1 | 0 | 0 | 0 | 0,1 |
| | O2 | 0 | 0 | 0 | 1,2 |
| | O3 | 3,6 | 0 | 0 | 0 |
| J2 | O1 | 0 | 0,1 | 0 | 0 |
| | O2 | 1,3 | 0 | 0 | 0 |
| | O3 | 0 | 3,4 | 0 | 0 |
| J3 | O1 | 0 | 0 | 0,3 | 0 |
| | O2 | 0 | 0 | 0 | 3,4 |

$P_2$

| Job | Operation | M1 | M2 | M3 | M4 |
|---|---|---|---|---|---|
| J1 | O1 | 0,1 | 0 | 0 | 0 |
| | O2 | 0 | 0 | 0 | 1,2 |
| | O3 | 3,6 | 0 | 0 | 0 |
| J2 | O1 | 0 | 0 | 0,1 | 0 |
| | O2 | 1,3 | 0 | 0 | 0 |
| | O3 | 0 | 3,4 | 0 | 0 |
| J3 | O1 | 0 | 0 | 1,4 | 0 |
| | O2 | 0 | 0 | 0 | 4,5 |

Construction of $e_1$

| Job | Operation | M1 | M2 | M3 | M4 |
|---|---|---|---|---|---|
| J1 | O1 | ?,? | 0 | 0 | 0 |
| | O2 | 0 | 0 | 0 | ?,? |
| | O3 | ?,? | 0 | 0 | 0 |
| J2 | O1 | 0 | ?,? | 0 | 0 |
| | O2 | ?,? | 0 | 0 | 0 |
| | O3 | 0 | ?,? | 0 | 0 |
| J3 | O1 | 0 | 0 | ?,? | 0 |
| | O2 | 0 | 0 | 0 | ?,? |

Construction of $e_2$

| Job | Operation | M1 | M2 | M3 | M4 |
|---|---|---|---|---|---|
| J1 | O1 | 0 | 0 | 0 | ?,? |
| | O2 | 0 | 0 | 0 | ?,? |
| | O3 | ?,? | 0 | 0 | 0 |
| J2 | O1 | 0 | 0 | ?,? | 0 |
| | O2 | ?,? | 0 | 0 | 0 |
| | O3 | 0 | ?,? | 0 | 0 |
| J3 | O1 | 0 | 0 | ?,? | 0 |
| | O2 | 0 | 0 | 0 | ?,? |

**Assignment exchange**

Final $e_1$

| Job | Operation | M1 | M2 | M3 | M4 |
|---|---|---|---|---|---|
| J1 | O1 | 0,1 | 0 | 0 | 0 |
| | O2 | 0 | 0 | 0 | 1,2 |
| | O3 | 3,6 | 0 | 0 | 0 |
| J2 | O1 | 0 | 0,1 | 0 | 0 |
| | O2 | 1,3 | 0 | 0 | 0 |
| | O3 | 0 | 3,4 | 0 | 0 |

Final $e_2$

| Job | Operation | M1 | M2 | M3 | M4 |
|---|---|---|---|---|---|
| J1 | O1 | 0 | 0 | 0 | 0,1 |
| | O2 | 0 | 0 | 0 | 1,2 |
| | O3 | 3,6 | 0 | 0 | 0 |
| J2 | O1 | 0 | 0 | 0,1 | 0 |
| | O2 | 1,3 | 0 | 0 | 0 |
| | O3 | 0 | 3,4 | 0 | 0 |

- **Mutation of One Operation's Assignment:** This operator selects randomly a multipurpose operation and assigns it randomly to another machine from its pool (Ghedjati, 1994; Mesghouni et al., 2004; Kacem, 2003).
- **Mutation of the Job With the Largest Effective Execution Time:** Find the job that has the largest effective execution time. Select randomly a multipurpose operation of that job and assigns it to another machine from its pool that is able to perform it faster than the current one.
- **Mutation of the Most Loaded Machine:** Reassign a multipurpose operation selected randomly from the sequence of the most loaded machine to the less loaded machine (Mesghouni et al., 2004; Kacem, 2003).

*Figure 17. Assignment crossover for the Tasks Sequencing List*

**Parent$_1$**

| $T_1$ | $T_2$ | $T_3$ | $T_4$ | $T_5$ | $T_6$ | $T_7$ | $T_8$ |
|---|---|---|---|---|---|---|---|
| (1,1,4) | (1,2,2) | (2,2,1) | (2,1,4) | (3,2,2) | (3,1,1) | (1,3,3) | (2,3,4) |

**Parent$_2$**

| $T_1$ | $T_2$ | $T_3$ | $T_4$ | $T_5$ | $T_6$ | $T_7$ | $T_8$ |
|---|---|---|---|---|---|---|---|
| (1,2,1) | (1,1,3) | (2,2,2) | (2,1,1) | (3,2,4) | (1,3,4) | (3,1,2) | (2,3,1) |

**Child$_1$**

| $T_1$ | $T_2$ | $T_3$ | $T_4$ | $T_5$ | $T_6$ | $T_7$ | $T_8$ |
|---|---|---|---|---|---|---|---|
| (1,1,3) | (1,2,2) | (2,2,2) | (2,1,4) | (3,2,2) | (3,1,2) | (1,3,3) | (2,3,1) |

**Child$_2$**

| $T_1$ | $T_2$ | $T_3$ | $T_4$ | $T_5$ | $T_6$ | $T_7$ | $T_8$ |
|---|---|---|---|---|---|---|---|
| (1,2,1) | (1,1,4) | (2,2,1) | (2,1,1) | (3,2,4) | (1,3,4) | (3,1,1) | (2,3,4) |

- **Mutation of the Last Finishing Job:** Given the most loaded machine and the job with the highest finishing time, if this machine executes one (or many) operation of the last finishing job, then remove and reschedule it on another machine of its pool.
- **Mutation by Job Rescheduling:** Extract all the operations of the last finishing job and reinsert them successively according to job precedence.
- **Swap Mutation on One Machine:** Select randomly one machine and two operations from its sequence than swap their positions in the sequence of the given machine when the swap is feasible.
- **Swap Mutation on Two Machines:** Choose randomly two machines and one operation from each of them. A swap of their assignment is performed when the two machines are part of the pool of each of them.

## LITERATURE REVIEW

In the literature two types of approaches have been used: hierarchical approaches where assignment and sequencing problems are treated separately and integrated approaches. (Brandimarte, 1993) was the first to use this decomposition for the FJSP. He solved the routing sub-problem using some existing dispatching rules and then focused on the scheduling sub-problem, which is solved using a tabu search heuristic. Much research has addressed the FJSP and most are based on metaheuristics and heuristics.

*Figure 18. Selective machine sequences crossover with the minimum completion time*

ParentP$_1$

| | | | | | | | | | | |
|---|---|---|---|---|---|---|---|---|---|---|
| $M_1$ | 2,3 | 1,6 | 2,4 | 2,1 | 5,3 | 3,2 | 4,6 | 4,5 | 5,5 | ***60*** |
| $M_2$ | 1,4 | 5,1 | 4,4 | 3,5 | | | | | | ***53*** |
| $M_3$ | 1,1 | 2,2 | 2,5 | 6,3 | | | | | | ***55*** |
| $M_4$ | 4,3 | 2,6 | 4,1 | 1,5 | 5,4 | 4,2 | | | | ***63*** |
| $M_5$ | 1,3 | 3,1 | 6,1 | 6,2 | 5,6 | 6,5 | | | | ***81*** |
| *$M_6$* | *1,2* | *3,3* | *3,4* | *3,6* | *5,2* | *6,6* | *6,4* | | | ***91*** |

ParentP$_2$

| | | | | | | | | | |
|---|---|---|---|---|---|---|---|---|---|
| $M_1$ | 2,1 | 1,6 | 2,4 | 4,1 | 2,3 | 3,3 | 4,6 | 4,3 | ***50*** |
| $M_2$ | 3,1 | 5,1 | 4,4 | 3,5 | 2,2 | 6,6 | 5,3 | 6,3 | ***58*** |
| $M_3$ | 1,1 | 1,3 | | | | | | | ***6*** |
| $M_4$ | 1,5 | 2,6 | 3,6 | 5,4 | 3,2 | 4,2 | 5,2 | 6,2 | ***75*** |
| $M_5$ | 6,1 | 2,5 | 6,5 | 5,6 | | | | | ***48*** |
| *$M_6$* | *1,4* | *1,2* | *3,4* | *4,5* | *5,5* | *6,4* | | | ***52*** |

List L

| | | | | | | | | |
|---|---|---|---|---|---|---|---|---|
| 3,2 | 2,2 | 6,6 | 6,3 | 6,2 | 3,1 | 5,3 | 5,2 | 3,6 |

Offspring

| | | | | | | | | |
|---|---|---|---|---|---|---|---|---|
| $M_1$ | 2,1 | 1,6 | 2,4 | 2,3 | 3,3 | 4,6 | | ***50*** |
| $M_2$ | 1,4 | 3,2 | 5,1 | 4,4 | 3,5 | | | ***60*** |
| $M_3$ | 1,1 | 1,3 | 2,2 | 6,6 | 6,3 | 6,2 | | ***86*** |
| $M_4$ | 4,3 | 2,6 | 4,1 | 1,5 | 5,4 | 4,2 | | ***71*** |
| $M_5$ | 3,1 | 5,3 | 6,1 | 2,5 | 6,5 | 5,6 | 5,2 | ***82*** |
| *$M_6$* | *1,2* | *3,6* | *3,4* | *4,5* | *5,5* | *6,4* | | ***76*** |

## Evolutionary Algorithm

In (Ghedjati, 1994) the author tried to solve the FJSP with the Operation/Rank/Machine and the Operation/priority rules/Heuristics representations. Varieties of genetic algorithms based on static and dynamic heuristics were designed to improve the solution quality. The representation chosen by (Kacem et al., 2002) is the tasks sequencing list. In (Mesghouni et al., 2004), the authors developed two genetic encodings, the parallel job and parallel machine representations of the chromosome and their associated genetic operators such as row crossover, column crossover and controlled mutation operator as it can balance the machine loads. In (Gao et al., 2008), a hybrid genetic and variable neighborhood descent (VND) algorithm in which the genetic algorithm (GA) uses two vectors to provide the machine assignment and operation sequence information where two representation methods are used: Gen et al.'s [9] representation and

*Figure 19. Mutation of one operation's assignment (Ghedjati, 1994)*

The chromosome

| Operation | **11** | 12 | 13 | 14 | 15 | 21 | 22 | 31 | 32 | 33 | 41 | 42 |
|---|---|---|---|---|---|---|---|---|---|---|---|---|
| Machine | **3** | 2 | 5 | 3 | 4 | 1 | 2 | 3 | 1 | 2 | 4 | 2 |
| Rang | **1** | 2 | 1 | 3 | 2 | 1 | 3 | 2 | 2 | 4 | 1 | 1 |

The mutated chromosome

| Operation | **11** | 12 | 13 | 14 | 15 | 21 | 22 | 31 | 32 | 33 | 41 | 42 |
|---|---|---|---|---|---|---|---|---|---|---|---|---|
| Machine | **1** | 2 | 5 | 3 | 4 | 1 | 2 | 3 | 1 | 2 | 4 | 2 |
| Rang | **1** | 2 | 1 | 3 | 2 | 1 | 3 | 2 | 2 | 4 | 1 | 1 |

The mutated chromosome after reparation

| Operation | **11** | 12 | 13 | **14** | 15 | 21 | 22 | **31** | **32** | 33 | 41 | 42 |
|---|---|---|---|---|---|---|---|---|---|---|---|---|
| Machine | **1** | 2 | 5 | **3** | 4 | 1 | 2 | **3** | **1** | 2 | 4 | 2 |
| Rang | **2** | 2 | 1 | **2** | 2 | 1 | 3 | **1** | **3** | 4 | 1 | 1 |

permutation representation. The VND is applied to each newly generated offspring in order to improve its quality before being injected into the population. In (Mitsuo et al., 2009), a multistage-based genetic algorithm with bottleneck shifting for the FJSP is proposed. A priority-based decoding is used before the chromosome involves Phenotype-based crossover and mutation operators. The bottleneck shifting worked over two neighborhoods, which use interchange of operation sequences and assignment of new machines for operations on the critical path. In order to enhance the search ability, the neighborhood structure can be adjusted dynamically in the local search heuristic. In their genetic algorithm (Zhang et al., 2011) have designed global selection (GS) and local selection (LS) methods in order to improve the quality of initial solution. An enhanced chromosome coding called "Machine Selection and Operation Sequence" is used to represent a solution of the FJSP. Then different suitable strategies for selection, crossover and mutation operators are adopted.

## CONCLUSION

The flexible job shop problem has attracted researches since it has many applications such as manufacturing and distributed computing. Only few works have treated real cases problems with further specific domains constraints such as the existence of more than one production plan or the management of many flexible systems. In this case, two assignment problems are considered job assignment and operation assignment. This variant has a direct application in distributed systems to dispatch workflows. The well-known makespan objective is always optimized by the resolution approaches but other criteria are interesting in presence of flexibility like the machines workloads and the job lateness. The genetic algorithms have generated powerful solutions to the benchmark test instances. Several resolution techniques are applied to the FJSP and neither one dominates the others on the well-known benchmark test sets. Genetic algorithms are competitive and the design of new adaptation is still a promising direction.

## REFERENCES

Bagchi, S., Uckun, S., Miyabe, Y., & Kawamura, K. (1991).Exploring problem-specific recombination operators for job-shop scheduling. In *Proceedings of the Fourth International Conference on Genetic Algorithms and their Applications* (pp. 10-17).

Chen, H., Ihlow, J., & Lehmann, C. (1999). A genetic algorithm for flexible job shop scheduling, In *Proceedings of IEEE International Conference on Robotics and Automation* (Vol. 2, pp. 1120-1125). doi:10.1109/ROBOT.1999.772512

Garey, M. R., Johnson, D. S., & Sethi, R. (1976). The Complexity of Flow Shop and Job-Shop Scheduling. *Mathematics of Operations Research, 1*(2), 117–129. doi:10.1287/moor.1.2.117

Ghedjati, F. (1994). Résolution par des heuristiques dynamiques et des algorithmes génétiques du problème d'ordonnancement de type Job-Shop généralisé (à machines non identiques en parallèle et contraintes de précédence). Unpublished doctoral dissertation, University of Paris.

Ho, N. B., & Tay, J. C. (2004, June 19-23). GENACE: An Efficient Cultural Algorithm for Solving the Flexible Job-Shop Problem. In *Proceedings of the IEEE Congress on Evolutionary Computation* (*Vol. 2*, pp. 1759-1766). doi:10.1109/CEC.2004.1331108

Kacem, I. (2003). Ordonnancement mutlticritère des Job-Shop Flexible: formulation, bornes inférieures et approche évolutionniste coopérative. Unpublished doctoral dissertation, University of Lille.

Kacem, I., Hammadi, F., & Borne, P. (2001a, October 7-10). Approach by Localisation And Genetic Manipulation algorithm for Flexible Job-shop Problems. In *Proceedings of the International IEEE Conference on systems & cybernetics*, Tucson, Arizona (pp. 2599-2604).

Kacem, I., Hammadi, F., & Borne, P. (2001b). Direct Chromosome Representation and Advanced Genetic Operators for Job-shop Problems. In *Proceedings of the International Conference on Computational Intelligence for Modelling, Control and Automation (CIMCA'01)*, Las Vegas, NV.

Kacem, I., Hammadi, F., & Borne, P. (2001c). Multi-objective Optimization for Flexible Job-Shop Scheduling Problem: Hybridization of Genetic Algorithms with Fuzzy Logic. In *Proceedings of IFDICON'2001, European Workshop on Intelligent Forecasting, Diagnosis and Control*, Santorini, Greece.

Kobayashi, S., Ono, I., & Yamamoura, M. (1995). An efficient genetic algorithm for Job-Shop scheduling Problem. In *Proceedings of the 6th International Conference on Genetic Algorithms* (pp. 506-511).

Lee, K.-M., & Yamakawa, T., & Lee, Keon-Myung., (1998).A genetic algorithm for general machine scheduling problems. In *Knowledge-Based Intelligent Electronic Systems, Proceedings KES '98. Second International Conference* (Vol. 2, pp. 60-66), 21-23 Apr 1998.

Mesghouni, K., Hammadi, S., & Borne, P. (2004). Evolutionary Algorithms for Job-Shop scheduling. *International Journal of Applied Mathematics and Computer Science, 14*(1), 91–103.

Naifar, F., Gzara, M., Moukrim, A., & Loukil, T. (2006). Hybrid Evolutionary Algorithm With Insertion Heuristics For The Flexible Job Shop Problem. In *Proceedings of the International Conference on Service Systems and Service Management*. doi:10.1109/ICSSSM.2006.320681

Paredis, J. (1992). *Exploiting constraints as background knowledge for genetic algorithms: A case-study for scheduling*. The Netherlands: Ever Science Publishers.

Roy, B., & Sussmann, B. (1964). *Les problèmes d'ordonnancement avec contraintes disjonctives (Note DS N°.9 bis)*. SEMA Montrouge.

Tay, J. C., & Wibowo, D. (2004). An Effective Chromosome Representation for Evolving Flexible Job Shop Schedules. In K. Deb et al. (Eds.), GECCO 2004, LNCS (Vol. 3103, pp. 210-221). Springer-Verlag Berlin Heidelberg 2004.

Uckun, S., Bagchi, S., Kawamura, K., & Miyabe, Y. (1993). Managing genetic search in Job-Shop scheduling. *IEEE Expert*, *8*(5), 15–24. doi:10.1109/64.236477

Yamada, T., & Nakano, R. (1992), A genetic algorithm applicable to a large scale Job-Shop problems. In R. Manner & B. Mandrick (Eds.), Proceedings of Parallel problem solving from nature (Vol. 2, pp. 281-290). Amsterdam: North Holland.

Zandieh, M., Mahdavi, I., & Bagheri, A. (2008). Solving a flexible job-shop scheduling problems by a Genetic Algorithm. *Journal of Applied Sciences*, *8*(24), 4650–4655. doi:10.3923/jas.2008.4650.4655

Zhang, H. P., & Gen, M. (2005). Multistage-based genetic algorithm for flexible job shop scheduling problem. *Journal of Complexity International*, *48*, 409–425.

## KEY TERMS AND DEFINITIONS

**Flexible Machine:** A machine that can execute different types of operations.

**Gantt Chart:** The Gantt-Chart is a convenient way of visually representing a solution of the scheduling problem.

**Genetic Algorithm:** A genetic algorithm is a metaheuristic inspired by the process of natural selection to solve optimization problems.

**Job Shop Scheduling:** The basic form of the problem of scheduling jobs with multiple (M) operations, over M machines, such that all of the first operations must be done on the first machine, all of the second operations on the second, etc.

**Metaheuristic:** A top-level general strategy which can be adapted to search for feasible solutions in domains where the task is hard.

**Scheduling:** Scheduling is the process of arranging, controlling and optimizing work and workloads in a production process or manufacturing process.

# Chapter 2
# Application and Evaluation of Bee-Based Algorithms in Scheduling:
## A Case Study on Project Scheduling

**Ayse Aycim Selam**
*Marmara University, Turkey*

**Ercan Oztemel**
*Marmara University, Turkey*

## ABSTRACT

*Scheduling is a vital element of manufacturing processes and requires optimal solutions under undetermined conditions. Highly dynamic and, complex scheduling problems can be classified as np-hard problems. Finding the optimal solution for multi-variable scheduling problems with polynomial computation times is extremely hard. Scheduling problems of this nature can be solved up to some degree using traditional methodologies. However, intelligent optimization tools, like BBAs, are inspired by the food foraging behavior of honey bees and capable of locating good solutions efficiently. The experiments on some benchmark problems show that BBA outperforms other methods which are used to solve scheduling problems in terms of the speed of optimization and accuracy of the results. This chapter first highlights the use of BBA and its variants for scheduling and provides a classification of scheduling problems with BBA applications. Following this, a step by step example is provided for multi-mode project scheduling problem in order to show how a BBA algorithm can be implemented.*

## INTRODUCTION

Scheduling, a complex optimization problem is an area of research that needs improvement using new methods and approaches. Solving this problem, which has a huge number of constraints; within acceptable levels of time and precision is a challenge. Scheduling problems can be solved up to some degree using traditional engineering models, algorithms, heuristics and meta-heuristics. Hence, methods employed

DOI: 10.4018/978-1-5225-2944-6.ch002

by living creatures in order to solve their problems for survival in nature; are inspiring researchers in many different ways which help to develop models for solving daily life problems. Optimization is one of the prominent fields where natural systems are providing foresight to generate acceptable solutions (Pham et al., 2005; Karaboga and Akay 2010).

Various natural systems (social insect colonies) such as bees or bacteria (Escherichia coli bacteria) indicate that very simple individual organisms can create systems which are able to perform highly complex tasks by dynamically interacting with each other and adopting social foraging behavior (Teodorovic et al., 2006; Tang, Nouri, and Motlagh, 2011). In recent years, a noticeable pattern has been observed in the area of academic scheduling where many complex problems were efficiently solved using the principles of meta-heuristics (Teoh, Wibowo, and Ngadiman, 2015). For instance, as an evolutionary computational approach, Passino (2002) is inspired by the foraging behavior of Escherichia coli bacteria in human intestines. Here, the bacterium striving to maximize the energy gained per unit of foraging time is seen as an optimization process (Passino, 2002). Tang, Nouri, and Motlagh, (2011) adapted this approach to the machine cell formation problem.

A recent algorithm based on the interesting breeding behavior such as brood parasitism of certain species of cuckoos combined with the Lévy flight behavior of some birds and fruit flies, is an example of this kind (Yang and Deb, 2009). Cuckoo Search Algorithm is thought to be a new and efficient population-based heuristic evolutionary algorithm for solving optimization problems with the advantages of simple implementation and few control parameters (Nguyen, Vo, and Truong, 2014). It is applied to the problems of hybrid flow shop scheduling (Marichevlam, Prabaharan, and Yang, 2014; Dasgupta and Das, 2015), short-term hydrothermal scheduling (Nguyen, Vo, and Truong, 2014; Nguyen and Vo, 2015; Nguyen and Vo, 2016), multi-objective scheduling (Chandrasekaran and Simon, 2012; Akbari and Rashidi, 2016) and 2-machine robotic cell scheduling (Majumder and Laha, 2016).

Fireflies are creatures which have also affected researchers with their bioluminescence abilities. The flashing light of fireflies attract mating partners (communication) and potential prey, and this light is formulated in such a way that it is associated with the objective function to be optimized (Yang, 2009). The researchers idealized some of the flashing characteristics of fireflies to develop a firefly inspired algorithm to solve optimization problems (Gandomi, Yang and Alavi, 2011; Ritthipakdee et al., 2014).

In another study, by using the navigation method of moths in nature which is called transverse orientation, the moth-flame optimization method is developed (Mirjalili, 2015). This method is also called moth swarm optimization and mimics to the orientation of moths towards moonlight to solve the constrained Optimal Power Flow problem (Mohamed et al., 2017).

Besides, the Bat Algorithm introduced to literature by Yang (2011) is inspired by the echolocation property of bats. This property is a type of sonar that guides bats in their flying and hunting behavior and provides them to move and distinguish different types of insects even in complete darkness (Yang, 2010). Topal and Altun introduced the dynamic virtual bat algorithm using only two bats to find the optimal solution (2016). Chakri et al., (2017) developed the directional bat algorithm to overcome premature convergence that can occur due to the low exploration ability of the standard bat algorithm. Some problems solved using bat algorithms are optimization problems (Yılmaz and Küçüksille, 2015), neural networks (Jaddi, Abdullah, and Hamdan, 2015), process planning (Wang et al., 2015), planning the sports training sessions (Fister et al., 2015), and visual tracking (Gao et al., 2016a).

Furthermore, the most recent bio-inspired algorithms are flower pollination algorithm (Yang, Karamanoğlu, and He, 2014) developed for applications in the domain of global optimization problems with multiple diverse criteria and multiple objectives and artificial plant algorithm (Cui and Cai, 2013) especially suitable for problems which are non-differential, multimodal, and high-dimensional in nature (Kar, 2016).

Apart from these, the natural bee system is one of those that can find promising food sources by communicating within the hive performing a complex real-life optimization problem. The Bees Algorithm is inspired by the food foraging behavior of honey bees and could be regarded as belonging to the category of "intelligent" optimization tools (Chong et al. 2006). It is a search algorithm capable of locating good solutions as efficiently as possible. Note that the bees are naturally searching for food and carrying the information outside the hive inside in order to find the best quality nectar and the maximum amount to fill the hive. Similarly, the BA is a search algorithm which is capable of locating good solutions efficiently.

The Bee-Based Algorithms (BBA) are applied on many scheduling problems, some of which are listed below.

- Single machine scheduling (Pham et al., 2007a).
- Flow shop scheduling (Pham and Koç, 2008; Pan et al., 2011; Taşgetiren et al., 2011; Liu and Liu, 2013; Taşgetiren et al., 2013).
- Open shop scheduling (Huang and Lin, 2011).
- Job shop scheduling (Chong et al., 2006; Wong et al., 2007; Zhang, 2011; Zhang et al., 2013).
- Project scheduling (Öztemel and Selam, 2010; Akbari et al., 2011; Tahooned and Ziarati, 2011; Ziarati et al., 2011).

The researchers have been influenced by the mechanism of swarms and how creatures in colonies (immigrating birds, fish flocks, ant colonies, bee colonies, etc.) interact with each other to survive in nature (Van Petegham and Vanhoucke, 2014; Salem and Hassine, 2015). This chapter describes one of these so called the Bee Colony Optimization (BCO) and its variants such as Artificial Bee Colony (ABC). Scheduling which is a manufacturing problem is known to be one of the promising application area of these algorithms as they may provide more effective schedules. In the preceding sections, the recent BBA are explained, following the scheduling problems defined and BBA applications from literature are first classified. Finally, a step by step example is provided for the multi-mode project scheduling problem showing how the algorithm is implemented and conclusions are drawn afterwards.

## AN OVERVIEW ON BEE-BASED ALGORITHMS

As mentioned above, solving especially np-hard problems within acceptable time duration and precision is a challenge. Finding the optimal solution for this type of complex multi-variable optimization problems with polynomial computation times is not or extremely hardly possible (Pham et al., 2005; Pham et al., 2006a). Complex problems of these nature can be solved up to some degree using traditional engineering models, algorithms, heuristics and meta-heuristics (Teodorovic et al., 2006). On the hand, the approaches employed by living creatures to solve their problems in the nature are inspiring the researchers in many different ways to develop problem solving algorithms especially for np-hard problems (Pham et al., 2005; Karaboga and Akay, 2010).

Bee based algorithms are inspired from food foraging bees in order to maximize the nectar intake to the hive. Researchers imitated this behavior to find optimal solutions for their problems. This section introduces the variants of BBA (BA, BCO, and ABC) and describes how they are utilized in optimization.

One of these initiatives, BCO approach is introduced by Nakrani and Tovey (2004), uses collective bee intelligence in solving combinatorial problems characterized by uncertainty. With this approach bees (also called artificial bees) communicate directly, and move in a way that the solution components are added to partial solution until creating a feasible one.

This BBA process consists of several iterations pseudo code of which is given in Table 1.

Here, forward and backward passes could be performed until some other stopping condition (the maximum total number of forward/backward passes, or the maximum total number of forward/backward passes between two objective function value improvements) is satisfied. In the related literature, this approach is being used for many applications including continuous optimization problems, training neural networks, mechanical and electronic components design optimization, combinatorial optimization problems such as job shop scheduling, internet server optimization problem, TSP, etc. (Nakrani and Tovey 2004). According to Teodorovic et al. (2006) various BCO algorithms can be developed describing the ways bees decide to act while creating and expanding solutions.

Secondly, BCO proposed by Pham et al. (2006a), is also inspired by the food foraging behavior of honey bees and could be regarded as belonging to the category of "intelligent" optimization tools. In nature, the bees are searching for food and carrying the information outside the hive to find the best quality and maximum amount of nectar in order to fill the hive. Similarly, the BCO is a search algorithm capable of locating good solutions efficiently. It is well-proven that the BCO is to be able solve complex optimization problems (Nakrani and Tovey, 2004; Pham et al., 2005; Pham et al., 2006a; Karaboga and Akay, 2010; Akay and Karaboga, 2012). Note that the food foraging system of honey bees is explained by Pham et al. as the following (Pham et al., 2006a): ABC is another approach introduced to literature by Karaboga and Baştürk (2008). In this approach, there are three types of bees: employed bees, onlookers and scouts. The colony is divided to two parts first of which are consists of the employed artificial bees and the second are the onlookers. Only one employed bee is assigned to every food source which means the number of employed bees is equal to the number of food sources. When a food source falls into disuse its employed bee becomes a scout. The search of the artificial bees can be summarized as follows;

*A colony of honey bees can extend itself over long distances (more than 10 km) and in multiple directions simultaneously to exploit a large number of food sources. The foraging colony sends scout bees to these flower patches. Scout bees move randomly from one patch to another. After returning to the hive the scout bees who found a promising patch go to the dance floor. On the dance floor scout bees perform a dance called waggle dance which gives information about the flower patch they found. The dance gives*

*Table 1. Pseudocode of bee colony optimization (Teodorovic et al., 2006)*

| (1) Initialization. Determine the number of bees B, and the number of iterations I. Select the set of stages $ST = \{st_1, st_2, ..., st_m\}$. Find any feasible solution $x$ of the problem. This solution is the *initial best solution*.<br>(2) Set $i: = 1$. Until $i = I$, repeat the following steps:<br>(3) Set $j = 1$. Until $j = m$, repeat the following steps:<br>*Forwardpass*: Allow bees to fly from the hive and to choose $B$ partial solutions from the set of partial solutions $S_j$ at stage $st_j$.<br>*Backwardpass*: Send all bees back to the hive. Allow bees to exchange information about quality of the partial solutions created and to decide whether to abandon the created partial solution and become again uncommitted follower, continue to expand the same partial solution without recruiting the nestmates, or dance and thus recruit the nestmates before returning to the created partial solution. Set, $j: = j + 1$.<br>(4) If the best solution $x_i$ obtained during the $i$-th iteration is better than the best- known solution, update the best known solution ($x: = x_i$).<br>(5) Set, $i: = i + 1$. |
|---|

*information about the quality rating, direction and distance of the flower patch. With this information, the colony sends the bees to the flower patches. The more the rating of the flower patch is, the more the follower bees are sent there. While harvesting from a patch, the bees monitor its food level. If the patch is still good enough as a food source then it will be advertised in the next waggle dance and more bees will be recruited to that source.*

While solving a scheduling problem with BCO, an initial set of "n" solutions (scout bees) are randomly chosen. Note that, each "scout bee" (n) represents a solution to the scheduling problem in question. The problem is solved using a neighborhood search approach. This search is carried out by changing the values of some of the variables in the objective function. In this way, some alternative but close solutions around the existing one can be generated. Once defined, "n" solutions are evaluated and ranked according to fitness for purpose (an objective function). The rating process ends with identifying "e" very best solutions, and "m" best solutions of the first iteration.

This process results with a series of neighborhood solutions around "(e) + (m-e)" points neighborhood search is respectively defined for the problem depending on the type and variables of the objective function. The number of neighborhood search around "e" solutions is "nep" and the number of neighborhood search around "m-e" solutions is "nsp". The best solution found during a neighborhood search is passed onto the next iteration and this process is repeated until the stopping criterion is met.

The BCO requires a number of parameters set as the following;

- Number of scout bees so called solutions (n), are randomly sent to the solution space, so that (n) sites (alternative solutions, points, vectors, etc.) are dealt with in the solution space.
- Number of sites (solutions) selected out of n alternatives (m), where each site is ranked and (m) site at the top is selected.
- Number of best sites out of m selected sites (e) so called elite sites, which are the number sites selected from (m) sites that are of minimum (or the maximum) solution values in the order.
- Number of bees recruited for best e sites (nep), which are the number of scout bees that will be seeking if there is a better solution around (e) solutions achieved up to that moment.
- Number of bees recruited for other (m-e) selected sites (nsp), which are responsible to find out if there is a better solution around the remaining (m-e) solutions.

ABC is another approach introduced to literature by Karaboga and Baştürk (2008). In this approach, there are three types of bees: employed bees, onlookers and scouts. The colony is divided to two parts of which the first consists of the employed artificial bees and the second are the onlookers. Only one employed bee is assigned to every food source which means the number of employed bees is equal to the number of food sources. When a food source falls into disuse its employed bee becomes a scout. The search of the artificial bees can be summarized as follows;

- *Employed bees determine a food source within the neighbourhood of the food source in their memory.*

- *Employed bees share their information with onlookers within the hive and then the onlookers select one of the food sources.*

- *Onlookers select a food source within the neighbourhood of the food sources chosen by themselves.*

*- An employed bee of which the source has been abandoned becomes a scout and starts to search a new food source randomly.*

The main steps of ABC algorithm is given in Table 2.

The experiments on some benchmark problems show that the BBA outperforms deterministic Simplex Method, Stochastic Simulated Annealing Optimization procedures, Genetic Algorithms (GA), Tabu Search (TS) and Ant Colony Optimization (ACO) etc. in terms of the speed of optimization and accuracy of the results (see Karaboğa and Baştürk, 2008; Pham et al., 2006a). In the initial research carried out by Pham et al. (2006a), the BCO is applied to some functional optimization problems such as DE Jong, Goldstein and Price, Branin, Martin and Gaddy, Rosenbrock, Hyper Sphere, and Griewank. They showed that the proposed algorithm has remarkable robustness producing 100% success rate in all cases. Similarly, Karaboğa and Baştürk (2008) showed that ABC algorithm performs better than Differential evolution (DE), Particle Swarm Optimization (PSO) and Evolutionary Algorithm (EA) for multi-dimensional and multimodal parameters, testing well known Griewank, Rastrigin and Rosenbrock functions with 50 parameters.

In general, the proposed bee methods handle constraints in a similar behavior and also follow a similar approach to evaluate their individuals and use a similar representation and decoding for them (Ziarati et al., 2011). Some other known algorithms based on Bee Swarm Intelligence are Virtual Bee, BeeAdHoc, the marriage in honeybees, the BeeHive, Bee System (Karaboğa et al. 2014). After the introduction of BCO, case studies and comparative studies appear in literature continually. BCO has been applied to several cases on training of neural networks, scheduling, TSP, data clustering, manufacturing cell formation, robotics, p-center problem, and large scale numerical problems, stochastic problems, multi-objective optimization, etc. Some application areas from literature (but not limited to) are provided in Table 3. The scheduling applications are evaluated in the proceeding section, so that they are not mentioned here. In the next section parameters of BCO (Based on Pham et al., 2005) is explained in detail.

*Table 2. Main steps of ABC algorithm (Karaboga and Baştürk 2008)*

| Initialize<br>REPEAT<br>• Move the employed bees onto their food sources and determine their nectar amounts.<br>• Move the onlookers onto the food sources and determine their nectar amounts.<br>• Move the scouts for searching new food sources.<br>• Memorize the best food source found so far.<br>UNTIL (requirements are met) |
|---|

*Table 3. Applications of BCO from literature*

| Research Area | Author(s) & Year |
|---|---|
| Training of neural networks | Pham et al. (2006b), Pham et al. (2006c), Pham et al. (2006e), Pham and Sholedolu (2008), Ozturk and Karaboga (2008), Pham and Darwish (2010) |
| Robotics | Pham et al. (2007a), Pham, Castellini, and Fahmy (2008), Pham et al. (2008b), Karaboğa a and Akay (2010) |
| TSP | Wong, Low, Chong, (2008), Wong, Low, Chong, (2010) |
| Clustering | Pham et al., (2007b), Pham et al., (2008c) |

## SCHEDULING PROBLEMS IN GENERAL

Since, scheduling problems are highly dynamic, complex and require extra resources as well as satisfy a set of various but inevitable constraints, it is obvious and well known that generating the solution algorithms is not easy. Most of the time, it is not even possible to find out the optimum solution. Some domain dependent and specifically designed solution algorithms as well as some procedures based on domain related assumptions are required for generating "good enough" solutions. Moreover, the difficulty of changing the schedules based upon the changes on resources, durations or respective activities is yet another problem requiring systematic attention and solution procedures. Intelligent algorithms and heuristics have therefore been a research interest for the related research.

Some of those tools and methods generated and employed in this respect can be listed as below.

- ACO (Li and Zhang, 2013).
- Swarm Intelligence (SI) (Pacini et al., 2014; Salem and Hassine, 2015).
- Genetic Algorithms (Okada et al., 2010; Van Petegham and Vanhoucke, 2010; Barrios et al., 2011; Afshar-Nadjafi et al., 2013).
- Priority rule-based heuristics (Buddhakulsomsiri and Kim, 2017).
- Classical and nonstandard meta-heuristics (Banaszak and Zaremba, 2006; Van Petegham and Vanhoucke, 2014).

Note that the scheduling problems in literature are classified according to the machine environment. The fundamentals of scheduling problems are single machine scheduling, parallel machine scheduling, flow shop scheduling, open shop scheduling, job shop scheduling, and finally a special case project scheduling.

As well-known in single machine scheduling, there is only one machine to process the jobs. This model is important in literature because some complicated cases can be degraded to single machine environment. Some objectives for this type of scheduling are total weighted completion time, maximum lateness, total tardiness, number of tardy jobs, etc.

On the other hand, parallel machine scheduling environment consists of 2 or more similar machines which have the ability of processing the same jobs. This way, a job can be processed on either one of the machines. The parallel machines may have the identical speed or different speeds (Allahverdi, 2015). In this environment, the target is to schedule which job will be processed on which machine aiming to, for example, minimize the makespan which is the total completion time, or to maximize lateness.

Similarly, flow shop machining is the environment when there is a series of machines to process the jobs in the same route. The waiting time between successive machines is one of the problems to be sorted out in generating good schedules in this case. Generally, the makespan is the most important objective of this type with limited and unlimited storage capacity of the machines. Flexible flow shops are the special case of flow shops where there are a set of parallel machines (multi-processor) at each step. The jobs have to be processed on one of these parallel machines at each step following the same route.

Unlike the flow shop, in the job shop scheduling problems, there is a fixed route, but it is not the same for each job. Branch-and-bound procedure and shifting bottleneck heuristic are some methods to solve this problem. A recent situation is recirculation where a job can be processed on machines more than once. The availability of recirculation affects the solution of the problem.

Furthermore; open shop machine environments are multi-operation models where the route is decided by the scheduler. The routes are not fixed for each job and the problem objects to makespan, maximum lateness or number of tardy jobs with/without preemption.

Other than above, there are cases of scheduling problems with specific conditions which reveal new problems. One of them, project scheduling problem (PSP), is a special case of parallel machine scheduling problem. In general, PSPs are defined having $n$ jobs, subjecting to precedence constraints in an unlimited parallel machine environment. The objective can be minimizing makespan, and/or minimizing the project cost. The well-known traditional techniques used to solve this problem are Critical Path Method (CPM) and Project Evaluation and Review Technique (PERT). It would not be underestimating the traditional methods by saying that the complexity of the real-world problems necessitates more capable methodologies such as BA and its variants.

## SCHEDULING USING BEE-BASED ALGORITHMS

As stated above, the BBA has been applied to several cases, some of which are on training of neural networks, production scheduling, traveling salesman problem, data clustering, manufacturing cell formation, robotics, p-center problem, and large-scale numerical problems, stochastic problems, multi-objective optimization, etc. However, scheduling is one of the optimization problems studied with BBA applications. In this section, the scheduling applications of BBA is provided according to scheduling environment. In a former study, single machine scheduling using BA was investigated by Pham et al., (2007a) showing that the BA performed more strongly than the existing techniques.

A frequent type of scheduling problem solved using BCO is permutation flow shop and lot streaming flow shop scheduling. Pham and Koc (2008) carried out a study on the flow shop sequencing tasks testing Talliard's benchmark problems. In another study, a discrete ABC algorithm is enhanced with a local search approach to solve the lot-streaming flow shop scheduling problems was proposed by Pan et al., (2011). The proposed algorithm obtained good results for total weighted earliness and tardiness criterion. Tasgetiren et al. (2011) presented a discrete ABC hybridized with a variant of iterated greedy algorithms to find the permutation that gives the smallest total flow time. Their method was compared against the best performing algorithms from the existing literature in terms of both solution quality and CPU times.

Similarly, Liu and Liu (2013) presented a hybrid discrete Artificial Bee Colony algorithm to minimize the makespan in the permutation flow shop scheduling problem. Their algorithm was tested on 21 problems by Reeves to prove its efficiency. Similarly, a discrete artificial bee colony algorithm was presented by Tasgetiren et al. (2013) to solve the no-idle permutation flow shop scheduling problems with the total tardiness. Highly competitive performance was achieved when compared to the genetic algorithm.

Pan et al. (2014) solved hybrid the flow shop scheduling problem using discrete ABC algorithm and compared the makespan with PSO and the artificial immune approach (AIS), which are the two best-performing algorithms for the hybrid flow shop with the makespan criterion in the literature.

A comparative study on job shop scheduling is performed by Chong et al. (2006), where BCO was compared to ACO and TS heuristics. Wong et al. (2008) improved BCO algorithm with Big Valley Exploitation and compared the performance of the method for job shop scheduling. Zhang (2011) solved job shop scheduling problems by an ABC algorithm. The computational results for different problem sizes showed that the proposed Algorithm was both effective and efficient. A novel and efficient ABC

Algorithm was also proposed by Zhang et al. (2013) aiming to minimize the total weighted tardiness in job shop scheduling problems.

Li, Pan and Taşgetiren (2014) studied a discrete ABC for multi objective flexible job shop scheduling showing the best performance from the literature. A two-stage ABC is applied to flexible job shop scheduling by Gao et al. (2015) and compared with parallel variable neighborhood search (PVNS), knowledge-based ant colony optimization (KBACO), tabu search algorithm with efficient neighborhood structure (TSPCB) effective artificial bee colony algorithm (EABC) and a simple and effective evolutionary algorithm (SEA). Also, for new job insertion three rescheduling strategies are proposed. In another study, Gao et al., (2016b) used an improved ABC algorithm for the flexible job-shop scheduling problem with fuzzy processing time. Sundar et al. (2016) proposed a hybrid ABC for job shop scheduling with no-wait constraint and compared with complete local search with limited memory (CLLM) and modified complete local search with memory (MCLM).

Huang and Lin (2011) proposed a BCO algorithm with idle-time-based filtering scheme for open shop scheduling problems. In their study, the filtering scheme could automatically stop searching a partial solution with insufficient profitability, while creating a new solution, and saving time.

One of the earlier study on utilizing BCO for PSP was carried out by Akbari et al. (2011). Here, an ABC Optimization was adapted to Resource Constrained PSP (RCPSP) by investigating its performance on several case studies from PSPLIB. However, the algorithm proposed was only able to solve the problem in single-mode resources. Each bee searched the space in n (which is the number of activities of the project) directions. The Artificial Bee Colony optimization was shown to be superior to the other algorithms used for the same purpose.

Minimizing project duration is a crucial problem for also stochastic RCPSP where ABC finds effective solutions as proposed by Tahooned and Ziarati (2011). In that study, the activity durations were modeled by uniform and exponential distributions. By taking the mean values of those, the project became deterministic. The expected makespans were then calculated according to the priority list and number of respective scenarios.

Three methods based on "scout bees" including BA, ABC, Bee Swarm Optimization and their variations empowered by new local search algorithms were also investigated for RCPSP by Ziarati et al. (2011). This produced a new constraint handling method to resolve infeasible solutions and performing better when compared with the other methods existing in the literature.

The applications of bee algorithms on the different types of scheduling problems from the literature is listed in Table 4. The classes of problems, which algorithm is used, and how it is compared is provided in detail. It can be clearly seen that same data sets used for comparison are big size scheduling problems. For example, j90 and j120 resource constrained problem cases can be counted as big size problems and the BBA generally outperforms other methods (Akbari et al., 2011, Ziarati et al., 2011; Tahooned et al., 2011).

## AN APPLICATION ON PROJECT SCHEDULING: A CASE STUDY

Project scheduling is a complex activity requiring computationally effective solutions. BCO algorithm is an alternative for this and proven to be able to cope with the respective computational complexity to some degree. In this section, BA is applied to a multi-mode project scheduling problem. Note that a very simple problem is explained here for the sake of better understanding. In literature, some big scale

*Table 4. Literature survey of scheduling using BBA*

| Author, Year | Type(s) of Scheduling Problem | Bee System(s) Used | Compared Methods | Compared Problems and / or Datasets | Major Findings |
|---|---|---|---|---|---|
| Chong et al., 2006 | Job shop | BCO | ACO, TDTS | 82 job shop problem instances ranging from 6 to 50 jobs and 5 to 20 machines | The performance of the algorithm is comparable to ACO algorithms, but gaps behind the efficient TS heuristics. |
| Pham et al., 2007a | Single machine | BA | GA, Particle swarm optimization (PSO), TS, and hybrids | Biskup and Feldmann's benchmark problems | Computational results show that the BA performed more strongly than the existing techniques. |
| Pham and Koç, 2008 | Permutation flow shop | BA | GA, PSO | Talliard's benchmark problems | The algorithm generally outperforms other techniques in terms of the quality of the job sequences produced. |
| Wong et al., 2008 | Job shop | BCO | Shifting bottleneck heuristic, TS and GA | Talliard's benchmark problems | Comparable to approaches provided. |
| Akbari et al., 2011 | Resource constrained project scheduling (RCPSP) | ABC | GA, PSO, ACO, ANGEL, GAPS, OOP-GA, PSO+, ACOSS, Neurogenetic | j30, j60, j90, and j120 problem cases of the PSPLIB | ABC provides an efficient way for solving RCPSP. |
| Huang and Lin, 2011 | Open shop | BCO | PSO | Talliard's benchmark problems | The average Cmax and time required are better. |
| Pan et al., 2011 | Lot-streaming flow shop | Discrete ABC | Discrete PSO, GA | 20 Problems are generated | Computational simulations and comparisons demonstrated the effectiveness and efficiency of the proposed Discrete ABC algorithm. |
| Tahooned et al., 2011 | Stochastic RCPSP | ABC | Genetic algorithm, GRASP algorithm | Instances from PSPLib | The ABC algorithm provides an efficient way to solve the stochastic RCPSP. |
| Taşgetiren et al., 2011 | Permutation flow shop | Discrete ABC | Hybrid discrete differential evolution (hDDE) algorithm, iterated greedy algorithm | Talliard's benchmark problems | Proposed algorithms were superior to the traditional IG_RS algorithm, and the performances of the Discrete ABC and hDDE algorithms are highly competitive. |
| Zhang, 2011 | Job shop | ABC | Hybrid particle swarm optimization | randomly generated different-scale jobshop instances | The computational results for problems of different sizes show that the proposed algorithm is both effective and efficient. |
| Ziarati et al., 2011 | RCPSP | BA, ABC, and bee swarm optimization | GA, TS, ACO, PSO and their hybrids | j30, j60, j90, and j120 problem cases of the PSPLIB | The proposed bee methods provide competitive results compared to the other methods investigated in this work. |
| Liu and Liu, 2013 | Permutation flow shop | Hybrid discrete ABC | PSO variable neighborhood search, PSO based memetic algorithm | 21 problems by Reeves | Computational results and comparisons demonstrate that the hybrid Discrete ABC is competitive. |
| | | | ACO, GA | Talliard's benchmark problems | |
| Taşgetiren et al., 2013 | Permutation flow shop | Discrete ABC | GA | Talliard's benchmark problems | The computational results show its highly competitive performance when compared to the genetic algorithm. |

*continued on following page*

*Table 4. Continued*

| Author, Year | Type(s) of Scheduling Problem | Bee System(s) Used | Compared Methods | Compared Problems and / or Datasets | Major Findings |
|---|---|---|---|---|---|
| Zhang et al., 2013 | Job shop | Hybrid ABC | Hybrid GA (genetic local search) | Pinedo & Singer's benchmark problems & instances from ORLib | Computational results verify the effectiveness and efficiency of the proposed approach, especially for larger-scale instances. |
| Li, Pan and Taşgetiren, 2014 | Flexible job shop | Discrete ABC | PVNS, KBACO, TSPCB, etc. | (1) five Kacem instances, (2) BR data, set of 10 problems by Brandimare. | The proposed DABC algorithm had the ability to obtain promising solutions for the problem considered. |
| Pan et al., 2014 | Hybrid flow shop | Discrete ABC | PSO, AIS | Carlier & Neron benchmark set of 77 instances | The results showed that the discrete ABC outperforms the other algorithms when solving the hybrid flow shop problem with the makespan criterion. |
| Gao et al., 2015 | Flexible job shop | Two-Stage ABC | PVNS, KBACO, TSPCB, EABC, and SEA | (1) five Kacem instances (2) BR data, set of 10 problems by Brandimare | The results and comparisons show that Two-Stage ABC is effective in both scheduling stage and rescheduling stage. |
| Gao et al., 2016b | Flexible job shop | Improved ABC | MinEnd heuristics | (1) 5 cases with fuzzy processing time (2) eight cases from a remanufacturing enterprise | The results by improved ABC algorithm is more competitive than the compared algorithms. |
| Sundar et al., 2016 | Job shop | Hybrid ABC | CLLM, MCLM | 21 small sized & 40 large sized instances from literature | Computational results demonstrate that overall hybrid ABC is better than both CLLM and MCLM on most of the instances in terms of solution quality and computational time. |

problems (j30, j60, j90, and j120 problem cases of the PSPLIB) are solved using BBA (Akbari et al., 2011; Tahooned et al., 2011; Ziarati et al 2011) and the solutions are compared with GA, TS, ACO, PSO and their hybrids. Each of the studies showed the competitiveness and efficiency of the proposed BBA. Here, a step by step solution procedure is provided in order to cover all solution space for the sake of validation.

Consider a scheduling problem of MPS8, ∞ | prec | Cmax with n=8 activities. The mathematical model of multi-mode project scheduling problem is provided below:

Objective Function:

$$Min = \sum_{j=1}^{m_{I+1}} \sum_{t=1}^{T} tx_{I+1,j,t} \tag{1}$$

Activity completion constraints:

$$S.t. \quad \sum_{j=1}^{m_i} \sum_{t=1}^{T} x_{ijt} = 1 \qquad \forall i \tag{2}$$

Precedence constraint:

$$Max_{e\in \mathrm{Pre}(i)}\left(\sum_{j=1}^{m_e}\sum_{t=1}^{T} tx_{ejt}\right)+\sum_{j=1}^{m_i}\sum_{t=1}^{T} p_{ij}x_{ijt} \leq \sum_{j=1}^{m_i}\sum_{t=1}^{T} tx_{ijt}, \qquad \forall i \tag{3}$$

Resource constraint:

$$\sum_{i=1}^{I}\sum_{j=1}^{m_i}\sum_{s=t}^{t+p_{ij}-1} r_{ijk}^{\rho}x_{ijs} \leq a_k^{\rho} \qquad \forall i \tag{4}$$

Decision variable:

$$x_{ijt} = 0 \qquad or \qquad 1, \qquad \forall i,j,t \tag{5}$$

Resource alternatives, predecessor and successor activities are provided for this project is shown in Table 5. Note that the second column illustrates the resource modes of each activity, where each activity can be completed using either one of the resources. Note that, only a single resource can be allocated to complete the activity. Predecessor activities in third column are precedence constraints of the problem.

Table 6 indicates activity times for each resource that the activity can be completed. Activities will be allocated to one of the resources and its duration will be assigned for that activity.

After providing this information, the project scheduling model can be executed as follows.

**Step 1:** Draw the project network using the precedence constraints.

Using the information provided, a precedency diagram of the project network is drawn as shown in Figure 1.

**Step 2:** Determine respective paths to complete the project.

*Table 5. Information for sample problem*

| Activity (Job) | Resource Alternatives | Predecessor Activity | Successor Activity |
|---|---|---|---|
| 1 | R1, R2 | - | 2, 3 |
| 2 | R3, R5 | 1 | 4, 5 |
| 3 | R6, R7 | 1 | 5 |
| 4 | R4, R8 | 2 | 6 |
| 5 | R2, R7 | 2, 3 | 6, 7 |
| 6 | R1, R8 | 4, 5 | 8 |
| 7 | R2, R3 | 5 | 8 |
| 8 | R6, R4 | 6, 7 | - |

*Table 6. Activity times for the resources*

| Activity | Resource | Duration | Resource | Duration |
|---|---|---|---|---|
| 1 | R1 | $t_{11} = 5$ | R2 | $t_{12} = 3$ |
| 2 | R3 | $t_{23} = 1$ | R5 | $t_{25} = 6$ |
| 3 | R6 | $t_{36} = 8$ | R7 | $t_{37} = 3$ |
| 4 | R4 | $t_{44} = 9$ | R8 | $t_{48} = 4$ |
| 5 | R2 | $t_{52} = 5$ | R7 | $t_{57} = 3$ |
| 6 | R1 | $t_{61} = 7$ | R8 | $t_{68} = 2$ |
| 7 | R2 | $t_{72} = 4$ | R3 | $t_{73} = 2$ |
| 8 | R6 | $t_{86} = 6$ | R4 | $t_{84} = 4$ |

*Figure 1. Project Network of the sample Problem*

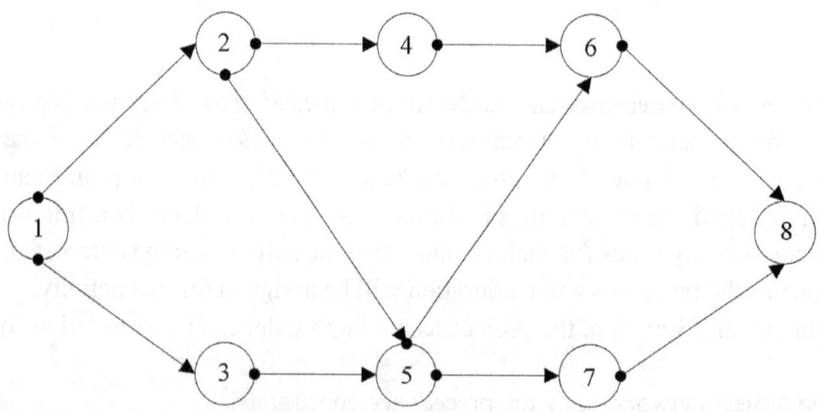


In this sample project, there are 5 paths to complete the overall project. These are:

- 1-2-4-6-8,
- 1-2-5-6-8,
- 1-2-5-7-8,
- 1-3-5-6-8,
- 1-3-5-7-8.

**Step 3:** Define the solution space.

As well known in project scheduling, the shortest completion time of the project is the resembled by the longest path duration so called *critical path.* In this case, to find out the optimum project completion time, each alternative path has to be traced in such a way that the longest path will be the shortest project completion time for each resource-activity combination. After finding the project completion time for each combination, the resources are allocated for the selected mode.

That presents a challenge especially in large scale projects. If $r_i$ represents the number of modes for activity i, and n (1, 2, 3, …, i, i+1, …, n) is the total number of activities to be completed; then there occurs ($r_1 * r_2 * \ldots * r_i * r_{i+1} * \ldots * r_n$) alternative cases.

For example, if there are 4 activities (jobs) in a project where job 1 can be completed using one of the 4 resources, job 2 can be completed using one of the 2 resources, job 3 can be completed using one of the 3 resources, job 1 can be completed using one of the 5 resources; then we there will be 4 * 2 * 3 * 5 = 120 alternative ways to complete the project.

In the sample problem here, each job has 2 alternative resources; this in turn creates $2^8 = 256$ alternative solutions. From these, the shortest completion time will be sought. This list of alternative solutions is considered to be the solution space (project pool) where the bees will scout for the best solution. Finally the stopping criterion of the algorithm for this case is set as to decrease the project pool to a traceable size. When the project pool is small enough to trace with the n scout bees, then there will not be further search for better a solution. This will be the final iteration remaining with the best solution.

**Step 4:** Restructure the problem as BA (problem mapping).

In this step project scheduling problem is transformed into the Bee foraging problem. This is called "problem mapping". A scout bee in the project scheduling problem will search a single path combination that can complete the project. There will be (n) scout bees in the project pool searching for a suitable path combination leading to the best solution. The project pool is formed by alternative solutions and represents the solution space of the respective scheduling problem. Similar to the visited sites in foraging problem, (n) will also indicate the number of varying paths (equal to number of scout bees) using alternative resources with different durations or having different costs for the specified tasks in the schedule.

Using the same kind of analogy, (m) will be the number of solutions selected out of n different path combinations whereas (e) will stand for the number of best solutions chosen among the m solutions.

The number of bees recruited for best (e) sites is referred as (nep) representing the number of sites to be searched around the best (e) solutions. Similarly (nsp) will denote the number of sites to be searched around the other (m-e) solutions.

The foraging allocation mechanism described in Table 7 is adopted from the pseudo code of Pham et. al., (2006).

*Table 7. Resource allocation mechanism*

| |
|---|
| 1. Initialize project pool with random schedules. |
| 2. Define completion times and evaluate alternative schedules. |
| 3. While (stopping criterion not met) // Forming new project pool. |
| a. Select schedules for neighborhood search. |
| b. Allocate number of schedules for the selected schedules (more searching area for best e schedules). |
| c. Select the most suitable schedule for the project pool. |
| d. Assign remaining schedules to search randomly and evaluate their fitness. |
| 4. End while. |

**Step 5:** Apply BCO to the problem mapped.

In this section, the proposed BCO based scheduling model is explained step by step. The project pool consists of alternative solutions (patches). The initial size of the patches (ngh) can be defined as the quantity of solutions in the project pool. In this sample problem, this is set to 256 alternative solutions.

Other parameters (n, m, e, nep, and nsp) are defined as the following.

- The number of visited sites (n) = 10.
- The number of selected sites (m) = 2.
- The number of best sites (e) = 1.
- The number of bees recruited for best (e) sites (nep) =10.
- The number of bees recruited for the other (m-e) sites (nsp) = 4.
- The stopping criterion = the total number of iteration.

(Note that the number of iteration is problem specific and defined as to be equal to 10. For larger projects a bigger number of iterations would be necessary.)

Following determining these parameters; the next step is to initialize and populate the project pool with random solutions. Based on the number of visited sites, 10 random solutions are selected for this purpose. Among those alternative solutions, the best 2 of them (m) are selected and the one very best solution (e) is identified. The completion times of best solutions are calculated as the following

- 20 time units for $m_{1.}$
- 22 time unit for $m_{2.}$

Note that, the best completion time among these is $m_1$ with 20 time units. This is in turn is set as the best solution (e) of the project at this iteration. Note that, $m_1$ and $m_2$ solutions were considered to be the focal points for further neighborhood search.

For the neighborhood of $m_1$ 10 new solutions (nep) are evaluated. Similarly, within the neighborhood of $m_2$ 4 new solutions (nsp) are assessed.

This process generated new set of best solutions which are;

- $m_1$ with 17 time units.
- $m_2$ with 20 time units.

This indicates an improvement of the completion time of the project from 20 to 17 time units. This completes the 1st iteration.

The second iteration is carried out in the same manner. A new project pool with 235 alternative solutions are generated. Note that only the very best solution with 17 time units are carried to the next pool and the remaining 13 solutions are left out of the solution space as they were already evaluated. In this iteration, the first assessment indicates that the following solutions are the best two solutions.

- $m_1$ with 17 time units.
- $m_2$ with 19 time units.

This indicates that the optimum solution for this generation is to be sought mainly around $m_1$. Implementing the same approach, the new solutions are generated and the optimum solution is found to be 15 units of completion time. This process is repeated until the $10^{th}$ iteration. It was experienced that the optimum solution was no longer possible to be improved. At the end, the optimum solution of project completion time is found to be 15 units. The respective resource combination for each activity can be identified easily as provided in Table 8.

## CONCLUSION AND DISCUSSIONS

In this chapter, scheduling, complex optimization problems that need improvement using new methods and approaches is studied. Solving the complex scheduling problems with huge numbers of constraints within acceptable levels of time and precision is still a big challenge. Many techniques and methods are generated and implemented. Some of them such as BA and its variants are considered to be intelligent optimization tools which are dealt with elaborately in this study.

The BA algorithms are capable of locating good solutions and mimicking the behavior of honey bees. Honey bees search for food and carry the information outside the hive inside to find the best quality nectar and the maximum amount in order to fill the hive. Analogous to this, in scheduling problems the solution space is searched by selected variables for identifying the best schedule.

It is also proven that the BBA gives promising solutions not only on best schedule but also on the time required to solve the problems. This was the case in this study as well. Nearly, in all cases provided in Table 4 the performance of BBA are comparable to existing methods (GA, ACO, TS, PSO, etc.). The computational results show the efficiency and effectiveness of the proposed BBA models. It is shown in literature that the BBA are superior to others in many ways (time, precision, performance and the number of iterations). Moreover, BBA can be enhanced by hybridizing various algorithms for the sake of better performance.

In the previous section, an example for a multi-mode project scheduling problem is solved. The size is kept minimum for the ease of understanding behind the algorithm. Also, the methods are applicable to large scale problems in literature (see section "Scheduling Using Bee Algorithms"). For the sake of validation, the proposed model is implemented on a small scale problem hence, it will be implemented

*Table 8. Best resource utilization for the sample problem*

| Activity | Resources | Duration |
|---|---|---|
| 1 | R2 | $t_{12}=3$ |
| 2 | R3 | $t_{23}=1$ |
| 3 | R7 | $t_{37}=3$ |
| 4 | R8 | $t_{48}=4$ |
| 5 | R7 | $t_{57}=3$ |
| 6 | R8 | $t_{68}=2$ |
| 7 | R3 | $t_{73}=2$ |
| 8 | R4 | $t_{84}=4$ |

on more complex problems in future studies. However, some adjustments and improvements are still required for further extensions of the scheduling problems. For example, it is known that in a scheduling problem, finding the makespan is not the only objective function. Tardiness, lateness, penalty costs are some other goals of objective functions in scheduling. Similar to the related research, it is claimed that BA can solve multi-objective problems efficiently. The scheduling problems can be can be adopted to solve competing objectives.

As well as finding an optimal schedule, adjusting the activities to sudden changes is also important in scheduling. Many interruptions such as breakdowns, can occur during the realization of activities. These affect the defined schedule and sometimes the activity durations change during realization. A flexible scheduling method is needed for these situations, where BBA seems to be promising in this area. The method is thought to be fast responding to sudden changes, which is another advantage in scheduling.

## REFERENCES


Afshar-Nadjafi, B., Rahimi, A., & Karimi, H. (2013). A genetic algorithm for mode identity and the resource constrained project scheduling problem. *Scientia Iranica*, *20*(3), 824–831.

Akay, B., & Karaboga, D. (2012). Artificial bee colony algorithm for large-scale problems and engineering design optimization. *Journal of Intelligent Manufacturing*, *23*(4), 1001–1014. doi:10.1007/s10845-010-0393-4

Akbari, M., & Rashidi, H. (2016). A multi-objectives scheduling algorithm based on cuckoo optimization for task allocation problem at compile time in heterogeneous systems. *Expert Systems with Applications*, *60*, 234–248. doi:10.1016/j.eswa.2016.05.014

Akbari, R., Zeighami, V., & Ziarati, K. (2011). Artificial Bee Colony for Resource Constrained Project Scheduling Problem. *International Journal of Industrial Engineering Computations*, *2*(1), 45–60. doi:10.5267/j.ijiec.2010.04.004

Allahverdi, A. (2015). The third comprehensive survey on scheduling problems with setup times/costs. *European Journal of Operational Research*, *246*(2), 345–378. doi:10.1016/j.ejor.2015.04.004

Banaszak, Z. A., & Zaremba, M. B. (2006). Project-Driven Planning and Scheduling Support for Virtual Manufacturing. *Journal of Intelligent Manufacturing*, *17*(6), 641–651. doi:10.1007/s10845-006-0034-0

Barrios, A., Ballestin, F., & Valls, V. (2011). A double genetic algorithm for the MRCPSP/max. *Computers & Operations Research*, *38*(1), 33–43. doi:10.1016/j.cor.2009.09.019

Buddhakulsomsiri, J., & Kim, D. S. (2007). Priority rule-based heuristic for multi-mode resource-constrained project scheduling problems with resource vacations and activity splitting. *European Journal of Operational Research*, *178*(2), 374–390. doi:10.1016/j.ejor.2006.02.010

Chakri, A., Khelif, R., Benouaret, M., & Yang, X. S. (2017). New directional bat algorithm for continuous optimization problems. *Expert Systems with Applications*, *69*, 159–175. doi:10.1016/j.eswa.2016.10.050

Chandrasekaran, K., & Simon, S. P. (2012). Multi-objective scheduling problem: Hybrid approach using fuzzy assisted cuckoo search algorithm. *Swarm and Evolutionary Computation*, *5*, 1–16. doi:10.1016/j.swevo.2012.01.001

Chong, C. S., Low, M. Y. H., Sivakumar, A. I., & Gay, K. L. (2006). A Bee Colony Optimization Algorithm to Job Shop Scheduling. In *Proceedings of the 2006 Winter Simulation Conference* (pp. 1954–1961). doi:10.1109/WSC.2006.322980

Cui, Z., & Cai, X. (2013). Artificial Plant Optimization Algorithm. In *Swarm Intelligence and Bio-Inspired Computation: Theory and Applications* (p. 351).

Dasgupta, P., & Das, S. (2015). A discrete inter-species cuckoo search for flowshop scheduling problems. *Computers & Operations Research*, *60*, 111–120. doi:10.1016/j.cor.2015.01.005

Fister, I., Rauter, S., Yang, X. S., Ljubič, K., & Fister, I. Jr. (2015). Planning the sports training sessions with the bat algorithm. *Neurocomputing*, *149*, 993–1002. doi:10.1016/j.neucom.2014.07.034

Gandomi, A. H., Yang, X. S., & Alavi, A. H. (2011). Mixed variable structural optimization using firefly algorithm. *Computers & Structures*, *89*(23), 2325–2336. doi:10.1016/j.compstruc.2011.08.002

Gao, K. Z., Suganthan, P. N., Chua, T. J., Chong, C. S., Cai, T. X., & Pan, Q. K. (2015). A two-stage artificial bee colony algorithm scheduling flexible job-shop scheduling problem with new job insertion. *Expert Systems with Applications*, *42*(21), 7652–7663. doi:10.1016/j.eswa.2015.06.004

Gao, K. Z., Suganthan, P. N., Pan, Q. K., Chua, T. J., Chong, C. S., & Cai, T. X. (2016b). An improved artificial bee colony algorithm for flexible job-shop scheduling problem with fuzzy processing time. *Expert Systems with Applications*, *65*, 52–67. doi:10.1016/j.eswa.2016.07.046

Gao, M. L., Shen, J., Yin, L. J., Liu, W., Zou, G. F., Li, H. T., & Fu, G. X. (2016a). A novel visual tracking method using bat algorithm. *Neurocomputing*, *177*, 612–619. doi:10.1016/j.neucom.2015.11.072

Huang, Y., & Lin, J. (2011). A new bee colony optimization algorithm with idle-time-based filtering scheme for open shop-scheduling problems. *Expert Systems with Applications*, *38*(5), 5438–5447. doi:10.1016/j.eswa.2010.10.010

Jaddi, N. S., Abdullah, S., & Hamdan, A. R. (2015). Optimization of neural network model using modified bat-inspired algorithm. *Applied Soft Computing*, *37*, 71–86. doi:10.1016/j.asoc.2015.08.002

Kar, A. K. (2016). Bio inspired computing–A review of algorithms and scope of applications. *Expert Systems with Applications*, *59*, 20–32. doi:10.1016/j.eswa.2016.04.018

Karaboga, D., & Akay, B. (2010). Proportional-Integral-Derivative Controller Design by Using Artificial Bee Colony, Harmony Search, and the Bees Algorithms. *Journal of Systems and Control Engineering*, *224*(1), 869–883.

Karaboga, D., & Baştürk, B. (2008). On the Performance of Artificial Bee Colony (ABC) Algorithm. *Applied Soft Computing*, *8*(1), 687–697. doi:10.1016/j.asoc.2007.05.007

Karaboga, D., Gorkemli, B., Ozturk, C., & Karaboga, N. (2014). A comprehensive survey: Artificial bee colony (ABC) algorithm and applications. *Artificial Intelligence Review*, *42*(1), 21–57. doi:10.1007/s10462-012-9328-0

Li, H., & Zhang, H. (2013). Ant Colony Optimization-Based Multi-Mode Scheduling Under Renewable and Nonrenewable Resource Constraints. *Automation in Construction*, *35*, 431–438. doi:10.1016/j.autcon.2013.05.030

Li, J. Q., Pan, Q. K., & Tasgetiren, M. F. (2014). A discrete artificial bee colony algorithm for the multi-objective flexible job-shop scheduling problem with maintenance activities. *Applied Mathematical Modelling*, *38*(3), 1111–1132. doi:10.1016/j.apm.2013.07.038

Liu, Y. F., & Liu, S. Y. (2013). A Hybrid Discrete Artificial Bee Colony Algorithm for Permutation Flow-shop Scheduling Problem. *Applied Soft Computing*, *13*(3), 1459–1463. doi:10.1016/j.asoc.2011.10.024

Majumder, A., & Laha, D. (2016). A new cuckoo search algorithm for 2-machine robotic cell scheduling problem with sequence-dependent setup times. *Swarm and Evolutionary Computation*, *28*, 131–143. doi:10.1016/j.swevo.2016.02.001

Marichelvam, M. K., Prabaharan, T., & Yang, X. S. (2014). Improved cuckoo search algorithm for hybrid flow shop scheduling problems to minimize makespan. *Applied Soft Computing*, *19*, 93–101. doi:10.1016/j.asoc.2014.02.005

Mirjalili, S. (2015). Moth-flame optimization algorithm: A novel nature-inspired heuristic paradigm. *Knowledge-Based Systems*, *89*, 228–249. doi:10.1016/j.knosys.2015.07.006

Mohamed, A. A. A., Mohamed, Y. S., El-Gaafary, A. A., & Hemeida, A. M. (2017). Optimal power flow using moth swarm algorithm. *Electric Power Systems Research*, *142*, 190–206. doi:10.1016/j.epsr.2016.09.025

Nakrani, S., & Tovey, C. (2004). On Honey Bees and Dynamic Server Allocation in Internet Hosting Centers. *Adaptive Behavior*, *12*(3-4), 223–240. doi:10.1177/105971230401200308

Nguyen, T. T., & Vo, D. N. (2015). Modified cuckoo search algorithm for short-term hydrothermal scheduling. *International Journal of Electrical Power & Energy Systems*, *65*, 271–281. doi:10.1016/j.ijepes.2014.10.004

Nguyen, T. T., & Vo, D. N. (2016). *An efficient cuckoo bird inspired meta-heuristic algorithm for short-term combined economic emission hydrothermal scheduling*. Ain Shams Engineering Journal.

Nguyen, T. T., Vo, D. N., & Truong, A. V. (2014). Cuckoo search algorithm for short-term hydrothermal scheduling. *Applied Energy*, *132*, 276–287. doi:10.1016/j.apenergy.2014.07.017

Okada, I., Zhang, F., Yang, Y., & Fujimura, S. (2010). A Random Key-based Genetic Algorithm for Resource-constrained Project Scheduling Problem with Multiple Modes. In *Proceedings of the International Multi Conference of Engineers and Computer Scientists 2010* (Vol. 1, pp. 106-111).

Öztemel, E., & Selam, A. A. (2010). Project Scheduling Using Bee Colony Optimization. In *Proceedings of the 7th International Conference on Intelligent and Manufacturing Systems* (pp. 709-716).

Özturk, C., & Karaboga, D. (2008, October). Classification by neural networks and clustering with artificial bee colony (abc) algorithm. In *Proceedings of the 6th International Symposium on Intelligent and Manufacturing Systems, Features, Strategies and Innovation* (pp. 110-124).

Pacini, E., Mateos, C., & Garino, C. G. (2014). Distributed job scheduling based on Swarm Intelligence: A survey. *Computers & Electrical Engineering*, *40*(1), 252–269. doi:10.1016/j.compeleceng.2013.11.023

Pan, Q. K., Tasgetiren, M. F., Suganthan, P. N., & Chua, T. J. (2011). A Discrete Artificial Bee Colony Algorithm for the Lot-Streaming Flowshop Scheduling Problem. *Information Sciences*, *181*(12), 2455–2468. doi:10.1016/j.ins.2009.12.025

Pan, Q. K., Wang, L., Li, J. Q., & Duan, J. H. (2014). A novel discrete artificial bee colony algorithm for the hybrid flowshop scheduling problem with makespan minimisation. *Omega*, *45*, 42–56. doi:10.1016/j.omega.2013.12.004

Passino, K. M. (2002). Biomimicry of bacterial foraging for distributed optimization and control. *IEEE control systems, 22*(3), 52-67.

Pham, D. T., Al-Jabbouli, H., Mahmuddin, M., Otri, S., & Darwish, A. H. (2008c). Application of the Bees algorithm to fuzzy clustering. In *Proceedings of the 4th International Virtual Conference on Intelligent Production Machines and Systems*, Dunbeath, Scotland.

Pham, D. T., Castellani, M., & Fahmy, A. A. (2008, July). Learning the inverse kinematics of a robot manipulator using the bees algorithm. In *Proceedings of the 6th IEEE International Conference on Industrial Informatics INDIN '08* (pp. 493-498). IEEE. doi:10.1109/INDIN.2008.4618151

Pham, D. T., Darwish, A. H., Eldukhri, E. E., & Otri, S. (2007a, July). Using the bees algorithm to tune a fuzzy logic controller for a robot gymnast. In *Proceedings of IPROMS 2007 conference*, Cardiff, UK.

Pham, D. T., & Darwish, H. A. (2010). Using the bees algorithm with Kalman filtering to train an artificial neural network for pattern classification. *Proceedings of the Institution of Mechanical Engineers. Part I, Journal of Systems and Control Engineering*, *224*(7), 885–892. doi:10.1243/09596518JSCE1004

Pham, D. T., Ghanbarzadeh, A., Koc, E., & Otri, S. (2006b, January). Application of the bees algorithm to the training of radial basis function networks for control chart pattern recognition. In *Proceedings of 5th CIRP international seminar on intelligent computation in manufacturing engineering (CIRP ICME'06), Ischia, Italy* (pp. 711-716).

Pham, D. T., Ghanbarzadeh, A., Koc, E., Otri, S., Rahim, S., & Zaidi, M. (2005). Bee Algorithm A Novel Approach to Function Optimization (Technical Note: MEC 0501). The Manufacturing Engineering Center, Cardiff University.

Pham, D.T., Ghanbarzadeh, A., Koc, E., Otri, S., Rahim, S., Zaidi, M. (2006a). The Bees Algorithm – A Novel Tool for Complex Optimization Problems. In *Intelligent Production Machines and Systems* (pp. 454–459).

Pham, D. T., & Koc, E. (2008). Flowshop Sequencing Using the Bees Algorithm. *In Proceedings of the IPROMS Innovative Production Machines and Systems Virtual Conference*, Cardiff, Wales, UK.

Pham, D. T., Koç, E., Kalyoncu, M., & Tınkır, M. (2008b). Hierarchical PID controller design for a flexible link robot manipulator using the bees algorithm. *methods, 25*, 32.

Pham, D. T., Koc, E., Lee, J. Y., & Phrueksanant, J. (2007a). Using the Bees Algorithm to Schedule Jobs for a Machine. In *Proceedings of the 8th International Conference on Laser Methodology, CMM and Machine Tool Performance* (pp. 430–439).

Pham, D. T., Otri, S., Afify, A., Mahmuddin, M., & Al-Jabbouli, H. (2007c, May). Data clustering using the bees algorithm. In *Proceedings of 40th CIRP international manufacturing systems seminar.*

Pham, D. T., Otri, S., Ghanbarzadeh, A., & Koc, E. (2006c, April). Application of the bees algorithm to the training of learning vector quantisation networks for control chart pattern recognition. In Proceedings of the Information and Communication Technologies ICTTA'06 (Vol. 1, pp. 1624-1629). IEEE. doi:10.1109/ICTTA.2006.1684627

Pham, D. T., & Sholedolu, M. (2008). Using a hybrid PSO-bees algorithm to train neural networks for wood defect classification. In *Proceedings of the 4th International Virtual Conference on Intelligent Production Machines and Systems, IPROMS.*

Pham, D. T., Soroka, A. J., Ghanbarzadeh, A., Koc, E., Otri, S., & Packianather, M. (2006d, August). Optimising neural networks for identification of wood defects using the bees algorithm. In *Proceedings of the 2006 IEEE International Conference on Industrial Informatics* (pp. 1346-1351). IEEE. doi:10.1109/INDIN.2006.275855

Ritthipakdee, A., Thammano, A., Premasathian, N., & Uyyanonvara, B. (2014). An Improved Firefly Algorithm for Optimization Problems. In *Proceedings of the 5th International Symposium on Advanced Control of Industrial Processes.*

Salem, H., & Hassine, A. B. (2015). Meeting Scheduling Based on Swarm Intelligence, *Procedia. Computer Science*, *60*, 1081–1091.

Sundar, S., Suganthan, P. N., Jin, C. T., Xiang, C. T., & Soon, C. C. (2016). A hybrid artificial bee colony algorithm for the job-shop scheduling problem with no-wait constraint. *Soft Computing.*

Tahooned, A., & Ziarati, K. (2011, June 12-15). Using Artificial Bee Colony to Solve Stochastic Resource Constrained Project Scheduling Problem, Advances in Swarm Intelligence. In *Proceedings of the Second International Conference, ICSI 2011, Chongqing*, China (pp. 293-302).

Tang, S. H., Nouri, H., & Motlagh, O. (2011). A bacteria foraging algorithm for solving multi-period cell formation and subcontracting production planning in a dynamic cellular manufacturing system. *South African Journal of Industrial Engineering*, *22*(2), 80–99. doi:10.7166/22-2-17

Tasgetiren, M. F., Pan, Q. K., Suganthan, P. N., & Chen, A. H.-L. (2011). A discrete artificial bee colony algorithm for the total flowtime minimization in permutation flowshops. *Information Sciences*, *181*(16), 3459–3475. doi:10.1016/j.ins.2011.04.018

Tasgetiren, M. F., Pan, Q. K., Suganthan, P. N., & Oner, A. (2013). A discrete artificial bee colony algorithm for the no-idle permutation flowshop scheduling problem with the total tardiness criterion. *Applied Mathematical Modelling*, *37*(10-11), 6758–6779. doi:10.1016/j.apm.2013.02.011

Teodorovic, D., Lucic, P., Markovic, G., & Dell'Orco, M. (2006). Bee Colony Optimization: Principles and Applications. In *Proceedings of the 8th Seminar on Neural Network Applications in Electrical Engineering.*

Teoh, C. K., Wibowo, A., & Ngadiman, M. S. (2015). Review of state of the art for metaheuristic techniques in Academic Scheduling Problems. *Artificial Intelligence Review*, *44*(1), 1–21. doi:10.1007/s10462-013-9399-6

Topal, A. O., & Altun, O. (2016). A novel meta-heuristic algorithm: Dynamic Virtual Bats Algorithm. *Information Sciences*, *354*, 222–235. doi:10.1016/j.ins.2016.03.025

Van Peteghem, V., & Vanhoucke, M. (2010). A Genetic Algorithm for the Preemptive and Non-Preemptive Multi-Mode Resource-Constrained Project Scheduling Problem. *European Journal of Operational Research*, *201*(2), 409–418. doi:10.1016/j.ejor.2009.03.034

Van Peteghem, V., & Vanhoucke, M. (2014). An Experimental Investigation of Metaheuristics for the Multi-Mode Resource-Constrained Project Scheduling Problem On New Dataset Instances. *European Journal of Operational Research*, *235*(1), 62–72. doi:10.1016/j.ejor.2013.10.012

Wang, J., Fan, X., Zhao, A., & Yang, M. (2015). A Hybrid Bat Algorithm for Process Planning Problem. *IFAC-PapersOnLine*, *48*(3), 1708–1713. doi:10.1016/j.ifacol.2015.06.332

Wong, L., Puan, C. Y., Low, M. Y. H., & Chong, C. S. (2008). Bee Colony Optimization Algorithm with Big Valley Landscape Exploitation for Job Shop Scheduling Problems. In *Proceedings of the 2008 Winter Simulation Conference* (pp. 2050–2058). doi:10.1109/WSC.2008.4736301

Wong, L. P., Low, M. Y. H., & Chong, C. S. (2008, May). A bee colony optimization algorithm for traveling salesman problem. In *Proceedings of the Second Asia International Conference on Modeling & Simulation AICMS '08* (pp. 818-823). IEEE. doi:10.1109/AMS.2008.27

Wong, L. P., Low, M. Y. H., & Chong, C. S. (2010). Bee colony optimization with local search for traveling salesman problem. *International Journal of Artificial Intelligence Tools*, *19*(03), 305–334. doi:10.1142/S0218213010000200

Yang, X. S. (2009, October). Firefly algorithms for multimodal optimization. In *Proceedings of the International symposium on stochastic algorithms* (pp. 169-178). Springer Berlin Heidelberg.

Yang, X. S. (2010). A new metaheuristic bat-inspired algorithm. In Proceedings of the Nature inspired cooperative strategies for optimization (NICSO 2010) (pp. 65-74). Springer Berlin Heidelberg. doi:10.1007/978-3-642-12538-6_6

Yang, X. S., & Deb, S. (2009, December). Cuckoo search via Lévy flights. In *Proceedings of the World Congress on Nature & Biologically Inspired Computing NaBIC '09* (pp. 210-214). IEEE.

Yang, X. S., Karamanoglu, M., & He, X. (2014). Flower pollination algorithm: A novel approach for multiobjective optimization. *Engineering Optimization*, *46*(9), 1222–1237. doi:10.1080/0305215X.2013.832237

Yılmaz, S., & Küçüksille, E. U. (2015). A new modification approach on bat algorithm for solving optimization problems. *Applied Soft Computing*, *28*, 259–275. doi:10.1016/j.asoc.2014.11.029

Zhang, R. (2011). An Artificial Bee Colony Algorithm Based on Problem Data Properties for Scheduling Job Shops. *Procedia Engineering*, *23*, 131–136. doi:10.1016/j.proeng.2011.11.2478

Zhang, R., Song, S., & Wu, C. (2013). A Hybrid Artificial Bee Colony Algorithm for the Job Shop Scheduling Problem. *International Journal of Production Economics*, *141*(1), 167–178. doi:10.1016/j.ijpe.2012.03.035

Ziarati, K., Akbari, R., & Zeighami, V. (2011). On the performance of Bee Algorithms for Resource Constrained Project Scheduling Problem. *Applied Soft Computing*, *11*(4), 3720–3733. doi:10.1016/j.asoc.2011.02.002

## KEY TERMS AND DEFINITIONS

**Ant Colony Optimization:** A probabilistic method that searches for the optimal path that mimics the ants while finding the way from food source to nest.

**Bee Colony Optimization:** A metaheuristic that mimics the food foraging bees in nature.

**Critical Path:** The longest path from start node to end node in a project network. Critical path determines the shortest time that a project can be completed.

**Metaheuristics:** A general concept of algorithms which can be applied to different types of optimization problems.

**Project:** A set of activities to accomplish a one-time effort that meets specifications under limited time, and budget.

**Scheduling:** The methods used to assign resources to a set of activities to be completed under different conditions.

**Swarm Intelligence:** The collective behavior of animal colonies who find organized solutions for their survival in nature.

# Chapter 3
# Metaheuristic Approaches for Extrusion Manufacturing Process:
## Utilization of Flower Pollination Algorithm and Particle Swarm Optimization

**Pauline Ong**
*Universiti Tun Hussein Onn Malaysia (UTHM), Malaysia*

**Desmond Daniel Vui Sheng Chin**
*Universiti Tun Hussein Onn Malaysia (UTHM), Malaysia*

**Choon Sin Ho**
*Universiti Tun Hussein Onn Malaysia (UTHM), Malaysia*

**Chuan Huat Ng**
*Universiti Tun Hussein Onn Malaysia (UTHM), Malaysia*

## ABSTRACT

*Optimization, basically, is a method used to find solutions for a particular problem without neglecting the existing boundaries or limitations. Flower Pollination Algorithm (FPA) is one of the recently developed nature inspired algorithms, based on the intriguing process of flower pollination in the world of nature. The main aim of this study is to utilize FPA in optimizing cold forward extrusion process in order to obtain optimal parameters to produce workpiece with the minimum force load. It is very important to find the most optimal parameters for an extrusion process in order to prevent waste from happening due to trial and error method in determining the optimal parameters and thus, FPA is used to replace the traditional trial and error method to optimize the cold forward extrusion process. The optimization performance of the FPA is then compared with the particle swarm optimization (PSO), in which the FPA shows comparable performance in this regard.*

DOI: 10.4018/978-1-5225-2944-6.ch003

## INTRODUCTION

Forming or better known as metal forming is a metalworking process to form or shape a raw material into desired geometry and shape. Metal working is also can be categorized as a process of forming a metal parts and objects through a mechanical deformation process whereby the workpiece will be reshaped without adding or removing of material. Various process parameters such as the shape of the workpiece, shape of the product, forming sequence, shapes of tools, shapes of the dies, coefficient of friction, and metal forming speed, working temperature and properties of the material as well as the properties of the tools used can be used to characterize a particular metal forming process. In order to improve the quality of the output product and contributes in reducing the production cost, it is very important and significant to determine and figure out the optimum forming parameters beforehand by utilizing optimization techniques (Byon & Hwang, 2003; Kuzman, 2001).

Currently, due to the development of applied mathematics, operational researches, informational computational methods, simulations and design of experiment (DOE) contributed in the improvement of forming technologies by employing the knowledge from the area such as modelling, optimizations, computer techniques and artificial intelligence previously. Today, there is plenty more different approach in optimization methods.

For this particular study, different optimization approaches, particularly, the classical mathematical approach, flower pollination algorithm (FPA) and particle swarm optimization (PSO) algorithm, are used to determine the optimum values of logarithmic strain, die angle as well as the friction factor of cold extrusion process. Experiment plans based on factorial design of experiment (DOE) and orthogonal array has been used to minimize the force required for cold extrusion process and then classical mathematical approach, FPA and PSO optimization have been performed, based on the response model of forming force for the cold extrusion process. By minimizing the extrusion force required, longer tool life, improve formability of workpiece and increment of product quality can be achieved.

The objectives of this study can be summarized as follows:

- To optimize the process parameters of cold forward extrusion, *viz* logarithmic strain, half-die angle and friction factor, such that the extrusion force is minimized.
- To compare the effectiveness of classical mathematical approach, FPA and PSO in optimization of the cold forward extrusion process.

## BACKGROUND

Extrusion is a plastic deformation process in which a block of metal, called the billet, is forced to flow or pass through the die opening of a smaller cross-sectional area than that of the original billet. Extrusion can be classified as hot and cold extrusion. However, for this particular study, only cold extrusion process will be considered. Selection of the optimal process parameters, including the shape of billet, extrude sequence, shapes of dies, friction, extrusion speed, temperature and material properties plays a determinant role to the success of the extrusion process. In this regard, different types of optimization techniques are continuously explored and applied in the process in order to reduce the production cost and to improve the quality of a product (Sadollah & Bahreininejad, 2012). The optimization techniques are used based on the required degree of accuracy for object modeling, and type of processes.

Attempting to minimize the tool load, research on the determination of the optimal cold forward extrusion parameters has been done by Jurkovic et al. (Jurković, Jurković, & Buljan, 2006). In this work, solving the derivation of the mathematical model of extrusion forming force, *F*, is being used to obtain the optimal parameters. Simply put, solving the derivation of the predicted mathematical model, i.e.,

$$\frac{dF}{dx_i} = 0, i = 1, 2, \ldots, n \tag{1}$$

was performed with the aim to find the optimal process parameters, $x_i$.

Classical mathematical optimization technique is a powerful tool to predict the response for any input parameters' values within the experiment domain. However, uses of derivative might be too complex and not easy to use, moreover if the number of process parameters increases. Besides, solving the derivation of the prediction model is tedious and requires very long steps (Jurković et al., 2006).

Solving the extrusion process optimization problem by using the classical mathematical techniques is not an easy task if many parameters are considered. In order to overcome this problem, Taguchi method has been used (Jurković, Brezočnik, Grizelj et al., 2009). In order to achieve the better optimization results, parameter design is the key step in the Taguchi method. In their work, orthogonal array is used where the experimental results are converted into S/N ratio as the measure of quality characteristic deviates from the preferred value. The S/N ratio is expressed as:

$$\frac{S}{N} = \eta = -10\log_{10}\left(\frac{1}{n}\sum_{i=1}^{n} y_i^2\right) \tag{2}$$

By using the minimum number of experimental trials, the Taguchi method is suitable for solving the optimization problems which involve a lot of parameters. By comparing to conventional method, Taguchi method is more easy and simple to be used in order to obtain optimal values of the process parameter (Jurković et al., 2006). In addition, the number of experiments conducted by Taguchi method in most cases is lesser than others which using statistical approach such as finite element method (Nuruddin & Bayuaji, 2009).

Ali Sadollah and Ardeshir Bahreininejad applied the GA in optimizing the die design of a cold forward extrusion process (Sadollah & Bahreininejad, 2012). In their work, minimization of tool load was set as the objective function. GA is very efficient and simple to be used in obtaining the optimal input parameters for variety of problems (Tabassum & Mathew, 2014). However, the convergence to the global optimum at all the time is not guaranteed. Moreover, many parameters in GA need to be fine-tuned, in addition to the slow convergence characteristic (Ong, 2014).

Apart from the classical mathematical approach, Taguchi method and GA, other optimization approaches have been applied in the extrusion process. The Finite Element Method (FEM) was used by Yanran et al. in order to determine the best result in the steady deformation stage of extrusion (Yanran, Wang, & Weimin, 1995). From the obtained deformation force, the optimal semi-code angle of the die was obtained by using FEM. Byon and Hwang have integrated thermo-mechanical finite element process model and derivation of an optimization scheme to optimize the cold and hot extrusion process (Byon & Hwang, 2003). Pathak and Ramakrishanan optimized the die angle and ram velocity by using GA and dynamic material modeling (DMM) (Pathak & Ramakrishnan, 2007).

# MAIN FOCUS OF THE CHAPTER

## Experimental Procedure

In this paper, the flow direction of metal is the same as the direction of action of the punch, where a solid workpiece based on the shape of the die opening as the final product, as shown in Figure 1.

In order to obtain the forward extrusion force value, experimental method such as definite measurement equipment and analytically according to known expression for total extrusion force can be applied (M. Jurković, Barišić, & Jurković, 2000; M. Jurković, Jurković, & Cukor, 2005; Kurt, 1985; Lange, 1985). Generally, forward extrusion force depends on die angle, friction factor, logarithmic strain, material properties as well as the initial geometry of workpiece (Bakhshi-Jooybari, 2002). Central composition design with five levels of the three main independent parameters which are the logarithmic strain (φ), friction factor (μ) and die angle (α) has been used to carried out the experiment, as shown in Table 1. The overall number of experiments conducted for this central composition design is $N = 2^3 + 6 + 6 = 20$ trials (Grizelj & Jurković, 2009). There is a total of eight ($2^3$) factorial designs with addition of six star points and center point which will be repeated six times in order to calculate the pure error of the experiment (Grizelj & Jurković, 2009).

Alloyed carbon steel (DIN16MnCr5) has been taken as the workpiece material for the forward extrusion process performed on hydraulic press. Different friction conditions have been applied to conduct the experiment using $MoS_2$, phosphate surface and oil, grease, oil, moist oil with five coefficients of frictions according to level parameters as the lubricants for the experiment. The initial diameter and height of the workpiece is 30mm and 37mm respectively and has been hold constant throughout the experiment. The mechanical properties and chemical composition of the workpiece are shown in Table 2.

*Figure 1. Extrusion die geometry with billet and final extruded product*

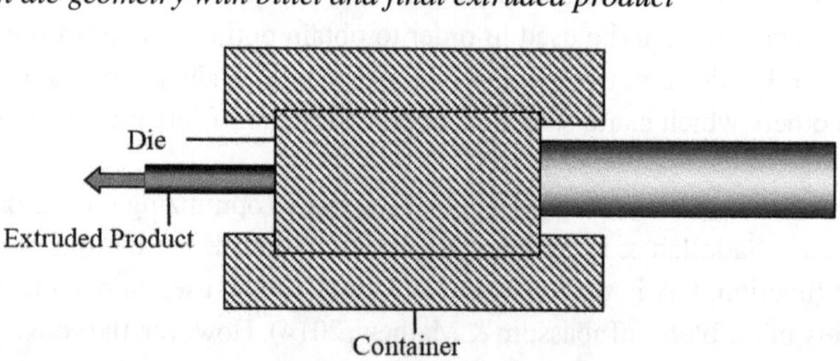


*Table 1. Levels of independent extrusion parameters*

| Symbol | Parameters / Levels | Lowest | Low | Centre | High | Highest |
|---|---|---|---|---|---|---|
| | Coding | -1.6817 | -1 | 0 | +1 | +1.6817 |
| A | Logarithmic strain φ | 0.112 | 0.308 | 0.596 | 0.884 | 1.080 |
| B | Half-die angle α (°) | 10 | 18 | 30 | 42 | 50 |
| C | Friction factor μ | 0.066 | 0.08 | 0.10 | 0.12 | 0.134 |

*Table 2. Mechanical properties and chemical composition of steel 16MnCr5 (DIN)*

| Mechanical Properties of Steel 16MnCr5 | | | | |
|---|---|---|---|---|
| **Tensile Strength MPA** | **Yield Strength MPa** | **Brinell Hardness HB** | **Elongation %** | **Reduction %** |
| 570 | 400 | 160 | 26 | 65 |
| Chemical Composition % | | | | |
| C | Si | Mn | Cr | S |
| 0.16 | 0.30 | 1.15 | 0.95 | 0.030 |

## Extrusion Force Model Prediction

Design of experiment (DOE) has been used for modelling and analysing purposes for this experiment. The experiment can be represented by the relationship between response of extrusion process, for this case the forward extrusion force, and the investigated independent parameters by the following polynomial form of mathematical model:

$$Y = \beta_0 + \sum_{i=1}^{k} \beta_i X_i + \sum_{i=1}^{k} \beta_{ii} X_i^2 + \sum_{i<j}^{k} \beta_{ij} X_i X_j \tag{3}$$

where $Y$ represents the response variable, in this case, the extrusion force. $k$ is the number of coded process parameters $X_i$, in this case, $k$ is equal to 3. $\beta$ is the approximated regression coefficient (RC) from the least square fitting, where the sign and magnitude of RC consider the relative effect of each process parameter for the extrusion force. $\beta_i$, $\beta_{ii}$ and $\beta_{ij}$ denote the relative influences of the linear effect of $X_i$, quadratic effect of $X_i^2$ and as well as the two variable interaction effect of $X_iX_j$, respectively.

In order to determine the unknown in Equation, Matlab R2013a Statistics (STAT) toolbox was used to simulate the polynomial model. The number of design factors must be at minimum in order to prevent the phenomenon of curse of dimensionality, whereby it will cause an equation to be extremely long and practically harder to solve and may cause algorithmic inefficiency as well. Therefore, the only design factor associated with a statistically significant coefficient at confidence level of 95% is considered (Grizelj & Jurković, 2009). The following equation shows the reduced prediction model of extrusion force:

$$F = 607.6453 + 170.403\varphi + 13.7993\alpha + 48.964\mu + 12.199\varphi^2 + 51.6246\alpha^2 \tag{4}$$

Table 3 shows the comparison of the experimental results and the results obtained by the prediction model from Equation 4. The obtained mean squared error (MSE) of 2.14 was considered satisfactorily, in which it can be concluded that the prediction model was able to explain the variation in the extrusion force with respect to the design factors effectively.

*Table 3. Design of experiments with experimental and model results (Grizelj & Jurković, 2009)*

| No | Parameters | | | Extrusion Force (kN) | |
|---|---|---|---|---|---|
| | Logarithmic Strain | Half-Die Angle | Friction Coefficient | Experiment (average F) | Prediction Model (Equation) |
| 1 | 0.308 | 18 | 0.080 | 445 | 438.30 |
| 2 | 0.884 | 18 | 0.080 | 790 | 779.11 |
| 3 | 0.308 | 42 | 0.080 | 478 | 465.90 |
| 4 | 0.884 | 42 | 0.080 | 770 | 806.71 |
| 5 | 0.308 | 18 | 0.120 | 560 | 536.23 |
| 6 | 0.884 | 18 | 0.120 | 860 | 877.04 |
| 7 | 0.308 | 42 | 0.120 | 566 | 563.83 |
| 8 | 0.884 | 42 | 0.120 | 905 | 904.64 |
| 9 | 0.596 | 30 | 0.100 | 610 | 607.65 |
| 10 | 0.596 | 30 | 0.100 | 614 | 607.65 |
| 11 | 0.596 | 30 | 0.100 | 605 | 607.65 |
| 12 | 0.596 | 30 | 0.100 | 611 | 607.65 |
| 13 | 0.596 | 30 | 0.100 | 606 | 607.65 |
| 14 | 0.596 | 30 | 0.100 | 597 | 607.65 |
| 15 | 0.112 | 30 | 0.100 | 338 | 355.58 |
| 16 | 1.080 | 30 | 0.100 | 963 | 928.71 |
| 17 | 0.596 | 10 | 0.100 | 725 | 730.44 |
| 18 | 0.596 | 50 | 0.100 | 799 | 776.85 |
| 19 | 0.596 | 30 | 0.066 | 556 | 525.30 |
| 20 | 0.596 | 30 | 0.134 | 711 | 689.99 |
| | MSE | | | 2.14 | |

## SOLUTIONS AND RECOMMENDATIONS

Optimization method has always been use in order to minimize the production cost of a manufacturing process while retaining or improving the existing quality of the product. The forming condition is the main factor that will impose any increase or decrease in the production cost and quality of the product. Therefore, it is very significant to determine the optimum value of parameters before the forming process, in this case, the extrusion process.

Extrusion is one of the most important processes in common manufacturing process. Optimization in extrusion process helps to ensure that the achievement of metal formation in a consistent thickness and within the allowed tolerance. Attempts to determine the optimal forming conditions eventually inspired the utilization of abundant optimization techniques, such as GA, response graph method, Taguchi's method, and ANNs. For this particular study, the extrusion process parameters were optimized by taking the advantage of the FPA and PSO for the sake of minimizing the forward extrusion force.

## Optimization of Extrusion Force Using Classical Mathematical Approach

The optimal forming conditions can be obtained by solving the first order derivative of the extrusion force model that has been obtained in Equation with respect to the corresponding process parameter by employing the classical mathematical approach. Due to the reason that reduction of extrusion force eventually causes low strain and low coefficient of friction, therefore it is logic that the only parameter to be optimized in this case is the half-die angle. By differentiating Equation, the obtained equation is:

$$\frac{dF}{d\alpha} = \frac{d[607.6453 + 170.403\varphi + 13.7993\alpha + 48.964\mu + 12.199\varphi^2 + 51.6246\alpha^2]}{d\alpha}$$

$$\frac{dF}{d\alpha} = 13.7993 + 103.2492\alpha$$

Taking $\frac{dF}{d\alpha} = 0$ and thus,

$$13.7993 + 103.2492\alpha = 0$$

$$\alpha = -0.1336504$$

Therefore, the obtained optimal half-die angle by the means of classical mathematical analysis is $\alpha_{opt} = 28.40°$. The minimum extrusion force obtained is 399.5549kN.

## Optimization of Extrusion Force Using Flower Pollination Algorithm

The FPA is used for the optimization of die design of cold direct extrusion process. The FPA is inspired by the natural pollination process of flower plant.

The main purpose of a flower is ultimately reproduction via pollination. Flower pollination is typically associated with the transfer of pollen, and such transfer is often linked with pollinators such as insects, birds, bats and other animals. In fact, some flowers and insects have co-evolved into a very specialized flower-pollinator partnership.

Pollination can be achieved by self-pollination or cross-pollination. Cross-pollination, or allogamy, means pollination can occur from the pollen of a flower of a different plant. While self-pollination is the fertilization of one flower, such as peach flowers, from the pollen of the same flower or different flowers of the same plant, which often occurs when there is no reliable pollinator available. Biotic, cross-pollination may occur at long distance, and the pollinators such as bees, bats, birds and flies can fly a long distance (Yang, 2012; Yang, Karamanoglu, & He, 2014). The FPA is designed from all these characteristics.

It is possible to idealize the characteristics of the pollination process, flower consistency and pollinator behaviour with the following rules:

1. Pollen carrying pollinators will perform the Lévy flights. The biotic and cross-pollination will be considered as a global pollination process.
2. Abiotic and self-pollination is considered as a local pollination process.
3. Flower consistency is proportional to the similarity of two flowers involved where flower consistency can be considered as reproduction probability.
4. Switch probability $p \in [0.1]$ is used to control local pollination and global pollination. This is due to the physical proximity and other external factors such as wind local pollination may have a significant fraction $p$ in overall pollination process (Yang, 2012; Yang et al., 2014).

The FPA has been applied to obtain the optimal forming condition of the extrusion process. Equation 4 was coded into the FPA code as the objective function by using Matlab and the obtained results is shown in Table 4.

## Optimization of Extrusion Force Using Particle Swarm Optimization

PSO is one of the population based stochastic optimization method which developed by Dr. Eberhart and Dr. Kennedy in 1995 (Kennedy & Eberhart, 1995). The formulation of PSO algorithm was attempted to simulate the group communication in social behaviour of organisms such as fish schooling and bird flocking. An individual in the swarm is termed as particle in the PSO, in which it represents the potential optimal solution of the problem. Each particle flies through the search space, and shares its individual best position to the swarm. Each particle has a velocity which directs the flying of the particle and its own fitness values which are calculated by the objective function to be optimized. Through such communication, all particles change their own position adaptively based on these shared experiences, eventually converging towards the optimality. This form of searching behaviour can be described as:

$$v_i^{(t+1)} = \omega \cdot v_i^{(t)} + c_1 \cdot rand \cdot \left(pbest_i - x_i^{(t)}\right) + c_2 \cdot rand \cdot \left(gbest - x_i^{(t)}\right) \quad (5)$$

$$x_i^{(t+1)} = x_i^{(t)} + v_i^{(t+1)} \quad (6)$$

where the position of each particle $x_i$ at time step $t+1$ is related to its velocity $v_i$ and $i = 1, 2, ..., N$ denotes the index of N particles. In every iteration, each particle is updated by following two best values

*Table 4. Results obtained by using FPA approach*

| Parameters | Result |
|---|---|
| Logarithmic strain, $\varphi$ | 0.308 |
| Half-die angle, $\alpha$ | 27.22 |
| Friction factor, $\mu$ | 0.08 |
| Extrusion Force, $f_{min}$ | 399.0430kN |

which are *pbest*, the best solution it has achieved so far and *gbest*, the global best position in the population. *rand* is a random number generated from a normal distribution from 0 to 1. The tuning parameters: $\omega$, $c_1$ and $c_2$ are constants which denote the inertia weight, weighting factor of acceleration towards $pbest_i$ and $gbest$, respectively.

The PSO was applied to find the optimal forming parameters in order to obtain the smallest objective value. The obtained extrusion polynomial model in Equation 4 was coded as the fitness function in PSO, and the results obtained are summarized in Table 5.

### Comparison of Classical Mathematical Approach, FPA and PSO Results in Optimization of Extrusion Force

Comparison of results obtained by different methods has been tabulated in Table 6. Based on the results obtained, it can be clearly seen that FPA and classical mathematical approach do not vary much in terms of the results obtained. FPA has the least extrusion force obtained compared to both PSO and classical mathematical approach for this particular study. However, in terms of performance, FPA and PSO outperformed the classical mathematical approach as it is deemed to be too complicated and hard to solve.

## FUTURE RESEARCH DIRECTIONS

This work hopefully provides another insight for the cold forward extrusion optimization using the metaheuristic approaches, specifically, the FPA and PSO. Although from the performance comparison shows that the FPA gives the best optimization performance, much effort needs to be done for further improvement. The searching process in the early stage of FPA may require a lot of time because of the random walk behaviour. To improve the effectiveness of FPA, the step size in the Lévy flight which related

*Table 5. Results obtained by using PSO approach*

| Parameters | Result |
|---|---|
| Logarithmic strain, $\varphi$ | 0.308 |
| Half-die angle, $\alpha$ | 28.76 |
| Friction factor, $\mu$ | 0.08 |
| Extrusion Force, $f_{min}$ | 406.3505kN |

*Table 6. Comparison of optimal results obtained by different approach*

| Parameters | Classical Mathematical Approach | FPA | PSO |
|---|---|---|---|
| Extrusion Force, F (kN) | 399.5549 | 399.0430 | 406.3505 |
| Optimal Half-Die Angle, α (°) | 28.40 | 27.22 | 28.76 |

to the scale of the search space, can be adjusted adaptively instead of using a constant value. With the adaptive step size, the searching process will search in the region with high probability containing the optimal solution. Moreover, hybridization of FPA with other algorithms might enhance the performance of FPA in solving the optimization problems. Through hybridization, the result obtained may be better.

Lastly, apart from the FPA, more recent developed metaheuristic algorithms, for instance, the cuckoo search algorithm (Yang & Deb, 2014), and moth flame algorithm (Mirjalili, 2015), can be utilized to find the optimal process parameters of the extrusion process.

## CONCLUSION

Different optimization approaches have been applied to find the optimal cold forward extrusion parameters with emphasis on the geometrical aspect of the process, which is the die angle. However, based on the analysis that has been done, it can be seen that the FPA performs better in order to obtain appropriate and acceptable results for the optimal extrusion force. This is due to the reason that FPA performance is more consistent and efficient as compared to the classical mathematical approach. In short, the FPA can be applied to improve initial process parameters or in the study case of the minimization of extrusion force by means of an optimal die angle with high accuracy.

## REFERENCES


Bakhshi-Jooybari, M. (2002). A theoretical and experimental study of friction in metal forming by the use of the forward extrusion process. *Journal of Materials Processing Technology*, *125*, 369–374. doi:10.1016/S0924-0136(02)00343-6

Byon, S., & Hwang, S. (2003). Die shape optimal design in cold and hot extrusion. *Journal of Materials Processing Technology*, *138*(1), 316–324. doi:10.1016/S0924-0136(03)00092-X

Grizelj, B., & Jurković, Z. (2009). Optimization of extrusion process by genetic algorithms and conventional techniques. *Tehnički vjesnik: znanstveno-stručni časopis tehničkih fakulteta Sveučilišta u Osijeku*, *16*(4), 27-33.

Jurković, M., Barišić, B., & Jurković, Z. (2000). *Mathematical modelling and optimization of machining processes.*

Jurković, M., Jurković, Z., & Cukor, G. (2005). Genetic algorithm application in optimization of extrusion forming process. *Paper presented at the 10th International Scientific Conference on Production Engineering CIM '05.*

Jurković, Z., Brezočnik, M., Grizelj, B., & Mandić, V. (2009). Optimization of extrusion process by genetic algorithms and conventional techniques. *Technical Gazette*, *16*(4), 27–33.

Jurković, Z., Jurković, M., & Buljan, S. (2006). Optimization of extrusion force prediction model using different techniques. *Journal of Achievements in materials and manufacturing engineering*, *17*(1-2), 353-356.

Kurt, L. (1985). *Handbook of metal forming*. Society of Manufacturing Engineers.

Kuzman, K. (2001). Problems of accuracy control in cold forming. *Journal of Materials Processing Technology, 113*(1), 10–15. doi:10.1016/S0924-0136(01)00688-4

Lange, K. (1985). Handbook of metal forming. McGraw-Hill Book Company.

Mirjalili, S. (2015). Moth-flame optimization algorithm: A novel nature-inspired heuristic paradigm. *Knowledge-Based Systems, 89*, 228–249. doi:10.1016/j.knosys.2015.07.006

Nuruddin, M., & Bayuaji, R. (2009). Application of Taguchi's approach in the optimization of mix proportion for Microwave Incinerated Rice Husk Ash foamed concrete. *International Journal of Cyber Ethics in Education, 9*, 121–129.

Ong, P. (2014). Adaptive Cuckoo Search Algorithm for Unconstrained Optimization. *TheScientificWorldJournal, 8*. doi:10.1155/2014/943403 PMID:25298971

Pathak, K., & Ramakrishnan, N. (2007). Optimization of die angle and ram velocity for rod extrusion using dynamic material modeling and genetic algorithm. *Indian Journal of Engineering and Materials Sciences, 14*(6), 399.

Sadollah, A., & Bahreininejad, A. (2012). Optimization of die design using metaheuristic methods in cold forward extrusion process. *Neural Computing & Applications, 21*(8), 2071–2076. doi:10.1007/s00521-011-0630-6

Tabassum, M., & Mathew, K. (2014). A genetic algorithm analysis towards optimization solutions. *International Journal of Digital Information and Wireless Communications, 4*(1), 124–142. doi:10.17781/P001091

Yang, X.-S. (2012). *Flower pollination algorithm for global optimization. In Unconventional computation and natural computation* (pp. 240–249). Springer. doi:10.1007/978-3-642-32894-7_27

Yang, X.-S., & Deb, S. (2014). Cuckoo search: Recent advances and applications. *Neural Computing & Applications, 24*(1), 169–174. doi:10.1007/s00521-013-1367-1

Yang, X.-S., Karamanoglu, M., & He, X. (2014). Flower pollination algorithm: A novel approach for multiobjective optimization. *Engineering Optimization, 46*(9), 1222–1237. doi:10.1080/0305215X.2013.832237

Yanran, Z., Wang, Z., & Weimin, C. (1995). Numerical simulations for extrusion and ironing and die-angle optimization. *Journal of Materials Processing Technology, 55*(1), 48–52. doi:10.1016/0924-0136(95)01811-5

## ADDITIONAL READING

Alam, M. S., Pathania, S., & Sharma, A. (2016). Optimization of the extrusion process for development of high fibre soybean-rice ready-to-eat snacks using carrot pomace and cauliflower trimmings. LWT -. *Food Science and Technology (Campinas.), 74*, 135–144. doi:10.1016/j.lwt.2016.07.031

Ansari, M. A., Behnagh, R. A., Narvan, M., Naeini, E. S., Givi, M. K. B., & Ding, H. (2015). Optimization of Friction Stir Extrusion (FSE) Parameters Through Taguchi Technique. *Transactions of the Indian Institute of Metals*.

Ashhab, M. S., Breitsprecher, T., & Wartzack, S. (2014). Neural network based modeling and optimization of deep drawing - Extrusion combined process. *Journal of Intelligent Manufacturing*, *25*(1), 77–84. doi:10.1007/s10845-012-0676-z

Bakhtiari, H., Karimi, M., & Rezazadeh, S. (2014). Modeling, analysis and multi-objective optimization of twist extrusion process using predictive models and meta-heuristic approaches, based on finite element results. *Journal of Intelligent Manufacturing*. doi:10.1007/s10845-014-0879-6

Barbara, R., Lorenzo, D., & Luca, T. (2016). Multi-goal optimization of industrial extrusion dies by means of meta-models. *International Journal of Advanced Manufacturing Technology*.

Breitsprecher, T., & Wartzack, S. (2014). Neural network based modeling and optimization of deep drawing–extrusion combined process. *Journal of Intelligent Manufacturing*, *25*(1), 77–84. doi:10.1007/s10845-012-0676-z

Chen, W. J., Su, W. C., Nian, F. L., Lin, J. R., & Chen, D. C. (2013). Application of ANOVA and taguchi-based mutation particle swarm algorithm for parameters design of multi-hole extrusion process. Research Journal of Applied Sciences. *Engineering and Technology*, *6*(13), 2316–2325.

Fourment, L., & Chenot, J. (1996). Optimal design for non-steady-state metal forming processes—i. shape optimization method. *International Journal for Numerical Methods in Engineering*, *39*(1), 33–50. doi:10.1002/(SICI)1097-0207(19960115)39:1<33::AID-NME844>3.0.CO;2-Z

Ghassemali, E., Tan, M.-J., Jarfors, A. E. W., & Lim, S. C. V. (2013). Optimization of axisymmetric open-die micro-forging/extrusion processes: An upper bound approach. *International Journal of Mechanical Sciences*, *71*, 58–67. doi:10.1016/j.ijmecsci.2013.03.010

Kim, N., Kang, C., & Kim, B. (2001). Die design optimization for axisymmetric hot extrusion of metal matrix composites. *International Journal of Mechanical Sciences*, *43*(6), 1507–1520. doi:10.1016/S0020-7403(00)00068-0

Kusiak, J., & Thompson, E. G. (1989). Optimization techniques for extrusion die shape design. *Paper presented at the Proc. 3rd Int. Conf. on Numer. Methods in Ind. Forming Processes*.

Lebaal, N., Schmidt, F., & Puissant, S. (2009). Design and optimization of three-dimensional extrusion dies, using constraint optimization algorithm. *Finite Elements in Analysis and Design*, *45*(5), 333–340. doi:10.1016/j.finel.2008.10.008

Lepadatu, D., Kobi, A., Hambli, R., & Barreau, A. (2005). Lifetime multiple response optimization of metal extrusion die. *Paper presented at the Annual Reliability and Maintainability Symposium*. doi:10.1109/RAMS.2005.1408335

Lin, Z., Juchen, X., Xinyun, W., & Guoan, H. (2003). Optimization of die profile for improving die life in the hot extrusion process. *Journal of Materials Processing Technology, 142*(3), 659–664. doi:10.1016/S0924-0136(03)00686-1

Narasimha, M., & Rejikumar, R. (2013). Plastic Pipe Defects Minimization. *International Journal of Innovative Research and Development, 2*(5), 1337–1351.

Ong, P., Zainuddin, Z., Sia, C. K., & Zain, B. A. M. (2015). *Adaptive Cuckoo Search Algorithm with Two-Parent Crossover for Solving Optimization Problems Advances in Swarm and Computational Intelligence* (pp. 427–435). Springer International Publishing.

Reggiani, B., Donati, L., & Tomesani, L. (2015). Multi-objective Optimization of the Extrusion Process. *Materials Today: Proceedings, 2*(10), 4847–4855. doi:10.1016/j.matpr.2015.10.031

Sharififar, M., & Akbari Mousavi, S. A. A. (2015). Simulation and optimization of hot extrusion process to produce rectangular waveguides. *International Journal of Advanced Manufacturing Technology, 79*(9-12), 1961–1973. doi:10.1007/s00170-015-6950-4

Sun, Y., Chen, Q., & Sun, W. (2015). Numerical simulation of extrusion process and die structure optimization for a complex magnesium doorframe. *International Journal of Advanced Manufacturing Technology, 80*(1-4), 495–506. doi:10.1007/s00170-015-7030-5

Ulysse, P. (2002). Extrusion die design for flow balance using FE and optimization methods. *International Journal of Mechanical Sciences, 44*(2), 319–341. doi:10.1016/S0020-7403(01)00093-5

Wang, J., Wu, Z., Gao, S., Lu, R., Qin, D., Yang, W., & Pan, F. (2015). Optimization of mechanical and damping properties of Mg–0.6 Zr alloy by different extrusion processing. *Journal of Magnesium and Alloys, 3*(1), 79–85. doi:10.1016/j.jma.2015.02.001

Yaghoobi, A., Bakhshi-Jooybari, M., Gorji, A., & Baseri, H. (2016). Application of adaptive neuro fuzzy inference system and genetic algorithm for pressure path optimization in sheet hydroforming process. *International Journal of Advanced Manufacturing Technology*. doi:10.1007/s00170-016-8349-2

Zhang, C., Zhao, G., Guan, Y., Gao, A., Wang, L., & Li, P. (2015). Virtual tryout and optimization of the extrusion die for an aluminum profile with complex cross-sections. *International Journal of Advanced Manufacturing Technology, 78*(5-8), 927–937. doi:10.1007/s00170-014-6691-9

Zhao, G., Chen, H., Zhang, C., & Guan, Y. (2013). Multiobjective optimization design of porthole extrusion die using Pareto-based genetic algorithm. *International Journal of Advanced Manufacturing Technology, 69*(5), 1547–1556. doi:10.1007/s00170-013-5124-5

Zhou, J., Lin, L., & Luo, Y. (2014). The multi-objective optimization design of a new closed extrusion forging technology for a steering knuckle with long rod and fork. *International Journal of Advanced Manufacturing Technology, 72*(9), 1219–1225. doi:10.1007/s00170-014-5742-6

## KEY TERMS AND DEFINITIONS

**Cold Extrusion:** The extrusion process which is carried out at room temperature or near room temperature.

**Extrusion:** A compression process where the material is pushed or forced to flow through a die orifice in order to produce a long continuous product.

**Flower Pollination Algorithm:** A type of bio-inspired optimization algorithm inspired by the flower pollination process.

**Forward Extrusion:** During the forward extrusion, a metal billet is put into a container. The material is compressed by a ram in which the material is forced to flow through a die.

**Particle Swarm Optimization:** A type of bio-inspired optimization algorithm inspired by how the fish school and birds fly.

# Chapter 4
# A Heuristic Approach for Car Sequencing Problem Including Assembly Ratio and Color Constraints

**Emek Gamze Köksoy Atiker**
*Intertech Information Technology and Marketing, Turkey*

**Fatma Betül Yeni**
*Istanbul Technical University, Turkey*

**Peiman A. Sarvari**
*Istanbul Technical University, Turkey*

**Emre Çevikcan**
*Istanbul Technical University, Turkey*

## ABSTRACT

*A car factory contains three main workshops; body shop, paint shop and assembly shop. Each of these three workshops has their set of constraints which have to be met in a production day by arranging the vehicles. The car sequencing problem is used to create a production sequence that meets these constraints. Car sequencing problem first handled in the literature by optimization of assembly constraints including ratio constraints. After that, color constraints are integrated to assembly constraints. At this chapter, the scenario in which high priority ratio constraints are primary, color constraints are secondary is tackled and a heuristic approach is proposed. For optimization of ratio constraints, an initial algorithm based on the greedy algorithm is used. The developed algorithm is coded and used on data set which is proposed by Renault at the ROADEF'2005 challenge. According to results, it is achieved the range of results which is achieved by ROADEF finalists.*

DOI: 10.4018/978-1-5225-2944-6.ch004

## INTRODUCTION

The car industry shows a rapid development in the late 19th century. The invention of the petroleum fuel powered engines and cars have occurred as a result of great efforts of European and American engineers. After these developments, some studies have performed in order to produce these commercial products. The first large-scale car production started at 1902 by Ransom Olds. At 1914, Henry Ford has moved up this mass production a step further with the developments on the assembly line. By latest developments, a car started to be produced in every 15 minutes.

Developments in the car industry continued both at Europa and America until the World War 2. Despite the pause in the growth during the war, it continued to grow at an increasing speed later. With the entry of new companies into the market, the competition increased, and it leads to the discovery of new targets such as manufacturing cars with less costly and better quality. Production systems have been improved in line with these objectives.

Nowadays, the car industry has become a challenging industry branch because of the strong competition, multi-product diversity and short product life. Increased consumer demand and quality expectations have become driving forces for the companies to improve their production methods. As a result of the studies about the productivity of the production line, a wide search area has emerged in the car industry. It has become mandatory for the companies to allow flexible operations to meet the demand of the customers. Optimal car sequencing is one of the many ways which provides this flexibility.

The standard car sequencing problem is known as a classical benchmark problem and has been widely studied since its first introduction in 1986 (Solnon, 2008). This problem involves scheduling cars using assembly shop constraints. In 2005 Renault proposed a car sequencing problem for ROADEF Challenge. The car sequencing problem which is proposed by Renault differs from the standard problem since, besides capacity constraints of the assembly shop, it also introduces color constraints to minimize the consumption of solvents in the paint shop and considers two categories of capacity constraints to take into account their priority.

In this chapter, the ROADEF scenario in which high priority ratio constraints are primary, color constraints are secondary is tackled and a heuristic approach is proposed. Firstly, a short information about standard car sequencing problem and ROADEF car sequencing problem are given. After the literature review, the proposed methodology is presented. The model is coded in C++, and the ROADEF'2005 data sets are used for evaluation.

## CAR SEQUENCING PROBLEM

A standard car factory consists of three main production workshops; a body shop where the body is built up by forming sheets, a paint shop where corrosion resistance of the body is increased, and the body is painted and an assembly shop, where different components of the vehicles are installed. Figure 1 depicts the stages of the production line.

The car sequencing problem is about deciding the best sequence which makes the production process in these three workshops easier (Estellon et al., 2007). Each production workshop has its constraints, and all these constraints can conflict with each other. Due to these constraints challenge in the problem, only assembly shops took into consideration in the earlier studies.

*Figure 1. Stages of the production line*

While the constraints of the body shop and assembly shop are quite similar each other, the constraints of paint shop are different (Prandstetter & Raidl, 2008). Some of the constraints that are defined for body and assembly shops are given below (Prandstetter, 2005):

- There must be found component c at maximum $l$ cars in a sliding window of $m$ vehicles.
- There must be exactly $l$ cars which have component $c$ in every subsequence with length $m$.
- None of the vehicles with component $c_1$ should be followed by vehicles with component $c_2$.
- The vehicle with component $c$ must be followed by at least $lmin$ cars are without component $c$.

Whereas, the paint shop defines only one constraint (Prandstetter, 2005):

- Maximum $s$ units of the vehicle which are same colored must be arranged consecutively.

While providing this constraint, the number of color changes has to be minimized. There are two reasons for this. Firstly; color changes cause environmental pollution, and they are quite expensive in both time and money because of the cleaning process of the injectors. Secondly, the paint injectors must be cleaned after a certain number of vehicles to preserve the quality.

If the color injector isn't cleaned often, the color will get sticky, and it would cause inaccurate painting results. Cleaning the injector is a difficult process, and it can be careless and sloppy if the same color is used over and over. Therefore, it is recommended to change the color after every cleaning.

Performing an arrangement with low cost and stable workload for the body shop will be costly with respect to paint shop. It likewise vice versa. In order to overcome this situation, stocks and buffers are located between the workshops, so that they allow to rearrange the cars during the production process.

Figure 2(a) shows the parallel buffer lines. At the end of the incoming line, a decision should be taken for the next car when selecting the buffer lines. The outgoing line is filled with cars taken from the parallel buffer lines. They use the First in, First out (FIFO) strategy. Figure 2(b) shows the loop buffer which enables the recirculation of cars. If the cars must be repaired, this method is usually used. If a car doesn't properly assemble, it is removed from the sequence. After the problem is solved, a new position is defined for the car. Figure 2(c) shows the layout which allows random access to all cars currently in the buffer (Prandstetter, 2005).

## ROADEF'2005 CAR SEQUENCING PROBLEM

ROADEF challenge is organized by the French Society of Operational Research. It aims to share the latest developments in the industry to give a chance to the researchers to face the real problems of the industry.

*Figure 2. Different buffer strategies: Rearrangement through merging of parallel lines (a); Recirculation (b) and Random access stock (c) (Prandstetter, 2005)*

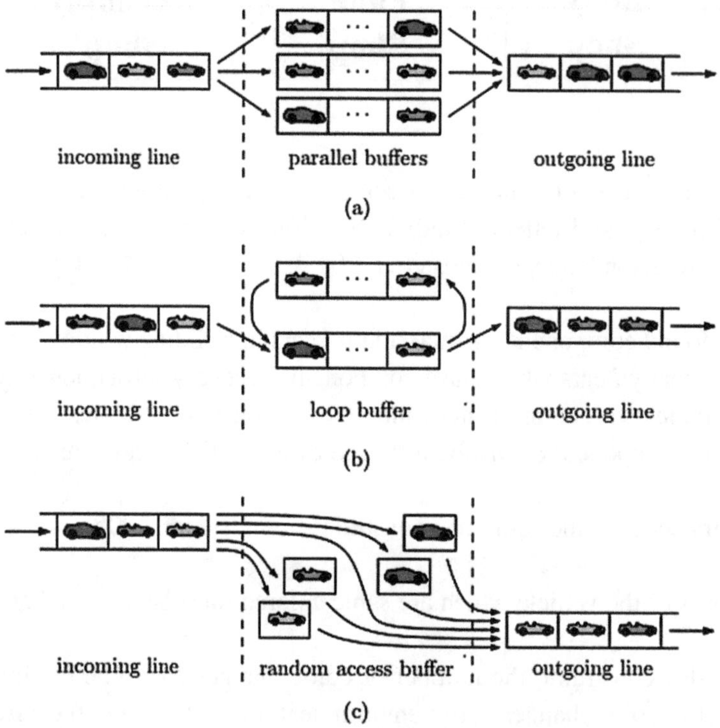


At the challenge in 2005, a new kind of car sequencing problem is proposed by Renault. This problem differs from the classic problems by including the color constraints as well as the capacity constraints imposed by the assembly shop, and it considers the capacity constraints in two categories. Detailed information about Renault car production problem is given at this section.

## Vehicle Production Planning and Sequencing at Renault

As it mentioned before, a car is produced by processing in the three different workshops. There are two main tasks for every workshop. The first one is to assign a one-day production period for every ordered vehicle, to take assembly line capacity constraints and client due dates into account. The second task is to sequence the vehicles for each production day while satisfying the best requirements of the production workshops. Vehicles are sent to the workshops due to the defined vehicle sequence.

For this study, also for the challenge, the following assumptions are made: Only the paint and assembly workshop constraints are discussed, body shop constraints are not critical for the schedule. The assignment decisions made at step 1, cannot go beyond the rules. The planning and scheduling process is performed by Renault using a software which uses linear programming for step 1 and simulated annealing for step 2.

The defined constraints for each workshop are as follows:

- **Paint Workshop Constraints:** The paint injection equipment has to be cleaned periodically for a good quality process. The main paint workshop objective is to minimize the consumption of a solvent which is used to clean the paint injection equipment in every color change.
- **Assembly Workshop Constraints:** The main assembly workshop objective is the work load balancing of the different work units on the assembly lines. To achieve this objective, the vehicles which require complex operations have to be sequenced with an adequate distance. In other words, the density of these "complicated produced vehicles" has to be limited in order to balance the workload of each station. This is provided with *p/q* ratio constraints.
- **Priority Classes for Assembly Workshop Constraints:** As it is mentioned before, a standard car sequencing problem considers only ratio constraints. Nevertheless, Renault always considers both the number of color changes and ratio constraints at the same time because of two major reasons. First, according to the Renault supply chain strategy, the paint and assembly workshops process the same vehicle sequence. Second, a common point can be found between these two objectives.

Depending on the labor cost, either satisfying the ratio constraints which allows to limit the workforce requirements of the assembly line or optimizing the color of the vehicle which minimizes the solvent consumption can be more advantageous.

Two kinds of ratio constraints, priority ratio constraints (PRC) and the nonpriority ratio constraints, are defined to make the balancing between assemblies and painting easier. While the priority ratio constraints are related to critical operations in the assembly shop, the nonpriority ones are related to less critical operations and are defined for workload smoothing process.

## LITERATURE

Car sequencing problem has been introduced into the literature by Parrello and Kabat (1986). It is about sequencing the cars which are produced in a day at the assembly line. Because the car sequencing problem is known to be NP-hard in the strong sense (Kis, 2004; Estellon & Gardi, 2013), different exact and heuristic solution approaches have been proposed widely in the literature. Among the exact approaches, Integer Linear Program (ILP) formulation is the most used method. Drexl and Kimms (2001) have proposed an integer program formulation to decide whether the car belongs any car class or not. Prandstetter and Raidl (2008) have presented an integer linear programming approach and a hybrid variable neighborhood search method to solve the car sequencing problem, in which the goal is to find an optimal arrangement of commissioned vehicles along a production line. Gravel et al. (2010) have proposed an Integer Linear Program avoiding symmetries by grouping cars with the same options. Duarte et al. (2012) have presented a new exact approach for car sequencing problems which considers limited capacity. Due to this problem, the cars for special markets should come first in the sequence, and the cars with the same color should be clustered in the sequence. Because of the complexity of this proposed model, they integrated it with a new heuristic. Estellon et al. (2008) presented two local search approaches which are integrated with a simple heuristic. The first one is a new approach to very large-scale neighborhood search and the second one is based on an original integer linear programming formulation. They compared and discussed these approaches through an extensive computational study on RENAULT's benchmarks. As it stated above, due to the difficulty of the problem solution, various heuristics are used in the literature besides these exact solution approaches. Gottlieb at al. (2003) have compared greedy, local search and ant colony optimization (ACO) approaches for the car sequencing problem. They obtained the best results by the ACO combined with a dynamic heuristic. Because of the success of ACO on solving hard combinatorial optimization problems, it has often used in the literature. (Gagne et al., 2006; Solnon, 2008; Morin et al., 2009). Gavronovic (2008) have proposed a solution approach for the car sequencing problem including color constraints. In this study, greedy algorithm for the initial sequence and local search for the problem solution is proposed. Withal, a tabu search heuristic is used to improve the results of the local search. Cordeau et al. (2008) also proposed an iterated tabu search based heuristic for the car sequencing problem in which a set of cars must be sequenced to satisfy the requirements of the paint shop and the assembly line. This presented heuristic combines a classical tabu search with perturbation operators which help to escape from local optima. Zufferey (2016) proposed Tabu search approach for two car sequencing problems involving smoothing constraints. The first one is denoted (P1) and was the subject of the ROADEF 2005 international Challenge proposed by the automobile manufacturer Renault, whereas the second one is denoted (P2) and extends some important features of (P1). The other heuristics which are commonly used in the literature are simulated annealing and genetic algorithm (Areal et al., 2011; Joly and Frein, 2008; Briant et al., 2008). Golle et al. (2014), examined and compared the solution quality (work overload) of car sequencing and mixed-model sequencing in their study. They derived the related car sequencing instances using different sequencing rule generation approaches and also applied various objective functions for car sequencing discussed in the literature. As it indicated in the literature review, heuristic approaches are used more often than the exact solution approaches, because of the difficulty of the solution of car sequencing problems. Also at this chapter heuristics methods are proposed too. For optimization of ratio constraints, an initial algorithm based on the greedy algorithm is used. After initial algorithm, the main ratio constraint optimization algorithm is proposed in two stages. The simulated annealing is integrated to this algorithm to improve the results. The last approach is related to color arrangement to recover color constraints violations.

## METHODOLOGY

In this study, a new heuristic approach for ROADEF '2005 car sequencing problem has been developed. The priority ratio constraints are considered as primary while nonpriority ones are considered as secondary. Firstly, an initial sequence is created using a starting algorithm before optimization of priority ratio constraints. Then the number of violations of priority ratio constraints through the defined sequence is tried to be minimized. Finally, the number of violations of color constraints is tried to be minimized while keeping the best result value. Figure 3 shows the flowchart of the algorithm.

### Starting Algorithm

The initial algorithm and heuristic approaches proposed by Gottlieb et al. (2003) are used as starting algorithm of this study. There are $q_{max}$ vehicles which belong to the previous day known as D-1 at the data set of car sequencing problem proposed by ROADEF '2005. The sequence of these cars is fixed and definitely cannot be changed. Figure 4 indicates the flowchart of starting algorithm.

Starting Algorithm consists two vehicle sets as scheduled and nonscheduled. Here, nonscheduled vehicles are evaluated one by one according to the minimum violation criteria and added to the set of scheduled vehicles. After each vehicle added to the scheduled vehicle set, the number of violation is

*Figure 3. Flowchart of the developed algorithm*

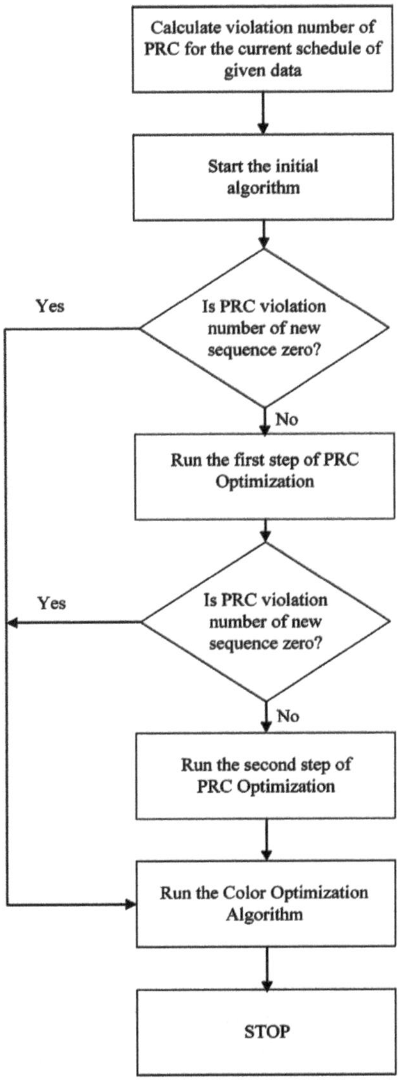


calculated. Firstly, each vehicle is considered to be added to the first line. The vehicle which causes minimum violation number is added to the first line at the scheduled vehicle set and deleted from da nonscheduled vehicle set. The same process is repeated until there is no vehicle at the nonscheduled vehicle set.

## Optimization of Priority Ratio Constraints

In this study, swapping move is used for searching the best solution that minimizing the priority ratio constraints. Two heuristic approaches are proposed to decide the swapping of the vehicles. Furthermore, simulated annealing algorithm is used at the first heuristic to expand the searching for candidate vehicles that will move.

*Figure 4. Flowchart of the starting algorithm*

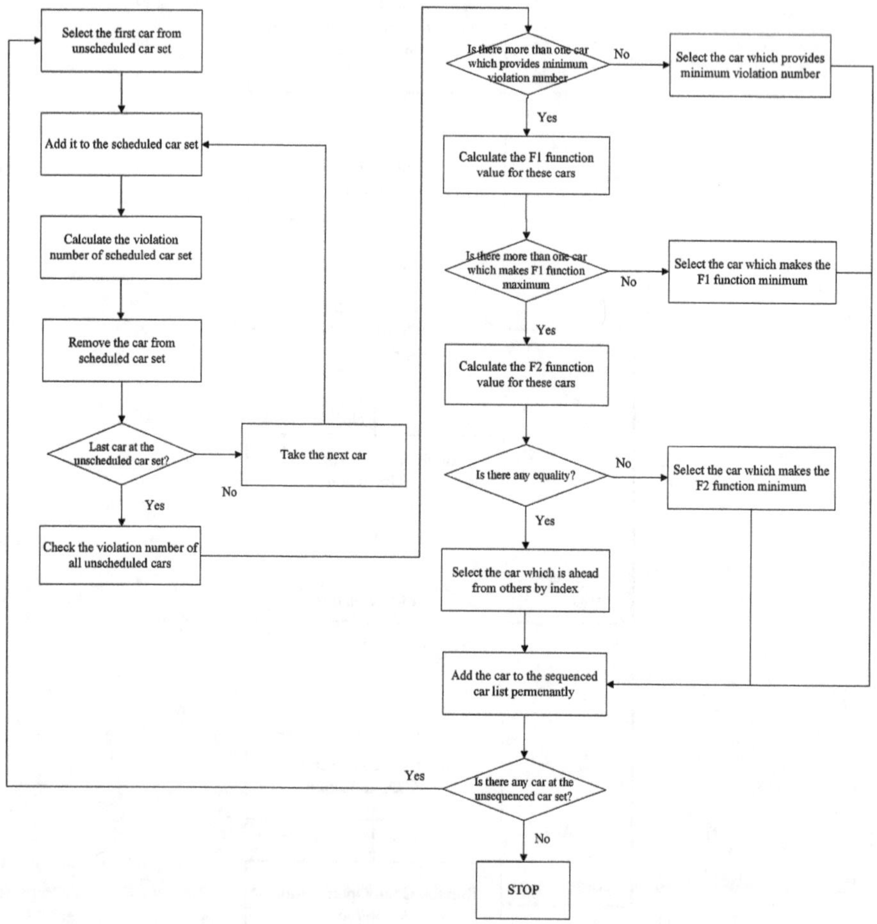


The first step of the optimization of priority ratio constraints is selecting vehicles which will swap, hence, replaced vehicles' degree of improvement in their own position is an important criterion to be evaluated. Although swapping of the vehicles can improve all sequence, the same process can also cause a local deterioration. In this study, vehicle swapping which will optimize a number of local violation is accepted as the certain change, and it is hoped that will contribute to an improvement in the long term.

Figure 5 shows the flowchart of this first step of the optimization. The symbols used in the flowchart and definitions are as follow;

S*: Best violation number.
$S_1$: Violation number before temporary relocation.
$S_2$: Violation number after temporary relocation.
bi: Local violation number of position i before a temporary car swapping between position i and j.
$b_j$: Local violation number of position j before a temporary car swapping between position i and j.
$c_i$: Local violation number of position j after a temporary car swapping between position i and j.
$c_j$: Local violation number of position j after a temporary car swapping between position i and j.
Δ: Sum of the difference in number of violation which occur by temporary relocations.

Δ*: The value of Δ obtained by relocation of cars which gives S*.
Poz*: Position of j which provides S* value after the loop j.

The algorithm used at this step consists of three phases. The reason of using these phases is giving priority to vehicle relocation which provides bilateral, regional improvement. By this way, in the case of failure to provide this condition, different alternatives can be evaluated. It is not needed to create separated loops for each phase because the phases get activated by evaluation of phase conditions in use. This general algorithm loop is continued until there isn't any opportunity for a relocation in all three phases.

*Figure 5. The flowchart of the first step of the optimization*

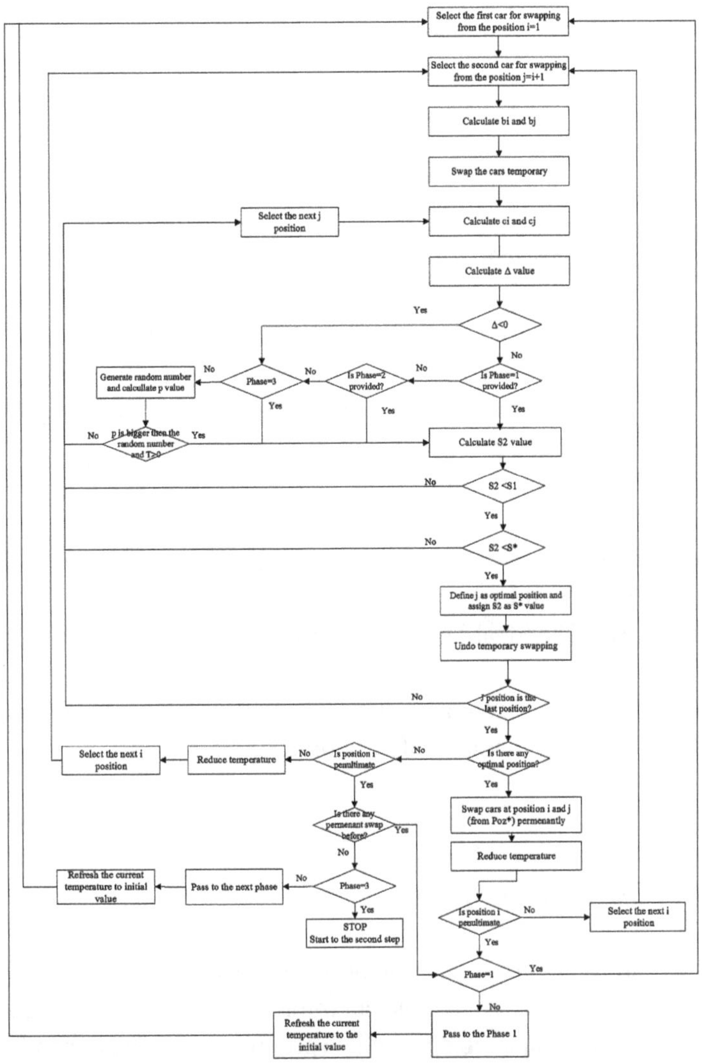

The aim of this second step is providing a rapid improvement for the solution by handling the cars which cause more violation primarily. This idea is based on Pareto analysis: "…20% of causes determine 80% of problems." Impacts of ratio constraints on the violation numbers of the first step are taking into account of the solution.

The framework of this step is given in Figure 6. The symbols used in the flowchart and definitions are as follows;

$S_1$: Violation number before temporary relocation.
$S_2$: Violation number after temporary relocation.
$\Delta_1$: Former $\Delta$ value. This value is updated if the new delta after the temporary relocation is better. It is reset after each permanent relocation.
$\Delta_2$: New delta value which is found after temporary relocation.
$E_1$: List of 20% of the vehicles with the highest number of local violation sorted by least number of local violation number.
$E_2$: List of the vehicles sorted by least number of local violation number.
k: Current position of the vehicle which is selected from the list $E_1$.
l: Current position of the vehicle which is selected from the list $E_2$.

## Simulated Annealing Approach

The conditions which are used for evaluation of the cars varies according to the phases; in a situation of $\Delta<0$, cars should be evaluated only in phase 3, not in phase 1 or 2. But in this case, the search space gets reduced; hence, these cars are considered as candidate cars by a certain probability according to the simulated annealing principal and evaluated at phase 1 and 2.

The parameters and functions are as follow:

- **Cooling Ratio ($\alpha$):** Cooling rate value gives the best result was found as 0.99 by the experimental design which will be described later.
- **Initial Temperature (t):** The ideal initial temperature was defined as 10000 at the experimental design.
- **Cooling Function:** Equation of cooling function is defined at below. F(t) means the current temperature, F(t+1) means the temperature after cooling.

$$F(t+1) = F(t) * \alpha$$

## Optimization of the Color Constraints

Color constraints are optimized after optimization of the priority ratio constraints while preserving previous solutions. The used approach aims to make an improvement by emplacing same-colored vehicle groups next to the other vehicle groups. Firstly, vehicle groups are created depending on color and color constraints through the sequence generated by priority ratio constraints optimization. Figure 7 indicates the flowchart of creating color group process.

*Figure 6. The flowchart of the second step of the optimization*

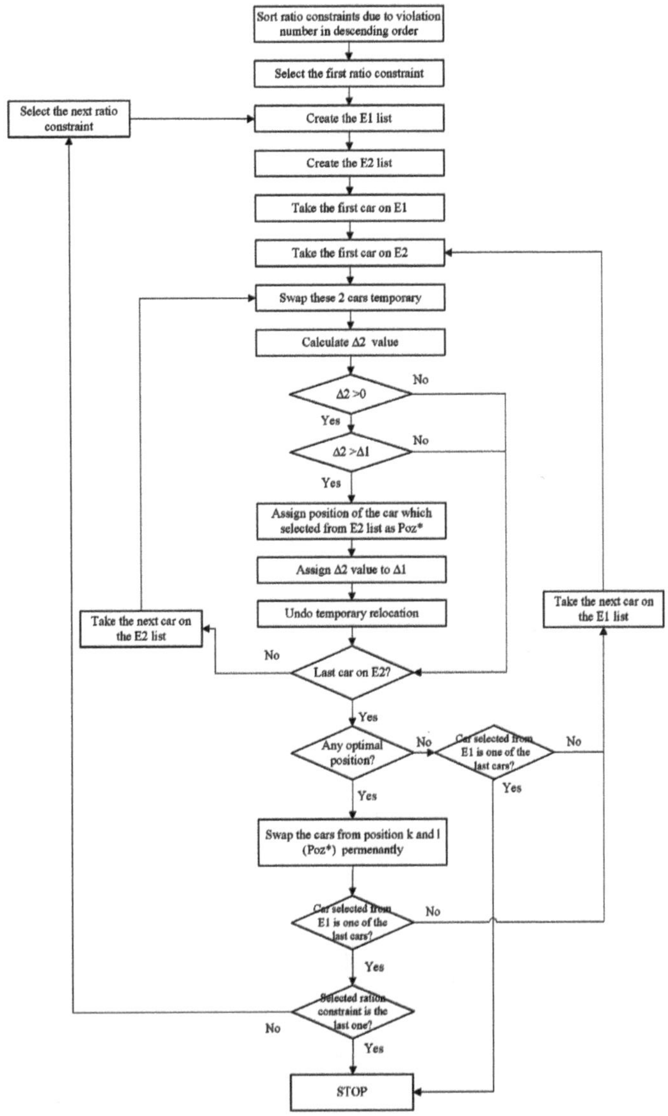


After creating vehicle groups, each vehicle group starting from the first one added to the next position respectively and the sequence assessed again in this structure. Figure 8 demonstrates the flowchart of color adjustment process.

## APPLICATION

The ROADEF'2005 data set is used to evaluate the model. The developed model is coded in C++ programming language and running time and results of ROADEF'2005 finalists are defined as efficiency scale.

*Figure 7. Creation of vehicle groups according to color*

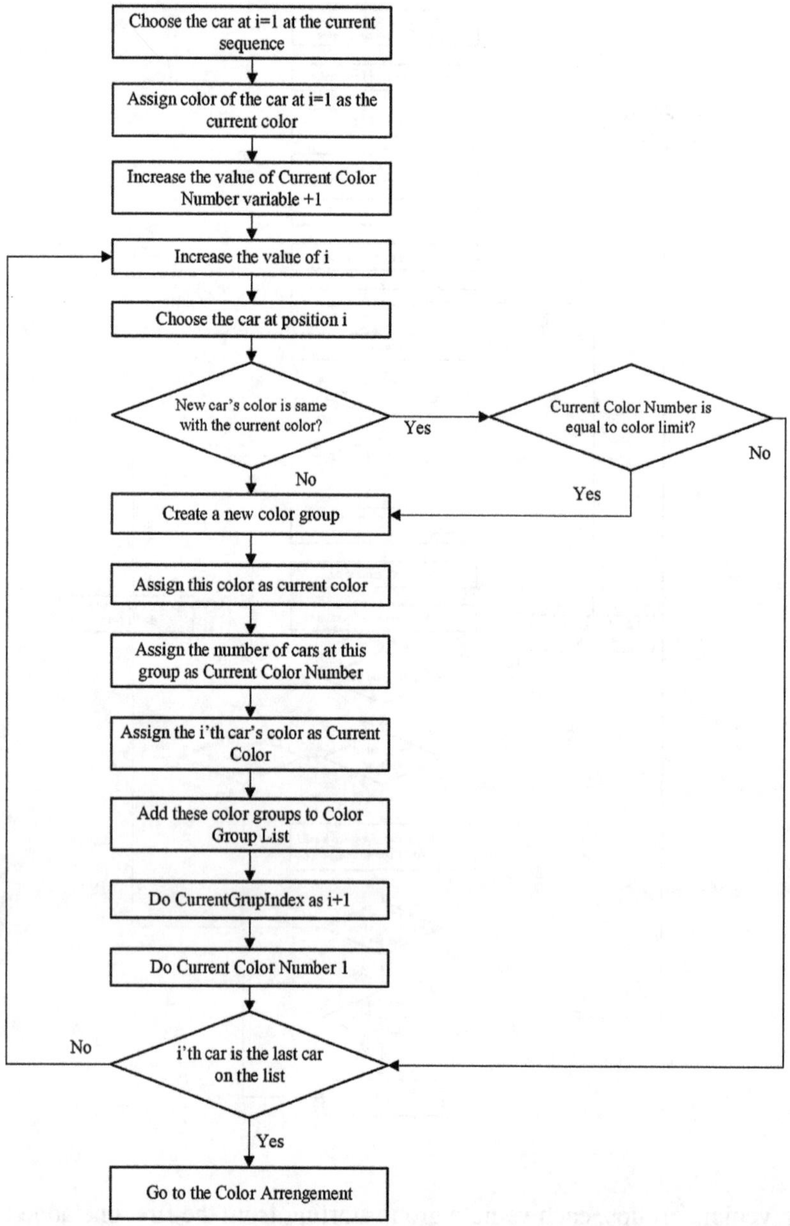


## Data Set

Three scenario and data set are generated by Renault. These data sets help the competitors to make their programs better.

- **Data Set A:** It is used to choose the finalists by the juri.
- **Data Set B:** It is used by finalists to control their programs before using data set X.

*Figure 8. Color adjustment process*

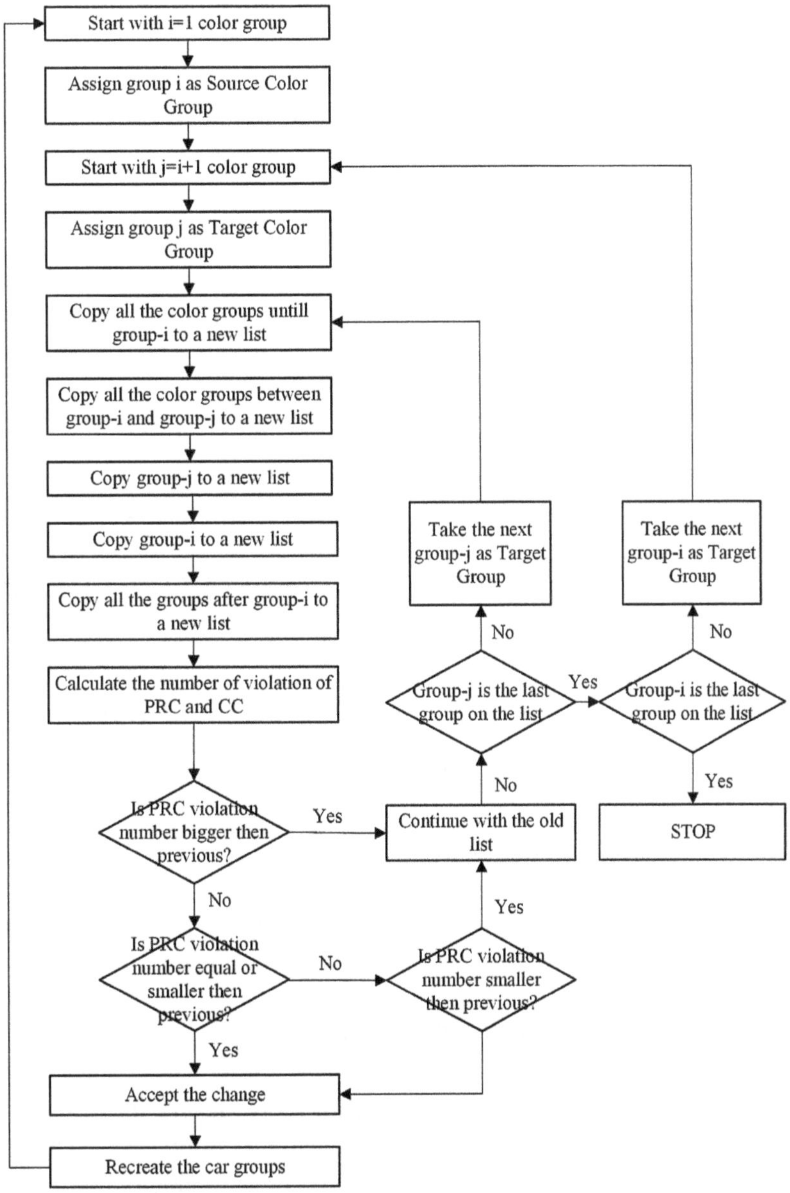


- **Data Set X:** It is the last data set used to sort the finalists.

In this study, only A and X data sets are used. The data sets where ratio constraints are primary and color constraints are secondary have been selected because the developed model aims to optimize ratio constraints firstly. As shown in Table 1, there are 16 production data where 9 of them belong to data set X and rest belong to data set A. This data set is divided into four groups according to the number of vehicles they contain; "Vehicle Group 1: Production data with the number of vehicles 0-349", "Vehicle Group 2: Production data with the number of vehicles 350-699", "Vehicle Group 3: Production data with

*Table 1. Properties of ROADEF Data*

| Group | Data Set | Vehicle Number | PRC Number | Easy to Do | Hard to Do | Color Limit | Data Set A | Data Set X |
|---|---|---|---|---|---|---|---|---|
| 1 | 655_CH2_EP_RAF_ENP_S52_J1_J2_S01_J1 (1) | 219 | 4 | x | | 10 | | x |
| | 034_VU_EP_RAF_ENP_S51_J1_J2_J3 (2) | 231 | 6 | x | | 30 | | x |
| | 064_38_2_EP_RAF_ENP_ch2 (3) | 335 | 4 | x | | 15 | x | |
| 2 | 048_CH2_EP_RAF_ENP_S49_J5 (4) | 459 | 8 | | x | 12 | | x |
| | 022_3_4_EP_RAF_ENP (5) | 485 | 3 | x | | 450 | x | |
| | 048_CH1_EP_RAF_ENP_S50_J4 (6) | 519 | 6 | x | | 12 | | x |
| | 048_39_1_EP_RAF_ENP (7) | 600 | 5 | | x | 10 | x | |
| 3 | 064_CH1_EP_RAF_ENP_S49_J1 (8) | 875 | 9 | x | | 15 | | x |
| | 064_38_2_EP_RAF_ENP_ch1 (9) | 875 | 7 | x | | 15 | x | |
| | 034_VP_EP_RAF_ENP_S51_J1_J2_J3 (10) | 921 | 3 | x | | 400 | | x |
| | 039_38_4_EP_RAF_ch1 (11) | 954 | 5 | | x | 20 | x | |
| | 039_CH3_EP_RAF_ENP_S49_J1 (12) | 1037 | 2 | x | | 20 | | x |
| 4 | 024_38_3_EP_RAF_ENP (13) | 1260 | 5 | | x | 10 | x | |
| | 023_EP_RAF_ENP_S49_J2 (14) | 1260 | 5 | x | | 40 | | x |
| | 024_38_5_EP_RAF_ENP (15) | 1315 | 5 | | x | 10 | x | |
| | 024_EP_RAF_ENP_S49_J2 (16) | 1319 | 7 | x | | 10 | | x |

the number of vehicles 700-1049" and "Vehicle Group 4: Production data with the number of vehicles 1050-1599".

## Experimental Design of Simulated Annealing

Experimental design is used to decide simulated annealing parameters like cooling ratio and initial temperature. Five value (0,99; 0,97; 0,94; 0,92; 0,9) evaluated for cooling ratio while four value (1000, 10000, 100000, 10000000) evaluated for initial temperature. Four production data is chosen according to the number of vehicles and assessed for improvement rate and time. Results are shown in Table 2.

According to this table; the higher initial temperature and cooling ratio, the better results. But it also causes long time solution. In this case, for the best improvement, the initial temperature is defined as 1000 and cooling ratio as 0,99.

## Evaluation of Starting Algorithm

Table 3 shows the result interval of ROADEF 2005 finale competitors according to the priority ratio constraints. As seen on the table, the higher number of the vehicle, the longer completion time. In some

*Table 2. Results of experimental design of simulated annealing*

| 048_CH2_EP_RAF_ENP_S49_J5 | | | | | | | | |
|---|---|---|---|---|---|---|---|---|
| **1** | **Initial Temperature** | | | | | | | |
| | **1000** | | **10000** | | **100000** | | **1000000** | |
| **Cooling Ratio** | **İmp. (%)** | **Time m:s** | **İmp. (%)** | **Time m:s** | **İmp. (%)** | **Time m:s** | **İmp. (%)** | **Time m:s** |
| 0.99 | 0.3016 | 03:00 | 0.317460317 | 05:02 | 0.317460317 | 06:56 | 0.317460317 | 07:27 |
| 0.97 | 0.3016 | 01:31 | 0.3016 | 02:31 | 0.3016 | 02:55 | 0.3016 | 03:03 |
| 0.94 | 0.238095238 | 01:10 | 0.2698 | 01:36 | 0.2698 | 01:53 | 0.3016 | 02:09 |
| 0.92 | 0.238095238 | 01:04 | 0.2698 | 01:19 | 0.2698 | 01:33 | 0.2698 | 01:48 |
| 0.9 | 0.2698 | 00:57 | 0.2698 | 01:10 | 0.2698 | 01:21 | 0.2698 | 01:32 |
| **034_VU_EP_RAF_ENP_S51_J1_J2_J3** | | | | | | | | |
| **2** | **Initial Temperature** | | | | | | | |
| | **1000** | | **10000** | | **100000** | | **1000000** | |
| **Cooling Ratio** | **İmp. (%)** | **Time m:s** | **İmp. (%)** | **Time m:s** | **İmp. (%)** | **Time m:s** | **İmp. (%)** | **Time m:s** |
| 0.99 | 0.555555556 | 01:01 | 0.555555556 | 01:16 | 0.555555556 | 01:18 | 0.555555556 | 01:19 |
| 0.97 | 0.380952381 | 00:34 | 0.571428571 | 00:39 | 0.555555556 | 00:37 | 0.555555556 | 00:55 |
| 0.94 | 0.380952381 | 00:27 | 0.380952381 | 00:31 | 0.365079365 | 00:34 | 0.523809524 | 00:30 |
| 0.92 | 0.380952381 | 00:24 | 0.380952381 | 00:28 | 0.380952381 | 00:31 | 0.380952381 | 00:38 |
| 0.9 | 0.380952381 | 00:23 | 0.380952381 | 00:27 | 0.380952381 | 00:28 | 0.380952381 | 00:30 |
| **039_38_4_EP_RAF_ch1** | | | | | | | | |
| **3** | **Initial Temperature** | | | | | | | |
| | **1000** | | **10000** | | **100000** | | **1000000** | |
| **Cooling Ratio** | **İmp. (%)** | **Time m:s** | **İmp. (%)** | **Time m:s** | **İmp. (%)** | **Time m:s** | **İmp. (%)** | **Time m:s** |
| 0.99 | 0.123 | 12:43 | 0.123 | 15:46 | 0.123 | 20:03 | 0.148 | 27:20 |
| 0.97 | 0.123 | 06:25 | 0.123 | 08:45 | 0.123 | 14:06 | 0.123 | 13:19 |
| 0.94 | 0.123 | 04:41 | 0.123 | 05:38 | 0.123 | 07:45 | 0.123 | 08:52 |
| 0.92 | 0.107 | 03:58 | 0.123 | 04:52 | 0.123 | 05:47 | 0.123 | 06:17 |
| 0.9 | 0.107 | 03:15 | 0.107 | 04:27 | 0.123 | 05:02 | 0.123 | 04:40 |
| **024_38_5_EP_RAF_ENP** | | | | | | | | |
| **4** | **Initial Temperature** | | | | | | | |
| | **1000** | | **10000** | | **100000** | | **1000000** | |
| **Cooling Ratio** | **İmp. (%)** | **Time m:s** | **İmp. (%)** | **Time m:s** | **İmp. (%)** | **Time m:s** | **İmp. (%)** | **Time m:s** |
| 0.99 | 0.194 | 52:55 | 0.194 | 51:06 | 0.194 | 47:06:00 | 0.194 | 44:13 |
| 0.97 | 0.194 | 50:45 | 0.194 | 35:10 | 0.194 | 38:08 | 0.194 | 36:23 |
| 0.94 | 0.194 | 54:34 | 0.194 | 49:14 | 0.194 | 40:67 | 0.194 | 38:53 |
| 0.92 | 0.194 | 40:25 | 0.194 | 33:22 | 0.194 | 30:02 | 0.194 | 32:34 |
| 0.9 | 0.194 | 39:55 | 0.194 | 37:12 | 0.194 | 32:11 | 0.194 | 30:12 |

*Table 3. Result of initial algorithm*

| Data | Given Violation Number of PRC | Violation number after SA | SA Time (mn:sc) | Total Result with SA | Total Result without SA | ROADEF'2005 Result |
|---|---|---|---|---|---|---|
| 1 | 180 | 206 | 00:04 | 159 | 167 | 153-306, insoluble |
| 2 | 63 | 65 | 00:07 | 17 | 21 | 8-12, insoluble |
| 3 | 28 | 0 | 00:21 | 0 | 0 | 0 |
| 4 | 63 | 72 | 00:39 | 34 | 43 | 31-34 |
| 5 | 2 | 0 | 00:43 | 0 | 0 | 0 |
| 6 | 8 | 1 | 02:28 | 1 | 8 | 0-2 |
| 7 | 51 | 83 | 02:50 | 18 | 34 | 0-27 |
| 8 | 77 | 101 | 07:35 | 63 | 70 | 61-67 |
| 9 | 2 | 0 | 10:51 | 0 | 0 | 0 |
| 10 | 22 | 1 | 07:42 | 1 | 7 | 0 |
| 11 | 122 | 142 | 04:03 | 6 | 72 | 13-84 |
| 12 | 8 | 7 | 05:04 | 7 | 7 | 0 |
| 13 | 73 | 92 | 08:43 | 57 | 57 | 4-63 |
| 14 | 6 | 1 | 08:45 | 0 | 0 | 0 |
| 15 | 98 | 93 | 08:59 | 57 | 78 | 4-75 |
| 16 | 73 | 42 | 09:02 | 14 | 23 | 0-49, insoluble |

data samples, starting algorithm gets really closer to the best solutions. For example, at the data 3,5 and 9 it is achieved to the optimal solution by only using starting algorithm.

## Evaluation of PRC Optimization

The results of optimizasion steps are evaluated based on a number of violations and processing time. At both steps, calculations are done for 'with SA' and 'without SA' separately and compared. When the second step continues after the first step, it has been observed that result remains stable. So, for a clear comparison, the second step also applied directly without the first step. Table 4 indicates the performance comparison of both steps of PRC optimization.

Although the results of the second step is worse than the first step, it still can be considered as good and besides it ends in a shorter time. This approach can be used in a comprehensive manner on further studies to obtain better results in less time.

## Evaluation of Color Constraint Optimization

The purpose of this approach is to improve the number of violation of color constraints while preserving the value of PRC optimization results. Table 5 shows the results of color constraint optimization for 16 data. The results obtained by 'with Sa' and 'without SA' are evaluated separately.

The results obtained from this approach are not as good as ROADEF'2005 finalist results, but they are acceptable as long as they get an improvement on the number of violation of PRC.

*Table 4. Performance comparison of steps of PRC optimization*

| Data | 1st Step With SA | Time (h:m:s) | 1st Step Without SA | Time (h:m:s) | 2nd Step With SA | Time (h:m:s) | 2nd Step With SA | Time (h:m:s) |
|---|---|---|---|---|---|---|---|---|
| 1 | 177 | 00:50 | 177 | 01:42 | 202 | 00:00:01 | 177 | 00:00:14 |
| 2 | 24 | 01:57 | 28 | 01:45 | 35 | 00:00:09 | 32 | 00:00:30 |
| 3 | 0 | 04:53 | 0 | 06:30 | 0 | 00:00:14 | 1 | 00:00:45 |
| 4 | 35 | 06:36 | 43 | 04:30 | 39 | 00:06:22 | 45 | 00:04:15 |
| 5 | 0 | 02:10 | 0 | 02:47 | 0 | 00:00:00 | 2 | 00:00:00 |
| 6 | 1 | 02:42 | 80 | 06:01 | 1 | 00:03:13 | 8 | 00:02:49 |
| 7 | 31 | 15:09 | 48 | 11:53 | 29 | 00:16:39 | 48 | 00:10:16 |
| 8 | 66 | 02:06:35 | 73 | 01:37:48 | 68 | 00:20:11 | 73 | 00:06:54 |
| 9 | 0 | 00:00:00 | 0 | 01:19:23 | 0 | 00:00:00 | 0 | 00:17:07 |
| 10 | 1 | 54:12 | 7 | 43:30 | 1 | 00:13:13 | 7 | 00:15:27 |
| 11 | 72 | 02:41:34 | 104 | 02:00:51 | 134 | 00:00:44 | 112 | 00:08:34 |
| 12 | 7 | 09:43 | 7 | 07:39 | 7 | 00:03:31 | 7 | 00:03:20 |
| 13 | 57 | 01:08:06 | 57 | 47:27 | 67 | 00:48:18 | 64 | 00:17:01 |
| 14 | 0 | 01:08:44 | 0 | 01:03:39 | 1 | 00:00:09 | 5 | 00:00:09 |
| 15 | 57 | 49:05 | 79 | 01:59:12 | 66 | 00:15:39 | 84 | 00:06:40 |
| 16 | 16 | 02:31:41 | 55 | 03:14:16 | 25 | 00:01:59 | 68 | 00:02:08 |

*Table 5. Results of color constraint optimization*

| Data | ROADEF Results | Color Imp. With SA | PRC Imp. With SA | Time With SA (m:s) | Color Imp. Without SA | PRC Imp. Without SA | Time Without SA (m:s) |
|---|---|---|---|---|---|---|---|
| 1 | 32-57 | 115-113 | 177-159 | 00:05 | 87-85 | 177-167 | 00:03 |
| 2 | 87-191 | 153-148 | 24-17 | 00:14 | 117-112 | 28-21 | 00:09 |
| 3 | 34-54 | 81-73 | 0-0 | 00:08 | 83-74 | 0-0 | 00:08 |
| 4 | 76-151 | 352-352 | 35-35 | 01:34 | 160-160 | 43-43 | 00:14 |
| 5 | 31-112 | 71-69 | 0-0 | 00:02 | 72-69 | 0-0 | 00:02 |
| 6 | 196-209 | 284-284 | 1-1 | 01:25 | 248-248 | 8-8 | 00:55 |
| 7 | 174-105 | 512-507 | 31-18 | 05:08 | 232-228 | 48-34 | 01:24 |
| 8 | 187-279 | 782-772 | 66-63 | 40:26 | 290-291 | 73-72 | 03:27 |
| 9 | 112-240 | 243-240 | 0-0 | 07:34 | 243-240 | 0-0 | 06:23 |
| 10 | 55-145 | 97-91 | 1-1 | 00:18 | 95-83 | 7-7 | 00:34 |
| 11 | 129-512 | 613-551 | 72-6 | 34:03 | 282-283 | 104-72 | 08:58 |
| 12 | 231-285 | 303-303 | 7-7 | 01:01 | 303-302 | 7-7 | 00:25 |
| 13 | 249-746 | 848-848 | 57-57 | 12:45 | 470-470 | 57-57 | 02:57 |
| 14 | 192-314 | 265-250 | 0-0 | 09:22 | 258-251 | 0-0 | 04:55 |
| 15 | 280-842 | 1003-1001 | 57-57 | 18:26 | 569-569 | 79-78 | 13:02 |
| 16 | 337-559, insoluble | 783-782 | 16-14 | 34:40 | 533-510 | 55-23 | 30:04 |

*Table 6. Improvement percentages over total*

| | 1st Step's Effect to the Total Improvement | | 2nd Step's Effect to the Total Improvement | | Color Constraint effect to the Total Improvement | |
|---|---|---|---|---|---|---|
| Data | With SA | Without SA | With SA | Without SA | With SA | Without SA |
| 1 | 61.70% | 23.08% | 0.00% | 0.00% | 38.30% | 76.92% |
| 2 | 85.42% | 83.33% | 0.00% | 0.00% | 14.58% | 16.67% |
| 3 | - | 100.00% | - | - | - | - |
| 4 | 97.37% | 100.00% | 0.00% | - | 2.63% | - |
| 5 | - | 100.00% | - | - | - | - |
| 6 | 0.00% | 0.00% | 0.00% | 0.00% | 0.00% | 0.00% |
| 7 | 80.00% | 17.65% | 0.00% | 0.00% | 20.00% | 82.35% |
| 8 | 92.11% | 57.14% | 0.00% | 0.00% | 7.89% | 42.86% |
| 9 | - | 100.00% | - | - | - | - |
| 10 | 0.00% | 100.00% | 0.00% | 0.00% | - | - |
| 11 | 51.47% | 36.00% | 0.00% | 0.00% | 48.53% | 64.00% |
| 12 | 0.00% | 100.00% | 0.00% | - | 0.00% | - |
| 13 | 100.00% | 100.00% | - | - | - | - |
| 14 | 100.00% | 100.00% | - | - | - | - |
| 15 | 100.00% | 95.00% | - | 0.00% | - | 5.00% |
| 16 | 92.86% | 36.00% | 0.00% | 0.00% | 7.14% | 64.00% |

Table 6 demonstrates the effects of these optimization approaches on the result of the algorithm. Mostly %80 of the improvement is obtained by the first step. The second step nearly no effect on the results while color constraint optimization approach is highly effective on especially some data.

## CONCLUSION

At this chapter, a heuristic approach for car sequencing problem including assembly ratio constraints and color constraints is proposed. For the application, car sequencing problem of ROADEF'2005 which considers minimization of violation numbers of priority ratio constraints and color constraints is used.

The obtained results of violation numbers of PRC are generally between the best and the worst results of ROADEF finalist, yet the results of color constraints optimization are not good at all. It is observed that the solution using starting algorithm always gives a better result for data, even though the starting algorithm causes a worsening at early results time to time.

The first step of PRC optimization constitutes 80% of the total improvement, but it is not considered successful in point of time performance. The second step of PRC optimization after the first step does not cause any difference. But, when it is used directly, it gives closer results to the first step's results in a shorter time. The results also showed that color constraint optimization is successful for only some data in reducing violation number of PRC.

For further studies;

- Performance enhancement can be performed to reduce the long processing time of the algorithm.
- New controls related to the color constraint can be added to the starting algorithm to avoid from worsening of the color constraints while scheduling cars according to violation numbers of PRC.
- The second step of the PRC optimization can improve in order to contribute to the total improvement.
- Color constraint optimization can be made more suitable for the PRC optimization with different changes.

## REFERENCES

Areal, J. J., Martín, R. M., & Campos, J. G. (2011). Simulated annealing vs. genetic algorithms applied using a new cost function for the car sequencing problem. *International Journal of Manufacturing Technology and Management, 23*(1-2), 113–136. doi:10.1504/IJMTM.2011.042111

Briant, O., Naddef, D., & Mounie, G. (2008), Greedy approach and multi-criteria simulated annealing for the car sequencing problem. *European Journal of Operational Research, 191*, 993–1003

Cordeau, J. F., Laporte, G., & Pasin, F. (2008). Iterated tabu search for the car sequencing problem. *European Journal of Operational Research, 191*, 945–956.

Drexl, A., & Kimms, A. (2001). Sequencing jit mixed-model assembly lines under station load and part-usage constraints. *Management Science, 47*(3), 480–491.

Duarte, C. S. R., Carvalho, J. V., & Barbosa-Póvoa, A. P. (2012). A new heuristic for car sequencing based on an integer programming approach. *IFAC Proceedings, 45*(6), 164-169. doi:10.3182/20120523-3-RO-2023.00073

Estellon, B., & Gardi, F. (2013). Car sequencing is NP-hard: A short proof. *The Journal of the Operational Research Society, 64*(10), 1503–1504. doi:10.1057/jors.2011.165

Estellon, B., Gardi, F., & Nouioua, K. (2008). Two local search approaches for solving real-life car sequencing problems. *European Journal of Operational Research, 191*, 928–944.

Gagne, C., Gravel, M., & Price, W. L. (2006). Solving real car sequencing problems with ant colony optimization. *European Journal of Operational Research, 174*, 1427-1448.

Gavronovic, H. (2008). Local search and suffix tree for car-sequencing problem with colors. *European Journal of Operational Research, 191*(3), 972–980. doi:10.1016/j.ejor.2007.04.051

Golle, U., Rothlauf, F., & Boysen, N. (2014). Car sequencing versus mixed-model sequencing: A computational study. *European Journal of Operational Research, 237*(1), 50–61. doi:10.1016/j.ejor.2014.01.012

Gottlieb, J., Puchta, M., & Solnon, C. (2003). A study of greedy, local search and ant colony optimization approaches for car sequencing problems. In Applications of Evolutionary Computing, LNCS (Vol. 2611, pp. 246–257).

Gravel, M., Gagne, C., & Price, W. L. (2005). Review and comparison of three methods for the solution of the car sequencing problem. *The Journal of the Operational Research Society, 56*(11), 1287–1295. doi:10.1057/palgrave.jors.2601955

Joly, A., & Frein, Y. (2008). Heuristics for an industrial car sequencing problem considering paint and assembly shop objectives. *Computers & Industrial Engineering*, *55*(2), 295–310. doi:10.1016/j.cie.2007.12.014

Kis, T. (2004). On the complexity of the car sequencing problem. *Operations Research Letters*, *32*(4), 331–335. doi:10.1016/j.orl.2003.09.003

Morin, S., Gagné, C., & Gravel, M. (2009). Ant colony optimization with a specialized pheromone trail for the car-sequencing problem. *European Journal of Operational Research*, *197*(3), 1185–1191. doi:10.1016/j.ejor.2008.03.033

Parello, B. D. & Kabat, W. C. (1986). Job-Shop Scheduling Using Automated Reasoning: A Case Study of the Car-Sequencing Problem. *Journal of Automated Reasoning*, *2*(1), 1-42.

Prandstetter, M. (2005). *Exact and heuristic methods for solving the car sequencing Problem.* Diplomarbeit, Viyana Technical Univercity.

Prandstetter, M. & Raidl, G. R. (2008). An integer linear programming approach and a hybrid variable neighborhood search for the car sequencing problem. *European Journal of Operational Research*, *191*, 1004-1022.

Solnon, C. (2008). Combining two pheromone structures for solving the car sequencing problem with Ant Colony Optimization. *European Journal of Operational Research*, *191*, 1043–1055

Solnon, C., Cung, V. D., Nguyen, A., & Artigues, C. (2008). The car sequencing problem: Overview of state-of-the-art methods and industrial case-study of the ROADEF2005 challenge problem. *European Journal of Operational Research*, *191*(3), 912–927. doi:10.1016/j.ejor.2007.04.033

Zufferey, N. (2016). Tabu Search Approaches for Two Car Sequencing Problems with Smoothing Constraints. In *Metaheuristics for Production Systems* (pp. 167–190). Springer International Publishing. doi:10.1007/978-3-319-23350-5_8

## KEY TERMS AND DEFINITIONS

**Heuristic:** A heuristic is a mental shortcut that allows people to solve problems and make judgments quickly and efficiently.

**Ratio Constraints:** In a standard car sequencing problem, the vehicles which require complex operations have to be sequenced with an adequate distance. This is provided with *p/q* ratio constraints.

**Roadef Challenge:** It is organized by the French Society of Operational Research and aims to share the latest developments in the industry.

**Sequencing Problem:** Selection of an appropriate order in which a number of jobs/operations can be assigned to a finite number of service facilities to optimize the outputs in term of time, cost or profit.

# Chapter 5
# Hub Location Allocation Problems and Solution Algorithms

**Peiman A. Sarvari**
*Istanbul Technical University, Turkey*

**Fatma Betül Yeni**
*Istanbul Technical University, Turkey*

**Emre Çevikcan**
*Istanbul Technical University, Turkey*

## ABSTRACT

*The Hub Location-Allocation Problem is one of the most important topics in industrial engineering and operations research, which aims to find a form of distribution strategy for goods, services, and information. There are plenty of applications for hub location problem, such as Transportation Management, Urban Management, locating service centers, Instrumentation Engineering, design of sensor networks, Computer Engineering, design of computer networks, Communication Networks Design, Power Engineering, localization of repair centers, maintenance and monitoring power lines, and Design of Manufacturing Systems. In order to define the hub location problem, the present chapter offers two different metaheuristic algorithms, namely Particle Swarm Optimization or PSO and Differential Evolution. The presented algorithms, then, are applied to one of the hub location problems. Finally, the performances of the given algorithms are compared in term of benchmarking.*

## INTRODUCTION

In this chapter, we discuss some services, such as database transaction, movements of people, commodities, information or unfinished parts that take place between an origin-destination pair of nodes. Such pairs of nodes can be found in the domain of a manufacturing site or spread along continents, as each origin-destination pair needs a service different from the other pairs.

DOI: 10.4018/978-1-5225-2944-6.ch005

Hub location problem is one of the most important topics of location problems. The facility location problem, also known as location analysis or k-center problem is a branch of operations research and computational geometry. Facility location problems try to reduce the costs of operations considering some set of constraints and relevant demands with locating different ranges of facilities. Making decisions for facility location are critically challenging regarding strategic planning for all types of business entities. Property acquisition and establishment are naturally costly so that one can consider facility location and relocation operations as long-term investments. Decision makers are challenging with different geographical, demographical and trending factors for selecting profitable sites. Thus, selection of robust facility locations is an important task, as far as future events are uncertain and unpredictable.

Hub location problem is an extension of the classical facility location problems. Hubs are facilities that operate as consolidating, connecting, and switching points for flows between the stipulated origins and destinations (Farahani et al., 2013). Hubs are also defined as special facilities that serve as switching, transshipping and sorting points in many-to-many distribution systems. The hub location problem is concerned with locating hub facilities and allocating demand nodes to hubs in order to route the traffic between origin–destination pairs (Alumur & Kara, 2008). Many applications are available for the hub location problem, and this section is primarily dedicated to introducing this problem to readers.

In this chapter, we have tried to fit what moves between an origin-destination pair of nodes, like information, people and commodities into the concept of HLP. Basically, each different pair of origin-destination node has to be serviced exclusively. For instance, people traveling from *i* to *j* are not interchangeable with those traveling from *j* to *i*. In order to have a fully connected network (a network in which all nodes are connected) with *N* nodes, in which each node can be either an origin or a destination, the number of pairs (*i-j* pairs which are different from *j-i* pairs) should be *N (N-1)*. Fig.1 illustrates a network composed of nodes and connections.

Assuming that we have different traffic services in this network and that each vehicle can service five origin-destination pairs every day, with 18 vehicles, we will be able to service ten nodes every day. If we set one of the nodes as a hub node and connect it to all the other nodes, which are introduced as spokes, we will have 2(n-1) connections to service all origin-destination node pairs. This network is presented in Figure 2 (Daskin, 1995).

Assume, if there are different traffic services, and if each vehicle can provide service for origin-destination pairs every day, with 18 vehicles, we will be able to service 46 nodes every day. Thus, with fixed traffic resources, we can service more cities with a hub network than with a completely connected network. Multi-hub network is another type of hub and spoke network that is a formation of two or several hubs and spoke networks in which all hubs are fully connected to each other.

This chapter is organized as follows. Section two, presents a technical and comprehensive literature review. In section three, the taxonomy of HLP is given, and section four, introduces some of the basic and fundamental models developed for the hub problem. In section five, an application with two metaheuristic solution algorithms is suggested together with its application in terms of performance evaluation of the proposed metaheuristics, and finally, the conclusion of the study is given in section six.

## LITERATURE REVIEW

The hub location problem has been studied for many years. With a glance at the related literature, one can find out the importance of the issue for the researchers. A panoramic view of its applications, research

*Figure 1. A regular full-connected network with connections between nodes*

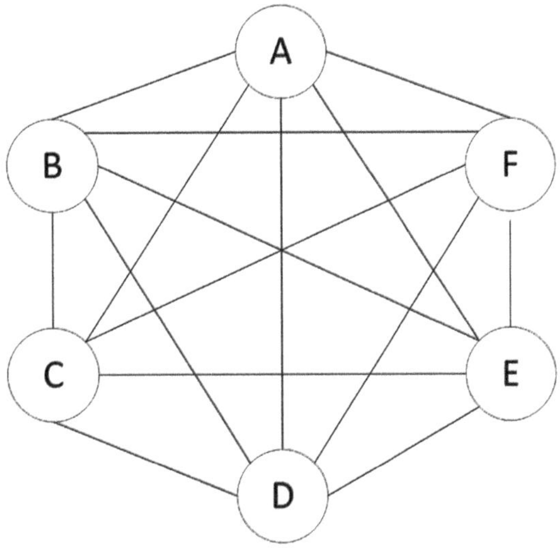


*Figure 2. A hub and spoke network*

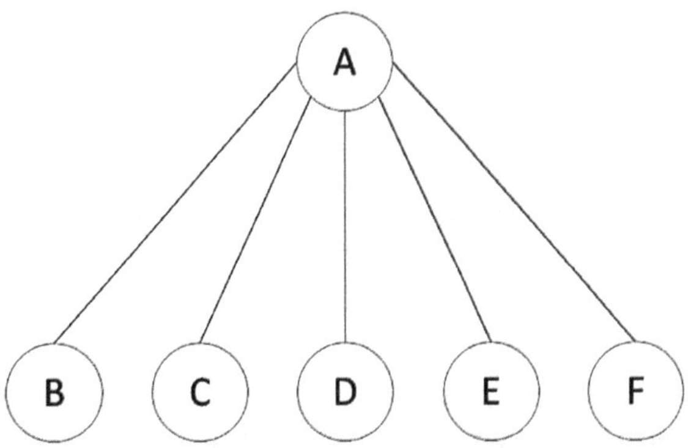


area, location and time, as illustrated in Figure 3, shows a boom of hub location subjected documents between the years 2000 and 2015. The U.S.A and China were pioneering regarding the number of the published documents more than half of which are mostly articles in engineering, mathematics, and computer sciences based areas.

Ostresh (1975) introduced a procedure to solve the two-center location-allocation problems. Likewise, O'Kelly (1987) formulated the P-hub median problem (P-HLMP) as a quadratic integer programming problem, which is a particular case of the None Linear Programming, when the objective function is in quadratic form, and the constraints are linear. Showing that the problem is NP-hard, he proposed two enumeration-based heuristics to solve it. Besides some exchange clustering methods presented by Klincewicz (1991), new heuristic approaches, such as Tabu search, genetic algorithm, and greedy search have also been used recently (Farahani et al., 2009).

*Figure 3. Research area and affiliation based systematic literature review*

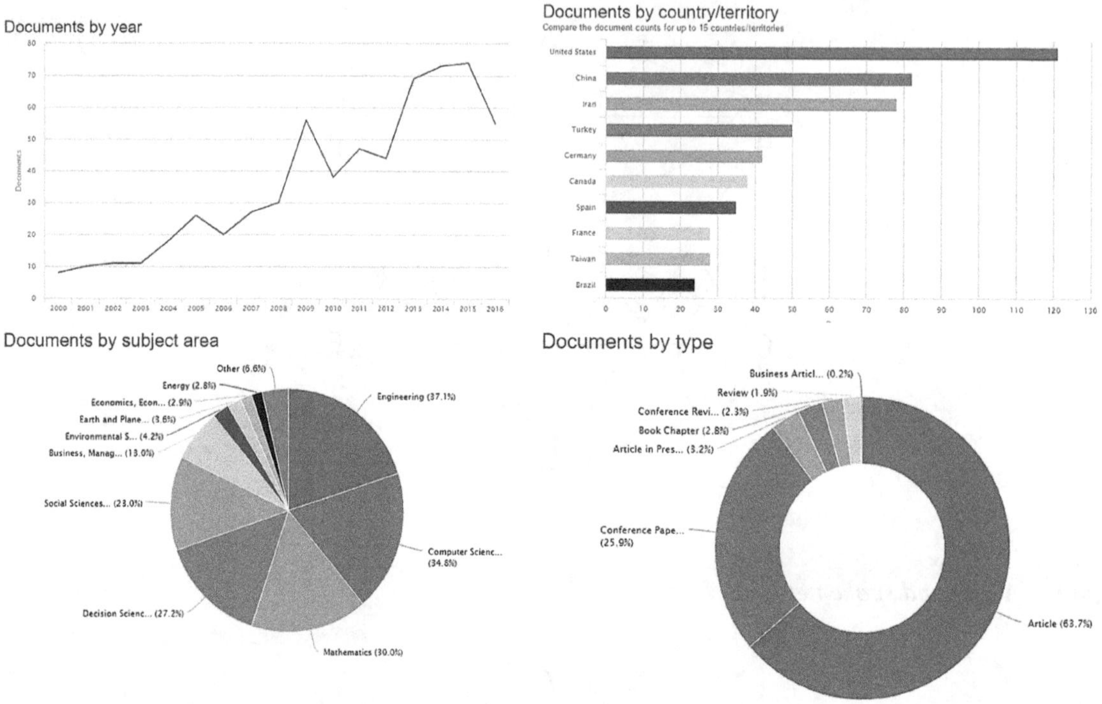


Sung and Jin (2001) proposed a dual-based solution algorithm which minimizes the total cost for hub location problems. They tested the effectiveness of the model with numerical examples among which the cost minimization objective is the most common one. For the same purpose, Martín and Román (2003) suggested a two-stage spatial competition game. Moreover, Rodríguez-Martín and Salazar-González (2008) and Contreras et al. (2010) have proposed a mixed integer programming (MIP) where the formulation is strengthened with valid inequalities. He et al. (2015) have also proposed an improved mixed integer programming (MIP) heuristic, which includes branch-and-bound, Lagrangian relaxation and linear programming relaxation. Thomadsen and Larsen (2007) represented branch-and-price algorithm or IP column generation (the combination of column generation and branch-and-bound algorithm) to solve a two-layered network (a hierarchical network) consisting of clusters of nodes, each defining an access network and a backbone network. Yaman et al. (2007) presented a generic mathematical model to solve the latest arrival hub location problem for cargo delivery systems. In their model, they aimed to minimize the longest delivery time. Sasaki et al. (2014) presented a general discrete Stackelberg hub location problem by adopting a multiple allocation hub-arc location models. They examined how the optimal solutions are affected by different customer allocation functions, different revenue sets, the number of hub arcs, and the degree of discount for hub arc travel. Rothenbächer et al. (2016) proposed branch-and-price-and cut algorithm for the solution of service network design hub location problems (SDNHLP). They conducted a computational experiment based on the combined road-rail transportation data. In addition to the mentioned studies, Marianov and Serra (2003) suggested a heuristic Tabu search algorithm for airline location in which each hub node is assumed to be an M/D/C queue. In this method, only a part of the feasible solution is surveyed, and the best neighborhood is selected as a new

*Table 1. Literature review matrix of Hub Location problem*

| Article | Objective | Type | Solution Algorithm |
|---|---|---|---|
| Ebery et al. (2000) | Cost minimization | CHLP-M | Linear programming, Heuristic approach |
| Sung and Jin (2001) | Cost minimization | HLP | Dual-based solution approach |
| Pamuk and Sepil (2001) | Maximum distance minimization | p-HLPC | Tabu search |
| Ebery (20019 | Cost minimization | USApHMP and pHAP | Mixed integer linear programming |
| Mayer and Wagner (2002) | Cost minimization | UHLP-M | Branch-and-bound method |
| Marianov and Serra (2003) | Cost minimization | HLP | Tabu search |
| Martin and Roman (2003) | Cost minimization | HLP | Spatial competition game |
| Bollapragada et al. (2005) | Maximizing expected total demand covered | CHMCLP-M | Greedy algorithm, Network-planning model |
| Aversa et al. (2005) | Cost minimization | p-HUB median | Mixed integer programming model |
| Marin (2005) | Cost minimization | SCHLP-M | Integer linear programming |
| Labbe et al. (2005) | Cost minimization | UHLP-S | Polyhedral analysis – Branch and cut algorithm |
| Topcouglu et al. (2005) | Cost minimization | UHLP-S | Genetic algorithm |
| Rodriguez et al. (2007) | Cost minimization | CHLP | Simulated annealing algorithm |
| Wagner (2007) | Cost minimization | CHLP | Tabu search |
| Berman et al. (2007) | Distance minimization | HLP | MiniSum model, MiniMax model |
| Thomadsen and Larsen (2007) | Cost minimization | HLP | Branch-and-price algorithm |

hub node with respect to continuous iterations on neighborhood nodes of the former hub, even the target function gains a worse solution. Furthermore et al. (2013) presented a hybrid heuristic approach based on simulated annealing and Tabu search algorithms to solve fully interconnected network design problem (FINDP) which is a specific application of hub location to network design.

Unlike the studies mentioned above, Berman et al. (2007) have analyzed three different kinds of the transfer point location problem in their study. They formulated each one of them based on two objective functions named Mini-Sum and Mini-Max.

P-hub location problems have also been studied in the literature. Besides some mathematical models aimed to minimize the total cost, maximum distance and total transportation time (Aversa et al., 2005; Campbell, 2009; Puerto et al., 2016), some heuristic algorithms have also been developed. Pamuk and Sepil (2001), for example, represented a single relocating heuristic by Tabu search to solve P-hub center problems (P-HLCP). They used two single-allocation schemes for the evaluation of the algorithm and employed a greedy local search to improve the resulting allocations.

Yaman (2008) represented a heuristic algorithm based on Lagrangian relaxation and local search to solve P-hub location median single allocation problems (P-HLMP-S). Mohammadi et al. (2016) have developed a bi-objective mixed integer non-linear model to study the bi-objective single allocation p-hub center median problems. They used a fuzzy queuing approach to model the uncertainties in the network and did several experiments besides a real transformation case to show the applicability of the proposed method.

*Table 2. Literature review matrix of Hub Location problem (Continues)*

| Article | Objective | Type | Solution Algorithm |
|---|---|---|---|
| Yaman, Kara and Tansel (2007) | Longest delivery time minimization | LAHLP | Generic mathematical model |
| Canovas, Garcia and Marin (2007) | Cost minimization | UHLP-M | Dual-ascent technique, Integer programming |
| Chen (2007) | Cost minimization | UHLP-S | Simulated annealing, Tabu list |
| Cunha and Silva (2007) | Cost minimization | UHLP-S | Hybrid genetic algorithm |
| Kratica et al. (2007) | Cost minimization | USApHMP | Genetic algorithm |
| Rodriguez and Salazar (2008) | Cost minimization | CHLP-M | Benders decomposition, Branch and cut algorithm |
| Costa, Captivo and Climaco (2008) | Service time minimization | CHLP-S | Bi-criteria model, Interactive methods |
| Alamur and Kara (2008) | Cost minimization | HLP | Survey |
| Yaman (2008) | Cost minimization | p-HLMP-S | Lagrangian relaxation based heuristic |
| Camargo, Miranda and Luna (2008) | Cost minimization | UHLP-S | Benders decomposition algorithm |
| Campbell (2009) | Cost minimization | p-HLMP-M | Mathematical model |
| Contreras, Fernandez and Marin (2010) | Cost minimization | THLP | Mixed integer programming (MIP) |
| Contreas, Cordeau and Laporte (2011) | Cost minimization | UHLP | Monte-carlo sampling, Benders decomposition |
| Alamur, Kara and Karasan (2012) | Cost minimization | MHLP | Linear mixed integer programming |
| Saboury et al. (2013) | Cost minimization | HLP | Tabu search and simulated annealing |
| Martin de Sa, Camargo and Miranda (2013) | Cost minimization | HLP-S | Benders decomposition method |

Bollapragada et al. (2005) represented a new network planning model and an efficient greedy solution heuristic to solve a model that is most closely related to the capacitated hub maximum-covering location problem with multi allocations (CHMCLP-M). The quality of the heuristic algorithms is evaluated by comparing its coverage with the optimal (for small problems) or with an upper bound obtained by solving a linear programming relaxation. Marn (2005) presented an integer linear programming formulation for splittable capacitated multiple allocation hub location problems and evaluated the model with a well-known data from the literature.

The uncapacitated hub location problems are the most common problem type at literature. Ebery (2001) proposed a new mixed integer linear programming to solve the uncapacitated single allocation p-hub median problems (USApHMP). Later, Mayer and Wagner (2002) used an aggregated branch and bound model. Labbe et al. (2005) represented a solution method based on branch and cut algorithm to solve uncapacitated single allocation hub location problem (UHLP-S). In this approach, the network connecting the hub nodes is called Backbone Network and the connected network of the terminal nodes is called access network. Cánovas et al. (2007) have also studied uncapacitated multiple allocation hub location problems (UHLP-M).

Focusing on the dual problem of a four-indexed formulation, they proposed a heuristic approach based on a dual ascent technique. They evaluated the obtained results based on two well-known data sets. Hsu and Chen (2007) have developed a hybrid heuristic approach based on Simulated Annealing

*Table 3. Literature review matrix of Hub Location problem (Continues)*

| Article | Objective | Type | Solution Algorithm |
|---|---|---|---|
| Rodriguez, Salazar and Yaman (2014) | Cost minimization | HLP | Mixed integer programming (MIP), Brunch-and-cut algorithm |
| Sasaki et al. (2014) | Cost minimization | HLP-M | Stackelberg competition model |
| He et al. (2015) | Cost minimization | HLP | Improved mixed integer programming (IMMIP) heuristic |
| Damgacıoğlu et al. (2015) | Cost minimization | UPHLP-S | Genetic algorithm |
| Mohammadi et al (2016) | Total transportation time minimization, Total system costs minimization | BpHCMP-S | Bi-objective mixed-integer non-linear programming (BMINLP), Fuzzy Queuing Approach, Game theory, Invasive weed optimization |
| Zhalechian et al. (2016) | Total transportation time minimization, Total system costs minimization | Multimodal HLP | Multi-objective mixed-integer non-linear mathematical model (MOMINLP) |
| Puerto et al. (2016) | Cost minimization | OMHLP-S | İnteger programming |
| Rothenbacher, Drexl and Irnich (2016) | Cost minimization | SNDHLP | Branch-and-price and Cut algorithm |

and Tabu Search. They tried to find a solution for uncapacitated single allocation hub location problems (USAHLP) with the hub-and-spoke network structure. Table 1, 2 and 3 indicate a summary of the related works, containing problem types and solution approaches, of different researchers. The examined studies are between 2000 and the first half of 2016.

## TAXONOMY OF HLPs

There are many applications of hub problems in real world. Here, we are going to give four major practices of hub location-allocation problems as following:

- **Energy Transfer:** Energy generator sites or drilling areas need to access to the best-located storages or transmission sites as hubs via pipelines or cables to transfer energy efficiently to the customers. As an example, we can name Waha Hub near Midland, Texas, the Katy Hub near Houston, Texas, and the Carthage Hub in East Texas.
- **Airlines and Airports:** As they aim to avoid empty direct flights and unreasonable flight fares, all airlines need to find the best-located airports as hubs for performing better operations and flight services. Frankfort airport, for example, is a hub in Europe that makes connection flights more reasonable than some direct flights.
- **Environmental Design:** It aims to handle the problems related to the transportation. For instance; it is concerned with finding the optimized land dump locations, with respect to garbage transportation stations.
- **Postal Logistics Network:** The strategic decisions for a hub based mail system include the following: the selection of suitable locations for consolidation, the assignment of customers to send-

ing and receiving depots, the determination of line-haul routes, and the choices of the types of transportation facilities. Operational decisions, which are based on strategic decisions, include the disposition of the number of vehicles for line-haul, and the planning of pick-up and delivery tours for parcels or part-loads to the customers from each depot (Zäpfel & Wasner, 2002).

Table 4 is giving a quick summary of the related works and the applications for location-allocation problem. The most common formulations, which have been widely applied in the literature, are introduced in the next section.

## FUNDAMENTAL HUB LOCATION MODELS

The problem of hub location has attracted many researchers who have worked on a variety of hub modeling problems. Since most of the applications of hub problems in the real world are discrete, the models developed so far are mostly discrete models. Sections 3.1-3.12 are introducing some most commonly used hub location problems in relevant literature and Tables 5, 6, 7 and 8 summarize the proposed mathematical models and their notations, model inputs and outputs (decision variables).

### Single Hub Location Problem

O'Kelly (1987) represented this problem with the following specifications:

- The total cost incurred by the location of hub nodes and allocation of non-hub nodes to hub nodes is minimized (Mini-Sum).
- The solution domain is all of the network nodes (network).
- The non-hub nodes are connected to the hub node.
- The number of hub nodes to locate is primarily specified (exogenous) and is equal to one.
- There is no cost for establishing the hub facility.
- The hub facility to locate is uncapacitated (capacity is not limited).
- The problem is the allocation of a non-hub node to just one hub (single allocation).

Considering the characteristics of this problem, its decision variables are binary (0 or 1). The mathematical formulation of single-HLP is depicted in Table 1.

### P-Hub Location Problem

In this problem, each non-hub node must be allocated to just one hub node. It is basically considered as a single allocation p-hub location problem. In this model:

- The total cost incurred by the location of hub nodes and allocation of non-hub nodes to hub nodes is minimized (criterion is Mini-Sum).
- The solution domain is all of the network nodes.
- The hub nodes are completely linked together.
- Every non-hub node is linked to a single hub node.

*Table 4. Applications of hub location problem and the related works*

| No | Article | Application | | | No | Article | Application | | |
|---|---|---|---|---|---|---|---|---|---|
| | | Yes | No | Field | | | Yes | No | Field |
| 1 | Wagner (2007) | x | | Transportation and Handling Problems | 21 | Campbell (2009) | x | | truck transportation |
| 2 | Ebery et al. (2000) | x | | Airline Passenger Transportation, Postal Logistic | 22 | Yaman (2008) | x | | Turkish network |
| 3 | Rodriguez and Salazar (2008) | x | | Telecommunications | 23 | Pamuk and Sepil (2001) | x | | airline passenger transportation |
| 4 | Costa, Captivo and Climaco (2008) | x | | Postal Logistic | 24 | Aversa et al. (2005) | x | | Ports |
| 5 | Bollapragada et al. (2005) | x | | Telecommunications | 25 | Contreas, et al (2011) | x | | ----- |
| 6 | Sung and Jin (2001) | x | | Transportation | 26 | Mayer and Wagner (2002) | x | | airline passenger transportation, postal logistic |
| 7 | Marianov and Serra (2003) | x | | Airlines and Airports | 27 | Canovas et al. (2007) | x | | airline passenger transportation, postal logistic |
| 8 | Martin and Roman (2003) | x | | Airlines and Airports | 28 | Labbe et al. (2005) | x | | Telecommunications |
| 9 | Beran et al. (2007) | | x | | 29 | Topcouglu et al. (2005) | x | | airline passenger transportation, postal logistic |
| 10 | Thomadsen and Larsen (2007) | | x | ------ | 30 | Chen (2007) | x | | airline passenger transportation, postal logistic |
| 11 | Alamur and Kara (2008) | | x | | 31 | Cunha and Silva (2007) | x | | transportation and handling problems |
| 12 | Saboury et al. (2013) | | x | | 32 | Camargo et al. (2008) | x | | airline passenger transportation, postal logistic |
| 13 | Rodriguez, Salazar and Yaman (2014) | x | | Airline Passenger Transportation, Postal Logistic | 33 | Damgacıoğlu et al. (2015) | x | | airline passenger transportation, postal logistic |
| 14 | He et al. (2015) | | x | ------- | 34 | Kratica et al. (2007) | x | | airline passenger transportation, postal logistic |
| 15 | Sasaki et al. (2014) | x | | Airline Passenger Transportation | 35 | Ebery (20019 | x | | postal logistic |
| 16 | Martin de Sa et al. (2013) | x | | Postal Logistic | 36 | Marin (2005) | x | | postal logistic |
| 17 | Alamur, Kara and Karasan (2012) | x | | Turkish Network | 37 | Rothenbacher et al. (2016) | x | | road-rail transportation |
| 18 | Zhalechian et al. (2016) | | x | | 38 | Contreras et al. (2010) | x | | airline passenger transportation, postal logistic |
| 19 | Puerto et al. (2016) | x | | Postal Logistic | 39 | Yaman et al. (2007) | x | | cargo logistic |
| 20 | Mohammadi et al (2016) | x | | Passenger Transportation | | | | | |

- The number of hub nodes to locate is primarily specified and is denoted by p, and at least one or at most two hub nodes have to be traversed for traveling between two non-hub nodes.
- The other features of p-Hub Location Problem are the same of the Single HLP.

## P-Hub Median Location Problem (Multiple Allocation p-HLP)

Because every non-hub node could be allocated to one hub node or more in p-hub median location problems, this model is named multiple allocation p-HLP. Most of the characteristics of this model are similar to those of the p-Hub LP except the following specifications:

- The problem tries to minimize the total transportation cost based on a nonlinear objective function.
- Its formulation is similar to the p-median formulation and is named p-hub median location problem.
- Non-hub nodes can be allocated to several hub nodes.

Flow rate between two nodes that has to be determined as part of the solution is a relaxed variable $\left(\geq 0 \geq 0\right)$.

## P-Hub Median Location Problem With Fixed Costs

Basically, the models mentioned above could be extended with fixed-link costs for connecting non-hub nodes to hub nodes. The following factors are the only differences between Multiple Allocation p-HLP and p-Hub Median LP with Fixed Costs:

- As the number of hubs to locate is not pre-specified so it must be considered both as a decision variable and as a part of the solution.
- A fixed cost related with links is incorporated into the model.

## Single Allocation p-Hub Location Problem

Unlike p-Hub Median Location model that allows the assignment of spokes to multiple hubs, sometimes we need to have each of the spoke nodes assigned to a single hub. Most of the assumptions of this model are similar to those of the median P-hub model except the following two features:

- Each non-hub node is assigned to only one hub.
- All of the outputs are binary variables (0–1).

## Minimum Value Flow on Any Spoke/Hub Connection Problem

Instead of arguing that each non-hub node should be allocated to a single hub node, we may contend that the flow between connections must be greater than or equal to some minimum flow threshold value. The assumptions of this model are similar to those of the median P-hub model except that there is a minimum flow for each spoke/hub connection.

*Table 5. Mathematical models of the most commonly used hub models*

| | Problem | Type | Model |
|---|---|---|---|
| **Single HLP** | $h_{ij}$ : Amount of flow between the *i*th node and *j*th node.<br>$C_{ij}$ : Cost amount between the *i*th node and *j*th node. | $Y_{ij} = \begin{cases} 1 & node\ i\ is\ allocated\ to\ hub\ j \\ 0 & otherwise \end{cases}$ | $\min \sum_i \sum_j \sum_k h_{ik} \left( C_{ij} + C_{jk} \right) Y_{ij} Y_{kj}$<br>Subject to<br>$\sum_j Y_{jj} = 1$<br>$Y_{ij} - Y_{jj} \leq 0 \ \forall i, j$<br>$Y_{ij} \in \{0,1\} \ \forall i, j$ |
| **p-Hub LP** | $h_{ij}$ : Amount of flow between the *i*th node and *j*th node.<br>$C_{ij}$ : Cost amount between the *i*th node and *j*th node.<br>$\alpha$ : Discount factor denoting economies of scale for transferring between hub nodes<br>$\left(0 \leq \alpha < 1\right)$ | $Y_{ij} = \begin{cases} 1 & node\ i\ is\ allocated\ to\ hub\ j \\ 0 & otherwise \end{cases}$ | $\min \sum_i \sum_k C_{ik} Y_{ik} \left( \sum_j h_{ij} \right)$<br>$+ \sum_k \sum_i C_{kj} Y_{ik} \left( \sum_J h_{ji} \right)$<br>$+ \alpha \sum_i \sum_j \sum_k \sum_m h_{ij} C_{km} Y_{ik} Y_{jm}$<br>Subject to<br>$\sum_j Y_{ij} = 1 \ \forall i$<br>$\sum_j Y_{jj} = P$<br>$Y_{ij} - Y_{jj} \leq 0 \ \forall i, j$<br>$y_{ij} \in \{0,1\} \ \forall i, j$ |
| **p-Hub Median LP** | $C_{ij}^{km}$ : The transportation cost between start node *i*, end node *j*, the *k*th hub nodes and the *m*th node. | $X_j = \begin{cases} 1 & a\ hub\ is\ located\ at\ node\ j \\ 0 & otherwise \end{cases}$<br>$Z_{ij}^{km} \geq o$; The non-hub nodes are allowed to be allocated to several hub nodes. | $\min \sum_i \sum_j \sum_k \sum_m C_{ij}^{km} h_{ij} Z_{ij}^{km}$<br>Subject to<br>$\sum_k X_k = P \ \sum_k \sum_m Z_{ij}^{km} = 1$<br>$\forall i, j$<br>$Z_{ij}^{km} \leq X_m \ \forall i, j, k, m$<br>$Z_{ij}^{km} \leq X_k \ \forall i, j, k, m$<br>$Z_{ij}^{km} \geq 0 \ \forall i, j, k, m$<br>$X_k \in \{0,1\} \ \forall k$ |

*continued on following page*

*Table 5. Continued*

| | | | |
|---|---|---|---|
| **Single HLP** | $h_{ij}$: Amount of flow between the *i*th node and *j*th node.<br>$C_{ij}$: Cost amount between the *i*th node and *j*th node. | $Y_{ij} = \begin{cases} 1 & node\ i\ is\ allocated \\ & to\ hub\ j \\ 0 & otherwise \end{cases}$ | $\min \sum_i \sum_j \sum_k h_{ik} \left( C_{ij} + C_{jk} \right) Y_{ij} Y_{kj}$<br>Subject to<br>$\sum_j Y_{jj} = 1$<br>$Y_{ij} - Y_{jj} \leq 0 \ \forall i, j$<br>$Y_{ij} \in \{0,1\} \ \forall i, j$ |
| **p-Hub LP** | $h_{ij}$: Amount of flow between the *i*th node and *j*th node.<br>$C_{ij}$: Cost amount between the *i*th node and *j*th node.<br>$\alpha$: Discount factor denoting economies of scale for transferring between hub nodes<br>$\left(0 \leq \alpha < 1\right)$ | $Y_{ij} = \begin{cases} 1 & node\ i\ is\ allocated \\ & to\ hub\ j \\ 0 & otherwise \end{cases}$ | $\min \sum_i \sum_k C_{ik} Y_{ik} \left( \sum_j h_{ij} \right) + \sum_k \sum_i C_{kj} Y_{ik} \left( \sum_J h_{ji} \right) + \alpha \sum_i \sum_j \sum_k \sum_m h_{ij} C_{km} Y_{ik} Y_{jm}$<br>Subject to<br>$\sum_j Y_{ij} = 1 \ \forall i$<br>$\sum_j Y_{jj} = P$<br>$Y_{ij} - Y_{jj} \leq 0 \ \forall i, j$<br>$y_{ij} \in \{0,1\} \ \forall i, j$ |
| **p-Hub Median LP** | $C_{ij}^{km}$: The transportation cost between start node *i*, end node *j*, the *k*th hub nodes and the *m*th node. | $X_j = \begin{cases} 1 & a\ hub\ is\ located \\ & at\ node\ j \\ 0 & otherwise \end{cases}$<br>$Z_{ij}^{km} \geq o;$ The non-hub nodes are allowed to be allocated to several hub nodes. | $\min \sum_i \sum_j \sum_k \sum_m C_{ij}^{km} h_{ij} Z_{ij}^{km}$<br>Subject to<br>$\sum_k X_k = P \ \sum_k \sum_m Z_{ij}^{km} = 1$<br>$\forall i, j$<br>$Z_{ij}^{km} \leq X_m \ \forall i, j, k, m$<br>$Z_{ij}^{km} \leq X_k \ \forall i, j, k, m$<br>$Z_{ij}^{km} \geq 0 \ \forall i, j, k, m$<br>$X_k \in \{0,1\} \ \forall k$ |

*Table 6. Mathematical models of the most commonly used hub models (continues)*

| Problem Type | Model Inputs | Model Outputs | Mathematical Model |
|---|---|---|---|
| **p-Hub Median LP with Fixed Costs** | $g_{ik}$ : The fixed cost of connecting non-hub node *i* to a hub facility located at node *k*. | $W_{ik}$ : The binary variable denoting selection of link *(i, k)* if it is equal to one. | $\min \sum_i \sum_j \sum_k \sum_m C_{ij}^{km} h_{ij} Z_{ij}^{km}$<br>Subject to<br>$\sum_k X_k = P$<br>$\sum_k \sum_m Z_{ij}^{km} = 1 \ \forall i, j$<br>$\sum_i \sum_k g_{ik} W_{ik}$<br>$Z_{ij}^{km} \leq X_m \ \forall i, j, k, m$<br>$Z_{ij}^{km} \leq X_k \ \forall i, j, k, m$<br>$Z_{ij}^{km} \geq 0 \ \forall i, j, k, m$<br>$X_k \in \{0,1\} \ \forall k$ |
| **Single Allocation p-Hub LP** | $C_{ij}^{km}$ : The transportation cost between start node *i*, end node *j*, the *k*th hub nodes and the *m*th node. | $Y_{ik}$ : The *i*th none-hub node is assigned to the *k*th hub node. | $Min \sum_i \sum_j \sum_k \sum_m C_{ij}^{km} h_{ij} Z_{ij}^{km}$<br>$\sum_k X_k = P$<br>$\sum_k \sum_m Z_{ij}^{km} = 1 \ \forall i, j$<br>$Y_{ik} \leq X_k \ \forall i, k$<br>$\sum_k Y_{ik} = 1 \ \forall i$<br>$Y_{ik} + Y_{jm} + 2Z_{ij}^{km} \geq 0 \ \forall i, j, k, m$<br>$X_k = 0,1 \ \forall k$<br>$Y_{ik} = 0,1 \ \forall i, k$<br>$Z_{ij}^{km} = 0,1 \ \forall i, j, k, m$ |

## P-HLP With Limited Capacity

If the incoming or outgoing flows in a network are limited to a fixed and certain value that is considered as hub capacity, we are faced with a p-HLP with Limited Capacity. The problem is formulated in a similar way to a general p-hub median location problem plus an extra capacity constraint.

## P-Hub Center Location Problem

Aiming to minimize the maximum travel time (or cost) between any origin–destination pair, the p-hub center problem is to locate p hubs and to allocate non-hub nodes to hub nodes. One may use this approach for decomposable or sensitive goods in a hub system. The characteristics of this model are similar to those of the median p-hub model except that some decision variables are relaxed (not necessary binary) and the objective function is *MiniMax*.

## Hub Set Covering Location Problem

The number of hubs in hub set covering problem is not determined. Hence, the objective function in this problem minimizes the establish cost of hubs. This issue is defined when all origin–destination are fully covered. Origin–destination can be allocated to one hub or more than one hub (Karimi & Bashiri, 2011). The assumptions of this model are similar to median P-hub model except that:

- The number of hubs are as decision variables and are not known before solving.
- A fixed cost of hub location is incorporated in the model.

## Hub Maximal Covering Location Problem

If the time (cost or distance) to cover all origin–destination pairs is greater than the available time (budget or distance), we can solve it by using a hub maximal covering problem, i.e., maximize the demand covered with a given number of hub facilities. The hub maximal covering objective function is maximizing the total flow between all origin–destination nodes which are allocated to the structured network (Karimi & Bashiri, 2011). The assumptions of this model are similar to those of the median P-hub model except that:

The number of hubs is known.
The fixed cost of hub location does not matter of consideration in the model.

## Multi-Objective p-Hub Location Problem

Costa et al. (2008) proposed a multi-objective HLP in which the first objective minimizes the total transportation cost, while the second one minimizes the maximum time that the hub nodes take to process the flow (i.e., minimizes the maximum service time of the hub nodes) (Farahani et al., 2013):

- In a similar manner to the p-HLP, each non-hub node in this problem is assigned to only one hub node.
- In this model, the criteria are Mini-Sum and Mini-Max.
- The solution domain is the nodes of the network.
- There is a full connection between hubs.
- Every non-hub node is linked to a single hub ultimately.
- The number of hubs to locate is pre-defined, and one or two hub nodes have to be traversed for traveling between two non-hub nodes.

*Table 7. Mathematical models of the most commonly used hub models (continues)*

| Problem Type | Model Inputs | Model Outputs | Mathematical Model |
|---|---|---|---|
| **Minimum Value Flow on any Spoke/Hub Connection Problem** | $L_{ik}$: The minimum flow between spoke $i$ and hub $k$ | $X_j = \begin{cases} 1 & a\ hub\ is\ located\ at\ node\ j \\ 0 & otherwise \end{cases}$<br>$Z_{ij}^{km} \geq o$; The non-hub nodes are allowed to be allocated to several hub nodes. | $\min \sum_i \sum_j \sum_k \sum_m C_{ij}^{km} h_{ij} Z_{ij}^{km}$<br>Subject to<br>$\sum_k X_k = P$<br>$\sum_k \sum_m Z_{ij}^{km} = 1 \ \forall i, j$<br>$Z_{ij}^{km} \leq X_m \ \forall i, j, k, m$<br>$Z_{ij}^{km} \leq X_k \ \forall i, j, k, m$<br>$Z_{ij}^{km} \geq 0 \ \forall i, j, k, m$<br>$X_k \in \{0,1\} \ \forall k$<br>$Y_{ik} + Y_{jm} - 2Z_{ij}^{km} \geq 0 \ \forall i, j, k, m$<br>$\sum_m \sum_j h_{ij} Z_{ij}^{km}$<br>$+\sum_p \sum_s h_{pi} Z_{pi}^{sk} \geq L_{ik} Y_{ik} \ \forall i, k$ |
| **Capacity Limitation of HLP** | $\theta_k$: The capacity of a hub at the $k$th candidate. | $X_j = \begin{cases} 1 & a\ hub\ is\ located\ at\ node\ j \\ 0 & otherwise \end{cases}$<br>$Z_{ij}^{km} \geq o$; The non-hub nodes are allowed to be allocated to several hub nodes. | $\min \sum_i \sum_j \sum_k \sum_m C_{ij}^{km} h_{ij} Z_{ij}^{km}$<br>Subject to<br>$\sum_k X_k = P$<br>$\sum_k \sum_m Z_{ij}^{km} = 1 \ \forall i, j$<br>$Z_{ij}^{km} \leq X_m \ \forall i, j, k, m$<br>$Z_{ij}^{km} \leq X_k \ \forall i, j, k, m$<br>$Z_{ij}^{km} \geq 0 \ \forall i, j, k, m$<br>$X_k \in \{0,1\} \ \forall k$<br>$\sum_m \sum_i \sum_j h_{ij} Z_{ij}^{km}$<br>$+\sum_s \sum_i \sum_j h_{ij} Z_{ij}^{sk} \leq \theta_k X_k \quad \forall k$ |

*continued on following page*

*Table 7. Continued*

| Problem Type | Model Inputs | Model Outputs | Mathematical Model |
|---|---|---|---|
| **p-Hub Center LP** | $C_{ij}^{km}$ : The transportation cost between start node *i*, end node *j*, the *k*th hub nodes and the *m*th node. | $X_j = \begin{cases} 1 & a\ hub\ is\ located\ at\ node\ j \\ 0 & otherwise \end{cases}$<br>$Z_{ij}^{km} \geq o;$ The non-hub nodes are allowed to be allocated to several hub nodes.<br>$X_k \in \{0,1\}\ \forall k$ | $\min \max_{i,j,k,m} \{C_{ij}^{km} h_{ij} Z_{ij}^{km}\}$<br>Subject to<br>$\sum_k X_k = p$<br>$\sum_k \sum_m Z_{ij}^{km} = 1 \quad \forall i, j$<br>$Z_{ij}^{km} \leq X_k\ \forall i, j, k, m$<br>$Z_{ij}^{km} \leq X_m\ \forall i, j, k, m$<br>$Z_{ij}^{km} \geq 0\ \forall i, j, k, m$ |

- There are not any fixed costs for hub nodes.
- The capacities of hubs are not limited.
- The decision variables are binary.

## Continuous p-HLP

In a hub location problem, sometimes we have to consider a continuous domain of solution that is not like a discrete set of nodes on a graph, yet like a plane or a sphere. The specifications of this model are:

- The criterion is Mini-Sum.
- The solution domain is a plane and is continuous.
- The hub nodes are completely linked, and every non-hub node is linked to only one hub facility.
- For traveling between two non-hub nodes, the number of hub nodes to locate is primarily specified as one or two.
- The fixed cost of opening hub facilities is not considered.
- The capacities of hubs are not limited.
- The decision variables are binary.

## Uncapacitated Single Allocation p-Hub Median Problem (USApHMP)

USApHMP belongs to the class of NP-hard problems. Even when the set of hubs is given, the assignment sub-problem of optimal allocation of non-hub nodes to hubs is also NP-hard (R.F. Love, J.G. Moris, 1988). The objective is to minimize the overall flow cost in a network under the following assumptions:

*Table 8. Mathematical models of the most commonly used hub models (continues)*

| Problem Type | Model Inputs | Model Outputs | Mathematical Model |
|---|---|---|---|
| **Hub Set Covering LP** | $F_K$ : The fixed cost in the *k*th candidate node. $V_{ij}^{km}$ : The node hubs of *m* and *k* cover the origin-destination of *I* and *j*. | $X_j = \begin{cases} 1 & a\ hub\ is\ located \\ & at\ node\ j \\ 0 & otherwise \end{cases}$ $Z_{ij}^{km} \geq o$; The non-hub nodes are allowed to be allocated to several hub nodes. | $\min \sum_k F_K X_K$ Subject to $\sum_k \sum_m V_{ij}^{km} Z_{ij}^{km} \geq 1 \ \forall i,j$ $Z_{ij}^{km} \leq X_k \ \forall i,j,k,m$ $Z_{ij}^{km} \leq X_m \ \forall i,j,k,m$ $Z_{ij}^{km} \geq 0 \ \forall k$ $X_k \in \{0,1\} \ \forall k$ |
| **Hub Maximal Covering LP** | $h_{ij}$ : The demand flow from origin *i* to destination *j*. | $X_j = \begin{cases} 1 & a\ hub\ is\ located \\ & at\ node\ j \\ 0 & otherwise \end{cases}$ $Z_{ij}^{km} \geq o$; The non-hub nodes are allowed to be allocated to several hub nodes. | $\max \sum_i \sum_j \sum_k \sum_m h_{ij} V_{ij}^{km} Z_{ij}^{km}$ Subject to $\sum_k X_k = p$ $\sum_k \sum_m Z_{ij}^{km} = 1 \ \forall i,j$ $Z_{ij}^{km} \leq X_k \ \forall i,j,k,m$ $Z_{ij}^{km} \leq X_m \ \forall i,j,k,m$ $Z_{ij}^{km} \geq 0 \ \forall k$ $X_k \in \{0,1\} \ \forall k$ |
| **Multi Objective p-HLP** | $T_k$ : The time unites that the hub node *k* takes to process one unit of flow. | $Y_{ij} = \begin{cases} 1 & node\ i\ is\ allocated \\ & to\ hub\ j \\ 0 & otherwise \end{cases}$ | $\min \sum_i \sum_j \sum_k \sum_m h_{ij} Y_{ik} Y_{jm} \left( C_{ik} + \alpha C_{km} + C_{jm} \right)$ $\min \max_k \left\{ T_k \sum_i \sum_j h_{ij} Y_{ik} + \sum_i \sum_j h_{ji} Y_{jm} Y_{ik} \right\}$ Subject to $\sum_k Y_{ik} = 1 \ \forall i$ $\sum_k Y_{kk} = p$ $Y_{ik} - Y_{kk} \leq 0 \ \forall i,k$ $Y_{ik} \in \{0,1\} \ \forall i,k$ |

- The number of hubs to be located is predetermined (p).
- There are no capacities or fixed costs involved.
- Each origin/destination node is assigned to a single hub.
- Direct transportation between non-hub nodes is not allowed.

The p-hub median formulation can sometimes lead to unsatisfactory results, for example, when the worst origin–destination distance (cost) is important. Difficulties of this kind can be avoided by using the p-hub center formulation, which minimizes the maximum distance between origin–destination pairs (Stanimirović, 2010).

As presented in Tables 1, 2 and 3 based on the relevant literature, there have been proposed a wide range of different solution algorithms to solve various types of HLPs. Even though integer programming optimization approaches are applied to solve small hub problems, larger instances of HLPs need to be solved by heuristic or meta-heuristic procedures. As a matter of fact, while large-size instances can be dealt with specialized exact methods, development of meta-heuristics has helped many real-world applications, in which optimal/near-optimal solutions can even be obtained in less computational time (Gelareh & Nickel, 2011).

In the past, few solving methods were proposed for hub location problems in which the number of hubs is a decision variable, and the fixed cost of establishing a hub is considered. Nevertheless, with the growth of meta-heuristic methods, the number of ways to solve such problems has been increased (Farahani et al., 2009). Although literature review of this episode is including most of the hub location-allocation problems and solution methods, the subsequent cases are, however, the major ones: a mixed method, Simulated Annealing and Tabu Search provided by Chen (2007); Genetic Algorithm method presented by Topcouglu et al. (2005); a Bi-criteria Integer Linear Programming to solve the capacitated single allocation hub location problem proposed by Costa et al. (2008); a heuristic algorithm based on Lagrangian Relaxation and Local Search to solve P-hub location median single allocation problems by Yaman (2008); and a Genetic Algorithm to solve UPHLP-S problem by Damgacioglu et al. (2015).

## PROPOSED HEURISTICS ALGORITHMS

In order to solve a general hub set covering location problem; that is fully introduced in Table 8, an application of hub location problem in manufacturing is addressed in this section by using two distinct metaheuristic algorithms, namely Differential Evolution (DE) and Particle Swarm Optimization (PSO) with different number of nodes, transportation charges, and fixed costs.

### Representation of Solution

The aim of this problem is to find the hub locations and the allocation of demand nodes to hubs. For presenting the given network, some discrete nodes are used here. The solutions are presented as a matrix. Each column shows a node in the network, in which its elements value explains the number of the hub and the nodes, which are allocated to them. Furthermore, when the value of each element on the entire column is equal to zero, the node is considered as a demand node. If the value of the item that is in the same row and the same column is equal to one, that node is considered as a hub and the rest of the elements show the demand nodes allocated to it. For example, a sample solution is obtained as in the following matrix (EghbaliZarch et al. 2013):

$$S = \begin{cases} 0 & 0 & 0 & 0 & 0 & 0 & 1 & 0 \\ 0 & 0 & 0 & 0 & 0 & 0 & 1 & 0 \\ 0 & 0 & 0 & 0 & 0 & 0 & 0 & 1 \\ 0 & 0 & 0 & 0 & 0 & 0 & 0 & 1 \\ 0 & 0 & 0 & 0 & 0 & 1 & 0 & 0 \\ 0 & 0 & 0 & 0 & 0 & 1 & 0 & 0 \\ 0 & 0 & 0 & 0 & 0 & 0 & 1 & 0 \\ 0 & 0 & 0 & 0 & 0 & 0 & 0 & 1 \end{cases}$$

In this solution, the matrix S is giving the set of connections between eight nodes. As it is clear, nodes 6, 7 and 8 are hubs. Also, node five is allocated to hub 6, nodes 1 and two are allocated to hub 7 and nodes 3 and four are assigned to hub 8.

The example mentioned above shows how a hub location problem should be solved. This solution is, however, inadequate for solving a very large set of nodes where some advanced solution techniques are needed. In such cases, metaheuristic algorithms should be used to receive an acceptable result in the reasonable amount of time. For that purpose, here we are going to introduce two popular metaheuristic algorithms namely Differential Evolution Algorithm and Particle Swarm Optimization Algorithm.

## Differential Evolution Algorithm

Differential Evolution is a Stochastic Direct Search as well as a Global Optimization algorithm that is an instance of an Evolutionary Algorithm from the field of Evolutionary Computation. It is related to sibling Evolutionary Algorithms such as the Genetic Algorithm, Evolutionary Programming, and Evolution Strategies. Differential Evolution (DE) was introduced by Ken Price and Rainer Storn in a series of papers that followed in quick succession (Storn, 1996; 2008, Storn & Price, 1996; 1997; 1995). DE is a population-based stochastic method for global optimization.

There are three kinds of vector in the literature of DE algorithm: a parent vector from the current generation that is called target vector; a mutant vector obtained through the differential mutation operation that is known as donor vector; and finally, an offspring formed by recombining the donor with the target vector namely trial vector (Das & Suganthan, 2011). The primary stages and algorithm steps of DE is illustrated in Flowchart 1. We are also giving details and descriptions of DE stages regarding algorithm installation. Moreover, the main stages and pseudocodes of DE algorithm are shown in Flowchart 1. The original version of DE can be defined by the following descriptions:

**Stage 1:** The population.

$$Pop_{x,y} = \left(X_{i,y}\right), \quad i = 0,1,\ldots,N_{Pop} - 1, \quad g = 0,1,\ldots,y_{\max}, \\ X_{i,y} = \left(x_{j,i,y}\right), \qquad j = 0,1,\ldots,Dim - 1. \tag{1}$$

where $N_{Pop}$ denotes the number of population vectors, y defines the generation counter, and *Dim* the dimensionality, i.e. the number of parameters.

**Stage 2:** The initialization of the population via x.

$$x_{j,i,0} = \alpha.\left(U_j - L_j\right) + L_j \tag{2}$$

For creating the initial population, the *j* numbers of hubs are firstly located randomly.

The dimensional initialization vectors, L and U, indicate the lower and upper bounds of the parameter vectors $x_{i,y}$. The random number generator α returns a uniformly distributed random number from within the range [0,1), i.e., $0 \leq \alpha < 1$. The subscript, j, indicates that a new random hub is generated for each parameter.

**Stage 3:** The perturbation of a base vector by using a difference vector based mutation to generate a mutation vector $m_{i,y}$.

$$M_{i,y} = g_{i,y} + \beta.\left(X_{r_1,y} - X_{r_2,y}\right) \tag{3}$$

The difference vector indices, and $r_1$, $r_2$ are randomly selected once per base vector. The setting $g_{i,y} = x_{r_0}$, g defines what is often called classic DE where the base vector is also a randomly chosen population vector. The random indexes $r_0$, $r_1$ and $r_2$ should be mutually exclusive. There are also variants of perturbations that are different to Eq. (3) and some of them will be described later. For example, setting the base vector to the current best vector or a linear combination of various vectors is also popular.

**Stage 4:** Diversity Enhancement.

The classic variant of diversity enhancement is a crossover, which mixes parameters of the mutation vector $m_{i,y}$ and the so-called target vector $x_{i,y}$ in order to generate the trial vector $v_{i,y}$. The most common form of crossover is uniform and is defined as:

$$V_{i,y} = v_{j,i,y} = \begin{cases} m_{j,i,y} & if\ \alpha \leq Cr \\ x_{j,i,y} & otherwise \end{cases} \tag{4}$$

**Stage 5:** Selection.

DE uses simple one-to-one survivor selection where the trial vector $v_{i,y}$ competes against the target vector $x_{i,y}$. The vector with the lowest objective function value survives into the next generation y+1.

$$X_{i,y+1} = \begin{cases} V_{i,y} & if\, f\left(V_{i,y}\right) \leq f\left(X_{i,y}\right) \\ X_{i,y} & otherwise \end{cases} \tag{5}$$

## Flowchart 1: DE Algorithm

```
Begin
Iteration=0
Create a random initial population
for each node i = 1 to Npop-1 do
    for j=1 to Dim do
                                 Equation 2.
    end for
end for
Evaluate Objective Function for each node of population
 for i=1 to Npop-1 do
                                 Fitness function
 end for
 Test vector generation
 for Iteration=1 to MaxIteration do
      for i=1 to Npop do
 Select randomly r1, r2, r3∈[1,Npop], r1≠r2≠r3≠i
 Mutation and Crossover Process
      jrand=α
      for j=1 to Dim do
              if(α<Cr or j==jrand) then
                       Equation 3.
              else
                       m(i,j)=x(i,j)
              end if
      end for
 end for
 Selection
        if f(m)≤f(x(i)) then
              x(i)=m(i,j)
        else
              x(i)= x(i) of prior Iteration
        end if
```

```
        end for
    end for
    End
```

In order to prevent the case $v_{i,y} = x_{i,y}$ at least one component is taken from the mutation vector $v_{i,y}$, a detail that is not expressed in Eq. (4). Other variants of crossover are described by Storn (2008).

## Particle Swarm Optimization

Inspired by the studies in neurosciences, cognitive psychology, social ethology and behavioral sciences, the concept of swarm intelligence (SI) was introduced in the domain of computing and artificial intelligence in 1989 as an innovative collective and distributed intelligent paradigm for solving problems, mostly in the domain of optimization, without centralized control or the provision of a global model (Marini & Walczak, 2015).

Particle Swarm Optimization (PSO) algorithm is a bio-inspired metaheuristic that is inspired by the swarm behavior in nature such as fish schooling or flocking of birds. A PSO consists of a pool of particles where a particle position in *n*-space represents a solution to a given problem. As each particle moves around in *n*-space, it remembers its best position so far, and it is also aware of the global best position found by all particles (Bailey et al., 2013). For the sake of simplification, the basic stages and algorithm steps of PSO are depicted in Flowchart 2.

### Flowchart 2: PSO Algorithm

```
for each particle i = 1, ..., N do
    Initialize the particle's position with a uniformly distributed random vector: x_i ~ U(L, U)
    Initialize the particle's best known position to its initial position: b(i) ← x(i)
    if f(b[i]) < f(g) then
        update the swarm's best-known position: g ← b(i)
    Initialize the particle's velocity: v(i) ~ U(-|U-L|, |U-L|)
while a termination criterion is not met do:
    for each particle i = 1, ..., N do
        for each dimension dim = 1, ..., n do
            Pick random numbers: r(i), r(g) ~ U(0,1)
            Update the particle's velocity: v(id) ← ω v(id) + φ(p) r(p) [p(id) -x(id)] + φ(g) r(g)
                                                                  [g(d)-x(id))
            Update the particle's position: x(i) ← x(i) + v(i)
            if f(x(i)) < f(p(i)) then
                Update the particle's best known position: p(i) ← x(i)
```

```
            if f(p(i)) < f(g) then
               Update the swarm's best-known position: g ← p(i)
```

To give more detailed information about stages of PSO, let $f\left(x\right) \mid \left\{\forall x \in \mathbb{R}\right\}$ be the objective function that must be minimized. The function takes a candidate solution as an argument in the form of a vector of real numbers and produces a real number as output which indicates the objective function value of the given candidate solution. The goal is to find a solution g for which, $f$(g) ≤ $f$(d) for all d in the search-space, which would mean g is the global minimum. Let $N$ be the number of particles in the swarm, each having a position $x_i \in \mathbb{R} \mid i \in N$ in the search space and a velocity $v_i \in \mathbb{R} \mid i \in N$. Let $b_i$ be the best-known position of particle $i$ and let g be the best-known position of the entire swarm. The values L and U are respectively the lower and upper boundaries of the search-space. The termination criterion can be a number of iterations performed, or a solution with adequate objective function value is found. The parameters $\omega$, $\varphi_p$, and $\varphi_g$ are selected by the practitioner and control the behavior and efficacy of the PSO method (Clerc, 2012).

## FINDINGS AND DISCUSSION

The DE and PSO were run on a predesigned data sets with fixed model parameters and different numbers of nodes, and the solution qualities were compared with each other. Before designing the DE and PSO, we first report the results with preset parameters for both DE and PSO algorithms. Table 9, describes empirically established parameter settings.

Regarding obtaining optimal solution, we applied an integer linear programming (ILP) algorithm for discrete space that was introduced and used by Marin (2005). We observed that the ILP algorithm was unable to manage more than 50 nodes as it just found *optimum values* of an objective function for the data sets with 10, 20, 30, 40 and 50 nodes. For the sake of comparison, we have used these optimum values for determining the best gaps of both DE and PSO using the formulation (5). The *best result* is obtained after the accomplishment of each metaheuristic at the end of 500[th] iterations or in 3600 seconds. The results summary of the analyses of a classic hub set covering location problem is given in Table 10. All analyses are fulfilled by a computer with 4770 (i7) CPU using MATLAB software.

*Table 9. Established parameter settings for DE and PSO in Matla*

| %% DE Parameters | %% PSO Parameters |
|---|---|
| MaxIt=500; % Maximum Number of Iterations | MaxIt=500; % Maximum Number of Iterations |
| $N_{Pop}$=200; % Population Size | $N_{Pop}$=200; % Population Size (Swarm Size) |
| beta_min=0.8; % Lower Bound of Scaling Factor | w=0.4; % Inertia Weight |
| beta_max=1.5; % Upper Bound of Scaling Factor | wdamp=1; % Inertia Weight Damping Ratio |
| pCR=0.2; % Crossover Probability | c1=0.3; % Personal(Cognitive) Learning Coefficient |
| | c2=0.9; % Global Learning Coefficient |
| | % Velocity Limits |
| | VelMax=0.1*(VarMax-VarMin); |
| | VelMin=-VelMax; |

$$\frac{\text{best cost of obtained by the metaheurestic - optimum value}}{\text{best result}} \times 100 \qquad (5)$$

Figure 4 and Figure 5 are graphically showing performances of both algorithms aiming to solve hub location-allocation problem for instance with 50 nodes. Proposed algorithms tried to minimize objective functions by assigning nodes to hubs. It is clear that both algorithms have found 11 hubs but with extremely different best costs. Moreover, the operation elapsed time for PSO and DE algorithms are behaving completely different.

In this certain example, the best cost and total CPU time to get the best cost by DE are respectively, 75663767 and 593 seconds, whereas these values for PSO are 71354969 and 537. Regarding preciseness and accuracy, gaps of PSO algorithm are lesser than DE.

In general, as it is depicted in Table 10, considering the first 10, 20 and 30 nodes, performances of DE are considerably better in terms of the best cost and CPU time.

With booming the numbers of nodes, the performance of PSO is getting better and better. As it is clear, the obtained best costs by PSO are very close to the optimum values obtained by an ILP tool. As illustrated in Table 10, ILP could not perform to catch optimum values of the experiments with more than 50 nodes and the results are not available (N.A). One might also consider the comparisons between the best costs and the total elapsed times for each experiment by using DE and PSO. Since minimization is the objective function of hub set covering location problem, so the best costs of PSO are significantly better than those of DE.

Gaps are metaheuristics variations that result from the difference between the best cost of metaheuristic algorithm and the optimum value. There are many possible reasons that cause variations, yet one of the most important factors is distribution type in the mutation phase for DE and the initialization stage for PSO. We used Uniform distribution for both metaheuristic algorithms as one can get better results with lower variations using different probabilistic distributions.

*Figure 4. Hub location allocation for an instance with 50 nodes using DE (left) and PSO (right) metaheuristics*

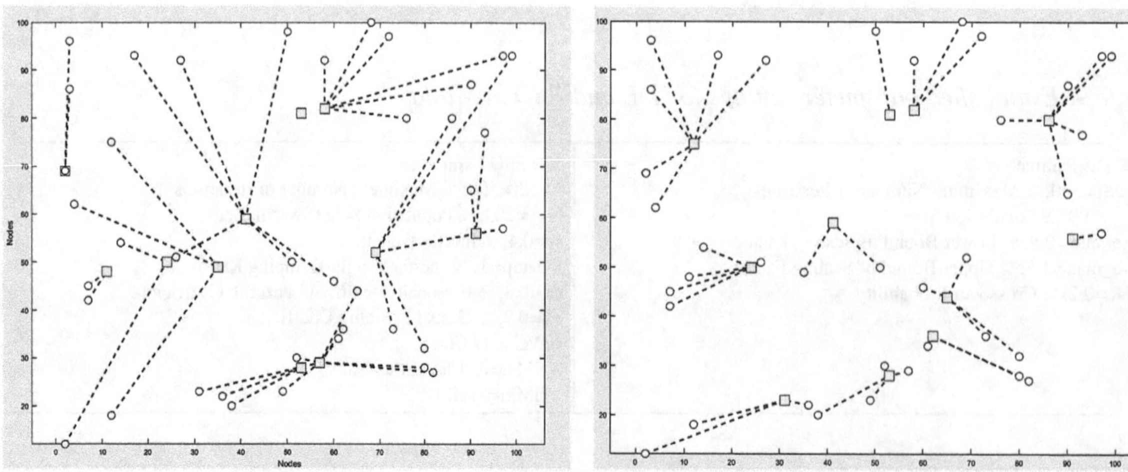

*Figure 5. Best cost (total cost) of objective function using DE (up) and PSO (down) metaheuristics*

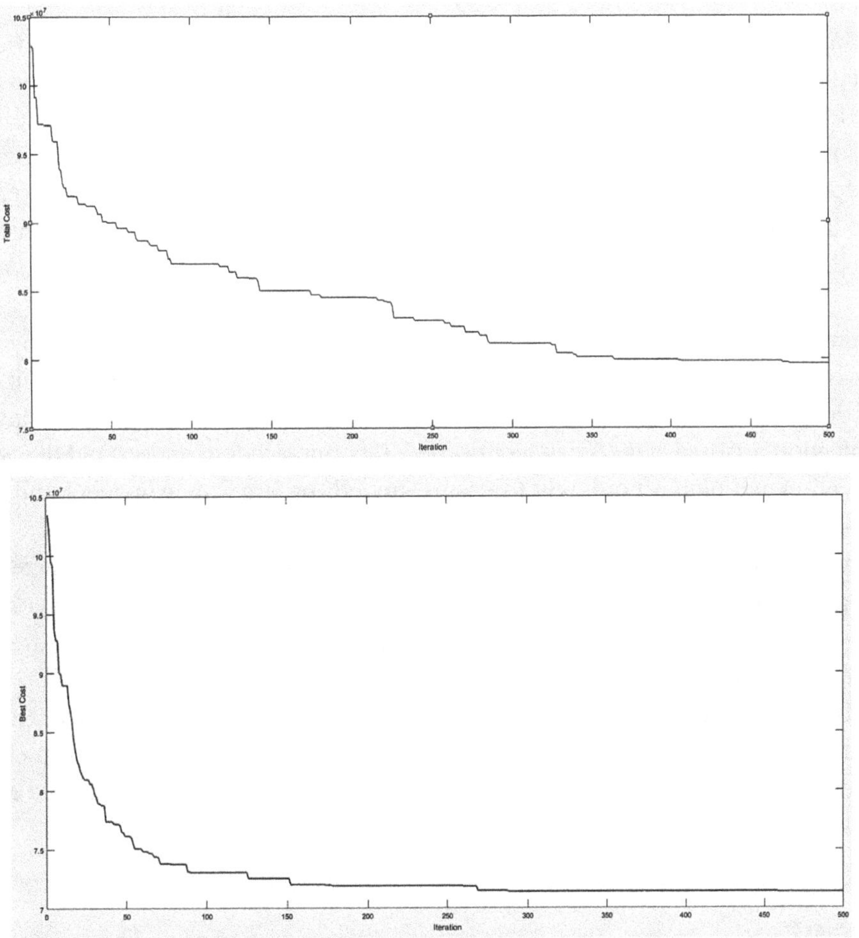

*Table 10. Analysis results for 10 experiments with different numbers of sets*

| | | Hub | | Best Cost | | CPU Total (Sec) | | Optimum Value | Lower Bound (%) or Gap (%) | |
|---|---|---|---|---|---|---|---|---|---|---|
| **No.** | **Node** | **DE** | **PSO** | **DE** | **PSO** | **DE** | **PSO** | | **DE** | **PSO** |
| **1** | 10 | 1 | 1 | 2106312 | 2106312 | 65 | 116 | 2106312 | 0 | 0 |
| **2** | 20 | 3 | 3 | 10813414 | 10813414 | 105 | 161 | 10813414 | 0 | 0 |
| **3** | 30 | 4 | 4 | 20981139 | 20981139 | 208 | 235 | 20981139 | 0 | 0 |
| **4** | 40 | 5 | 8 | 49605297 | 48284833 | 453 | 331 | 48165245 | 2.9 | 0.24 |
| **5** | 50 | 11 | 11 | 75663767 | 71354969 | 593 | 537 | 71168452 | 5.9 | 0.26 |
| **6** | 60 | 14 | 15 | 95287163 | 92245522 | 856 | 616 | N/A | N/A | N/A |
| **7** | 70 | 14 | 16 | 167840678 | 144327597 | 845 | 763 | N/A | N/A | N/A |
| **8** | 80 | 14 | 16 | 204174556 | 182860891 | 1250 | 925 | N/A | N/A | N/A |
| **9** | 90 | 15 | 15 | 254737053 | 226993504 | 1531 | 1126 | N/A | N/A | N/A |
| **10** | 100 | 27 | 27 | 343083222 | 303949181 | 2709 | 1382 | N/A | N/A | N/A |

## FUTURE RESEARCH DIRECTIONS

With the development of technology, the logistics sector has found faster and more cost-effective ways of shipping freight. The hub-and-spoke model was born from industry's efforts to develop more efficient networks. The functionality of the hubs and spokes differ according to the industry. Today, using solution results of hub location-allocation problems can lead managers and researchers to new horizons. We faced with the concept of hub location allocation not just as a transportation problem optimizer, but also as a new paradigm in industry. As managerial applications to move into the future, one can consider some challenges, as dependent variables or constraints, for hub location-allocation problem:

- **Differences in Local Regulatory Environment, Culture and Time Zones:** With centers in various parts of the world, one company may face differences in regulatory environment, cultural preferences, time zone, etc. which may pose a challenge. Due to differences in time zones and cultures, communication between centers can be impeded. It is important to understand the 'softer' aspects of the location and plan accordingly to ensure smooth program management and open bilateral communication.
- **Integration of Resource Pools to Provide Seamless Services:** The solution results of a hub location problem can help to synchronize operations between the Hubs and Spokes and tightly integrate their resource pools to provide a seamless service offering to clients. Failure to do so may defeat the entire purpose of running this business model.
- **Addressing Tax Issues:** Different countries have different tax structures; understanding and complying with them can be an arduous task for companies. Thus, firms need to formulate a plan and seek expert advice to optimize tax treatment, minimize uncontrolled tax risks, and ensure ongoing compliance with laws.

## CONCLUSION

Even though integer programming optimization approaches are applied to solve small hub problems, larger instances of HLPs need to be solved by heuristic procedures or meta-heuristic procedures as one can clearly observe that it is tough to solve such problems effectively with the conventional approaches. While large-size instances can be dealt with specific exact methods, development of meta-heuristics has, in fact, helped many real-world applications, in which optimal/near-optimal solutions can even be obtained in less computational time. With a glance at the literature of HLP, we can see that the trend of heuristic and metaheuristic algorithms in HLPs is similar to the exact solution algorithms. Two points are clear in the related literature: first, the majority of studies have dealt with the uncapacitated cases of HLPs. Second, most of the capacitated HLPs have been investigated in recent years.

In this chapter, we tried to introduce a hub location problem based on the already existing models proposed by the studies. Thus, we reviewed over 40 papers between the years 2000 and the first half of 2016 dealing with or related to hub location problem. In this regard, we mentioned applications, apprehensions and the definitions of the hub location problem. Moreover, we explained the basic classifications and fundamental mathematical models and formulations for different variants of hub location problem. Then, we delivered a categorization of solution approaches and algorithms including exact methods as well as heuristics and meta-heuristics. Afterward, we reviewed an application of HLP and analyzed the shapes of different experimental designs with a different number of nodes. We finally introduced, in Table 8, DE and PSO metaheuristics for a general hub set covering location problem that is one of the popular hub location-allocation problems. We tested their performance comparing them with an exact solution technique and also with each other. The comparisons showed that for booming the engineering competence and control, we need to handle more advanced metaheuristics. We have concluded that we need to use up-to-date and valid parameters of algorithms in order to improve the efficacy and influence of metaheuristics.

## REFERENCES

Aversa, R., Botter, R. C., Haralambides, H. E., & Yoshizaki, H. T. Y. (2005). A Mixed Integer Programming Model on the Location of a Hub Port in the East Coast of South America. *Maritime Economics & Logistics*, *7*(1), 1–18. doi:10.1057/palgrave.mel.9100121

Bailey, A., Ornbuki-Berrnan, B., & Asobiela, S. (2013). Discrete PSO for the uncapacitated single allocation hub location problem. In *Proceedings of the 2013 IEEE Symposium on Computational Intelligence in Production and Logistics Systems, CIPLS 2013 - 2013 IEEE Symposium Series on Computational Intelligence SSCI '13* (pp. 92–98). doi:10.1109/CIPLS.2013.6595205

Berman, O., Drezner, Z., & Wesolowsky, G. O. (2007). The transfer point location problem. *European Journal of Operational Research*, *179*(3), 978–989. doi:10.1016/j.ejor.2005.08.030

Bollapragada, R., Camm, J., Rao, U. S., & Junying, W. (2005). A two-phase greedy algorithm to locate and allocate hubs for fixed-wireless broadband access. *Operations Research Letters*, *33*(2), 134–142. doi:10.1016/j.orl.2004.05.007

Campbell, J. F. (2009). Hub location for time definite transportation. *Computers & Operations Research*, *36*(12), 3107–3116. doi:10.1016/j.cor.2009.01.009

Cánovas, L., García, S., & Marín, A. (2007). Solving the uncapacitated multiple allocation hub location problem by means of a dual-ascent technique. *European Journal of Operational Research*, *179*(3), 990–1007. doi:10.1016/j.ejor.2005.08.028

Clerc, M. (2012). Standard Particle Swarm Optimisation.

Contreras, I., Fernández, E., & Marín, A. (2010). The Tree of Hubs Location Problem. *European Journal of Operational Research*, *202*(2), 390–400. doi:10.1016/j.ejor.2009.05.044

Das, S., & Suganthan, P. N. (2011). Differential evolution: A survey of the state-of-the-art. *IEEE Transactions on Evolutionary Computation*, *15*(1), 4–31. doi:10.1109/TEVC.2010.2059031

EghbaliZarch, M., Abedzadeh, M., & Setak, M. (2013). Differential evolution algorithm for multi-commodity and multi-level of service hub covering location problem. *International Journal of Industrial Engineering Computations*, *4*(1).

Farahani, R., Abedian, M., & Sharahi, S. (2009). *Dynamic Facility Location Problem.*

Farahani, R. Z., Hekmatfar, M., Arabani, A. B., & Nikbakhsh, E. (2013). Hub location problems: A review of models, classification, solution techniques, and applications. *Computers & Industrial Engineering*, *64*(4), 1096–1109. doi:10.1016/j.cie.2013.01.012

Gelareh, S., & Nickel, S. (2011). Hub location problems in transportation networks. *Transportation Research Part E, Logistics and Transportation Review*, *47*(6), 1092–1111. doi:10.1016/j.tre.2011.04.009

He, Y., Wu, T., Zhang, C., & Liang, Z. (2015). An improved MIP heuristic for the intermodal hub location problem. *Omega (United Kingdom)*, *57*, 203–211.

Hsu, C.-C., & Chen, Y.-C. (2007). Mining of mixed data with application to catalog marketing. *Expert Systems with Applications*, *32*(1), 12–23. doi:10.1016/j.eswa.2005.11.017

Karimi, H., & Bashiri, M. (2011). Hub covering location problems with different coverage types. *Scientia Iranica*, *18*(6), 1571–1578. doi:10.1016/j.scient.2011.09.018

Love, R. F., & Moris, J. G. G. W. (Ed.). (1988). Facility location: Models and methods Publication in Operations Research (7th ed.). New York: Elsevier.

Marianov, V., & Serra, D. (2003). Location models for airline hubs behaving as M/D/c queues. *Computers & Operations Research*, *30*(7), 983–1003. doi:10.1016/S0305-0548(02)00052-7

Marín, A. (2005). Formulating and solving splittable capacitated multiple allocation hub location problems. *Computers & Operations Research*, *32*(12), 3093–3109. doi:10.1016/j.cor.2004.04.008

Marini, F., & Walczak, B. (2015). Particle swarm optimization (PSO). A tutorial. *Chemometrics and Intelligent Laboratory Systems*, *149*, 153–165. doi:10.1016/j.chemolab.2015.08.020

Martín, J. C., & Román, C. (2003). Hub location in the South-Atlantic airline market A spatial competition game. *Transportation Research Part A, Policy and Practice*, *37*(10), 865–888. doi:10.1016/S0965-8564(03)00060-0

Mohammadi, M., Tavakkoli-Moghaddam, R., Siadat, A., & Rahimi, Y. (2016). A game-based meta-heuristic for a fuzzy bi-objective reliable hub location problem. *Engineering Applications of Artificial Intelligence*, *50*, 1–19. doi:10.1016/j.engappai.2015.12.009

Pamuk, F. S., & Sepil, C. (2001). A solution to the hub center problem via a single-relocation algorithm with tabu search. *IIE Transactions*, *33*(5), 399–411. doi:10.1080/07408170108936838

Puerto, J., Ramos, A. B., Rodriguez-Chia, A. M., & Sanchez-Gil, M. C. (2016). Ordered median hub location problems with capacity constraints. *Transportation Research Part C, Emerging Technologies*, *70*, 142–156. doi:10.1016/j.trc.2015.05.012

Rodríguez-Martín, I., & Salazar-González, J. J. (2008). Solving a capacitated hub location problem. *European Journal of Operational Research*, *184*(2), 468–479. doi:10.1016/j.ejor.2006.11.026

Rothenbächer, A.-K., Drexl, M., & Irnich, S. (2016). Branch-and-Price-and-Cut for a Service Network Design and Hub Location Problem. *European Journal of Operational Research, 255*(3), 935–947. doi:10.1016/j.ejor.2016.05.058

Saboury, A., Ghaffari-Nasab, N., Barzinpour, F., & Saeed Jabalameli, M. (2013). Applying two efficient hybrid heuristics for hub location problem with fully interconnected backbone and access networks. *Computers & Operations Research, 40*(10), 2493–2507. doi:10.1016/j.cor.2013.01.008

Sasaki, M., Campbell, J. F., Krishnamoorthy, M., & Ernst, A. T. (2014). A Stackelberg hub arc location model for a competitive environment. *Computers & Operations Research, 47*, 27–41. doi:10.1016/j.cor.2014.01.009

Stanimirović, Z. (2010). A genetic algorithm approach for the capacitated single allocation p-hub median problem. *Computing and Informatics, 29*(1), 117–132.

Storn, R. (1996). On the usage of differential evolution for function optimization. In *Proceedings of the1996 Biennial Conference of the North American Fuzzy Information Processing Society* (pp. 519–523).

Storn, R. (2008). Differential evolution research—Trends and open questions BT - Advances in Differential Evolution. *Advances in Differential Evolution, 143*(1), 1–31. doi:10.1007/978-3-540-68830-3_1

Storn, R., & Price, K. (1996). Minimizing the real functions of the ICEC'96 contest by differential evolution. In *Proceedings of IEEE International Conference on Evolutionary Computation* (pp. 842–844). doi:10.1109/ICEC.1996.542711

Storn, R., & Price, K. (1997). Differential Evolution: A Simple and Efficient Heuristic for global Optimization over Continuous Spaces. *Journal of Global Optimization, 11*(4), 341–359. doi:10.1023/A:1008202821328

Storn, R., & Price, K. V. (1995). Differential Evolution - A simple and efficient adaptive scheme for global optimization over continuous spaces.

Sung, C. S., & Jin, H. W. (2001). Dual-based approach for a hub network design problem under non-restrictive policy. *European Journal of Operational Research, 132*(1), 88–105. doi:10.1016/S0377-2217(00)00114-4

Thomadsen, T., & Larsen, J. (2007). A hub location problem with fully interconnected backbone and access networks. *Computers & Operations Research, 34*(8), 2520–2531. doi:10.1016/j.cor.2005.09.018

Yaman, H. (2008). Star p-hub median problem with modular arc capacities. *Computers & Operations Research, 35*(9), 3009–3019. doi:10.1016/j.cor.2007.01.014

Yaman, H., Kara, B. Y., & Tansel, B. Ç. (2007). The latest arrival hub location problem for cargo delivery systems with stopovers. *Transportation Research Part B: Methodological, 41*(8), 906–919. doi:10.1016/j.trb.2007.03.003

## KEY TERMS AND DEFINITIONS

**Database Transaction:** A unit of work performed within a database management system (Sarvari et al., 2016).

**Hub:** The facilities that are servicing many origin-destination pairs as transformation and tradeoff nodes.

**Hybrid Manufacturing System:** A system is one in which functional layout (generally job-shop type) and cells (manufacturing or assembly) coexist.

**Metaheuristics:** In computer science and mathematical optimization, a metaheuristic is a higher-level procedure or heuristic designed to find, generate, or select a heuristic (partial search algorithm) that may provide a sufficiently good solution to an optimization problem, especially with incomplete or imperfect information or limited computation capacity (Bianchi et al., 2009).

**Quadratic Integer Programming:** A special case of the None Linear Programming, when the objective function is in quadratic form and the constraints are linear.

**Spoke:** A none hub node that is connecting to a hub.

**Uncapacitated Single Allocation p-Hub Median Problems (USApHMP):** In the classical USApHMP, transportation costs are modeled as linear functions of the transport volume, where a fixed discount factor on hub-hub connections is introduced to simulate economies of scale.

# Chapter 6
# Heuristic Approaches in Clustering Problems

**Onur Doğan**
*Istanbul Technical University, Turkey*

## ABSTRACT

*Clustering is an approach used in data mining to classify objects in parallel with similarities or separate according to dissimilarities. The aim of clustering is to decrease the amount of data by grouping similar data items together. There are different methods to cluster. One of the most popular techniques is K-means algorithm and widely used in literature to solve clustering problem is discussed. Although it is a simple and fast algorithm, there are two main drawbacks. One of them is that, in minimizing problems, solution may trap into local minimum point since objective function is not convex. Since the clustering is an NP-hard problem and to avoid converging to a local minimum point, several heuristic algorithms applied to clustering analysis. The heuristic approaches are a good way to reach solution in a short time. Five approaches are mentioned briefly in the chapter and given some directions for details. For an example, particle swarm optimization approach was used for clustering problem. In example, iris dataset including 3 clusters and 150 data was used.*

## INTRODUCTION

Researchers have studied for years to find optimal solutions to the problems (Reeves, 1995). The problem searching the optimum solution according to decision variable is called *optimization problem*. The main purpose of an optimization problem is to maximize or minimize a function which is called *objective function*. The objective function can sometimes be maximizing profit or minimizing total cost of transportation. "*Mathematical models*" are a bridge between the mathematic and real world (Meerschaert, 2013). If $x$ refers to decision variables vector, the objective function of the problem depending on the decision variables is $f(x)$. A mathematical model of a minimization problem can be shown as follows:

$$\text{minimize } f(x) \tag{1}$$

DOI: 10.4018/978-1-5225-2944-6.ch006

subject to $g_0(x) \leq b_i \; i = 1,\ldots,m$ (2)

$h_j(x) = c_j \; j = 1,\ldots,n$ (3)

Most of the real world problems contain multiple objectives and contradictory criteria (Singh & Yadav, 2015). If a mathematical model consists of more than one objective function, it is called *multi-objective programming*. In this case, the objective function is $f(x) = g(x) + h(x) - l(x)$ where g(x), h(x) and l(x) show objective functions.

Optimization problems are classified two main categories, continuous and discrete optimization. While decision variables can take any value in continuous optimization problems, in discrete problems, they can take predefined values in solution space such as taking integer values (Cura T., 2008).

In real life problems, finding the best solution is usually take so much time due to infinite solution space. It is expected to find a result near to the best solution in an acceptable time. Using some rules and some solutions instead of all of them, heuristic algorithms reach near to an optimum solution. It does not guarantee to find the best solution. They reduce the time consumption and give the flexibility (Bassett, 2000; Yavuz, Inan, & Fığlalı, 2008).

Although, in operation research problem, searching a solution to a problem, the first step is to create a model, in this chapter, it is considered independent from mathematical model.

Due to clustering problem is an NP-hard problem (Aloise, Deshpande, & Hansen, 2009; Dasgupta, 2007; Drineas, Frieze, Kannan, Vempala, & Vinay, 2004), heuristic algorithms can be used to solve it. Clustering analysis is a type of data mining methods. The goal of the clustering is to create groups according to similarities among the individuals and dissimilarities among the groups. It uses distances to calculate similarities or dissimilarities. There are different ways to calculate it, Euclidean, Pearson, Manhattan, Minkowski etc. Euclidean distance between two objects is commonly used in literature.

Clustering problem is represented mathematically as a set of subsets $C = C_1,\ldots,C_n$ of $S$ such that $S = \bigcup_{i=1}^{n} C_i$ and $C_i \cap C_j$ for $i \neq j$ (Rokach & Maimon, 2005). Therefore, one object can belong to one and only one group.

It is very useful since the process for separation of data in a large solution space is critical for making right decisions. Scope of the clustering analysis can be classified as determining appropriate category, establishing modeling, forecasting according to groups, hypothesis tests, data analysis, etc. (Ball, 1971).

The chapter has five main sections, clustering problems, heuristic algorithms, experiments, future directions and conclusions. After the introduction, clustering problem will be introduced. Its mathematical background will be mentioned. In the second section, heuristic algorithms, the meaning of heuristic and classification of heuristic algorithms can be found. One may find a literature review about heuristic methods and more details. As a subtitle in the second section, five of heuristic algorithms, simulated annealing, tabu search, genetic algorithms, ant colony algorithm and particle swarm optimization, will be explained. Then, in the third section, iris dataset will be introduced as a clustering problem. K-means and particle swarm optimization algorithm will be applied to the problem, respectively. In addition, their results will be compared. One can find the whole algorithm for K-means and a pseudo code for the particle swarm optimization algorithm due to the length of the codes. In future directions, some improvements will be discussed to obtain better solutions. Conclusion section will summarize the whole chapter.

## CLUSTERING PROBLEMS

Numerous data gathering from different information tools has some important opportunities to take advantage. Because of the huge data point, in many examples, finding relationship among data cannot be clear. To effectively divide the information, we must firstly define a criterion for creating groups, and secondly, find an optimal grouping based on that criterion. In the almost wholly analytical problem, it is possible to encounter some situations related to clustering due to many features (Gungor & Unler, 2008).

Clustering is the process of assigning discrete objects to groups according to similarities such as the grouping plants of different species in order to find species having most similarities. Clustering problem is represented mathematically as a set of subsets $C = C_1,...,C_n$ of S such that $S = \bigcup_{i=1}^{n} C_i$ and $C_i \cap C_j$ for $i \neq j$ (Rokach & Maimon, 2005). Therefore, one object can belong to one and only one group.

One and mostly used similarity criterion is the distance between two objects. The formula of Euclidean distance mostly referred in the literature to calculate distances between objects o and z is Equation 1.

$$D(o,z) = D(z,o) = \sqrt{\sum_{i=1}^{n}\left(o_i - z_i\right)^2} \tag{4}$$

The mathematical formulation of the problem can be express as follows (Shelokar, Jayaraman, & Kulkarni, 2004).

$$\min \sum_{k=1}^{K}\sum_{i=1}^{n} w_{ik} D\left(o_i, z_i\right) \tag{5}$$

$$\sum_{k=1}^{K} w_{ik} = 1,\ i = 1,2,...,n \tag{6}$$

$$\sum_{i=1}^{n} w_{ik} \geq 1,\ k = 1,2,...,K \tag{7}$$

where

$z_k$: the center of cluster k
$w_{ik} \in \{0,1\}$, if object i belongs to cluster k, $w_{ik} = 1$

Steps of the K-means algorithm are below.

- Place K points into the space represented by the objects that are being clustered. These points represent initial group centroids.
- Assign each object to the group that has the closest centroid.

- When all objects have been assigned, recalculate the positions of the K centroids.
- Repeat Steps 2 and 3 until the centroids no longer move. It produces a separation of the objects into groups from which the metric to be minimized can be calculated.

One of the most popular techniques is K-means algorithm and widely used in literature to solve clustering problem. Although it is a simple and fast algorithm, there are two main drawbacks. One, it is highly depending on initial values of centroids. Two, in minimizing problems, solution may trap into local minimum point since objective function is not convex (Selim & Ismail, 1984).

To avoid converging to a local minimum point, several heuristic algorithms applied to clustering analysis. Khaled Al-Sultan (1995) and Sung and Jin (2000) used Tabu and simulating annealing algorithms, Rahman and Islam (2014) and Kashan, et al. (2013) modified grouping genetic algorithm. Gungor and Unler (2008), Che (2012) used simulated annealing. While Shelokar and friends (2004) studied ant colony optimization, Cura (2012) studied particle swarm optimization method to solve clustering problem.

## HEURISTIC ALGORITHMS

In original of the word, heuristic means to explore or discover. This term derives from Greek *heuriskein*. Archimedes screamed after realizing buoyancy of water as "eureka" which means "I have found (it)" is the past form of heuristic (Reeves, 1995). Although this is associated with finding the best solution and assuring that it is the best. Reeves (1995) defined more appropriate definition for heuristic:

*A heuristic is a technique which seeks good (i.e. near optimum) solutions at a reasonable computational cost without being able to guarantee either feasibility or optimality, or even many cases to state how close to optimality or a particle feasible solution is.*

Different scientists categorize heuristic techniques in various ways. Table 1 shows a separation of heuristics according to Talbi. Although it is separated more than two groups, in this chapter, single-

*Table 1. Topology of the heuristic algorithms (Talbi, 2009)*

| 1. Single-Solution Based Metaheuristics | 2. Population-Based Metaheuristics |
|---|---|
| 1.1. Simulated Annealing | 2.1. Evolutionary Algorithms |
| 1.2. Tabu Search | 2.1.1. Genetic Algorithms |
| 1.3. Iterated Local Search | 2.1.2. Evolutionary Programming |
| 1.4. Variable Neighborhood Search | 2.1.3. Genetic Programming |
| 1.5. Guided Local Search | 2.2. Scatter Search |
| 1.6. Smoothing Methods | 2.3. Swarm Intelligence |
| 1.7. Noisy Method | 2.3.1. Ant Colony Optimization |
| 1.8. GRASP | 2.3.2. Particle Swarm Optimization |
| | 2.4. Bees Colony |
| | 2.5. Artificial Immune Systems |

solution based and population based algorithms are focused. Simulated annealing and tabu search from single-solution based heuristics and genetic algorithm, ant colony optimization and particle swarm optimization from population based heuristics are the main topics of the chapter.

Single-solution based metaheuristics (S-metaheuristics) generate one solution. They select one of candidate solutions and a new candidate solution is created from it. This process is gone on until a given condition is satisfied. It is generally a number of iteration or an acceptable error percentage. The generation phase may be memory or memoryless. Figure 1 shows the main principles of the S-metaheuristics.

On the other hand, as seen in Figure 2, population-based metaheuristics (P-metaheuristics) search the solution around a sample of candidate solutions rather than a single solution. They first initialized a population to find a solution and then generates a new population. Finally, the new and initialized population is combined using some selection rules. The iterations may have memory or memoryless. This process is stopped when a given condition, a number of iteration or an acceptable error percentage, is satisfied.

A comparison between S-metaheuristics and P-metaheuristics can be shown in Figure 3. There are two criteria in designing a metaheuristic: exploration (diversification) and exploitation (intensification). While S-metaheuristics are good intensification algorithms by searching at the local area, P-metaheuristics are a good choose to diversify because of the random search. Due to some disadvantages of the S-metaheuristics and P-metaheuristics, hybrid algorithms can be used to reach near to the best solution. Some scientists consider hybrid metaheuristics as a class of the topology of the metaheuristics (Talbi, 2009; Sörensen & Glover, 2013).

*Figure 1.*

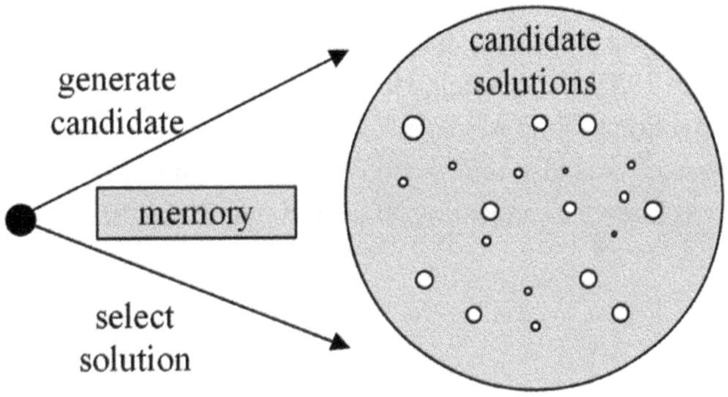


*Figure 2.*

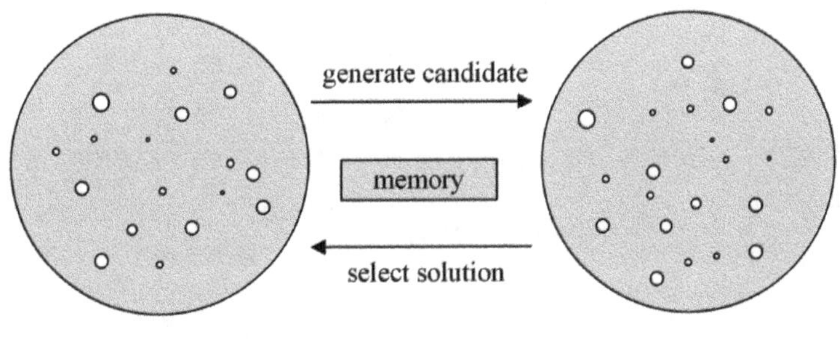

*Figure 3.*

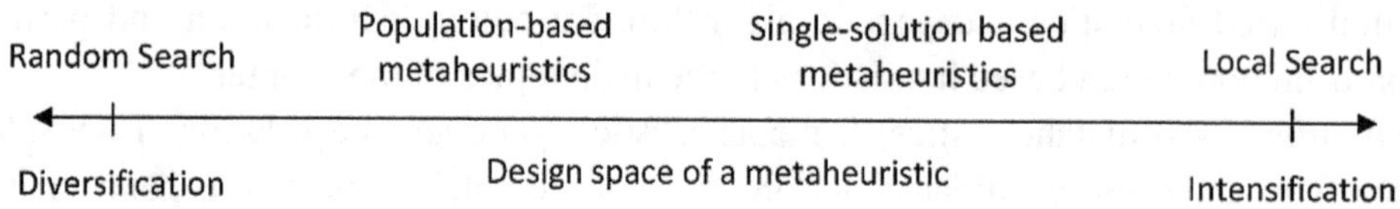


Many problems in real life, solution (design) space is infinite or too big to evaluate all candidate solutions. For that reason, within an acceptable time and under an acceptable error, finding a solution is an obligation. Mentioned techniques in the chapter are used to find a solution near to optimum by using some of the candidate solutions instead of all solutions. Since some candidate solutions are used in the algorithm, it is possible to find a different solution at different times even though the same algorithm is used.

Many heuristic methods have been proposed in literature. In this chapter, five of them are mentioned briefly and given some references for anyone who wants to go details. Figure 4 demonstrates timeline of the various metaheuristics. The original studies of the SA (Simulated Annealing) (Cerny, 1985; Kirkpatrick, Gelatt, & Vecchi, 1983), TS (Tabu Search) (Glover, 1986; Hansen, 1986), GA (Genetic Algorithm) (Holland, 1962; Holland, 1975), ACO (Ant Colony Optimization) (Dorigo, 1992) and PSO (Particle Swarm Optimization) (Kennedy & Eberhart, 1995) are related to optimization or machine learning or both.

## Simulated Annealing (SA)

Simulated annealing was developed by various researchers in the mid-1980s (Luke, 2009). It is a technique for discrete optimization (Dowsland, 1991). The idea of the basis of SA algorithm was first proposed by Metropolis and friends in 1953. Kirkpatrick, Gelatt and Vecchi (1983) showed Metropolis algorithm could be applied to optimization problems. Cerny (1985) modified the original algorithm by letting the temperature decrease to zero. As is also understood from the name, SA works as the cooling of material in a heat bath, a process known as annealing. If a solid material is heated until its melting point and

*Figure 4.*

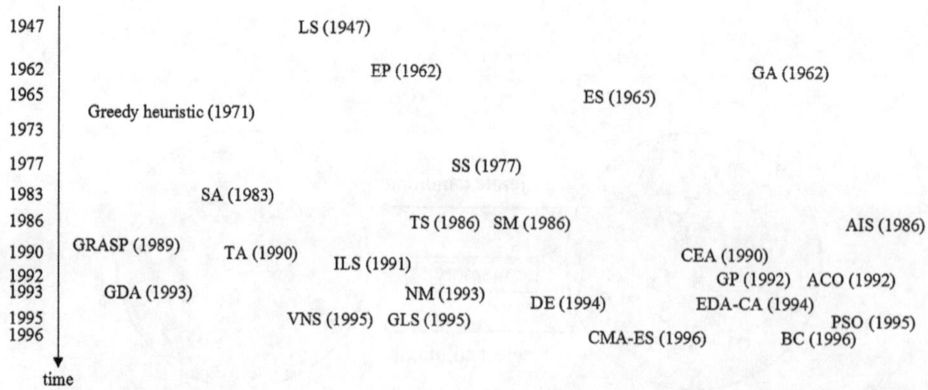

then cooled back into the solid state, the structural of the material depends on the rate of cooling. For example, if a material is grown by fast cooling rate, it will contain some imperfections (Dowsland, 1991).

The most important advantage of the method is to get rid of trapping in a local point. SA searches the solution according to not only improvements but also deteriorations. If at a new position objective function has a deterioration, it is saved with a probability and either it goes to next position or previous position. Because of this property, it can be said SA has a memory. While searching the solution, SA starts finding a local optimum point. It aims to reach a global optimum solution using the local search point.

For details, it may be studied (Selim & Alsultan, 1991) and (Merendino & Celebi, 2013)

## Tabu Search (TS)

Tabu (taboo) search (TS) algorithm was developed by Glover in 1986. Hansen (1986) proposed a similar approach named steepest ascent/mildest descent. Tabu means something that you should not say or do because people generally think it is morally wrong, unpleasant, or embarrassing in Cambridge Dictionary Online. The algorithm adopts as the main idea very close meaning with the definition.

While searching the solution, solutions previously visited could be selected again. For example, in a minimization problem, the algorithm finds a local minimum point like in Figure 5. If the next neighbor solution is larger than the local minimum point, it could immediately return to same local minima. Finally, the algorithm could not reach global minima. To avoid the danger, Tabu search uses a list named "tabu list" and the local minima is hold in the memory as a taboo. If a candidate is "good," but it is in tabu list, sometimes it could not reject. The tabu list may be too restrictive; a nongenerated solution may be forbidden. For some conditions called "aspiration criteria", solutions in the tabu list may be accepted. A candidate in the tabu list is released after some iterations named "tabu tenure". Tabu tenure, in fact, is a cycle time and it is used to avoid moving the same point in the tabu list.

Al-Sultan (1995) used standard tabu search algorithm for clustering problems. Let A be an array of dimension m whose *i*th element ($A_i$) is a number represents the cluster to which the *i*th object is allocated. Evidently, given A, all ($w_{ij}$) are defined as binary in the following manner:

*Figure 5.*

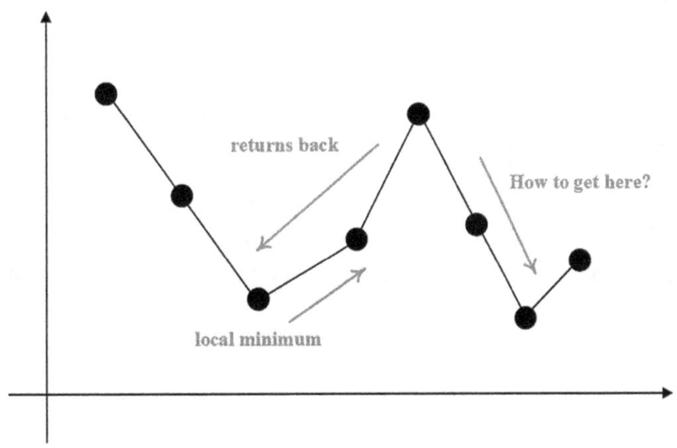

$$w_{ij} = \begin{cases} 1 & if\ A_i = j \\ 0 & otherwise \end{cases} \tag{8}$$

for all $i = 1, 2, ..., m$ and $j = 1, 2, ..., c$ . *i* refers to the number of data objects and *c* is the number of clusters. In the iris dataset $m = 150$ and $c = 3$ . To give an instance for $m = 10$ .

1 – 2 – 1 – 3 – 3 – 1 – 3 – 2 – 2 -3

It shows that the first object is assigned to the first cluster because of $w_{1,1} = 1$, $w_{1,2} = 0$, $w_{1,3} = 0$ and the second object is assigned to second cluster because of $w_{2,1} = 0$, $w_{2,2} = 1$, $w_{2,3} = 0$ etc.

Given a set of all ($w_{ij}$)s, the center of each cluster ($z_j$) can be computed as the center of the data objects allocated to that cluster.

$$z_j = \frac{\sum_{i=1}^{m} w_{ij} x_{ij}}{\sum_{i=1}^{m} w_{ij}} \tag{9}$$

Given the array (*A*) and the centers ($z_j$), the objective function *J(w,z)* can be defined as:

$$J(w, z) = \sum_{i=1}^{m} \sum_{j=1}^{c} w_{ij} \left\| x_i - z_i \right\|^2 \tag{10}$$

Thus, it is obvious for the array A (solution), a specific value is the fitness function is denoted by J. Let $A_b$, $A_t$ and $A_c$ denote the best, trial and current solutions and $J_b$, $J_t$ and $J_c$ denote the corresponding best, trial and current objective fitness values, respectively. The algorithm works with a current solution $A_c$ and then by applying neighborhood structures, Al-Sultan (1995) generated a trial solution $A_t$. The best solution found so far which is denoted by $A_b$ will always be kept throughout the searching process.

## Genetic Algorithms (GA)

Genetic algorithms are one of the evolutionary algorithms and were developed firstly by Holland and his friends at The University of Michigan in 1962 to understand the adaptive processes of natural systems as a doctoral thesis (Holland, 1962). Then they have been applied to machine learning and optimization (Goldberg, 1989; Jong, 1985) GA are based on an evolutionary principle and have been successfully applied to several NP-hard combinatorial optimization problems (Leu, Matheson, & Rees, 1996; Watanabe, Ida, & Gen, 2005; Onwubolu & Mutingi, 2001). GA can evaluate more than one point at the same time and generate more than one solutions. They create a set which consists of the best points instead of finding one solution. To create new solutions, GA usually applies a crossover operator to critical two solutions and a mutation operator that means a random modification of the chromosome (Talbi, 2009).

Exactly how the GAs work is not clear (Glover & Kochenberger, 2003). An explanatory study can be found in (Revees & Rowe, 2003).

## Ant Colony Optimization (ACO)

Ant colony optimization (Dorigo & Caro, 1999; Dorigo, Caro, & Gambardella, 1999) is a population-based stochastic search method for solving combinatorial problems. The inspiring source of ACO is that most of the ant species have trial laying and trial following behavior during the foraging as a communication medium. Each ant deposits pheromone trails, which is a chemical substance, on the paths they move. In other words, an ant is influenced to move towards a food source by the pheromone released by another ant (Açıl, 2008). Released pheromone warns other ants to follow a path. Pheromone level on a path is proportional to the number of ants which followed the path. Hence, ants are not interested with length of the path, but the pheromone level on the path (Bonabeau, Dorigo, & Theraulaz, 1999). For details, (Shelokar, Jayaraman, & Kulkarni, 2004; Gao, 2016) and also it can be visited (URL-1) for investigating some codes and application.

## Particle Swarm Optimization (PSO)

PSO is another stochastic population-based metaheuristic method inspired from swarm intelligence (Kennedy, Eberhart, & Yuhui, 2001). It imitates the social behavior of bird flocking or fish schooling to find a place with enough food. It has successfully applied for continuous problems (Talbi, 2009).

In PSO, each particle is a potential solution. At each iteration, particles move according to their best previous position and best previous position visited any particle in the swarm. So, it is a method with memory. A particle has its own position and velocity, which means flying direction and step of the particle (Figure 6).

*Figure 6.*

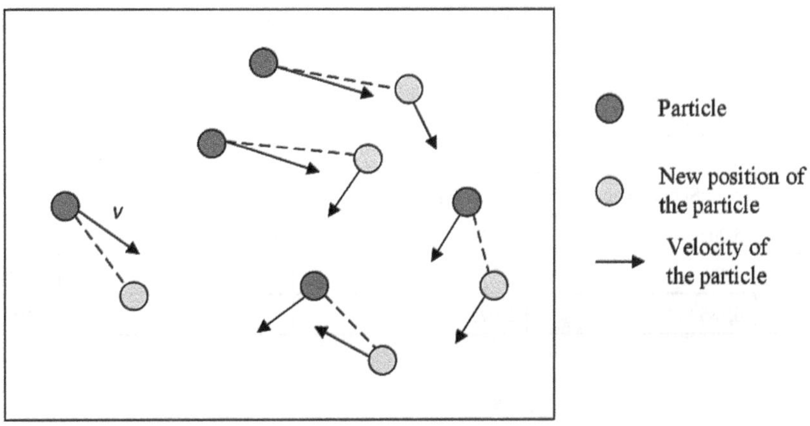

## EXPERIMENTS

This is perhaps the best known database to be found in the pattern recognition literature. The data set contains 3 classes of 50 instances each, where each class refers to a type of iris plant. One class is linearly separable from the other 2; the latter are not linearly separable from each other. The data set has 4 attributes, sepal length in cm, sepal width in cm, petal length in cm, petal width in cm (URL-2). Table 2 shows a part of the data set. The goal of the experiments is to assign objects to the cluster with the exact number.

In fact, the centers of each class are shown in Table 3.

*Table 2. A part of the iris data set*

| Sepal Length | Sepal Width | Petal Length | Petal Width | Class |
|---|---|---|---|---|
| 5,1 | 3,5 | 1,4 | 0,2 | Iris-setosa |
| 4,9 | 3 | 1,4 | 0,2 | Iris-setosa |
| 4,7 | 3,2 | 1,3 | 0,2 | Iris-setosa |
| 5 | 3,6 | 1,4 | 0,2 | Iris-setosa |
| 5,4 | 3,9 | 1,7 | 0,4 | Iris-setosa |
| 4,6 | 3,4 | 1,4 | 0,3 | Iris-setosa |
| 5,7 | 2,8 | 4,5 | 1,3 | Iris-versicolor |
| 6,3 | 3,3 | 4,7 | 1,6 | Iris-versicolor |
| 4,9 | 2,4 | 3,3 | 1 | Iris-versicolor |
| 6,5 | 2,8 | 4,6 | 1,5 | Iris-versicolor |
| 5,7 | 2,8 | 4,5 | 1,3 | Iris-versicolor |
| 6,3 | 3,3 | 4,7 | 1,6 | Iris-versicolor |
| 6,3 | 2,9 | 5,6 | 1,8 | Iris-virginica |
| 6,3 | 2,9 | 5,6 | 1,8 | Iris-virginica |
| 6,5 | 3 | 5,8 | 2,2 | Iris-virginica |
| 7,6 | 3 | 6,6 | 2,1 | Iris-virginica |
| 4,9 | 2,5 | 4,5 | 1,7 | Iris-virginica |

*Table 3. Real center points for classes*

| | Sepal Length | Sepal Width | Petal Length | Petal Width |
|---|---|---|---|---|
| **Iris-Setosa** | 5,006 | 3,418 | 1,464 | 0,244 |
| **Iris-Versicolor** | 5,936 | 2,770 | 4,260 | 1,326 |
| **Iris-Virginica** | 6,586 | 2,973 | 5,514 | 2,016 |

## K-Means Algorithm in Clustering

Firstly, K-means algorithm is applied. Whole Matlab code is presented in Appendix 1 for anyone who wants to try.

The algorithm whose steps were mentioned before is run for 30 times and according to mean of runs, the average center points of each dimension for cluster 1, cluster 2 and cluster 3 are {5,8683 2.7400 4,3817 1,4350}, {5.0060 3,4280 1,4620 0,2460} and {6,8525 3.0700 5,6925 2,0375}, respectively. The number of data in each cluster is 47, 52 and 51, respectively. Mean square errors (MSE) of each class are {3,3369 3,2083 0,3358}. When considered the number of data assigned to each class, mean square errors are support them. Table 4 summarizes K-means algorithm results.

## Particle Swarm Optimization for Clustering

Figure 7 demonstrates the encoding of the four-dimensional problem with three clusters. The encoding is used to define all particles.

Due to the length of the codes, pseudo code for PSO is used briefly.

```
Initialize the swarm;
Initialize centers randomly for each dimension;
while iteration<iter_no;
for i=1:swarmsize;
```

*Table 4. K-means Algorithm Results*

| | Center points | | | | Number of Data | MSE |
|---|---|---|---|---|---|---|
| | Sepal Length | Sepal Width | Petal Length | Petal Width | | |
| **Iris-Setosa** | 5,8683 | 2,7400 | 4,3817 | 1,4350 | 47 | 3,3369 |
| **Iris-Versicolor** | 5,0060 | 3,4280 | 1,4620 | 0,2460 | 52 | 3,2083 |
| **Iris-Virginica** | 6,8525 | 3,0700 | 5,6925 | 2,0375 | 51 | 0,3358 |

*Figure 7.*

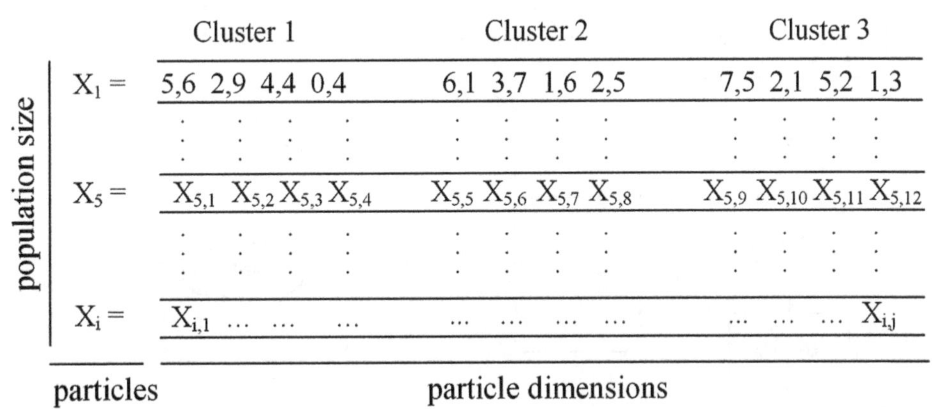

```
Finding the minimum distance;
Finding the new center;
end.
Update positions;
Update pbest;
Update sbest;
end.
```

A basic flowchart as seen in Figure 8 for particle swarm optimization in clustering problems is drawn for better understanding. For details and formulas (Kennedy & Eberhart, 1995; Kennedy, Eberhart, & Yuhui, 2001) can be studied.

The algorithm is run for 30 times and according to mean of runs, the average center points of each dimension for cluster 1, cluster 2 and cluster 3 are {5,0986 3,8660 2,7893 1,5419}, {5.9272 2.7507 3.6323 1.4839} and {6.4412 2.8979 5.0494 2.4687}, respectively. The number of data in each cluster is 50, 47 and 53, respectively. Mean square errors of each class are {1,9105 0,6476 0,6687}. There is an interesting result. Although the first class has the exact number of data, MSE is higher than other two classes, which have near to zero error. It shows that the number of data assigned to the first class is correct but there are some mistakes at assigned data. In our study, since the aim is to assign data to each cluster, we can rule out the mistake. To assign right object to right class can be another study. Table 5 summarizes PSO algorithm results.

*Figure 8.*

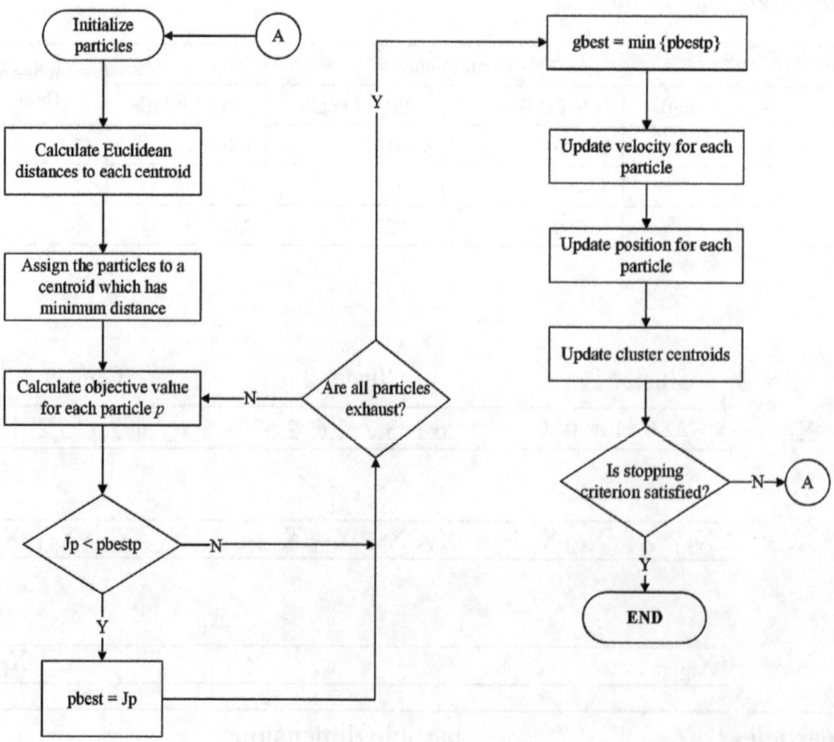

*Table 5. PSO algorithm results*

| | Center Points | | | | Number of Data | MSE |
|---|---|---|---|---|---|---|
| | Sepal Length | Sepal Width | Petal Length | Petal Width | | |
| **Iris-Setosa** | 5,098613 | 3,865953 | 2,7893 | 1,54189 | 50 | 1,910544 |
| **Iris-Versicolor** | 5,927233 | 2,750729 | 3,632293 | 1,48389 | 47 | 0,647606 |
| **Iris-Virginica** | 6,441227 | 2,897853 | 5,049447 | 2,46865 | 53 | 0,668668 |

## FUTURE DIRECTIONS

This chapter explains some solutions to clustering problems using some of the heuristic methods. An experimental example including iris data set, which has 150 data, was used to show how to apply a heuristic method to the problem. The first further research will include different datasets and it will be controlled whether the heuristic method used gives better results for each dataset. If it gives better results then as the second improvement, it will be applied to a dataset including big data to demonstrate that the method can be used for real life problems. If it gives worse results, then some hybrid algorithms will be tried to obtain better results. For instance, combining particle swarm and tabu search algorithm can result in better solutions. Tabu search may be used to determine a local minimum point at any iteration during PSO algorithm. Some other hybrid heuristic algorithms can also give better results. After establishing hybrid algorithm, it will be used for a dataset including big data.

## CONCLUSION

In the chapter, firstly, clustering problems were discussed. Then some heuristic approaches are introduced briefly. Heuristics are useful methods to solve NP-hard problems. We also said that because clustering is an NP-hard problem, heuristic approaches can be used. There are many ways to reach clustering solution. Using different methods such as simulated annealing, tabu search, genetic algorithm, ant colony optimization and particle swarm optimization, it was mentioned that how to obtain a solution. Some directions for details were given. Applying all mentioned methods to clustering might be another study. K-means and particle swarm optimization algorithms were applied to the same data set. A whole code was given to apply the K-means algorithm. A pseudo code and a flowchart were used to make understandable main logic of PSO. According to results, particle swarm optimization gave a better solution. However, it does not mean that it is the best solution due to the operating logic of heuristic approaches. PSO reaches faster than the K-means algorithm. In iris dataset, the number of data is not too large but when you use big data, time is a critical factor. The parameters of PSO ($c_1$, $c_2$, w) can be changed to study about whether there may be better solutions.

## REFERENCES

Açıl, A. (2008). Solving the profitable tour problem using ant colony system [Doctoral Thesis].

Al-Sultan, K. (1995). A Tabu search approach to the clustering problem. *Pattern Recognition*, *9*(28), 1443–1451. doi:10.1016/0031-3203(95)00022-R

Aloise, D., Deshpande, A., Hansen, P., & Popat, P. (2009). NP-hardness of Euclidean sum-of-squares clustering. *Machine Learning*, *75*(2), 245–248. doi:10.1007/s10994-009-5103-0

Ball, G. H. (1971). Classification analysis (Technical Report SRI Project 5533). Stanford Research Institute.

Bassett, M. (2000). Assigning projects to optimize the utilization of employees time and expertise. *Computers & Chemical Engineering*, *24*(2-7), 1013–1021. doi:10.1016/S0098-1354(00)00534-2

Bonabeau, E., Dorigo, M., & Theraulaz, G. (1999). *Swarm intelligence: from natural to artificial systems*. New York, NY, USA: Oxford University Press, Inc.

Cerny, V. (1985). Thermodynamical approach to the traveling salesman problem an efficient simulation algorithm. *Journal of Optimization Theory and Applications*, *45*(1), 41–51. doi:10.1007/BF00940812

Che, Z. H. (2012). Clustering and selecting suppliers based on simulated annealing algorithms. *Computers & Mathematics with Applications (Oxford, England)*, *1*(63), 228–238. doi:10.1016/j.camwa.2011.11.014

Cura, T. (2008). *Modern sezgisel teknikler ve uygulamaları*. İstanbul: Papatya Yayıncılık.

Cura, T. (2012). A particle swarm optimization approach to clustering. *Expert Systems with Applications*, *I*(39), 1582–1588. doi:10.1016/j.eswa.2011.07.123

Dasgupta, S. (2007). *The hardness of k-means clustering* (Technical Report CS2007-0890). University of California, San Diego.

Dorigo, M. (1992). *Optimization, learning and natural algorithms* [PhD thesis]. Politecnico di Milano, Italy.

Dorigo, M., & Caro, G. D. (1999). The ant colony optimization meta-heuristic. In New ideas in optimization (pp. 11-32). Maidenhead, UK: McGraw-Hill Ltd.

Dorigo, M., Caro, G. D., & Gambardella, L. M. (1999). Ant algorithms for discrete optimization. *Artificial Life*, *5*(2), 137–172. doi:10.1162/106454699568728 PMID:10633574

Dowsland, K. A. (1991). *Simulated annealing. C. R. Revees içinde, Modern heuristic techniques (s. 20)*. New York: John Wiley&Sons, Inc.

Drineas, P., Frieze, A., Kannan, R., Vempala, S., & Vinay, V. (2004). Clustering large graphs via the singular value decomposition. *Machine Learning, 2*(9-33), 56.

Gao, W. (2016). Improved ant colony clustering algorithm and its performance study. *Computational Intelligence and Neuroscience, 2016*, 1–14. doi:10.1155/2016/4835932 PMID:26839533

Glover, F. (1986). Future paths for integer programming and links to artificial intelligence. *Computers & Operations Research, 13*(5), 533–549. doi:10.1016/0305-0548(86)90048-1

Glover, F., & Kochenberger, G. A. (2003). *Genetic algorithms. Hand book of metaheuristics (s. 59). içinde New York*. Boston, Dordrecht, London, Moscow: Kluwer Academic Publishers.

Goldberg, D. E. (1989). *Genetic algorithms in search, optimization, and machine learning*. Boston, MA, USA: Addison-Wesley Longman Publishing Co., Inc.

Gungor, Z., & Unler, A. (2008). K-Harmonic means data clustering with tabu-search method. *Applied Mathematical Modelling*, *6*(32), 1115–1125. doi:10.1016/j.apm.2007.03.011

Hansen, P. (1986). The steepest ascent mildest descent heuristic for combinatorial programming. In Proceedings of the Congress on Numerical Methods in Combinatorial Optimization, Capri, Italy.

Holland, J. H. (1962). Outline for a logical theory of adaptive systems. *Journal of the ACM*, *9*(3), 297–314. doi:10.1145/321127.321128

Holland, J. H. (1975). *Adaptation in natural and artificial systems*. Ann Arbor, MI: University of Michigan Press.

Jong, K. A. (1985). Genetic algorithms: A 10 year perspective. In *Proceedings of the 1st International Conference on Genetic Algorithms* (pp. 169-177). Hillsdale, NJ, USA: L. Erlbaum Associates Inc.

Kashan, H. (2013). A particle swarm optimizer for grouping problems. *Information Sciences*, *252*, 81–95. doi:10.1016/j.ins.2012.10.036

Kennedy, J., & Eberhart, R. C. (1995). A new optimizer using particle swarm theory. In *Proceedings of the Sixth International Symposium on Micromachine and Human Science* (pp. 39-43).

Kennedy, J., Eberhart, R. C., & Yuhui, S. (2001). *Swarm Intelligence*. San Francisco, CA: Morgan Kaufmann.

Kirkpatrick, S., Gelatt, C. D., & Vecchi, M. P. (1983). Optimization by simulated annealing. *Science*, *220*(4598), 671–680. doi:10.1126/science.220.4598.671 PMID:17813860

Leu, Y.-Y., Matheson, L., & Rees, L. (1996). Sequencing mixed-model assembly lines with genetic algorithms. *Computers & Industrial Engineering*, *30*(4), 1027–1036. doi:10.1016/0360-8352(96)00050-2

Luke, S. (2009). *Simulated annealing. S. Luke içinde, Essentials of metaheuristics (s. 23)*. Raleigh, North Carolina: Lulu.

Meerschaert, M. M. (2013). *Mathematical Modeling*. East Lansing: Academic Press.

Merendino, S., & Celebi, M. E. (2013). A simulated annealing clustering algorithm based on center perturbation using gaussian mutation. In *Proceedings of the Twenty-Sixth International Florida Artificial Intelligence Research Society Conference*, Florida.

Onwubolu, G., & Mutingi, M. (2001). A genetic algorithm approach to cellular manufacturing systems. *Computers & Industrial Engineering*, *39*(1-2), 125–144. doi:10.1016/S0360-8352(00)00074-7

Rahman, M. A., & Islam, M. Z. (2014). A hybrid clustering technique combining a novel genetic algorithm with K-Means. *Knowledge-Based Systems*, *71*, 345–365. doi:10.1016/j.knosys.2014.08.011

Reeves, C. R. (1995). *Modern heuristic techniques for combinatorial problems*. London: McGraw-Hill.

Revees, C. R., & Rowe, J. E. (2003). *Genetic algorithmsp principles and perspectives*. Dordrecht: Kluwer Academic Publishers.

Rokach, L., & Maimon, O. (2005). Clustering methods. In Data mining and knowledge discovery handbook (pp. 321-352). Springer US. doi:10.1007/0-387-25465-X_15

Selim, S. Z., & Alsultan, K. (1991). A simulated annealing algorithm for the clustering problem. *Pattern Recognition*, *24*(10), 1003–1008. doi:10.1016/0031-3203(91)90097-O

Selim, S. Z., & Ismail, M. A. (1984). K-means-type algorithms: A generalized convergence theorem and characterization of local optimality. *IEEE Transactions on Pattern Analysis and Machine Intelligence*, *I*(6), 81–87. doi:10.1109/TPAMI.1984.4767478 PMID:21869168

Shelokar, P., Jayaraman, V., & Kulkarni, B. (2004). An ant colony approach for clustering. *Analytica Chimica Acta*, *509*(2), 187–195. doi:10.1016/j.aca.2003.12.032

Singh, S. K., & Yadav, S. P. (2015). Modeling and optimization of multi objective non-linear programming problem in intuitionistic fuzzy environment. *Applied Mathematical Modelling*, *39*(16), 4617–4629. doi:10.1016/j.apm.2015.03.064

Sörensen, K., & Glover, F. (2013). Metaheuristics. In S. I. Gass & M. C. Fu (Eds.), *Encyclopedia of Operations Research and Management Science* (3rd ed., pp. 960–970). London: Springer. doi:10.1007/978-1-4419-1153-7_1167

Sung, C., & Jin, H. (2000). A tabu-search-based heuristic for clustering. *Pattern Recognition*, *V*(33), 849–858. doi:10.1016/S0031-3203(99)00090-4

Talbi, E.-G. (2009). *Metaheuristics: From Design to Implementation*. Lille: John Wiley & Sons, Inc. doi:10.1002/9780470496916

Turing Finance. (2016, 09 29). *Clustering using Ant Colony Optimization*. Retrieved 10 19, 2016, from http://www.turingfinance.com/ant-colony-optimization-finance/

UCI Machine Learning Repository. (2015). Iris datasets. Retrieved from https://archive.ics.uci.edu/ml/datasets/Iris

Watanabe, M., Ida, K., & Gen, M. (2005). A genetic algorithm with modified crossover operator and search area adaptation for the job-shop scheduling problem. *Computers & Industrial Engineering*, *48*(4), 743–752. doi:10.1016/j.cie.2004.12.008

Yavuz, M., Inan, U. H., & Fığlalı, A. (2008). Fair referee assignments for professional football leagues. *Computers & Operations Research*, *35*(9), 2937–2951. doi:10.1016/j.cor.2007.01.004

# APPENDIX

```
% Initializing the dataset of flowers
load irisdataset.txt
Data=irisdataset;
Dim=size(Data);
% Selecting 3 random centers
Selection=rand(1,3);
Selection=Selection*Dim(1,1);
Selection=ceil(Selection); %Selecting the row number.
% 3 random centers are found
Center1=Data(Selection(1),:);
Center2=Data(Selection(2),:);
Center3=Data(Selection(3),:);
% K means Algorithm
for j=1:1:n
count1=0;
Mean1=zeros(1,4);
count2=0;
cluster1=[];
Mean2=zeros(1,4);
cluster2=[];
count3=0;
cluster3 =[];
Mean3=zeros(1,4);
%Finding the minimum distance for each cluster center
for i=1:1:Dim(1,1)
Pattern1(i)=sqrt((Center1(1,1)-Data(i,1))^2+(Center1(1,2)-
Data(i,2))^2+(Center1(1,3)-Data(i,3))^2+(Center1(1,4)-Data(i,4))^2);
Pattern2(i)=sqrt((Center2(1,1)-Data(i,1))^2+(Center2(1,2)-
Data(i,2))^2+(Center2(1,3)-Data(i,3))^2+(Center1(1,4)-Data(i,4))^2);
Pattern3(i)=sqrt((Center3(1,1)-Data(i,1))^2+(Center3(1,2)-
Data(i,2))^2+(Center3(1,3)-Data(i,3))^2+(Center1(1,4)-Data(i,4))^2);
LessDist=[Pattern1(i) Pattern2(i) Pattern3(i)];
Minimum=min(LessDist);
%Finding the new centre
if (Minimum==Pattern1(i))
    count1=count1+1;
    Mean1=Mean1+Data(i,:);
    cluster1=[cluster1 i];
```

```
else if (Minimum==Pattern2(i))
        count2=count2+1;
        Mean2=Mean2+Data(i,:);
        Cluster2=[cluster2 i];
    else count3=count3+1;
        Mean3=Mean3+Data(i,:);
        cluster3=[cluster3 i];
    end
end
end
```

# Chapter 7
# An Integrated Methodology for Order Release and Scheduling in Hybrid Manufacturing Systems
## Considering Worker Assignment and Utility Workers

**Ömer Faruk Yılmaz**
*Istanbul Technical University, Turkey & Yalova University, Turkey*

**Mehmet Bülent Durmuşoğlu**
*Istanbul Technical University, Turkey*

## ABSTRACT

*There are three main problems that could impact the performance of a Hybrid Manufacturing System (HMS): (1) order release (OR), (2) batch scheduling and (3) worker assignment. This paper deals with these three main problems hierarchically for an HMS. Three different mathematical models are developed to describe the problems more clearly. A novel methodology is proposed to adopt a holistic approach to these problems and find an effective solution. Implementation of the proposed methodology permits integrating batch scheduling and worker timetabling. Feasible solutions in the best-known Pareto front are evaluated as alternative solutions. The goal is to select a preferred solution that satisfies worker constraints, creates effective worker teams in cells, minimizes the number of utility workers, and the average flow time. The study also presents several improvements, which are made following the application of the proposed methodology to a real company that produces expansion joints.*

DOI: 10.4018/978-1-5225-2944-6.ch007

## INTRODUCTION

In today's manufacturing systems, a high variety of parts are produced to meet customer expectations. Some of the parts produced belong to the early stages of life cycle, whereas others belong to later stages. Demand variability is usually high for parts in the early stages and low for parts in the later stages. A manufacturing system, which can produce both types of parts, is the hybrid manufacturing system (HMS). HMS is a manufacturing system comprised of both cells and functional area. Parts with low and erratic demand are produced in the functional area, whereas parts with high and stable demand are produced in the cells (Durmusoglu and Satoglu, 2011). Studies conducted in real manufacturing environments show that many cellular manufacturing environments are designed to have a hybrid structure. That is why studies on HMS are highly relevant for industrial applications.

The problems in an HMS can be classified mainly into the design and the operational classes. Design problems include HMS formation and layout decisions whereas operational problems include order release, scheduling of batches/parts and worker assignment decisions (Aglan and Durmusoglu, 2015). Operational problems have not been considered extensively in the literature as compared to design problems (Satoglu and Suresh, 2009). This chapter deals with the problems of order release, batch scheduling and worker assignment in the HMS consisting of a number of parallel independent manufacturing cells and a functional layout.

Decisions on the order release, the batch scheduling, and the worker assignment problems are typically made independently. However, the efficiency of the HMS can be increased when these decisions are made concurrently due to the interrelation among these problems. The objective of this study is to propose a methodology to adopt a holistic approach to the order release, the batch scheduling and the worker assignment problems in the HMS.

The proposed methodology consists of four stages. In the first stage, an optimization model is developed for the OR problem. In the second stage, the multi-objective mathematical model developed by Yılmaz and Durmusoglu (2017) is used for the batch scheduling problem in HMS. In the third stage, the goal programming model is used to make the worker assignment decisions. Decision rules to select a preferred solution are used in the fourth stage.

The first stage of the proposed methodology concerns order release (job release). The goal is to improve system performance by using OR to control the order flow. It decides which parts (jobs) be allowed to be released to the shop floor, at what time and under what conditions they are to be released (Cevikcan and Durmusoglu, 2014). Answers to these questions make OR one of the main components of the workload control (WLC) mechanism. Hence, it is possible to release parts to the shop floor in a controlled way and make efficient use of the capacity of the cells and the machines.

In the OR stage, which is the first stage of the proposed methodology, the aim of the optimization model is to release the maximum number of parts (jobs) to the shop floor. No attention is given to the batch scheduling problem. The multi-objective optimization model used in the scheduling stage, which is the second stage of the proposed methodology, has the following objectives: (1) minimizing the maximum number of workers, (2) minimizing the maximum number of worker transfer and (3) minimizing the average flow time. In the third stage of the methodology, the optimization model has the following objectives: (1) minimizing deviations from the number of workers assigned to operations in the scheduling stage, (2) minimizing utility workers and (3) minimizing deviations from the desired team synergy.

In what follows, the rationale behind the second objective of the worker assignment goal programming model is explained in detail.

When workers fail to complete the tasks assigned to them in their permissible work zone, utility work emerges. Then, utility workers are assigned to perform the utility work (Cevikcan and Durmusoglu, 2011). In this study, a utility worker is defined as a worker that can be assigned to different cells on a need basis to reduce the cell cycle times.

The concepts of utility work and utility worker are commonly used in the context of assembly lines. However, they can also be effectively utilized in studies on other manufacturing systems. That is why the present study uses the concepts of utility work and utility worker in the context of HMS.

The hierarchical application of the second and third stages of the proposed methodology solves the problems of batch scheduling and worker assignment. Therefore, an integration mechanism is developed to address the second and the third stages concurrently. A heuristic algorithm is proposed within the integration mechanism, which combines a (multi-objective evolutionary algorithm) FastPGA-based heuristic and a local search heuristic.

In the fourth stage of the proposed methodology, a preferred solution is selected. Decision rules are developed to select a preferred solution from the best-known feasible Pareto set, which is obtained by applying the integration mechanism to the second and the third stages.

The motivation of this study is the integration of the order release, the batch scheduling and the worker assignment problems for a real-world HMS. Therefore, the study has the potential of adding value to industry by the way of effectively raising engineering control for production planning activities in the HMS. In this context, this study has the originality of proposing a novel methodology with the aid of the mathematical models and heuristic approach via the hierarchical consideration of the order release, scheduling and worker assignment problems in the HMS.

The rest of the chapter includes the following plan: The second section is divided into three sub-sections. The first sub-section reviews OR studies, the second sub-section reviews studies on sequencing and scheduling in hybrid and cellular manufacturing systems, and the third sub-section reviews studies on worker allocation and assignment issues. The third section provides a detailed presentation of the proposed methodology, and the developed mathematical models. The fourth section explains the proposed algorithm which is used in the integration mechanism. The fifth section deals with real life application of the proposed methodology. The sixth section presents conclusions and recommendations for future research directions.

## LITERATURE REVIEW

Since this study adopts a holistic approach to the order release, scheduling and worker assignment issues, the literature on these issues is reviewed in this section.

### Order Release Studies

Order release (OR) production control policy can be a step that precedes WLC, or it can be a main part of WLC. To examine these two cases of OR, some studies in the literature use the simulation method, whereas others use mathematical models. In this section, studies using simulation techniques are not covered.

Liao (1992) focused on the problem of accepting or rejecting arriving jobs (parts) to a job shop in which flow times of jobs are stochastic. Ashby and Uzsoy (1995) combined OR, group scheduling and order sequencing for a make-to-order manufacturing facility. Choi et al. (1997) examined the multi-job OR problem, using a heuristic they developed, which was very effective in minimizing work-in-process (WIP), flow time and tardy jobs. Fowler et al. (2002) developed a heuristic for WLC, which they implemented in the semiconductor industry. In their experiments on make-to-order manufacturing systems, Haskose et al. (2004) examined the effects of two different WLC mechanisms (job release and order acceptance) on manufacturing lead time (MLT). Rogers and Nandi (2007) examined the effects of two-stage input control systems on make-to-order manufacturing systems. Betrand and Van Ooijen (2008) examined the OR problem in terms of order lead time, order tardiness and work-in-process objectives. They developed an equation to identify the maximum number of orders that are allowed to be in process. Moreira and Alves (2009) proposed a multiple decision-making scheme for the WLC. They argue that four different decisions (order acceptance, due date setting, job release and dispatching) have to be evaluated simultaneously. Thürer et al. (2014) evaluated four different OR mechanisms that are widely used in the literature together with sequence dependent setup times. They emphasized that usage of sequence dependent setup times in OR does not improve WLC performance. Cevikcan and Durmusoglu (2014) focused on scheduling and OR problems in parallel machines. They developed separate mathematical models for each problem, and used them for small-sized problems. For large-sized problems, they developed a single heuristic that covers both problems.

The present study develops a mathematical model for the OR problem, and proposes a methodology that integrates the OR and the batch scheduling in the HMS.

## Batch Scheduling Studies for Hybrid and Cellular Manufacturing Systems

There are a limited number of studies on HMS in the literature. Some of the studies on HMS are simulation-based studies conducted to evaluate system performance (Suresh, 1991; Burgess et al., 1993; Shambu and Suresh, 2000; Kher and Jensen, 2002, Zolfaghari and Lopez Roa, 2006). Design studies, on the other hand, make up a large portion of all studies on HMS (Murthy and Srinivasan, 1995; Harhalakis et al., 1996; Gravel et al., 2000; Venkataramanaiah and Krishnaiah, 2002; Viguier and Pierreval, 2004; Ioannou, 2006; Feyzioglu and Pierreval, 2009; Satoglu and Suresh 2009; Satoglu et al., 2010; Torabi and Amiri, 2012; Durmusoglu and Satoglu, 2011; Durmusoglu and Kaya, 2012).

There is also a study (Aglan and Durmusoglu, 2015) on production control methods for HMS. Another study (Yilmaz and Erbiyik, 2016) examined the use of optimization methods in HMS.

To the best knowledge of the authors, batch scheduling problem in the HMS has not been the subject of any published study.

Because the cellular manufacturing system (CMS) is one of the major components of HMS, this section also reviews studies on batch scheduling problem in CMS. Some of the findings from these studies also apply to HMS.

Little research has been conducted on batch scheduling problem in CMS in the literature. The following is a review of studies that examine the batch scheduling problem in multi cell manufacturing system.

Studies on batch scheduling problem in CMS are reviewed below:

Das and Canel (2005) developed a branch and bound model to solve the problem of scheduling of batches in the multi cell flexible manufacturing system (MCFMS). Celano et al. (2008) used simulation method to analyze the problem of scheduling batches within a manufacturing system consisting of multiple cells. Balaji and Porselvi (2014) developed a mathematical model for batch scheduling problem in a MCFMS having sequence dependent batch setup times with flowline characteristics. Yilmaz et al. (2016) proposed a mathematical model for the bath scheduling problem in a specific type of CMS. They developed a heuristic method for solving the mathematical model.

## Worker Allocation and Assignment Studies in CMS and HMS

This section reviews studies on worker allocation and assignment in CMS and HMS. In this section, studies using simulation techniques are not covered.

Askin and Huang (2001) developed a mixed integer goal programming model to improve the fitness of individual workers to tasks performed in cells, and to create effective teams. Norman et al. (2002) examined the problem of assigning workers to manufacturing cells to improve organizational effectiveness. They treated organizational effectiveness as a function of productivity, output quality and training cost. Suer and Dagli (2005) examined cell loading and labor allocation problems. They created a three-stage structure and looked for solutions to sequencing, labor allocation and cell loading problems. Cesani and Steudel (2005) used the simulation method to examine the effects of different labor allocation strategies on system performance. Suer and Tummaluri (2008) studied the problem of assigning operators to operations in labor intensive cells. They developed a three-stage approach for the solution of the problem. Fowler et al. (2008) examined differences between workers, in terms of their general cognitive ability (GCA), and developed a mixed integer mathematical model to minimize worker-related costs over multiple periods. Satoglu and Suresh (2009) adopted a three-stage approach for HMS design. In the third stage of this approach, they developed a goal programming mathematical model for labor allocation. Fan et al. (2010) developed a Dual Resource Constraint (DRC) multi-objective mathematical model for cell formation and operator assignment problems. Suer and Alhawari (2011) examined the use of two different operator assignment strategies (Max-Min and Max) in labor intensive manufacturing cells. Their results showed that the max-min strategy is more successful in improving the skill levels of operators. Egilmez et al. (2014) examined the problem of stochastic skill-based manpower allocation in a cellular manufacturing environment where both operation times and demand are uncertain. They developed three different stochastic non-linear mathematical models inside a hierarchical, four-stage methodology. Azadeh et al. (2015) proposed bi-objective mathematical model for clustering parts, machines and workers simultaneously in cellular manufacturing. They applied ε-constraint method to obtain non-dominated pareto solutions and then used common weighted multi-criteria decision analysis (MCDA)-data envelopment analysis method to select a preferred solution from Pareto optimal solutions. Niakan et al. (2016) developed a new bi-objective mathematical model of the dynamic cell formation problem to handle worker assignment and environmental and social criteria. Due to the NP-hard nature of the problem, a hybrid algorithm called NSGA-II-MOSA (Multi-Objective Simulated Annealing) was developed.

The present study introduces the concept of utility worker for HMS. The following is a review of studies that utilize the concept of utility worker.

The concept of utility work is often used in the problem of mixed model assembly line (MMAL) sequencing. In addition, many studies in the literature have multi objective, and minimizing utility work is one of the objectives in these studies (Hyun et al., 1998; Ponnambalam et al., 2003; Tavakkoli-Moghaddam and Rahimi-Vahed, 2006; Akgunduz and Tunali, 2010; Chutima and Naruemitwong, 2014). These studies generally use meta-heuristic methods to find solutions to multi-objective problems.

In addition to multi-objective studies where utility work is one of the objectives, there are also studies that exclusively focus on utility workers (Tsai, 1995; Kotani et al., 2004; Kim and Jeong, 2007; Giard and Jeunet, 2010; Cevikcan and Durmusoglu, 2011). These studies use both heuristic and meta-heuristic methods.

More detailed information on utility work and utility workers can be found in Cevikcan and Durmusoglu (2011).

There are studies in the literature that focus on the order release, the batch scheduling or the worker assignment problems, as well as studies that combine two of these problems. However, to the best of author's knowledge, there are no studies that examine two or more of these problems in a holistic manner for HMS. Therefore, the purpose of the present study is to fill this gap in the literature by adopting a holistic approach and finding a common solution to these problems.

## PROPOSED METHODOLOGY

This study proposes a four-stage, hierarchical methodology in order to find a holistic solution to the order release, the batch scheduling and the worker assignment problems in the HMS. The first stage of the proposed methodology deals with OR problem, the second stage deals with batch scheduling problem, the third stage deals with worker assignment problem, and the fourth stage deals with decision rules. Figure 1 shows the proposed methodology.

*Figure 1. Roadmap for the proposed methodology*

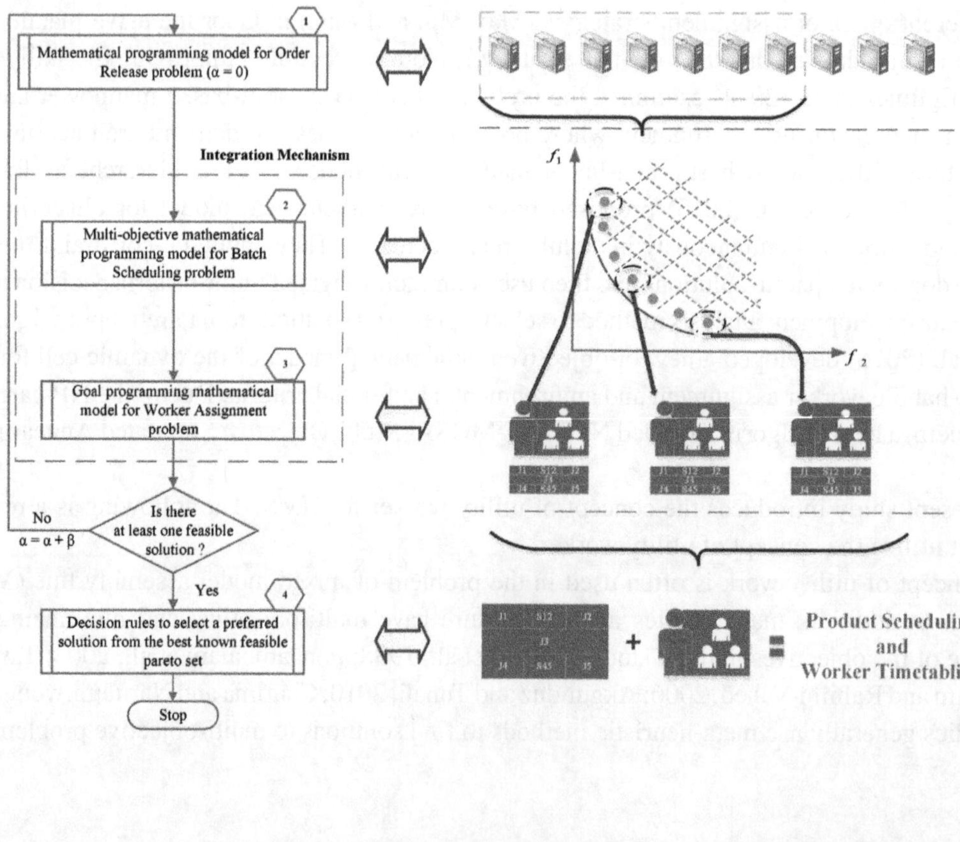

The general working principle of the methodology that integrates scheduling and worker timetabling for HMS is explained below.

In the first stage, the OR mathematical model (where $\alpha = 0$) is used to find a solution for the OR problem. In this model, the objective is to release the maximum number of parts to the shop floor. No attention is paid to scheduling. The released batches can be scheduled, and workers can be assigned by using the integration mechanism.

In the second stage of the proposed model, the focus is on the multi-objective batch scheduling problem, and the mixed integer non-linear multi-objective mathematical model, developed by Yilmaz and Durmusoglu (2017), is reformulated and used. The objectives used in the mathematical model are as follows: minimizing the average flow time, the maximum number of workers and the maximum number of workers transfer. Since HMS has structure that combine the cells and functional area, finding the optimal solution for this type of system requires a high computational time even for small-sized problem instances.

The third stage of the proposed methodology is the stage of worker assignment. In this stage, a mixed integer linear goal programming model is developed to evaluate three different objectives within the same objective function. Objectives in this stage are as follows: minimizing deviations from the number of workers assigned to operations ($x_{i,k}$) in the second stage, minimizing utility workers, and minimizing deviations from the desired team synergy. The first objective ensures that operations are completed in time, and thus prevents the creation of infeasible schedules. The second objective minimizes quality problems. The third objective ensures the formation of effective teams in cells.

## Integration Mechanism

The integration mechanism that combines the second and third stages is shown in Figure 2.

The integration mechanism consists of two parts, one for small-sized problems and the other for large-sized problems. If the total number of cells and machines is smaller than five and the scheduling horizon is equal to or smaller than eight hours (one shift), it is a small-sized problem. For this type of problem, it is possible to find an optimal solution for the worker assignment problem in a reasonable amount of computational time.

Finding the optimal solution for the batch scheduling problem requires high computational time even when the problem at hand is a small-sized illustrative example.

Therefore, the proposed algorithm is applied in the second stage if the problem is small-sized and applied in both second and third stages when the problem is large-sized.

When the problem is small-sized, feasible solutions are selected from the Pareto set obtained at the end of the second stage, and used in the third stage. Two rules are taken into consideration when deciding whether a solution is feasible or not. The first rule is that the makespan of the solution should not exceed the determined scheduling horizon. The second rule is that the total number of workers for the solution should not exceed the number of existing workers. Solutions that meet both of these rules are accepted as feasible solutions.

In the third stage, it is possible to find an optimal solution for small-sized problems using the worker assignment mathematical model. In this stage, the model is solved for each of the feasible solutions which are found at the end of the second stage.

If the problem at hand is large-sized, the second and the third stages are combined and common solutions are obtained for the batch scheduling and the worker assignment problems using the proposed algorithm. This is followed by the fourth stage, in which the goal is to select a preferred solution from these solutions.

In the fourth stage, a preferred solution is selected from the best-known feasible Pareto set using the decision rules created.

If no feasible solutions are obtained by the end of the third stage, parameter α is increased by β, and the process is restarted from the order release stage.

## Order Release

In production systems with high part variety and demand variability, using pure pull production control methods, such as Kanban is not possible. In these kinds of production systems, OR can be used as an alternative production control method (Cevikcan and Durmusoglu, 2013).

In the first stage of the proposed methodology, the OR mathematical model developed is used to decide which batches will be released to the shop floor. In the OR mathematical model, the objective is to release the maximum number of parts to the shop floor by taking capacity utilization into consideration. Unreleased batches have priority to be released in the next scheduling horizon.

The zero-one integer mathematical model developed for the OR stage is presented in Figure 1.

*Figure 2. Mechanism for integrating second and third stages of proposed methodology*

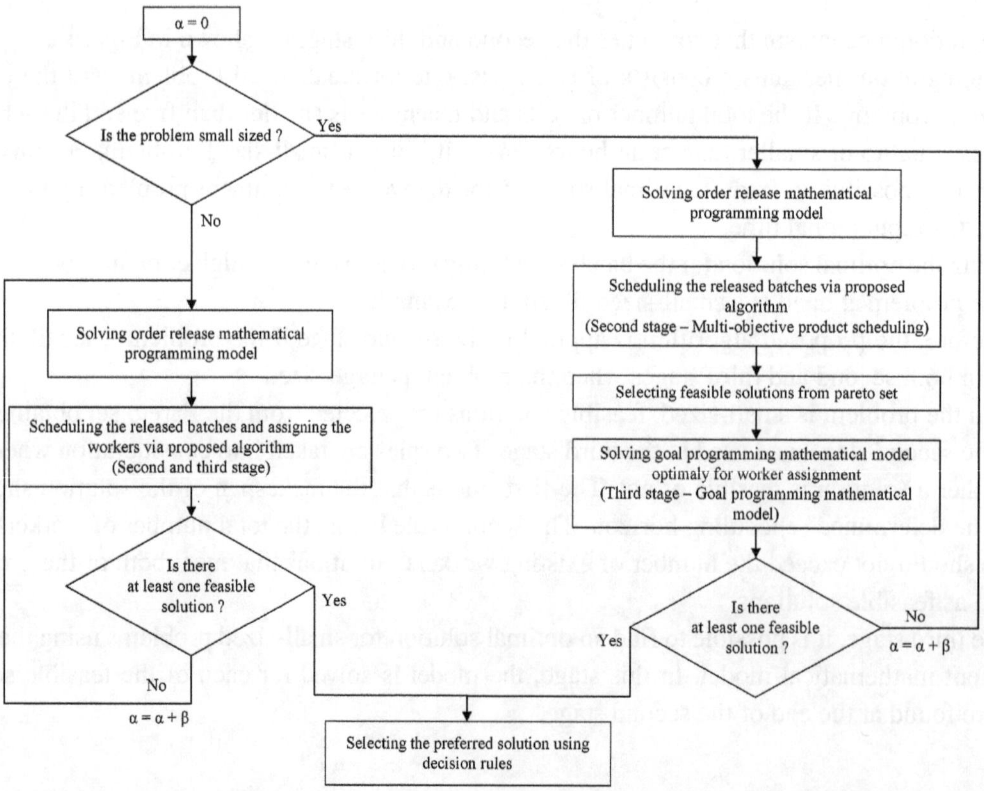

Indices:

$i$: Index set of batches ($i = 1, \ldots, NP$).
$k$: Index set of the cells and the machines in the functional area ($k = 1, \ldots, K$).

Parameters:

$\alpha$: Cycle time changing ratio
$avesetup_{i,k}$: Average setup time of batch $i$ on cell/machine $k$
$q_i$: Number of parts in batch $i$
$capacity_k$: Capacity of cell/machine $k$
$ak_{i,k}$: If batch $i$ is allocated to cell/machine $k$, 1; if not, 0
$FLT_{i,k}$: Completion time of the first part of batch $i$ on cell/machine $k$
$cycmin_{i,k}$: Minimum cycle time for batch $i$ on cell/machine $k$
$cycmax_{i,k}$: Maximum cycle time for batch $i$ on cell/machine $k$

Variables:

$cyc_{i,k}$: Cycle time for batch $i$ on cell/machine $k$
$p_{i,k}$: Processing time of batch $i$ on cell/machine $k$

Decision variables:

$release_i$: If batch i is released, 1; if not, 0.

Mathematical Model:
Objective Function:

$$Maximize \quad f\left(x\right) = \sum_{i=1}^{NP} release_i * q_i \tag{1}$$

Constraints:

$$\sum_{i=1}^{NP} pr_{i,k} \leq capacity_k \quad \forall\, k \tag{2}$$

$$p_{i,k} = \left(cyc_{i,k} * \left(q_i - 1\right) + FLT_{i,k} + avesetup_{i,k}\right) \quad \forall i, k \tag{3}$$

$$p_{i,k} - pr_{i,k} \leq M * \left(1 - release_i\right) \quad \forall i, k \tag{4}$$

$$pr_{i,k} \leq M * release_i \quad \forall i, k \tag{5}$$

$$cyc_{i,k} = cyc\min_{i,k} + \left(cyc\max_{i,k} - cyc\min_{i,k}\right) * \alpha \quad \forall i, k \tag{6}$$

$$release_i 0 \ or \ 1 \tag{7}$$

Objective function (1) maximizes the total number of parts released to the shop floor. Equation (2) deals with cell/machine capacity constraint. Equations (3)-(5) are used to calculate processing times of the released batches. Equation (6) is used to change the cell cycle times. Varying parameter $\alpha$ between 0 and 1 can change cycle times. When parameter $\alpha$ is equal to zero, cycle time is at its minimum, and the maximum number of parts is released to shop floor. Equation (7) enforces the binary restriction on decision variable. The constant $M$ in the equations should be sufficiently large.

The solution of the mathematical models is performed on a 2.4 GHz Intel(R) Core™ i7-3630QM CPU with 16 GB of RAM via GAMS® optimization software. The average computational time for the OR mathematical model is 15 seconds.

## Batch Scheduling

In the second stage of the proposed methodology, the mathematical model developed Yilmaz and Durmusoglu (2017) for the batch scheduling problems is used. This multi-objective mathematical model has three objectives: minimizing average flow time, minimizing maximum number of workers in the HMS and minimizing maximum number of workers transfer within the cells.

The first objective attempts to decrease manufacturing lead time (MLT). The second objective attempts to minimize the maximum number of workers. The third objective attempts to minimize intercell utility worker transfers. Figure 3 presents the objectives and the variables which are used in the mathematical model.

In Figure 3, $O_{iz}$ represents the $z^{th}$ operation of batch $i$. The vertical axis shows the number of workers assigned to the operations. The horizontal axis shows starting and finishing times of operations and setups for batches.

Changing the number of workers in cells causes changes in cell cycle times, which in turn changes the flow times of batches. An increase in the number of workers in cells results in a decrease in the flow times of batches, and vice versa.

As Figure 3 shows, the first two objectives are conflicting objectives.

The third objective is to keep the numbers of workers in cells constant. The variables $wa_{i,k}$, $wa1_{i,k}$, and $wa2_{i,k}$ in Figure 4 are explained using an illustrative example.

Figure 3 shows the variables of $wa_{i,k}$, $wa1_{i,k}$, and $wa2_{i,k}$.

We provide an illustrative example to emphasize the importance of the batch scheduling problems in the CMS and HMS.

As an illustration, consider a CMS consisting of two manufacturing cells. Batches 1 and 2 are produced in the cell 1, and batches 3 and 4 are produced in the cell 2. The number of workers used to produce the

*Figure 3. Scheduling operations with objectives*

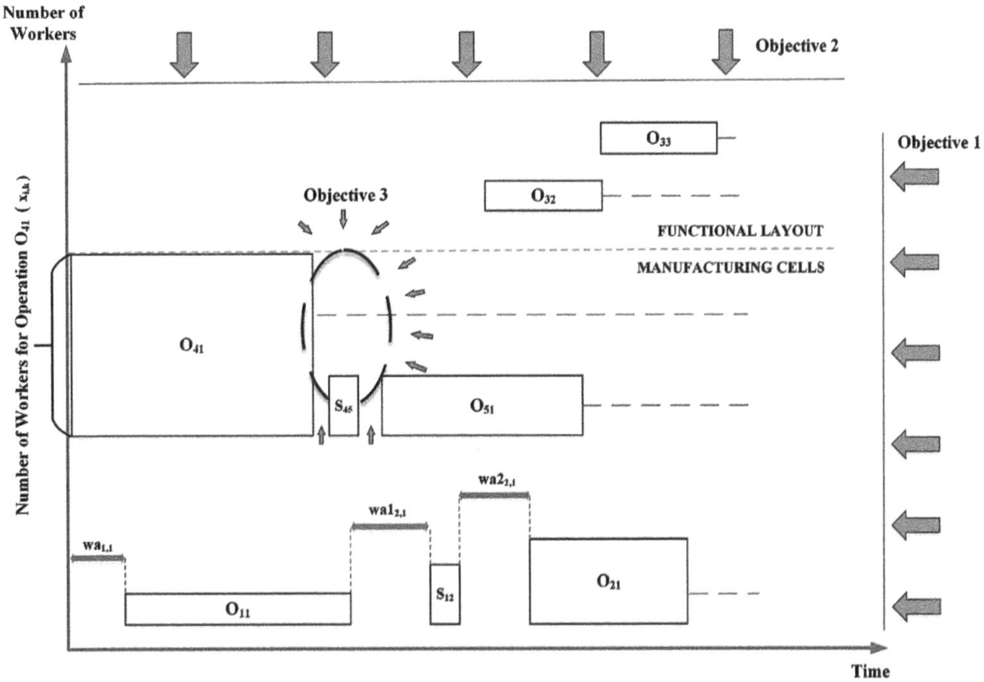


four batches, are three, two, three and four, respectively. When the number of workers working in the production of a batch decreases, the flow time of batch increases and the time limit is exceeded. Figure 4 (a) shows one of the feasible schedules that may result when variables $wa_{i,k}$, $wa1_{i,k}$, and $wa2_{i,k}$ are not used. With this schedule, the total number of workers required is seven. Figure 4 (b) shows one of the feasible schedules that may result when variables $wa1_{i,k}$, and $wa2_{i,k}$ are used. With the schedule in Figure 4 (b), the total number of workers decreases to six. This is because variables $wa1_{i,k}$, and $wa2_{i,k}$ are used in scheduling. These variables play an important role in reducing the total number of workers in DRC manufacturing systems.

The multi-objective mixed integer non-linear mathematical model used in second stage is presented below (Figure 1). The purpose of the mathematical model is to contribute to the explanation of the batch scheduling problem addressed in the study.

The multi-objective problem, notations, assumptions and mathematical model were introduced in Yılmaz and Durmusoglu (2017).

The following assumptions have been made in this study.

- The order of operations for each batch is predefined.
- Preemption of operations is not allowed.
- Each cell and machine can process only one operation at a time.
- There are no precedence constraints among the operations of different batches.
- Batches are available for processing at time zero.
- The number of workers in cells may change according to time and operation. Each worker has the same multi-skills.

*Figure 4. The importance of the variables* $wa_{i,k}$, $wa1_{i,k}$, $wa2_{i,k}$

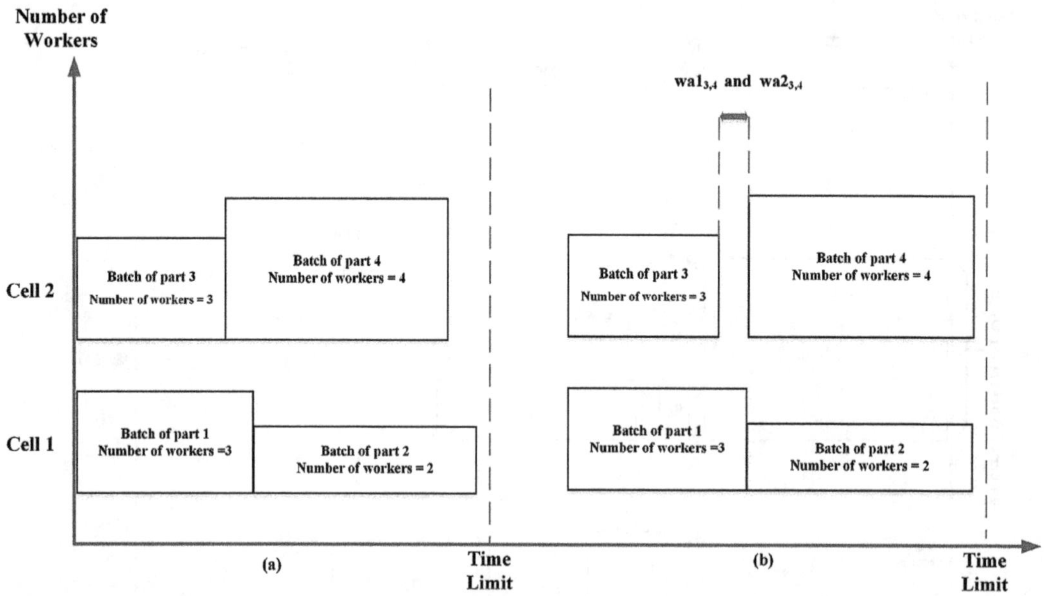


- The setup times are sequence-dependent and symmetrical for both cells and machines.
- The batches pass through the cell towards the functional area. Backflow is not allowed.

The modeling, objectives, and constraint are presented in the following section.

## Notations

The following terms were defined:

Indices:

*i, j*: Index set of batches (*i, j* = 1, …, *N*).
*k, l*: Index set of the cells and the machines in the functional area (*k*, l = 1, …, *K*).
*t*: Index set of workloads' changing times (t = 1,2)

Parameters:

$cycmin_{i,k}$: Minimum cycle time of batch *i* in cell/machine *k*
$cycmax_{i,k}$: Maximum cycle time of batch *i* in cell/machine *k*
$last_{i,k}$: If the last operation of batch *i* is in cell/machine *k*, 1; if not, 0
$d_{i,l,k}$: If batch *i* is allocated to cell/machine *k* following its operation in the cell/machine l, 1; if not, 0
$ak_{i,k}$: If batch *i* is allocated to cell/machine *k*, 1; if not, 0
$FLT_{i,k}$: Completion time of the first part in the batch *i* in cell/machine *k*
$q_i$: Size of batch *i*

$s_{i,j,k}$: Sequence-dependent setup time created by setup of batch $i$ following batch $j$ is processed in cell/machine $k$

$setupworker_{i,j,k}$: Sequence-dependent number of workers for the operation of batch $i$ after operation of batch $j$ is processed in cell/machine $k$

$sw_{i,k}$: Necessary number of workers for the setup of batch $i$ in cell/machine $k$

Variables:

$p_{i,k}$: Processing time of batch $i$ in cell/machine $k$

$c_{i,k}$: Completion time of batch $i$ in cell/machine $k$

$setupstart_{i,k}$: Starting time of the setup before operation of batch $i$ in cell/machine $k$

$setupfinish_{i,k}$: Finishing time of the setup before operation of batch $i$ in cell/machine $k$

$time_{j,l,t}$: Starting time of operation/setup of batch $j$ in cell/machine $l$ (number of workers changing time)

$workload_{j,l,t}$: Total number of workers at the start of setup/operation of batch $j$ in cell/machine $l$

$workload1_{j,l,t}$: Total number of workers at the start of operation of batch $j$ in cell/machine $l$

$workload2_{j,l,t}$: Total number of workers at the start of setup of batch $j$ in cell/machine $l$

$cyc_{i,k}$: Cycle time for batch $i$ in cell/machine $k$

Decision Variables:

$b_{i,j,k}$: If batch j precedes batch $i$ in cell/machine $k$, 1; if not, 0

$x_{i,k}$: Number of workers for operation of batch $i$ in cell/machine $k$

Problem Formulation:

Three objectives used in the problem are as follows.

Objective Functions:

$$\text{Minimize } f(x) = \left[f_1(x), f_2(x), f_3(x)\right] \tag{8}$$

$$f_1(x) = \frac{\sum_{i=1}^{N}\sum_{k=1}^{K} c_{i,k} * last_{i,k}}{N} \tag{9}$$

$$f_2(x) = \max_{\forall j,l,t}\left(workload_{j,l,t}\right) \tag{10}$$

$$f_3(x) = \max_{\forall k}\left[\max_{\forall i}\left(x_{i,k}\right) - \min_{\forall j}\left(x_{j,k} > 0\right)\right] \tag{11}$$

The constraints used in the problem are stated below.
Constraints:

$$c_{i,k} - p_{i,k} = \left( \begin{array}{l} \sum_{j=1}^{N} b_{i,j,k} * \left( setupfinish_{i,k} + wa2_{i,k} \right) \\ + \left( 1 - \sum_{j=1}^{N} b_{i,j,k} \right) * \left( wa_{i,k} + \sum_{l=1}^{K} d_{i,l,k} * c_{i,l} \right) \end{array} \right) * ak_{i,k} \quad \forall i,k \tag{12}$$

$$c_{i,k} - p_{i,k} \geq \sum_{l=1}^{K} \left( c_{i,l} * d_{i,l,k} \right) \quad \forall i,k,l \quad k \neq l \tag{13}$$

$$setupstart_{i,k} = \sum_{j=1}^{N} b_{i,j,k} * \left( c_{j,k} + wa1_{i,j,k} \right) \quad \forall i,k \tag{14}$$

$$setupfinish_{i,k} = setupstart_{i,k} + \sum_{j=1}^{N} \left( b_{i,j,k} * s_{i,j,k} \right) \quad \forall i,k \tag{15}$$

$$b_{i,j,k} \leq ak_{i,k} * ak_{j,k} \quad \forall i,j,k \quad i \neq j \tag{16}$$

$$\sum_{i=1}^{N} ak_{i,k} - 1 = \sum_{i=1}^{N} \sum_{j=1}^{N} b_{i,j,k} \quad \forall k \tag{17}$$

$$b_{i,j,k} + b_{j,i,k} \leq 1 \quad \forall i,j,k \quad i \neq j \tag{18}$$

$$\sum_{i=1}^{N} b_{i,j,k} \leq 1 \quad \forall j,k \quad i \neq j \tag{19}$$

$$\sum_{j=1}^{N} b_{i,j,k} \leq 1 \quad \forall i,k \quad i \neq j \tag{20}$$

$$time_{j,l,t} = setupstart_{j,l} \quad \forall j,l,t = 1 \tag{21}$$

$$time_{j,l,t} = c_{j,l} - p_{j,l} \quad \forall j,l,t = 2 \tag{22}$$

$$time_{j,l,t} - \left(c_{i,k} - p_{i,k}\right) < M * ka_{i,k,j,l,t} \quad \forall i,k,j,l,t \tag{23}$$

$$c_{i,k} - time_{j,l,t} \leq M * kb_{i,k,j,l,t} \quad \forall i,k,j,l,t \tag{24}$$

$$g1_{i,k,j,l,t} = ka_{i,k,j,l,t} * kb_{i,k,j,l,t} \quad \forall i,k,j,l,t \tag{25}$$

$$workload\ \ 1_{j,l,t} = \sum_{i=1}^{N}\sum_{k=1}^{K}\left(x_{\ i,k} * g1_{\ i,k,j,l,t}\right) \ \forall j,l,t \tag{26}$$

$$time_{j,l,t} - setupstart_{i,k} < M * kc_{i,k,j,l,t} \quad \forall i,k,j,l,t \tag{27}$$

$$setupfinish_{i,k} - time_{j,l,t} \leq M * kd_{i,k,j,l,t} \quad \forall i,k,j,l,t \tag{28}$$

$$g2_{i,k,j,l,t} = kc_{i,k,j,l,t} * kd_{i,k,j,l,t} \quad \forall i,k,j,l,t \tag{29}$$

$$workload\ \ 2_{j,l,t} = \sum_{i=1}^{N}\sum_{k=1}^{K}\left(sw_{i,k} * g2_{i,k,j,l,t}\right) \ \forall j,l,t \tag{30}$$

$$workload_{j,l,t} = workload\ \ 1_{j,l,t} + workload\ \ 2_{j,l,t} \quad \forall\ j,l,t \tag{31}$$

$$sw_{i,k} = \sum_{j=1}^{N} \left( setupworker_{i,j,k} * b_{i,j,k} \right) \quad \forall i,k \quad i \neq j \tag{32}$$

$$p_{i,k} = cyc_{i,k} * \left( q_i - 1 \right) + FLT_{i,k} \quad \forall i,k \tag{33}$$

$$cyc_{i,k} = \max \left( \frac{cyc\max_{i,k}}{x_{i,k}} ; cyc\min_{i,k} \right) \quad \forall\, i,k \tag{34}$$

$$x_{i,k} = \left( \frac{cyc\max_{i,k}}{cyc_{i,k}} \right)^{+} \quad \forall i,k \tag{35}$$

$$b_{i,j,k} = 0 \quad or \quad 1; cyc_{i,k} \geq 0; wa_{i,k} \geq 0; wa1_{i,k} \geq 0; wa2_{i,k} \geq 0 \tag{36}$$

In this chapter, three objectives are considered. The first objective (9) minimizes the average flow time of batches. The second objective (10) minimizes the maximum number of workers in HMS. The third objective (11) minimizes the maximum number of workers changing in cells. When the number of workers changing in cells is decreased, the inter-cellular movements of workers will also decrease.

Constraints (12) and (13) represent if batch *j* precedes batch *i* then the start time of batch *i* on its cell/machine depends on the finishing time of batch *j*, finishing time of setup and elapsed time after setup ($wa2_{i,k}$) on the same cell/machine. If the operation of batch *i* is the first operation then the start time of batch *i* on its cell/machine depends on the elapsed time after the starting time ($wa_{i,k}$).

Constraint (14) and (15), respectively, represent starting and finishing time of setup. If batch *j* precedes batch *i*, then starting time of setup depends on finishing time of batch *j* and elapsed time before setup ($wa1_{i,k}$). Finishing time of setup depends on starting time of setup and setup time.

Constraints (16), (17), (18), (19), and (20) represent the sequencing of the operations in the cells and machines. There can only be a single operation before and after any operation.

Constraints (21) and (22) represent the time points where the number of workers might be changed. These time points include the starting time of setup and the operation of batches. The total number of workers is computed for every time point determined using constraints 18 and 19.

Constraints (23) and (24) are formed to identify the time points (as computed by using constraints (21) and (22)) with which the operations coincide.

Constraints (25) and (26) are used to determine the total number of workers required in the operations for each time point computed by using constraints (21) and (22).

Constraints (27) and (28) are formed to identify the time points (as computed by using constraints (21) and (22)) with which the setups coincide.

Constraints (29) and (30) are used to determine the total number of workers required for the setups for each time point computed by using constraints (21) and (22).

Constraint (31) is used to determine the total number of workers for each time point computed by using constraints (21) and (22).

Constraint (32) is used to determine the number of workers for the sequence-dependent setup time.

Constraint (33) shows the processing time of operations in the cells and machines.

Constraint (34) and (35) show the variation in the cycle time depending on the number of workers required for the operations in the cells. Long cycle times require small numbers of workers, while short cycle times require large numbers of workers.

Constraint (36) enforces the binary and non-negative restrictions on the variables.

## Worker Assignment

The worker assignment goal programming mathematical model developed for the third stage of the proposed methodology is used to assign workers to operations. Worker capabilities are taken into account in the development of the mathematical model.

The mathematical model has three different objectives: minimizing deviations from the number of workers assigned to operations ($x_{i,k}$) in the second stage, minimizing utility workers and minimizing deviations from desired team synergy level.

The binary integer linear goal programming mathematical model used in the third stage is presented below (Figure 1).

Indices:

$i,j$: Index set of batches ($i = 1, \ldots, N$).
$z$: Index set of workers ($z = 1, \ldots, Z$).
$m$: Index set of the machines in the cells and functional area ($m = 1, \ldots, M$).
$k,l$: Index set of the cells and the machines in the functional area ($k = 1, \ldots, K$).
$y$: Index set of traits ($y = 1, 2, 3, 4$)
$n$: Index set of modes ($n = 1, 2, 3$)

Parameters:

$w_1$: Weight of the first objective
$w_2$: Weight of the second objective
$w_3$: Weight of the third objective
$using1_{i,k,m}$: If machine $m$ is used for operation of batch $i$ on cell/machine $k$, 1; if not, 0.
$using2_{i,k,m}$: If machine $m$ is used for setup of batch $i$ on cell/machine $k$, 1; if not, 0.
$s_{z,m}$: If worker $z$ is currently capable to perform operations/setups on machine type $m$, 1; if not, 0.
$sw_{i,k}$: Necessary number of workers for the setup of batch $i$ on cell/machine $k$
$x_{i,k}$: Number of workers for operation of batch $i$ on cell/machine $k$
$a_{z,y,n}$: If worker $z$'s mode of operation is at level $n$ for trait $y$, 1; if not, 0.
$ak_{i,k}$: If batch $i$ is allocated to cell/machine $k$, 1; if not, 0
$avewalking_k$: Average walking time of workers in cell $k$
$manualtask_{i,k,m}$: Manual task time for batch $i$ on machine $m$ in cell $k$

$lmct_{i,k}$: Longest machine cycle time for batch $i$ on cell $k$
$ka1_{i,k,j,l}$: If the operation of the batch $i$ on cell/machine $k$ and the operation of batch $j$ in cell/machine $l$ overlap in time, 1; if not, 0
$ka2_{i,k,j,l}$: If the setup of the batch $i$ on cell/machine $k$ and the setup of batch $j$ on cell/machine $l$ overlap in time, 1; if not, 0
$ka3_{i,k,j,l}$: If the operation of the batch $i$ on cell/machine $k$ and the the setup of batch $j$ in cell/machine $l$ overlap in time, 1; if not, 0

Variables:

$woo_{z,i,k}$: If worker z is assigned to operation of batch $i$ on cell/machine $k$, 1; if not, 0
$wooo_{z,k}$: If worker $z$ is assigned to cell/machine $k$, 1; if not, 0
$wss_{z,i,k}$: If worker $z$ is assigned to setup of batch $i$ on cell/machine $k$, 1; if not, 0

Deviational Variables:

$d1_{i,k}$ - $d2_{i,k}$: Positive and negative deviational variables for the first objective
$d3_z$ - $d4_z$: Positive and negative deviational variables for the second objective
$d5_{n,k}$ - $d6_{n,k}$: Positive and negative deviational variables for the third objective

Decision Variables

$wo_{z,i,k,m}$: If worker $z$ is assigned to machine $m$ for operation of batch $i$ on cell/machine $k$
$ws_{z,i,k,m}$: If worker $z$ is assigned to machine $m$ for setup of batch $i$ on cell/machine $k$

Mathematical Model:
Objective Function:

$$Minimize \quad w_1 * \sum_{i=1}^{N}\sum_{k=1}^{K}\left(d1_{i,k} + d2_{i,k}\right) + w_2 * \sum_{z=1}^{Z} d3_z + w_3 * \sum_{n=1}^{3}\sum_{k=1}^{K}\left(d5_{n,k} + d6_{n,k}\right) \tag{37}$$

Constraints:

$$\sum_{z=1}^{Z} woo_{z,i,k} - d1_{i,k} + d2_{i,k} = x_{i,k} \quad \forall i,k \tag{38}$$

$$\sum_{k=1}^{K} wooo_{z,k} - d3_z + d4_z = 1 \quad \forall z \tag{39}$$

$$\sum_{z=1}^{Z}\sum_{y=1}^{4}\left(a_{z,y,n} * wooo_{z,k}\right) - d5_{n,k} + d6_{n,k} = \sum_{z=1}^{Z} wooo_{z,k} \quad k = 1,..,\text{number of cells}; n = 1,3 \tag{40}$$

$$\sum_{z=1}^{Z}\sum_{y=1}^{4}\left(a_{z,y,n} * wooo_{z,k}\right) - d5_{n,k} + d6_{n,k} = 2 * \sum_{z=1}^{Z} wooo_{z,k} \quad k = 1,..,\text{number of cells}; n = 2 \tag{41}$$

$$wo_{z,i,k,m} \leq s_{z,m} \quad \forall z,i,k,m \tag{42}$$

$$\sum_{z=1}^{Z} wo_{z,i,k,m} = using1_{i,k,m} \quad \forall\, i,k,m \tag{43}$$

$$ws_{z,i,k,m} \leq s_{z,m} \quad \forall z,i,k,m \tag{44}$$

$$\sum_{z=1}^{Z} ws_{z,i,k,m} = using2_{i,k,m} \quad \forall i,k,m \tag{45}$$

$$\sum_{z=1}^{Z} wss_{z,i,k} = sw_{i,k} \quad \forall i,k \tag{46}$$

$$\sum_{m=1}^{M} ws_{z,i,k,m} \leq wss_{z,i,k} * M \quad \forall z,i,k \tag{47}$$

$$\sum_{m=1}^{M} wo_{z,i,k,m} \leq woo_{z,i,k} * M \quad \forall z,i,k \tag{48}$$

$$\sum_{i=1}^{N}\left(woo_{z,i,k} + wss_{z,i,k}\right) \leq wooo_{z,k} * M \quad \forall z,k \tag{49}$$

$$woo_{z,i,k} + wss_{z,i,k} \leq ak_{i,k} * M \quad \forall z,k \tag{50}$$

$$\max\left(\max_{\forall z}\left(\sum_{m=1}^{M} wo_{z,i,k,m} * \begin{pmatrix} avewalking_{k} \\ + manualtask_{i,k,m} \end{pmatrix}; lmct_{i,k}\right)\right) \leq cyc_{i,k} \quad k = 1,.., \text{number of cells} \quad \forall i \tag{51}$$

$$\left(woo_{z,i,k} + woo_{z,j,l}\right) * ka1_{i,k,j,l} \leq 1 \quad \forall z,i,k,j,l \quad i \neq j, k \neq l \tag{52}$$

$$\left(wss_{z,i,k} + wss_{z,j,l}\right) * ka2_{i,k,j,l} \leq 1 \quad \forall z,i,k,j,l \quad i \neq j, k \neq l \tag{53}$$

$$\left(woo_{z,i,k} + wss_{z,j,l}\right) * ka3_{i,k,j,l} \leq 1 \quad \forall z,i,k,j,l \quad i \neq j, k \neq l \tag{54}$$

$$wo_{z,i,k,m} 0 \quad or \quad 1; ws_{z,i,k,m} 0 \quad or \quad 1; d1_{i,k} \geq 0; d2_{i,k} \geq 0; d3_{z} \geq 0; d4_{z} \geq 0; d5_{n,k} \geq 0; d6_{n,k} \geq 0 \tag{55}$$

The objectives function (37) presents a weighted average of the three objectives that are assigned weights of $w_1$ through $w_3$. Equation (38) is used to assign the determined number of workers to operations. For this soft constraint, the target level is set at $x_{i,k}$. Equation (39) is created to minimize the number of utility workers. To prevent the over-assignment of workers to more than one cell, the target level is set at one. Equations (40) and (41) are used to form effective teams in each cell. In these equations, the target level is set at ideal team synergy level (25% resistor, 25% initiator, 50% accommodator). (For more information on forming effective teams, please refer to Askin and Huang, 2001 and Kolbe, 1994). Equation (42) ensures that each worker-operation assignment is within the available capabilities.

Equation (43) is used to assign workers to machines where the operations are performed.

Equation (44) ensures that each worker-setup assignment is within capabilities made available.

Equation (45) is used to assign workers to machines where the setups are performed.

Equation (46) ensures the assignment of workers to setups.

Equations (47), (48) and (49) ensure that workers assigned to a machine for setup or operation are also assigned to the cell that contains the machine in question. The letter $M$ in the equations represents a large number.

Equation (50) expresses the necessary condition for the assignment of a worker to an operation or setup.

Equation (51) ensures that cycle times obtained in the case of the dedicated assignment of workers are smaller than the cycle times obtained in the second stage; otherwise, the schedule in the preceding model becomes infeasible.

Because of the non-linear structure of Equation (51), finding the optimal solution requires high computational time even for small-sized problems. Therefore, instead of Equation (51), its linear version, Equation (56) is used in the solution of the model.

$$\sum_{m=1}^{M} wo_{z,i,k,m} * \left(avewalking_k + manualtask_{i,k,m}\right) \leq cyc_{i,k} \quad k = 1,.., \text{number of cells} \quad \forall i, z \tag{56}$$

Equation (52), (53), and (54) prevents the overlapping assignment of workers to the operations and setups.

Equation (55) enforces the binary, integer and non-negative restrictions on the variables.

## Decision Rules

From the best-known feasible Pareto set, which is obtained using the integration mechanism, a preferred solution is selected. The selection is made on the basis of pre-set decision rules.

Decision Rules:

If (there is one feasible solution)
Select the solution as preferred solution
Else (there are two or more feasible solutions)
If (there is one solution that has the biggest value of ALTAMLT (Equation 57))
Select the solution as preferred solution
Else (there are two or more solutions that have the biggest value of ALTAMLT)
Select a random solution as preferred solution
End
End

# PROPOSED ALGORITHM

The proposed algorithm is used inside the integration mechanism that combines the second and third stages of the methodology. The proposed algorithm is developed on the basis of the FastGPA algorithm, and a local search heuristic is developed and employed inside the proposed algorithm.

The FastPGA (Fast Pareto Genetic Algorithm; Eskandari et al., 2007) is a population-based MOEA (Multi-Objective Evolutionary Algorithm) and it is very similar to the NSGA-II algorithm regarding computational complexity. Eskandari et al. (2007) introduced a new fitness assignment and ranking strategy with the FastPGA algorithm. In addition, they used a population regulation operator to make it effective in the solution of MOPs. Results of the experiments they conducted on test problems showed that the FastPGA algorithm performs well in the solution of MOPs, regarding the fast convergence, the diversity and the distance metrics. Therefore, the FastPGA algorithm is used in the proposed algorithm for the current study. (For detailed information concerning the FastPGA algorithm, please refer to Eskandari et al., 2007).

A flowchart of the proposed algorithm is shown in Figure 5. The proposed algorithm consists of two parts, one for small-sized problems and the other for large-sized problems. If the problem at hand is small-sized, the proposed algorithm is used to find a solution only for the batch scheduling problem. If the problem is large-sized, it is used to find solutions to both the batch scheduling and worker assignment problems. In small-sized problems, an optimal solution is obtained in the worker assignment stage after the proposed algorithm is used. In large-sized problems, worker assignment decisions are made using the proposed algorithm.

Sub-chromosome is used for worker assignment decisions to be made in large-sized problems. The second and the third objectives in the mathematical model (equations 39, 40 and 41) are normalized and transformed into a single objective function. Solutions are evaluated considering the weighted sum of fitness functions.

In the scheduling stage of the proposed algorithm, a local search heuristic is used.

Local search
Input: P and population_size of P
Output: Local population L
(1) Set S = Random [(population_size of P)*plocal]$^{+}$ solutions from population P and
(2) For each $x_k \in$ S do
Generate Number_neigh neighborhood of $x_k$, N($x_k$) (Using polynomial mutation)
For each y$\in$ N($x_k$) do
If y is not dominated by $x_k$ then

*Figure 5. Flowchart of the proposed algorithm*

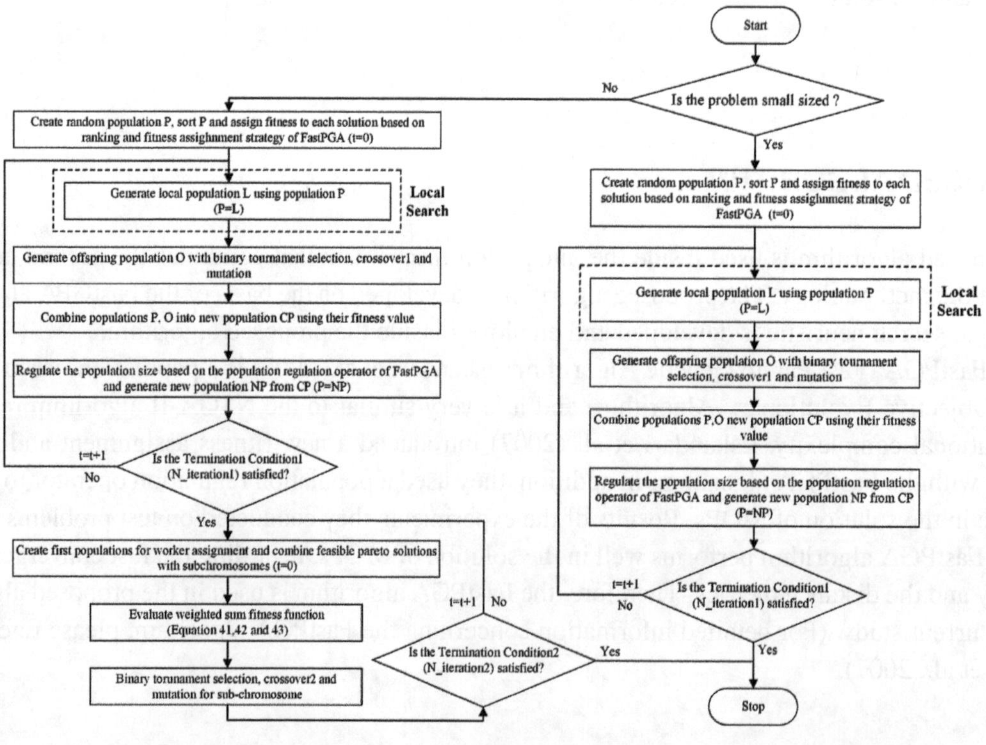

$x_k \leftarrow y$
End if
End for
End for
(3) Combine S and P to form local population L
(4) Rank local population L and regulate population_size of L according to the number of nondominated solution
(5) Output L

For the first step of the local search heuristic, a sample of $[(\text{population\_size of P})*\text{plocal}]^+$ solutions are randomly selected from population P. At the second step, neighborhood solutions are generated and compared with the existing solutions. At the third step, population S, which consists of the best neighborhood solutions, and the existing population P are combined to create population L. At step four, dominated solutions within the population L are eliminated to create a population that consists of non-dominated solutions.

## Chromosome and Sub-Chromosome Representation

The following chromosome structure proposed by Yılmaz and Durmusoglu (2017) is adopted in this chapter. Serial scheduling scheme (SSS) is applied to obtain feasible schedules.

It is important to take into account both the encoding-decoding and the chromosome structures that are used for representing the solution of the problems. The chromosome used in this study consists of five lines. The first line ($b_{i,j,k}$) represents the sequencing of the batches within the cells and machines. The second line ($wa_{i,k}$) represents the elapsed time before the start of operations of first batches within the cells and machines. The third line ($wa1_{i,k}$) represents the elapsed times before the start of setups. The fourth line ($wa2_{i,k}$) represents the elapsed times following the end of setups. The fifth line ($x_{i,k}$) shows the number of workers for the cells and machines. Another important point in this context is that the second, the third and the fourth lines in the chromosome structure are dependent on the first and last lines. The decisions are made based on the first and last lines. The other lines show the elapsed times which are caused by decision variables. Figure 6 shows the chromosome structure.

After the sequence of batches on cells and machines ($b_{i,j,k}$) is determined, the assignment of workers are supposed to be determined for both cells and functional areas. Each machine in a cell corresponds to a task, and these tasks combine to form operations/setups. In the functional area, each task corresponds to an operation/setup.

Figure 7 shows the structure of the sub-chromosome developed for worker assignment decisions. The upper part of the figure shows assignments to setups, and the lower part shows assignments to operations. In the sub-chromosome structure, workers are first assigned to operations, and then to tasks.

## Selection, Crossover and Mutation Operators

The binary tournament selection approach (Beyer and Deb, 2001), which is commonly used in the literature, is applied to both algorithms in this study.

*Figure 6. Chromosome structure*

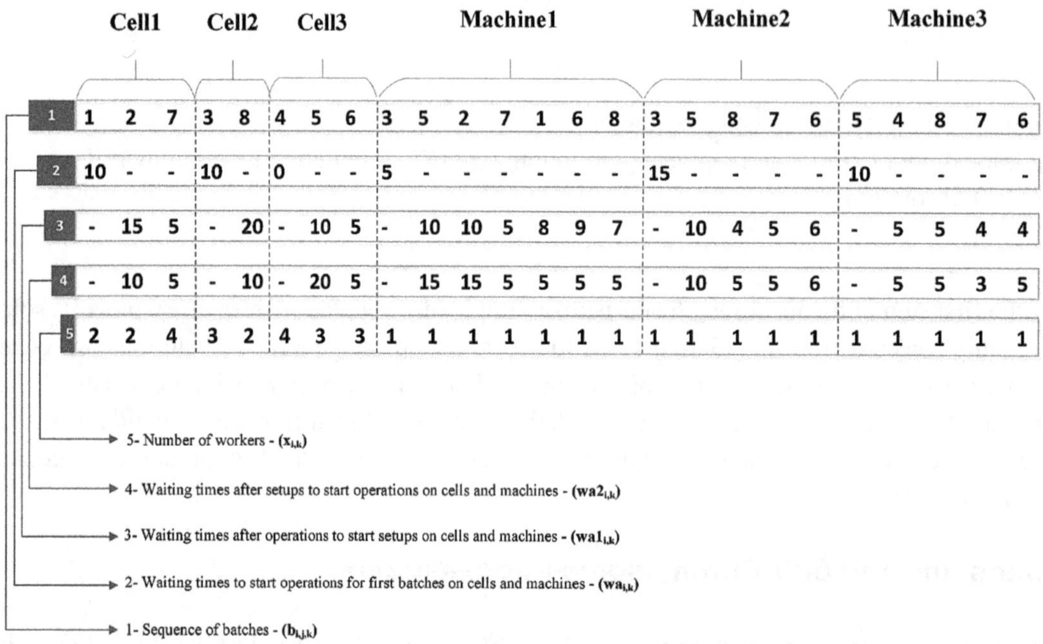


*Figure 7. Sub-chromosome structure*

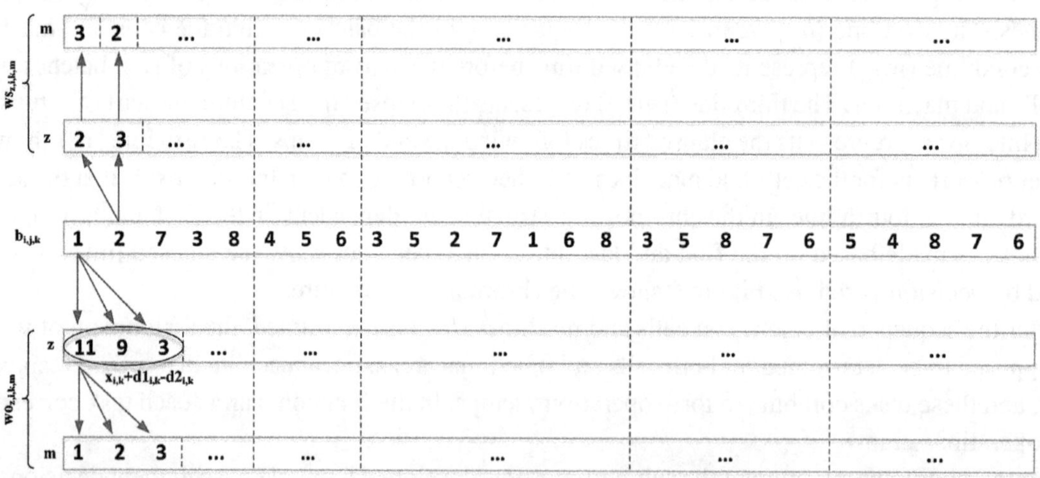


The crossover operator is formed by adapting the crossover operator and is used for both algorithms. This operator consists of two stages: In the first stage, the cells and machines are divided into two different groups for two parents, while in the second stage, one of these groups is selected and the operation sequencing is swapped according to a certain probability and two children are produced.

Since the chromosomes in this study are formed through real-coding, polynomial mutation (Hamdan, 2012) is used for both algorithms.

## INDUSTRIAL APPLICATION OF THE PROPOSED METHODOLOGY

To test its validity, the proposed methodology was applied to a real-life expansion joints manufacturing system. The company produces many types of expansion joints, and also fills custom orders.

In the selected scheduling horizon, only axial metal bellowed expansion joint parts were produced, and, as a result, the present study examined the producing of these parts. These parts are divided into two main categories of fixed and floating flange joints.

### Manufacturing System of the Company

The manufacturing system of the company consists of three different manufacturing cells and a functional area. Figure 8 shows these cells and the functional area.

There are seven different types of machines in the cells. In the functional area, there are two types of welding machines, two types of flange machines (floating and fixed), and one testing machine. The HMS has thirteen different types of machine.

After the operations are performed on the cells, the resulting semi-finished parts are called bellows, and the finished parts are called expansion joints.

Table 2 shows maximum and minimum cell cycle times for parts, final orders of parts, batch sizes of parts, processing times of tasks, and the order of operations (Yilmaz et al. 2016).

Table 2 also provides, under the names of the machines, information about the operation orders and processing times. As this table shows, the order of operations performed on cells is equal to one for each task. This is because each operation performed on the cells consists of seven separate tasks.

Table 3 provides information on sequence dependent setup times and workers in cell 1. Sequence-dependent setup times and workers in other cells and machines are not reported.

Table 4 shows the types of machines, and the modes of workers' levels for traits.

Production takes place over a single shift in a day, which lasts 540 minutes. Thus, the scheduling horizon for this study is set at 540 minutes.

### Applying Proposed Methodology

The OR mathematical model was solved for a 540-minute shift. Final orders and batch sizes are reported in Table 2. At this stage, the selected batches released to the shop floor for scheduling. In Table 2, batches released to the shop floor are shown in yellow. Using the integration mechanism, scheduling and worker assignment decisions are made.

The proposed algorithm is used inside the integration mechanism. Values of the parameters used in the algorithm are selected on the basis of previous studies (Deb et al., 2002; Eskandari and Geiger, 2008), and repetitive experiments are carried out to select appropriate parameter combination. Table 1 reports these parameter values. The algorithm is stopped when the values generated by the proposed algorithm do not improve after a certain number of iterations.

A total of 12 workers are employed in the production of axial metal bellowed expansion joints. Figure 7 shows the schedule that results when the proposed methodology is applied. The schedule shows starting and finishing times of operations and provides information on workers and operations. In Figure 7, the scheduling horizon is divided into 33 equal parts, each of which equals 1000 seconds. Workers assigned to operations are shown in parentheses.

*Figure 8. Hybrid Manufacturing System*

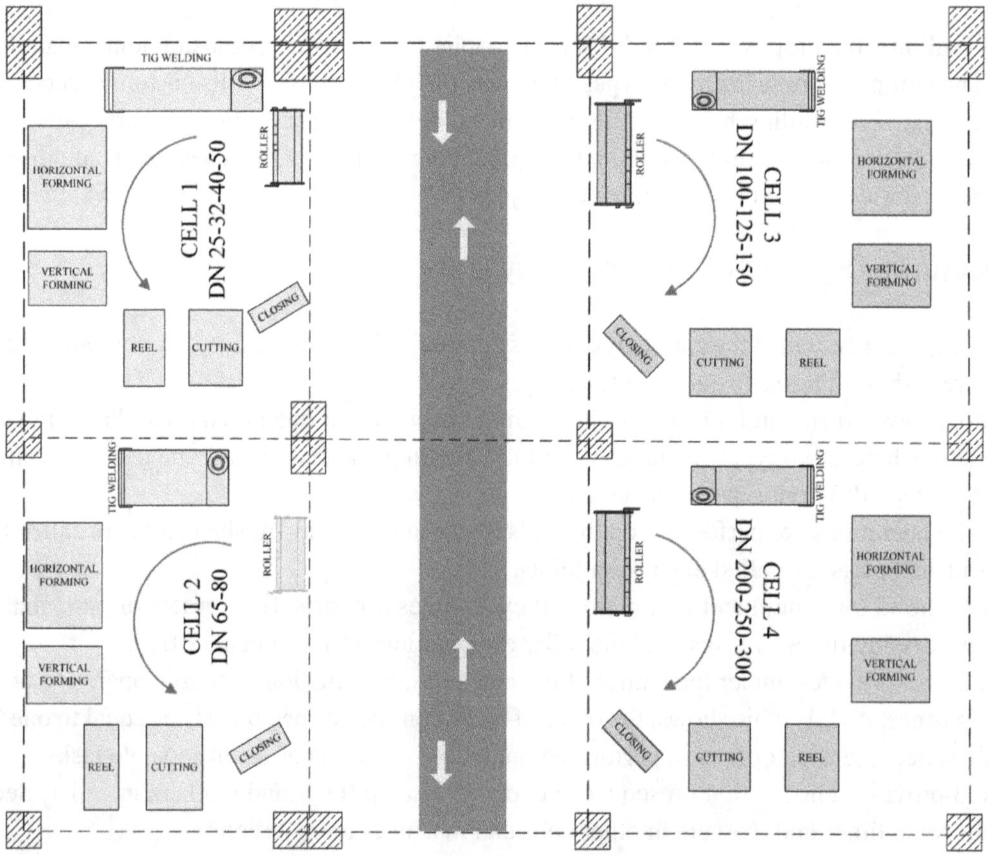


*Table 1. Parameter values for proposed algorithm*

| | Scheduling Stage | Worker Assignment Stage |
|---|---|---|
| Initial Population Size | 200 | - |
| Maximum Population Size | 200 | - |
| Population Size | - | 10 |
| plocal | 0.6 | - |
| $\eta_m$ | 20 | - |
| Mutation Probability | 0.9 | 1 |
| Mutation Probability for Local Search | 1 | - |
| Number_neigh | 5 | - |
| Termination Condition | 20 | 5 |

Application of the methodology to the HMS reduced the number of utility workers from three to two. Third and seventh workers are assigned to operations in cells to complete operations that would otherwise not finish in time. This makes it possible to decrease the cell cycle time to the pre-set level, and operations are finished in time. While the third worker is assigned to Cell 1 and Cell 3 as a utility worker; the seventh worker is assigned to Cell 1 and Cell 2 as a utility worker. A reduction in the number of utility workers indicates that congestion is reduced, a more effective production control is performed, and the work is standardized for each worker. This also decreases quality problems encountered in the cells.

Application of the proposed methodology increases the percentage of the labor time in MLT. This increase is shown in Figure 10. Equation (57) is developed to calculate the exact amount of this increase and named as average labor time of workers within average manufacturing lead time (ALTAMLT).

$$ALTAMLT = \left( \left( \sum_{z=1}^{Z} \sum_{i=1}^{N} \sum_{k=1}^{K} \left( p_{i,k} * woo_{z,i,k} \right) \Big/ Z \right) \Big/ \left( \sum_{i=1}^{N} \sum_{k=1}^{K} \left( c_{i,k} * last_{i,k} \right) \Big/ N \right) \right) \tag{57}$$

An increase in the ALTAMLT ratio indicates that the number of parts produced within the scheduling horizon has increased. Figure 10 (c) shows the ALTAMLT value obtained in the HMS without using the proposed methodology, Figure 10 (b) shows the ALTAMLT value obtained when the proposed algorithm is used without local search, and Figure 10 (c) shows the ALTAMLT value obtained when the proposed methodology is applied. As the figure shows, the proposed methodology leads to 35% increase in the ALTAMLT value.

## Sensitivity Analysis

This section presents the sensitivity analyses conducted to examine the effects of variables *wa*, *wa1* and *wa2* and deviational variables *d1* and *d2* on schedules and assignments. Effects of these variables on the schedules are shown in Figure 11.

For sensitivity analysis, three schedules are evaluated over four different variables. These variables are the makespan, the average flow time, the total number of workers and the total number of workers transfer. Figure 11 reports the values of these variables obtained from three different schedules. The first schedule (1) is created using the variables *wa*, *wa1* and *wa2* (Figure 9). The second schedule (2) is created without using the variables *wa*, *wa1* and *wa2*. The third schedule (3) is created without using variables *wa*, *wa1* and *wa2*, and by assigning the minimum number of workers to each operation.

Makespan and average flow time values obtained from the second schedule are lower compared to the first schedule, but the total number of workers in the second schedule exceeds the current number of workers in HMS, which means that the second schedule is effective but not feasible. The first and second schedules have the same value regarding the total number of workers transfer.

In the third schedule, the total number of workers is six, but this schedule cannot be used because its makespan value exceeds the determined scheduling horizon.

The results of the sensitivity analysis demonstrate the importance of using variables *wa*, *wa1* and *wa2* in generating an effective and feasible schedule.

In the third stage of the proposed methodology, deviational variables *d1* and *d2* are used in the worker assignment model. Effects of these variables on assignments are shown in Figure 12.

*Table 2. The production data from real hybrid manufacturing system*

| | Parts | | Cycle Times | | Orders | Batch Sizes | Cells | | | | | | | Functional Area | | | | | | | |
|---|---|---|---|---|---|---|---|---|---|---|---|---|---|---|---|---|---|---|---|---|---|
| | | | Max | Min | | | Mac.1 | Machine 2 | Machine 3 | Machine 4 | Mac. 5 | Mac. 6 | Mac. 7 | Machine 8 | Machine 9 | Mac. 10 | Mac. 11 | Mac. 12 | Mac. 13 | Testing |
| **Cell 1** | **1** | Flo. Fla. | 957 | 450 | 5 | 5 | 1-16 | 1-150 | - | 1-450 | 1-155 | 1-108 | 1-62 | | | | 2-102 | 3-193 | | 4-48 |
| | | Fix. Fla. | 622 | 468 | 4 | 4 | 1-16 | 1-150 | 1-468 | - | 1-160 | 1-108 | 1-62 | | 2-232 | 3-225 | | | | 4-36 |
| | **2** | Flo. Fla. | 915 | 432 | 10 | 10 | 1-15 | 1-144 | - | 1-432 | 1-148 | 1-105 | 1-55 | | | | 2-105 | 3-194 | | 4-52 |
| | | Fix. Fla. | 590 | 455 | 5 | 5 | 1-15 | 1-144 | 1-455 | - | 1-152 | 1-105 | 1-58 | | 2-258 | 3-252 | | | | 4-38 |
| | **3** | Flo. Fla. | 887 | 420 | 15 | 15 | 1-15 | 1-140 | - | 1-420 | 1-145 | 1-96 | 1-55 | | | | 2-111 | 3-214 | | 4-64 |
| | | Fix. Fla. | 555 | 438 | 8 | 8 | 1-15 | 1-140 | 1-438 | - | 1-145 | 1-96 | 1-55 | | 2-293 | 3-275 | | | | 4-47 |
| | **4** | Flo. Fla. | 741 | 319 | 15 | 15 | 1-12 | 1-135 | - | 1-319 | 1-130 | 1-84 | 1-45 | | | | 2-104 | 3-207 | | 4-70 |
| | | Fix. Fla. | 508 | 355 | 5 | 5 | 1-12 | 1-135 | 1-355 | - | 1-138 | 1-84 | 1-48 | | 2-331 | 3-325 | | | | 4-46 |
| **Cell 2** | **5** | Flo. Fla. | 626 | 250 | 10 | 10 | 1-9 | 1-120 | - | 1-250 | 1-117 | 1-71 | 1-44 | | | | 2-108 | 3-211 | | 4-76 |
| | | Fix. Fla. | 436 | 272 | 10 | 10 | 1-9 | 1-120 | 1-272 | - | 1-117 | 1-71 | 1-44 | | 2-368 | 3-411 | | | | 4-58 |
| | **6** | Flo. Fla. | 472 | 155 | 15 | 15 | 1-8 | 1-108 | - | 1-155 | 1-88 | 1-66 | 1-32 | | | | 2-132 | | 3-204 | 4-88 |
| | | Fix. Fla. | 363 | 186 | 9 | 9 | 1-8 | 1-108 | 1-186 | - | 1-102 | 1-66 | 1-38 | 2-383 | | 3-490 | | | | 4-64 |
| **Cell 3** | **7** | Flo. Fla. | 525 | 182 | 12 | 12 | 1-8 | 1-115 | - | 1-182 | 1-95 | 1-71 | 1-40 | | | | 2-151 | | 3-211 | 4-91 |
| | | Fix. Fla. | 414 | 210 | 8 | 8 | 1-8 | 1-115 | 1-210 | - | 1-112 | 1-71 | 1-44 | 2-398 | | 3-600 | | | | 4-68 |
| | **8** | Flo. Fla. | 617 | 226 | 14 | 14 | 1-11 | 1-131 | - | 1-226 | 1-102 | 1-85 | 1-48 | | | | 2-156 | | 3-220 | 4-112 |
| | | Fix. Fla. | 469 | 258 | 8 | 8 | 1-11 | 1-131 | 1-258 | - | 1-120 | 1-85 | 1-50 | 2-436 | | 3-721 | | | | 4-84 |
| | **9** | Flo. Fla. | 755 | 318 | 10 | 10 | 1-14 | 1-145 | - | 1-318 | 1-115 | 1-94 | 1-55 | | | | 2-169 | | 3-290 | 4-140 |
| | | Fix. Fla. | 541 | 360 | 5 | 5 | 1-14 | 1-145 | 1-360 | - | 1-134 | 1-94 | 1-60 | 2-467 | | 3-890 | | | | 4-105 |

*Table 3. Sequence dependent setup times and workers for cell 1*

| Cell1 | 1 | 2 | 3 | 4 | 5 | 6 | 7 | 8 |
|---|---|---|---|---|---|---|---|---|
| 1 | - | 210-1 | 525-2 | 238-1 | 564-2 | 265-1 | 605-2 | 294-1 |
| 2 | 564-2 | - | 549-2 | 278-1 | 556-2 | 286-1 | 582-2 | 300-1 |
| 3 | 525-2 | 215-1 | - | 212-1 | 542-2 | 230-1 | 580-2 | 267-1 |
| 4 | 564-2 | 278-1 | 549-2 | - | 556-2 | 280-1 | 582-2 | 295-1 |
| 5 | 564-2 | 232-1 | 542-2 | 225-1 | - | 214-1 | 562-2 | 245-1 |
| 6 | 564-2 | 286-1 | 549-2 | 280-1 | 556-2 | - | 582-2 | 286-1 |
| 7 | 605-2 | 248-1 | 580-2 | 234-1 | 562-2 | 220-1 | - | 232-1 |
| 8 | 540-2 | 300-1 | 549-2 | 295-1 | 556-2 | 286-1 | 582-2 | - |

*Table 4. Trait levels and capabilities for each worker*

| | | Traits | | | | Machine Types | | | | | | | | | | | | | |
|---|---|---|---|---|---|---|---|---|---|---|---|---|---|---|---|---|---|---|---|
| | | 1 | 2 | 3 | 4 | 1 | 2 | 3 | 4 | 5 | 6 | 7 | 8 | 9 | 10 | 11 | 12 | 13 | 14 |
| Workers | 1 | 1 | 1 | 2 | 2 | 1 | | 1 | 1 | 1 | 1 | 1 | | | 1 | 1 | | | 1 |
| | 2 | 3 | 3 | 2 | 2 | 1 | | 1 | 1 | 1 | 1 | 1 | | | 1 | 1 | | | 1 |
| | 3 | 2 | 3 | 2 | 2 | | 1 | | | | | | 1 | 1 | | | 1 | 1 | |
| | 4 | 1 | 2 | 2 | 3 | 1 | | | | 1 | 1 | 1 | | | | | | | 1 |
| | 5 | 3 | 2 | 2 | 1 | | 1 | | | | | | 1 | 1 | | | 1 | 1 | |
| | 6 | 1 | 3 | 2 | 2 | | 1 | | | | | | 1 | 1 | | | 1 | 1 | |
| | 7 | 2 | 2 | 2 | 3 | 1 | | 1 | | 1 | 1 | 1 | | | 1 | | | | 1 |
| | 8 | 1 | 2 | 2 | 2 | 1 | 1 | | 1 | 1 | 1 | 1 | | | | 1 | | | 1 |
| | 9 | 2 | 1 | 3 | 2 | 1 | 1 | 1 | | 1 | 1 | 1 | | | 1 | | | | 1 |
| | 10 | 2 | 2 | 2 | 2 | 1 | 1 | 1 | 1 | 1 | 1 | 1 | | | 1 | 1 | | | 1 |
| | 11 | 2 | 2 | 2 | 2 | 1 | | 1 | 1 | 1 | 1 | 1 | | | 1 | 1 | | | 1 |
| | 12 | 2 | 3 | 2 | 2 | 1 | | 1 | 1 | 1 | 1 | 1 | | | 1 | 1 | | | 1 |

Figure 12 evaluates three different worker assignment strategies using two variables. The variables used are ALTAMLT (Equation 57) and the number of utility workers.

In the first strategy, both of the deviational variables *d1* and *d2* are used to determine the length of the sub-chromosome ($x_{i,k}+d1_{i,k}-d2_{i,k}$). In the second strategy, only the deviational variable *d2* is used to determine the length of the sub-chromosome ($x_{i,k}-d2_{i,k}$). The third strategy represents the current situation in real life manufacturing system.

In the first strategy, the goal is to make the number of workers assigned to operations equal to $x_{i,k}$. In the second strategy, the goal is to make the number of workers assigned to operations equal to or smaller than $x_{i,k}$.

Figure 12 shows the worker assignment decisions. The first strategy outperforms the second strategy in terms of the ALTAMLT. Regarding the number of utility workers, the first and the second strategies are equal. These results indicate that changing the $x_{i,k}$ values obtained in the first stage of the proposed

*Figure 9. Scheduling and worker timetabling*

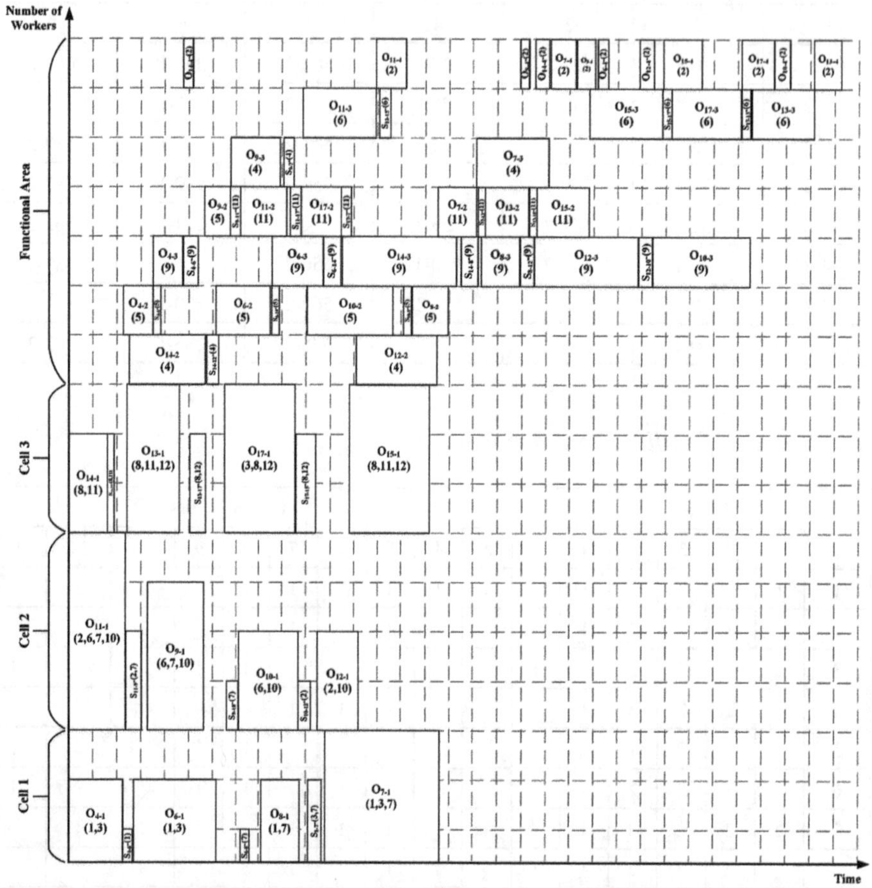


algorithm should be avoided. They also show that an effective schedule is created in the scheduling stage. What is more, the first two strategies outperform the third strategy regarding the number of utility workers.

## CONCLUSION

Studies on HMS offer a more realistic and comprehensive view of manufacturing systems. In particular, the use of holistic methodologies in HMS studies offer solutions to many problems encountered in real manufacturing systems. With these considerations in mind, instead of solving the order release, the batch scheduling, and the worker assignment problems independently, they all became the levels of the proposed holistic methodology in this study.

This study makes a valuable contribution to the literature, primarily, by proposing a holistic methodology. A second contribution is that it develops novel mathematical models, algorithms and decisions rules, and applies them to novel problems.

In this chapter, a holistic methodology is proposed to find solutions to the OR, the batch scheduling and the worker assignment problems in HMS. Integration of the mathematical models developed for these problems with the proposed methodology allows examining the problems in relation to one another.

*Figure 10. Average labor time of workers in average manufacturing lead time*

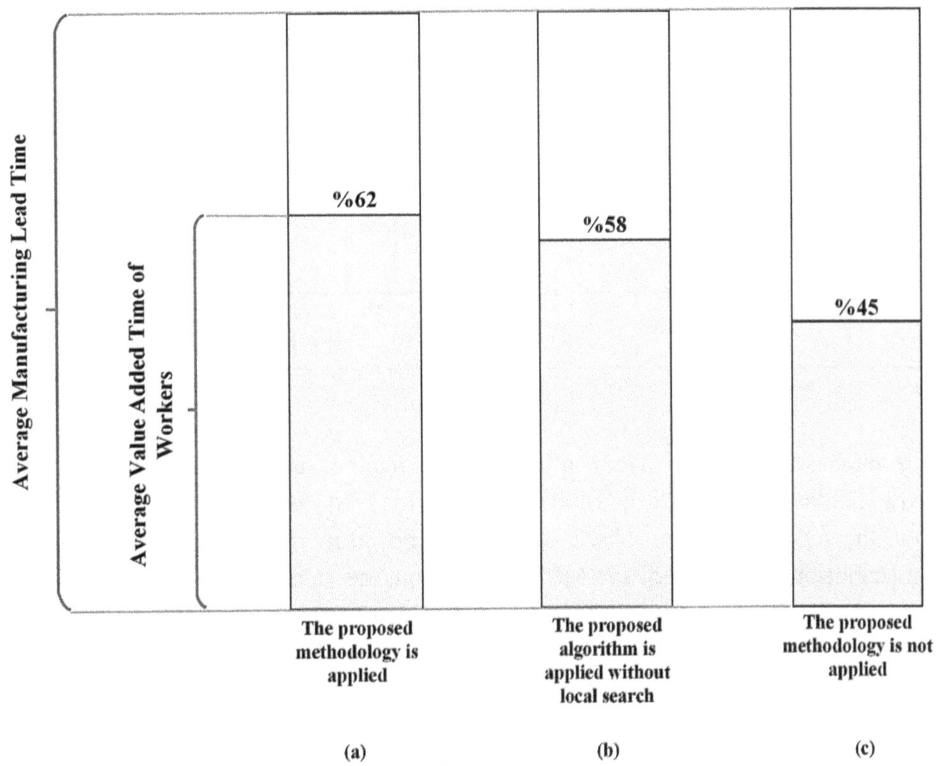


*Figure 11. Sensitivity analysis 1*

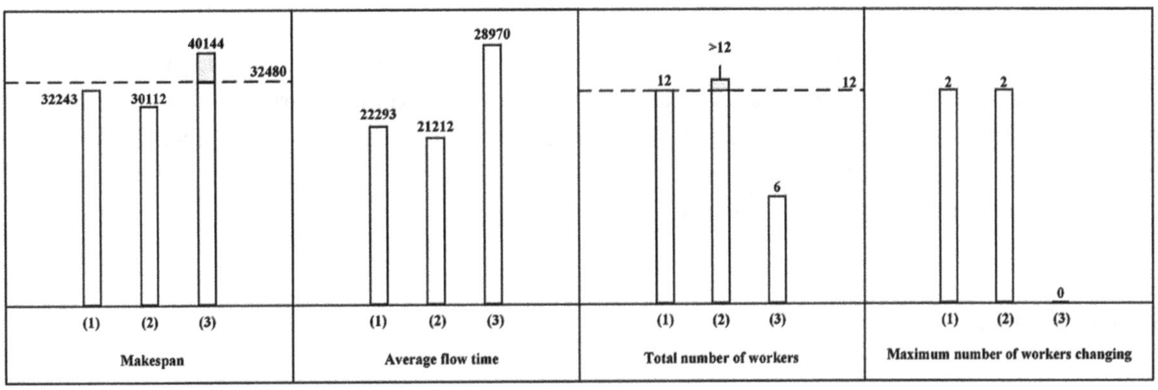


The integration mechanism developed makes it possible to combine scheduling and worker assignment problems. Since the proposed algorithm is FastGPA-based, effective and efficient solutions are found for the problems regarding the quality and the computational time.

The proposed methodology aims to maximize the number of parts produced without adding new workers while keeping MLT under control. Therefore, the WLC mechanism is initiated in the OR stage of the proposed methodology and is also kept running during the other stages. In the OR stage, custom-

*Figure 12. Sensitivity analysis 2*

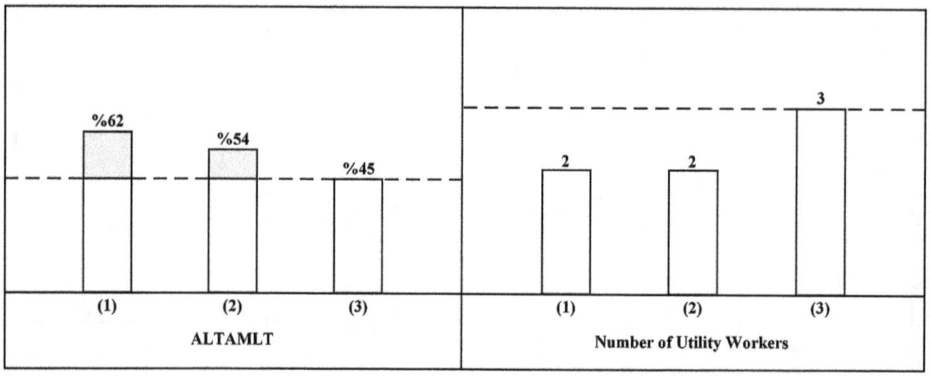


ers' orders are released to the shop floor in a controlled manner, and in the scheduling stage, batches are scheduled to achieve a balanced distribution of workload. In the worker assignment stage, workers are assigned to the scheduled operations. Use of the proposed methodology reduces congestion, WIP, unbalanced distribution of workload, and MLT. In addition, the reduction in the number of utility workers and the creation of effective teams in cells prevent quality problems and improve worker efficiency.

Because the problems handled in the second and third stages of the methodology are multi-objective, the methodology employs a Pareto front structure. The methodology makes it possible to obtain feasible non-dominated solutions by the end of the third stage, and a preferred solution is selected from the alternative solutions. To be able to make this selection, decision rules are developed in the fourth stage of the methodology. ALTAMLT value, which is shown in the decision rules, demonstrates the effectiveness of the proposed methodology.

A sensitivity analysis is conducted in the final section of the study to examine the effects of variables *wa*, *wa1* and *wa2* on schedules, and the effects of deviational variables *d1* and *d2* on assignments.

The proposed methodology offers a holistic perspective on the problems at hand, and makes it possible to obtain effective solutions. In future studies, the methodology and the algorithm used in the integration mechanism could be improved considering different meta-heuristic or heuristic methods to obtain more effective solutions. Last but not least, variations of the mathematical models could be developed and implemented for different manufacturing systems.

## REFERENCES

Aglan, C., & Durmusoglu, M. B. (2015). Lot-splitting approach of a hybrid manufacturing system under CONWIP production control: A mathematical model. *International Journal of Production Research*, *53*(5), 1561–1583. doi:10.1080/00207543.2014.957873

Akgündüz, O. S., & Tunalı, S. (2010). An adaptive genetic algorithm approach for the mixed-model assembly line sequencing problem. *International Journal of Production Research*, *48*(17), 5157–5179. doi:10.1080/00207540903117857

Ashby, J. R., & Uzsoy, R. (1995). Scheduling and order release in a single-stage production system. *Journal of Manufacturing Systems*, *14*(4), 290–306. doi:10.1016/0278-6125(95)98881-6

Askin, R. G., & Huang, Y. (2001). Forming effective worker teams for cellular manufacturing. *International Journal of Production Research*, *39*(11), 2431–2451. doi:10.1080/00207540110040466

Azadeh, A., Sheikhalishahi, M., & Koushan, M. (2015). An integrated fuzzy DEA–Fuzzy simulation approach for optimization of operator allocation with learning effects in multi products CMS. *Applied Mathematical Modelling*, *37*(24), 9922–9933. doi:10.1016/j.apm.2013.05.039

Balaji, A. N., & Porselvi, S. (2014). Artificial Immune System Algorithm and Simulated Annealing Algorithm for Scheduling Batches of Parts based on Job Availability Model in a Multi-cell Flexible Manufacturing System. *Procedia Engineering*, *97*, 1524–1533. doi:10.1016/j.proeng.2014.12.436

Betrand, J. W. M., & Van Ooijen, H. P. G. (2008). Optimal work order release for make-to-order job shops with customer order lead-time costs, tardiness costs and work-in-process costs. *International Journal of Production Economics*, *116*(2), 233–241. doi:10.1016/j.ijpe.2008.08.055

Beyer, H. G., & Deb, K. (2001). On self-adaptive features in real-parameter evolutionary algorithms. *Evolutionary Computation. IEEE Transactions on*, *5*(3), 250–270.

Burgess, A. G., Morgan, I., & Vollmann, T. E. (1993). Cellular manufacturing: Its impact on the total factory. *International Journal of Production Research*, *31*(9), 2059–2077. doi:10.1080/00207549308956844

Celano, G., Costa, A., & Fichera, S. (2008). Scheduling of unrelated parallel manufacturing cells with limited human resources. *International Journal of Production Research*, *46*(2), 405–427. doi:10.1080/00207540601138452

Cesaní, V. I., & Steudel, H. J. (2005). A study of labor assignment flexibility in cellular manufacturing systems. *Computers & Industrial Engineering*, *48*(3), 571–591. doi:10.1016/j.cie.2003.04.001

Cevikcan, E., & Durmusoglu, M. B. (2011). Minimising utility work and utility worker transfers for a mixed-model assembly line. *International Journal of Production Research*, *49*(24), 7293–7314. doi:10.1080/00207543.2010.537385

Cevikcan, E., & Durmusoglu, M. B. (2014). An integrated job release and scheduling approach on parallel machines: An application in electric wire-harness industry. *Computers & Industrial Engineering*, *76*, 318–332. doi:10.1016/j.cie.2014.08.012

Choi, K., Kim, S., Lee, H., & Kwon, I. (1997). An operation scheme for make-to-order job-shop production systems. *Computers & Industrial Engineering*, *33*(3), 765–768. doi:10.1016/S0360-8352(97)00248-9

Chutima, P., & Naruemitwong, W. (2014). A Pareto biogeography-based optimisation for multi-objective two-sided assembly line sequencing problems with a learning effect. *Computers & Industrial Engineering*, *69*, 89–104. doi:10.1016/j.cie.2014.01.001

Das, S. R., & Canel, C. (2005). An algorithm for scheduling batches of parts in a multi-cell flexible manufacturing system. *International Journal of Production Economics*, *97*(3), 247–262. doi:10.1016/j.ijpe.2004.07.006

Deb, K., Agrawal, S., Pratap, A., & Meyarivan, T. (2002). A fast elitist non-dominated sorting genetic algorithm for multi-objective optimization: NSGA-II. In *Proceedings of the International Conference on Parallel Problem Solving From Nature* (pp. 849-858). Springer Berlin Heidelberg.

Durmusoglu, M. B., & Kaya, G. (2012). Lean Thinking Based Investment Planning at Design Stage of Cellular/Hybrid Manufacturing Systems. In *Industrial Engineering: Concepts, Methodologies, Tools, and Applications: Concepts, Methodologies, Tools, and Applications* (p. 409).

Durmusoglu, M. B., & Satoglu, S. I. (2011). Axiomatic design of hybrid manufacturing systems in erratic demand conditions. *International Journal of Production Research, 49*(17), 5231–5261. doi:10.1080/00207543.2010.510487

Egilmez, G., Erenay, B., & Süer, G. A. (2014). Stochastic skill-based manpower allocation in a cellular manufacturing system. *Journal of Manufacturing Systems, 33*(4), 578–588. doi:10.1016/j.jmsy.2014.05.005

Elmi, A., Solimanpur, M., Topaloglu, S., & Elmi, A. (2011). A simulated annealing algorithm for the job shop cell scheduling problem with intercellular moves and reentrant parts. *Computers & Industrial Engineering, 61*(1), 171–178. doi:10.1016/j.cie.2011.03.007

Eskandari, H., Geiger, C., & Lamont, G. (2007). FastPGA: A dynamic population sizing approach for solving expensive multiobjective optimization problems. In Evolutionary Multi-Criterion Optimization (pp. 141-155). Springer Berlin/Heidelberg. doi:10.1007/978-3-540-70928-2_14

Eskandari, H., & Geiger, C. D. (2008). A fast Pareto genetic algorithm approach for solving expensive multiobjective optimization problems. *Journal of Heuristics, 14*(3), 203–241. doi:10.1007/s10732-007-9037-z

Fan, J., Cao, M., & Feng, D. (2010, June). Multi-objective dual resource-constrained model for cell formation problem. In *Proceedings of the 2010 IEEE International Conference on Management of Innovation and Technology (ICMIT)* (pp. 1031-1036). IEEE. doi:10.1109/ICMIT.2010.5492881

Feyzioglu, O., & Pierreval, H. (2009). Hybrid organization of functional departments and manufacturing cells in the presence of imprecise data. *International Journal of Production Research, 47*(2), 343–368. doi:10.1080/00207540802425898

Fowler, J. W., Hogg, G. L., & Mason, S. J. (2002). Workload control in the semiconductor industry. *Production Planning and Control, 13*(7), 568–578. doi:10.1080/0953728021000026294

Fowler, J. W., Wirojanagud, P., & Gel, E. S. (2008). Heuristics for workforce planning with worker differences. *European Journal of Operational Research, 190*(3), 724–740. doi:10.1016/j.ejor.2007.06.038

Giard, V., & Jeunet, J. (2010). Optimal sequencing of mixed models with sequence-dependent setups and utility workers on an assembly line. *International Journal of Production Economics, 123*(2), 290–300. doi:10.1016/j.ijpe.2009.09.001

Gravel, M., Price, W., & Gagne, C. (2000). An interactive tool for designing manufacturing cells for an assembly job-shop. *International Journal of Production Research, 38*(2), 309–322. doi:10.1080/002075400189428

Hamdan, M. (2012). On the disruption-level of polynomial mutation for evolutionary multi-objective optimisation algorithms. *Computing and Informatics*, *29*(5), 783–800.

Harhalakis, G., Lu, T., Minis, I., & Nagi, R. (1996). A practical method for design of hybrid-type production facilities. *International Journal of Production Research*, *34*(4), 897–918. doi:10.1080/00207549608904942

Haskose, A., Kingsman, B. G., & Worthington, D. (2004). Performance analysis of make-to-order manufacturing systems under different workload control regimes. *International Journal of Production Economics*, *90*(2), 169–186. doi:10.1016/S0925-5273(03)00052-5

Hyun, C. J., Kim, Y., & Kim, Y. K. (1998). A genetic algorithm for multiple objective sequencing problems in mixed model assembly lines. *Computers & Operations Research*, *25*(7), 675–690. doi:10.1016/S0305-0548(98)00026-4

Ioannou, G. (2006). Time-phased creation of hybrid manufacturing systems. *International Journal of Production Economics*, *102*(2), 183–198. doi:10.1016/j.ijpe.2005.03.005

Kher, H. V., & Jensen, J. B. (2002). Shop Performance Implications of Using Cells, Partial Cells, and Remainder Cells*. *Decision Sciences*, *33*(2), 161–190. doi:10.1111/j.1540-5915.2002.tb01641.x

Kim, S., & Jeong, B. (2007). Product sequencing problem in Mixed-Model Assembly Line to minimize unfinished works. *Computers & Industrial Engineering*, *53*(2), 206–214. doi:10.1016/j.cie.2007.06.011

Kolbe, K. Pure Instinct, 1994.

Kotani, S., Ito, T., & Ohno, K. (2004). Sequencing problem for a mixed-model assembly line in the Toyota production system. *International Journal of Production Research, 42*(23), 4955-4974.

Liao, C. J. (1992). Optimal control of jobs for production systems. *Computers & Industrial Engineering*, *22*(2), 163–169. doi:10.1016/0360-8352(92)90043-J

Moreira, M. R. A., & Alves, R. A. F. (2009). A methodology for planning and controlling workload in a job-shop: A four-way decision-making problem. *International Journal of Production Research*, *47*(10), 2805–2821. doi:10.1080/00207540701725083

Murthy, C. V., & Srinivasan, G. (1995). Fractional cell formation in group technology. *International Journal of Production Research*, *33*(5), 1323–1337. doi:10.1080/00207549508930212

Niakan, F., Baboli, A., Moyaux, T., & Botta-Genoulaz, V. (2016). A bi-objective model in sustainable dynamic cell formation problem with skill-based worker assignment. *Journal of Manufacturing Systems*, *38*, 46–62. doi:10.1016/j.jmsy.2015.11.001

Norman, B. A., Tharmmaphornphilas, W., Needy, K. L., Bidanda, B., & Warner, R. C. (2002). Worker assignment in cellular manufacturing considering technical and human skills. *International Journal of Production Research*, *40*(6), 1479–1492. doi:10.1080/00207540110118082

Ponnambalam, S. G., Aravindan, P., & Rao, M. S. (2003). Genetic algorithms for sequencing problems in mixed model assembly lines. *Computers & Industrial Engineering*, *45*(4), 669–690. doi:10.1016/j.cie.2003.09.001

Rogers, P., & Nandi, A. (2007). Judicious order acceptance and order release in make-to-order manufacturing systems. *Production Planning and Control*, *18*(7), 610–625. doi:10.1080/09537280701582422

Satoglu, S. I., Durmusoglu, M. B., & Ertay, T. (2010). A mathematical model and a heuristic approach for design of the hybrid manufacturing systems to facilitate one-piece flow. *International Journal of Production Research*, *48*(17), 5195–5220. doi:10.1080/00207540903089544

Satoglu, S. I., & Suresh, N. C. (2009). A goal-programming approach for design of hybrid cellular manufacturing systems in dual resource constrained environments. *Computers & Industrial Engineering*, *56*(2), 560–575. doi:10.1016/j.cie.2008.06.009

Shambu, G., & Suresh, N. C. (2000). Performance of hybrid cellular manufacturing systems: A computer simulation investigation. *European Journal of Operational Research*, *120*(2), 436–458. doi:10.1016/S0377-2217(98)00371-3

Süer, G. A., & Alhawari, O. (2011). Operator assignment decisions in a highly dynamic cellular environment. In *Operations Management Research and Cellular Manufacturing Systems: Innovative Methods and Approaches: Innovative Methods and Approaches* (p. 258).

Süer, G. A., & Dagli, C. (2005). Intra-cell manpower transfers and cell loading in labor-intensive manufacturing cells. *Computers & Industrial Engineering*, *48*(3), 643–655. doi:10.1016/j.cie.2003.03.006

Süer, G. A., & Tummaluri, R. R. (2008). Multi-period operator assignment considering skills, learning and forgetting in labour-intensive cells. *International Journal of Production Research*, *46*(2), 469–493. doi:10.1080/00207540601138551

Suresh, N. C. (1991). Partitioning work centres for group technology: Insights from an analytical model. *Decision Sciences*, *22*(4), 772–791. doi:10.1111/j.1540-5915.1991.tb00364.x

Tavakkoli-Moghaddam, R., & Rahimi-Vahed, A. R. (2006). Multi-criteria sequencing problem for a mixed-model assembly line in a JIT production system. *Applied Mathematics and Computation*, *181*(2), 1471–1481. doi:10.1016/j.amc.2006.02.033

Thürer, M., Silva, C., Stevenson, M., & Land, M. (2014). Controlled order release: A performance assessment in job shops with sequence-dependent set-up times. *Production Planning and Control*, *25*(7), 603–615. doi:10.1080/09537287.2012.735801

Torabi, S. A., & Amiri, A. S. (2012). A possibilistic approach for designing hybrid cellular manufacturing systems. *International Journal of Production Research*, *50*(15), 4090–4104. doi:10.1080/00207543.2011.590827

Tsai, L. H. (1995). Mixed-model sequencing to minimize utility work and the risk of conveyor stoppage. *Management Science*, *41*(3), 485–495. doi:10.1287/mnsc.41.3.485

Venkataramanaiah, S., & Krishnaiah, K. (2002). Hybrid heuristic for design of cellular manufacturing systems. *Production Planning and Control*, *13*(3), 274–283. doi:10.1080/09537280110073978

Viguier, F., & Pierreval, H. (2004). An approach to the design of a hybrid organization of workshops into functional layout and group technology cells. *International Journal of Computer Integrated Manufacturing*, *17*(2), 108–116. doi:10.1080/09511920310001593092

Yilmaz, O. F., Cevikcan, E., & Durmusoglu, M. B. (2016). Scheduling batches in multi hybrid cell manufacturing system considering worker resources: A case study from pipeline industry. *Advances in Production Engineering & Management*, *11*(3), 192–206. doi:10.14743/apem2016.3.220

Yilmaz, O. F., & Durmusoglu, M.B. (in press). A Multi-objective scheduling of hybrid manufacturing systems with walking workers.

Yilmaz, O. F., & Erbiyik, H. (2016). *Effective Applications of Optimization Methods in the Manufacturing Environment in Turkey. Comparative Economics and Regional Development in Turkey*. Hershey, PA: Business Science Reference; doi:10.4018/978-1-4666-8729-5.ch015

Zolfaghari, S., & Lopez Roa, E. V. (2006). Cellular manufacturing versus a hybrid system: A comparative study. *Journal of Manufacturing Technology Management*, *17*(7), 942–961. doi:10.1108/17410380610688250

## KEY TERMS AND DEFINITIONS

**Batch Scheduling in Hybrid Manufacturing:** A type of operations research problem involving sequencing and scheduling of batches on cells and machines in functional area.

**Evolutionary Algorithm:** An evolutionary algorithm is inspired by biological evolution and uses crossover, mutation, selection mechanisms.

**Hybrid Manufacturing System:** A manufacturing system that contains both cells and a functional area. It provides flexibility and allows the production of many kinds of products.

**Multi-Objective Evolutionary Algorithm:** A multi-objective evolutionary algorithm uses Pareto front and deals with multiple objective functions simultaneously.

**Multi-Objective Optimization:** A topic of multiple-criteria decision-making dealing with mathematical optimization problems which consists of more than one objective to be optimized at the same time.

**Order Release:** Batches are sent to manufacturing system for processing.

**Pareto Front:** The set that corresponds to Pareto set and is composed of all Pareto optimal decision vectors is called Pareto front.

**Pareto Optimal Solution:** A solution is called Pareto optimal, if none of the objective functions can be improved without degrading some of the other objective values.

**Worker Assignment:** Assignment of employees to operations for the processing of parties.

# Chapter 8
# Evolutionary Algorithms for Multi-Objective Scheduling in a Hybrid Manufacturing System

**Ömer Faruk Yılmaz**
*Istanbul Technical University, Turkey & Yalova University, Turkey*

**Mehmet Bülent Durmuşoğlu**
*Istanbul Technical University, Turkey*

## ABSTRACT

*Problems encountered in real manufacturing environments are complex to solve optimally, and they are expected to fulfill multiple objectives. Such problems are called multi-objective optimization problems(MOPs) involving conflicting objectives. The use of multi-objective evolutionary algorithms (MOEAs) to find solutions for these problems has increased over the last decade. It has been shown that MOEAs are well-suited to search solutions for MOPs having multiple objectives. In this chapter, in addition to comprehensive information, two different MOEAs are implemented to solve a MOP for comparison purposes. One of these algorithms is the non-dominated sorting genetic algorithm (NSGA-II), the effectiveness of which has already been demonstrated in the literature for solving complex MOPs. The other algorithm is fast Pareto genetic algorithm (FastPGA), which has population regulation operator to adapt the population size. These two algorithms are used to solve a scheduling problem in a Hybrid Manufacturing System (HMS). Computational results indicate that FastPGA outperforms NSGA-II.*

## INTRODUCTION

There are two important difficult to handle features of real-world problems. One is the size of the problem which is quite larger than the hypothetical problems and the other is presence of hard constraints to be satisfied. Therefore, finding solutions to such problems through classical optimization methods is difficult regarding the excessive computational time. It may take hours or even days to find a feasible solution to such problems. Metaheuristic or problem-specific heuristic methods are proposed to solve these types of problems. In this study, the evolutionary algorithms (EAs) that provide good solutions especially to

DOI: 10.4018/978-1-5225-2944-6.ch008

large-sized problems are used (Das and Panigrahi, 2009). EAs are stochastic population-based algorithms which can be implemented to many real-world problems easily and inspire their structures from the original mechanism of nature. EAs are the most utilized population-based algorithms and have proved their effectiveness in many fields such as continuous or combinatorial optimization, system modeling and identification, planning and control, engineering design, data mining and machine learning (Talbi, 2009).

There are different optimization methods such as goal programming and variations, and min-max algorithms, which are used for solving multi-objective optimization problems. However, it is seen that multi-objective evolutionary algorithms have been extensively used in recent years including this study (Yilmaz and Erbiyik, 2016; Zhou et al., 2011; Tang and Wang, 2013). Moreoever, multi-objective evolutionary algorithms have proved their effectiveness to solve MOPs. (Yilmaz and Erbiyik, 2016; Deb, 2001; Deb et al., 2002; Zhou et al., 2011; Tang and Wang, 2013; Eskandari and Geiger, 2008).

It must also be noted that MOEAs are different than single-objective optimization algorithms regarding the fitness function, diversity preserving, and elitism and these ensure that many missing points faced in single-objective optimization algorithms are fulfilled (Tang and Wang, 2013).

In this study, two different MOEAs (NSGA-II and FastPGA) are used to solve multi-objective scheduling problem in HMS and their solutions are compared with one another. In this context, five different hypothetical cases are designed, each having different problem sizes. The NSGA-II algorithm has already demonstrated its effectiveness in many expensive MOPs. That is why it is considered suitable for MOPs. The FastPGA algorithm, on the other hand, is used in this study because it employs the adaptive population sizing strategy, which is known as effective to solve expensive MOPs.

One of the most important points of the current study is the multi-objective mathematical model that was developed for these types of problems (Yılmaz and Durmusoglu, 2017). In addition, the similarity of HMSs with real manufacturing environments increases the applicability of both the developed mathematical model and the study results in industrial applications. In this respect, the current study will also contribute to industrial applications. The developed mathematical model within this study may also be adapted and used for other manufacturing systems, including assembly lines, and cellular manufacturing systems (CMS).

The rest of the study is organized as follows: In the second section, the evolutionary algorithms for multi-objective optimization problems are given. In the third section, literature review is presented for the related studies. The fourth section covers the definition of the problem and its mathematical model. The detailed structures of MOEAs are addressed in the fifth part. The algorithms are run through hypothetical examples and the NSGA-II and the FastPGA are compared in the sixth section. The conclusions and future research directions are given in the sixth section.

## EVOLUTIONARY ALGORITHM FOR MULTI-OBJECTIVE OPTIMIZATION

### Multi-Objective Optimization

MOPs involve two or more conflicting objectives. That is why Pareto optimality and dominance approaches are utilized. A MOP can be expressed as follows:

Minimize $f(x) = \left[f_1(x), \ldots f_m(x)\right]^T$ (1)

$$\text{s.t } g_i(x) \geq 0 \left(i = 1, 2, ..., n\right) \tag{2}$$

$$h_i(x) = 0 \left(i = 1, 2, ..., p\right) \tag{3}$$

$$x \in \Omega \tag{4}$$

Expression $\Omega$ in the fourth equation is a decision space and $x \in \Omega$ is a decision variable vector. The objectives conflict and the first equation has *m* different objective function. While the second equation expresses the inequality constraints, the third one expresses the equality constraints.

Let us assume that $x = \left[x_1, ..., x_m\right]^T$ and $y = \left[y_1, ..., y_m\right]^T$ are two separate decision variable vectors. If the conditions $f_i\left(x\right) \leq f_i\left(y\right)\ \forall i \in \left\{1, ..., m\right\}$ and $f_j\left(x\right) < f_j\left(y\right)\ \exists j \in \left\{1, ..., m\right\}$ are satisfied, it can be said that $x$ dominates $y$ and symbolically shown as $x \prec y$. If there are no vectors to satisfy the condition $x \prec x^*$ for the decision variable vector $x$, vector $x^*$ is Pareto optimal. The set composed of all Pareto optimal vectors are called Pareto set and expressed with $P$. The set that corresponds to Pareto set and is composed of all Pareto optimal decision vectors is called Pareto front and expressed with $PF$. Pareto front is the appearance of Pareto set within the *m*-size objective function space. Four kinds of Pareto front structures are shown in Figure 1 (Luke, 2013).

The Pareto front can inherently be convex or concave as well as non-convex. In addition, the Pareto front can be discontinuous in cases where there are areas that cannot be accessed by individuals on the front.

There are two goals that multi-objective optimization algorithms try to achieve. These are convergence and diversity. The results obtained through the algorithm cannot possibly reflect the true Pareto optimal set or the whole Pareto optimal front. Konak et al. (2006) indicated and detailed three important features which multi-objective algorithms are supposed to provide.

- The best known Pareto front should be as close as possible to the true Pareto front. An ideal situation is that the best know Pareto set is the subset of the Pareto optimal set.
- The solutions in the best-known Pareto set should be distributed properly on the Pareto front. This way, the true picture of trade-offs can be seen in a better way.
- The best-known Pareto front should cover the area covered by the true Pareto front. This ensures that extreme solutions within the objective function space can be evaluated.

While the first one of these items is more related to the convergence, the other two items are closely related to the diversity. Figure 2 shows different cases concerning the concept of convergence and diversity.

Two different structures introduced by Goldberg (1989) are used to investigate the convergence of the algorithms. The first one is domination and the sequence of a solution is determined by how many solutions it is dominated with. The second approach is non-dominated sorting. While any solution has a sequence, the approach in which solutions with the same rank do not dominate each other is used. Hence, the solutions that do not dominate each other have the same rank when an evaluation is made for each solution.

*Figure 1. Four kinds of Pareto fronts (Luke, 2013)*

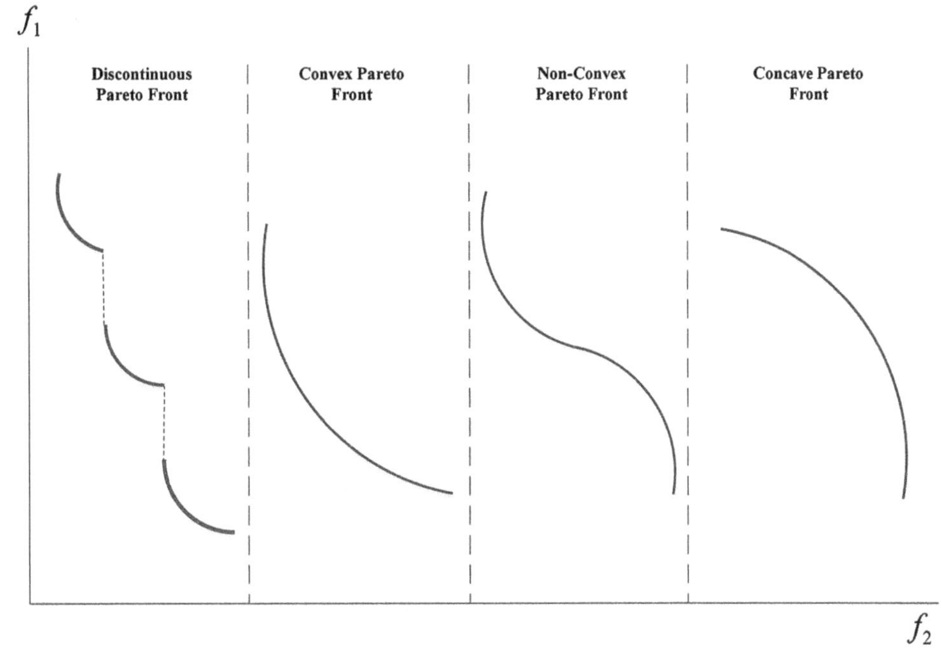


*Figure 2. Convergence and diversity concept in multi-objective optimization algorithms*

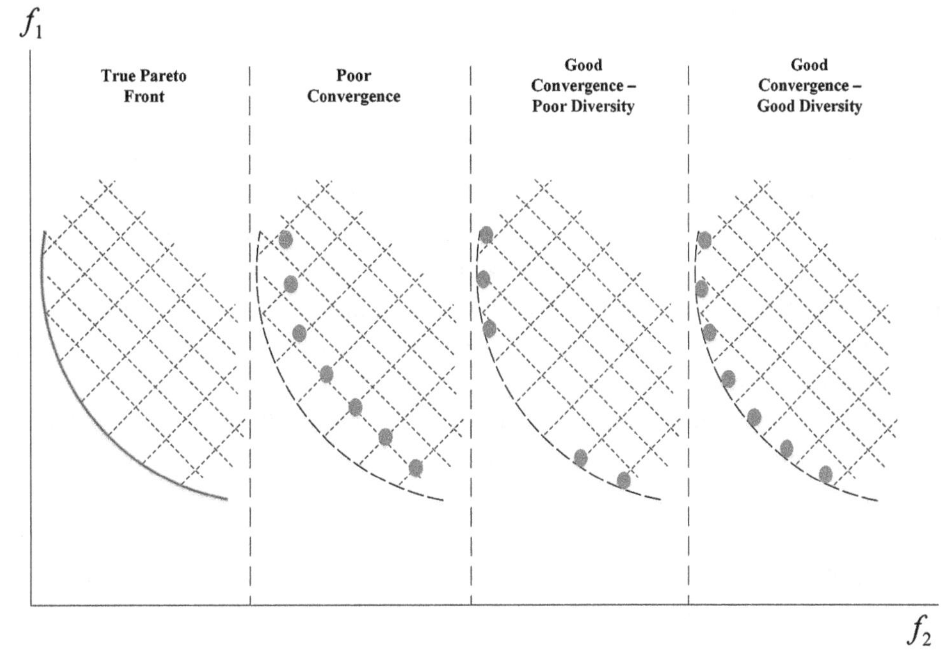

Three different structures are used while evaluating the algorithms in terms of diversity. One of them is bounding hypercube (Deb et al., 2002). This structure is based on the crowding distance calculation. The distance of any solution to its two neighbors is compute and added for each objective function, and therefore the distance value is obtained for that solution. The distance value is accepted as being infinite for solutions on the corners on the Pareto front. The high crowding distance value provides priority for that solution. In the second approach (Knowles and Corne, 2000), a hyper grid is created through the objective function and a density calculation is made. The third approach (Zitzler et al., 2002) is based on determining each solution's $k^{th}$ nearest neighbor. More detailed information on these approaches can be found in the related references and in the study by Das and Panigrahi (2009).

## Most Known Multi-Objective Evolutionary Algorithms

In this section, the multi-objective evolutionary algorithms commonly used in the literature are presented. Since these algorithms have proven their efficiency, they have been used in many studies as comparison algorithms. The use of multi-objective evolutionary algorithms in the literature was provided extensively in the study by Zhou et al. (2011).

The NSGA-II (Deb et al., 2002) algorithm promotes the solutions in each iteration by using two $N$-size populations, namely $P$ and $Q$, simultaneously. While population $P$ provides elite preservation, population $Q$ which is formed as a result of selection, crossover, and mutation facilitates new solutions. The use of crowding distance only in the objective space can be regarded as the disadvantage of this algorithm. SPEA2 (Strength Pareto Evolutionary Approach) was developed by Zitzler et al. (2002) and is quite similar to the NSGA-II algorithm. MOPSO (Multi-objective particle swarm optimization) was developed by Coello et al. (2004). It is an algorithm based on the Particle Swarm Optimization (PSO) and developed to find solutions to multi-objective optimization problems. PAES (The Pareto Archive Evolutionary Strategy) is an algorithm which is easy to implement and not computationally expensive, however, its performance depends on the cell size which can be considered its disadvantage (Knowles and Corne, 2000). PESA (Pareto Envelope based Selection Algorithm) has similarites to PAES in terms of advantages and disadvantages (Corne et al., 2001). MOGA (Multi-objective Genetic Algorithms) is a genetic-based multi-objective algorithm. Slow convergence can be considered a disadvantage of this algorithm (Fonseca et al., 1993). NPGA (Niched Pareto Genetic Algorithm) has a simple search process with the tournament selection operator. However, its need for an extra parameter for the search process can be accepted as the disadvantage of this algorithm (Horn et al., 1994). As can be seen, multi-objective evolutionary algorithms have their advantages as well as their disadvantages. Konak et al. (2006) and Kanthababu (2013) mentioned the advantages and disadvantages of these algorithms in detail in their studies. In addition, Camara et al. (2013) addressed the comparisons of multi-objective evolutionary algorithms for the optimization problems in their study.

Since the NSGA-II and FastPGA algorithms are used in this study, implementation areas of NSGA-II and FastPGA are presented in the next section.

## LITERATURE REVIEW

The literature is reviewed in two subsections. The first subsection describes studies on HMS and CMS.

There are numerous studies in the literature regarding the NSGA-II algorithm but not all of these are covered in the literature section below. The emphasis is mainly placed on those carried out in recent years. On the other hand, only a few studies involve FastPGA algorithm are found in the literature.

The studies of NSGA-II and FastPGA are addressed in the second subsection.

### Hybrid and Cellular Manufacturing System Studies

In this section, studies using simulation techniques are not covered.

Murthy and Srinivasan (1995) developed a non-linear mathematical model to form an HMS. The objective of this mathematical model is the minimization of inter-cellular movements. Harhalakis et al. (1996) developed a mathematical model with a single objective function to minimize inter-cellular movements. Venkataramanaiah and Krishnaiah (2002) developed a hybrid heuristic for the design of an HMS to minimize the intercell flow. The performance of the developed heuristic was evaluated with standard problems in the literature. Viguier and Pierreval (2004) developed a mathematical model to design HMS, where parts have several routes. The problem was solved by an evolutionary programming approach. Ioannou (2006) proposed a linear mathematical model for redesigning functional layout into HMS, where the layout of each cell is designed. Feyzioglu and Pierreval (2009) formulated the HMS design problem as a constrained fuzzy multi-objective optimization problem and proposed an evolutionary algorithm to solve it. Satoglu and Suresh (2009) proposed a two-stage goal-programming model for the design of an HMS and the worker assignment in a dual resource constraint environment. Optimum solutions were found using GAMS optimization program. Satoglu et al. (2010) developed the mathematical model and a heuristic approach to design HMS and facilitate one-piece flow. Durmusoglu and Kaya (2012) provided a complete methodology using an axiomatic design for the lean thinking based investment planning in the HMS. Aglan and Durmusoglu (2014) developed a linear CONWIP control mathematical model in the case of lot splitting to minimize average flow time in HMS.

Different from the aforementioned studies, Yilmaz and Erbiyik (2016) proposed an effective way to use optimization methods for HMS in their study.

Since CMS underlies the HMS, the sequencing and scheduling studies for cellular manufacturing environments are addressed in this section.

Schaller (2000) proposed efficient heuristic algorithms and lower bounds on the makespan for finding permutation schedule for part families and jobs within each part family in the flowline manufacturing cell where setup times are sequence-dependent in order to minimize the makespan. Das and Canel (2005) developed a branch and bound solution method for the scheduling batches of parts in a flexible manufacturing system consisting of flow-type cells. Tavakkoli-Moghaddam et al. (2008) proposed two evolutionary algorithms (genetic algorithm (GA) and memetic algorithm (MA)) to find optimal permutation schedule for job-shop type cells. They show that the proposed MA outperforms the proposed GA with respect to the average elapsed time to obtain makespan. Venkataramanaiah (2008) developed a simulated annealing based algorithm to minimize the weighted sum of makespan, the flow time and the idle time for the scheduling of flow-shop type manufacturing cell. Tavakkoli-Moghaddam et al. (2010) proposed a non-linear mathematical model to minimize the makespan, intracellular movement, and tardiness for the scheduling of a job-shop type manufacturing cell. A meta-heuristic algorithm based on scatter search (SS) was developed to solve the scheduling problem. Wang et al. (2010) addressed the joint decision problem of cell formation and batch scheduling problem in a cellular manufacturing environment. A non-linear mixed integer mathematical model was developed to minimize the tardiness for scheduling and formation problems. The SS with dispatching rules was proposed to solve the model for the real-world problems. Akturk (2011) developed a multi-objective heuristic algorithm and pricing mechanism for the cell loading, lot size determination, and cell scheduling problems. The linkage between problems is achieved using the developed pricing mechanism. Lin et al. (2011) developed an effective multi-start simulated annealing algorithm to minimize the makespan for the flowline manufacturing cell scheduling problem. Solimanpur and Elmi (2013) developed a mixed integer linear programming model to minimize the makespan for the cell scheduling problem. Nested tabu search (NTS) heuristic algorithm was proposed to solve the formulated problem. Li et al. (2015) proposed hybrid harmony search algorithm to minimize the tardiness and average makespan for multi-objective flowline manufacturing cell scheduling problem. Yilmaz et al. (2016) proposed a mathematical model for the bath scheduling problem in a specific type of CMS. They developed a heuristic method to find solution for the problem.

It is seen in these studies that workers related issues are not included in the scheduling objectives. The objectives used in the study can easily be applied to cellular manufacturing environments easily as well.

## NSGA-II and FastPGA

The NSGA-II algorithm is a metaheuristic algorithm commonly used in the literature. This algorithm has been applied to various fields and some of these studies are given in this section.

NSGA-II algorithm was developed by Deb et al. (2002) to overcome the shortcomings of NSGA and introduced fast non-dominated sorting approach and selection operator. The NSGA-II is one of the widely used MOEAs for the MOPs. Murugan et al. (2009) presented an application of the NSGA-II algorithm to the multi-objective expansion planning problem. Ramesh et al. (2012) presented an application of Modified NSGA-II algorithm for the multi-objective reactive power planning problem. Rabiee et al. (2012) applied the NSGA-II algorithm for solving a bi-objective partial flexible job-shop scheduling problem for comparison purposes. Pires et al. (2012) suggested the NSGA-II algorithm with local search for the multi-objective reactive power compensation problem. Yu et al. (2013) proposed an approach of combining local search into NSGA-II and presented an application to the multi-objective line cell conversion problem. Han et al. (2014) proposed an improved NSGA-II algorithm and applied it to lot streaming flow-shop scheduling problem. Xu et al. (2015) used an improved NSGA-II algorithm for the cross-trained workers scheduling problem.

As can be seen from these studies, the NSGA-II algorithm has been applied to different problem types over the last two decades.

FastPGA algorithm was developed by Eskandari et al. (2007). The FastPGA method introduced population regulation operator and new ranking strategy. Eskandari et al. (2007) compared FastPGA with NSGA-II and computational results for a number of test problem revealed that FastPGA is a promising approach for real-world optimization problems. Iranmanesh et al. (2009) used FastPGA algorithm for the multi-objective problem in project scheduling. Eskandari and Geiger (2009) extended FastPGA approach to solve MOPs in stochastic environments by incorporating a stochastic non-domination based solution ranking procedure.

Both NSGA-II and FastPGA algorithms are adapted for the multi-objective scheduling problem in this study.

## MULTI-OBJECTIVE SCHEDULING PROBLEM

The multi-objective scheduling problem, notations, assumptions and mathematical model are presented in below.

### Problem Definition

The objective is to find sequence of batches that minimize the average flow time of batches (objective 1), the maximum number of workers (objective 2) in the system and the maximum number of workers changing (objective 3) in manufacturing cells. The above three objectives are considered simultaneously since all these measures are important and affect the HMS performance. With the second objective, it is aimed to reduce the total number of workers. The third objective aims to reduce the unnecessary movement such as inter-cellular movement by decreasing the number of workers switching cells. The reason for using these objectives is to overcome the problems faced in a real manufacturing environment.

Figure 3 shows the objectives along with the scheduling horizon. In this figure, $O_{iz}$ represents the $z^{th}$ operation of batch *i*. Figure 3 also illustrates the scheduling for the operation of batches in an HMS consisting of a cell and functional area (the latter comprises two machines). The $O_{11}$ and $O_{12}$ operations are performed on the first cell, while the $O_{31}$ operation is performed in the second cell, and the $O_{41}$ and $O_{51}$ operations are performed on the third cell. The $O_{32}$ operation is performed in the first machine, while the $O_{33}$ operation is performed on the second machine.

The vertical axis shows the number of workers performing operations, while the horizontal axis shows the start and finishing time of each operation and setup. The first two objectives in Figure 3 reveal that the total number of workers can be reduced by compressing the upper part of the vertical axis, while the average flow time can be reduced by compressing the right side of the horizontal axis. Since it is not possible to minimize all three objectives simultaneously, trade-off solutions are proposed to be used instead of a unique solution. That is why we attemp to obtain non-dominated solutions with good convergence and diversity performances.

The following assumptions have been made in this study.

- The order of operations for each batch is predefined.
- Preemption of operations is not allowed.
- Each cell and machine can process only one operation at a time.
- There are no precedence constraints among the operations of different batches.
- Batches are available for processing at time zero.
- The number of workers in cells may change according to time and operation. Each worker has the same multi-skills.
- The setup times are sequence-dependent and symmetrical for both cells and machines.
- The batches pass through the cell towards the functional area. Backflow is not allowed.

The modeling, objectives, and constraint are presented in the following section.

### Notations

The following terms were defined:

*Figure 3. Scheduling operations with objectives*

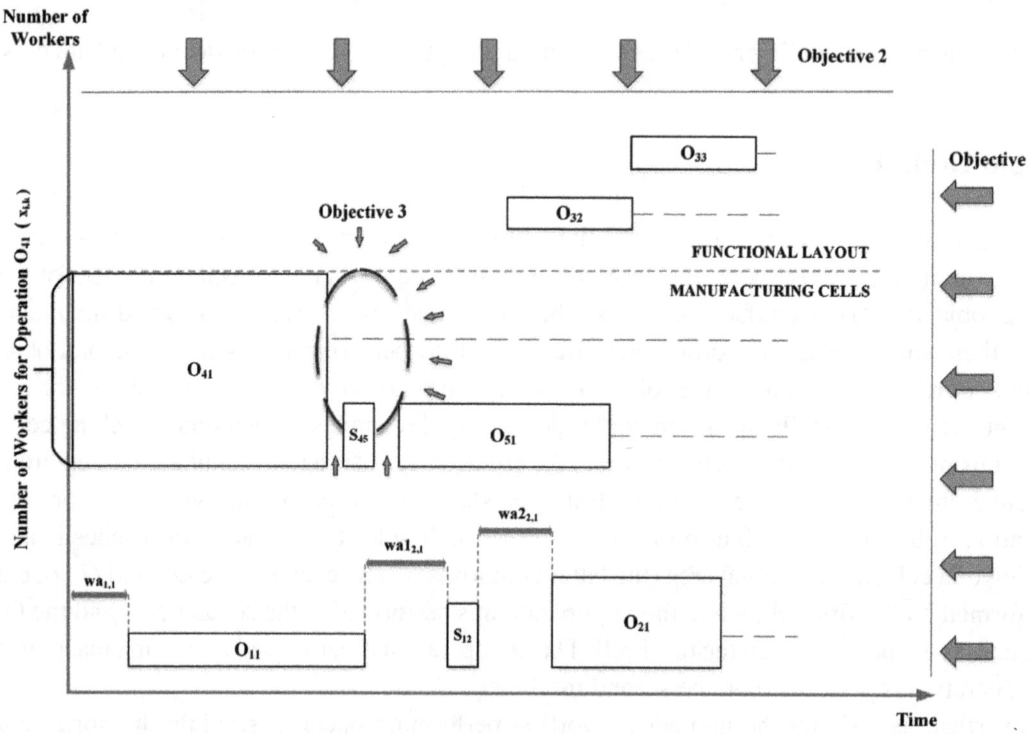


Indices:

*i,j*: Index set of batches (*i,j*=1,…,*N*).
*k,l*: Index set of the cells and the machines in the functional area (*k*,l=1,…,*K*).
*t*: Index set of workloads' changing times (t=1,2)

Parameters:

$cycmin_{i,k}$: Minimum cycle time of batch *i* in cell/machine *k*
$cycmax_{i,k}$: Maximum cycle time of batch *i* in cell/machine *k*
$last_{i,k}$: If the last operation of batch *i* is in cell/machine *k*, 1; if not, 0
$d_{i,l,k}$: If batch *i* is allocated to cell/machine *k* following its operation in the cell/machine l, 1; if not, 0
$ak_{i,k}$: If batch *i* is allocated to cell/machine *k*, 1; if not, 0
$FLT_{i,k}$: Completion time of the first part in the batch *i* in cell/machine *k*
$q_i$: Size of batch *i*
$s_{i,j,k}$: Sequence-dependent setup time created by setup of batch *i* following batch *j* is processed in cell/machine *k*
$setupworker_{i,j,k}$: Sequence-dependent number of workers for the operation of batch *i* after operation of batch *j* is processed in cell/machine *k*
$sw_{i,k}$: Necessary number of workers for the setup of batch *i* in cell/machine *k*

Variables:

$p_{i,k}$: Processing time of batch $i$ in cell/machine $k$
$c_{i,k}$: Completion time of batch $i$ in cell/machine $k$
$setupstart_{i,k}$: Starting time of the setup before operation of batch $i$ in cell/machine $k$
$setupfinish_{i,k}$: Finishing time of the setup before operation of batch $i$ in cell/machine $k$
$time_{j,l,t}$: Starting time of operation/setup of batch $j$ in cell/machine $l$ (number of workers changing time)
$workload_{j,l,t}$: Total number of workers at the start of setup/operation of batch $j$ in cell/machine $l$
$workload1_{j,l,t}$: Total number of workers at the start of operation of batch $j$ in cell/machine $l$
$workload2_{j,l,t}$: Total number of workers at the start of setup of batch $j$ in cell/machine $l$
$cyc_{i,k}$: Cycle time for batch $i$ in cell/machine $k$

Decision Variables:

$b_{i,j,k}$: If batch j precedes batch $i$ in cell/machine $k$, 1; if not, 0
$x_{i,k}$: Number of workers for operation of batch $i$ in cell/machine $k$

## Problem Formulation

Three objectives used in the problem are as follows.
Objective Functions:

$$\text{Minimize } f(x) = \left[f_1(x), f_2(x), f_3(x)\right] \tag{5}$$

$$f_1(x) = \frac{\sum_{i=1}^{N}\sum_{k=1}^{K} c_{i,k} * last_{i,k}}{N} \tag{6}$$

$$f_2(x) = \max_{\forall j,l,t}\left(workload_{j,l,t}\right) \tag{7}$$

$$f_3(x) = \max_{\forall k}\left|\max_{\forall i}\left(x_{i,k}\right) - \min_{\forall j}\left(x_{j,k} > 0\right)\right| \tag{8}$$

The constraints used in the problem are stated below.
Constraints:

$$c_{i,k} - p_{i,k} = \left( \sum_{j=1}^{N} b_{i,j,k} * \begin{pmatrix} setupfinish_{i,k} \\ +wa2_{i,k} \end{pmatrix} + \left(1 - \sum_{j=1}^{N} b_{i,j,k}\right) * \left( wa_{i,k} + \sum_{l=1}^{K} d_{i,l,k} * c_{i,l} \right) \right) * ak_{i,k} \quad \forall i,k \tag{9}$$

$$c_{i,k} - p_{i,k} \geq \sum_{l=1}^{K} \left( c_{i,l} * d_{i,l,k} \right) \quad \forall i,k,l \quad k \neq l \tag{10}$$

$$setupstart_{i,k} = \sum_{j=1}^{N} b_{i,j,k} * \left( c_{j,k} + wa1_{i,j,k} \right) \quad \forall i,k \tag{11}$$

$$setupfinish_{i,k} = setupstart_{i,k} + \sum_{j=1}^{N} \left( b_{i,j,k} * s_{i,j,k} \right) \quad \forall i,k \tag{12}$$

$$b_{i,j,k} \leq ak_{i,k} * ak_{j,k} \quad \forall i,j,k \quad i \neq j \tag{13}$$

$$\sum_{i=1}^{N} ak_{i,k} - 1 = \sum_{i=1}^{N} \sum_{j=1}^{N} b_{i,j,k} \quad \forall k \tag{14}$$

$$b_{i,j,k} + b_{j,i,k} \leq 1 \quad \forall i,j,k \quad i \neq j \tag{15}$$

$$\sum_{i=1}^{N} b_{i,j,k} \leq 1 \quad \forall j,k \quad i \neq j \tag{16}$$

$$\sum_{j=1}^{N} b_{i,j,k} \leq 1 \quad \forall i,k \quad i \neq j \tag{17}$$

$$time_{j,l,t} = setupstart_{j,l} \quad \forall j,l,t = 1 \tag{18}$$

$$time_{j,l,t} = c_{j,l} - p_{j,l} \quad \forall j, l, t = 2 \tag{19}$$

$$time_{j,l,t} - \left(c_{i,k} - p_{i,k}\right) < M * ka_{i,k,j,l,t} \quad \forall i, k, j, l, t \tag{20}$$

$$c_{i,k} - time_{j,l,t} \leq M * kb_{i,k,j,l,t} \quad \forall i, k, j, l, t \tag{21}$$

$$g1_{i,k,j,l,t} = ka_{i,k,j,l,t} * kb_{i,k,j,l,t} \quad \forall i, k, j, l, t \tag{22}$$

$$workload1_{j,l,t} = \sum_{i=1}^{N} \sum_{k=1}^{K} \left(x_{i,k} * g1_{i,k,j,l,t}\right) \quad \forall j, l, t \tag{23}$$

$$time_{j,l,t} - setupstart_{i,k} < M * kc_{i,k,j,l,t} \quad \forall i, k, j, l, t \tag{24}$$

$$setupfinish_{i,k} - time_{j,l,t} \leq M * kd_{i,k,j,l,t} \quad \forall i, k, j, l, t \tag{25}$$

$$g2_{i,k,j,l,t} = kc_{i,k,j,l,t} * kd_{i,k,j,l,t} * \forall i, k, j, l, t \tag{26}$$

$$workload2_{j,l,t} = \sum_{i=1}^{N} \sum_{k=1}^{K} \left(sw_{i,k} * g2_{i,k,j,l,t}\right) \quad \forall j, l, t \tag{27}$$

$$workload_{j,l,t} = workload1_{j,l,t} + workload2_{j,l,t} \quad \forall\, j, l, t \tag{28}$$

$$sw_{i,k} = \sum_{j=1}^{N} \left(setupworker_{i,j,k} * b_{i,j,k}\right) \quad \forall i, k \quad i \neq j \tag{26}$$

$$p_{i,k} = cyc_{i,k} * \left(q_i - 1\right) + FLT_{i,k} \quad \forall i,k \tag{30}$$

$$cyc_{i,k} = \max\left(\frac{cyc\max_{i,k}}{x_{i,k}}; cyc\min_{i,k}\right) \quad \forall\, i,k \tag{31}$$

$$x_{i,k} = \left(\frac{cyc\max_{i,k}}{cyc_{i,k}}\right)^{+} \quad \forall i,k \tag{32}$$

$$b_{i,j,k} = 0 \quad or \quad 1; cyc_{i,k} \geq 0; wa_{i,k} \geq 0; wa1_{i,k} \geq 0; wa2_{i,k} \geq 0 \tag{33}$$

In this chapter, three objectives are considered. The first objective (6) minimizes the average flow time of batches. The second objective (7) minimizes the maximum number of workers in HMS. The third objective (8) minimizes the maximum number of workers changing in cells. When the number of workers changing in cells is decreased, the inter-cellular movements of workers will also decrease.

Constraints (9) and (10) represent if batch *j* precedes batch *i* then the start time of batch *i* on its cell/machine depends on the finishing time of batch *j*, finishing time of setup and elapsed time after setup ($wa2_{i,k}$) on the same cell/machine. If the operation of batch *i* is the first operation then the start time of batch *i* on its cell/machine depends on the elapsed time after the starting time ($wa_{i,k}$).

Constraint (11) and (12), respectively, represent starting and finishing time of setup. If batch *j* precedes batch *i*, then starting time of setup depends on finishing time of batch *j* and elapsed time before setup ($wa1_{i,k}$). Finishing time of setup depends on starting time of setup and setup time.

Constraints (13), (14), (15), (16), and (17) represent the sequencing of the operations in the cells and machines. There can only be a single operation before and after any operation.

Constraints (18) and (19) represent the time points where the number of workers might be changed. These time points include the starting time of setup and the operation of batches. The total number of workers is computed for every time point determined using constraints 18 and 19.

Constraints (20) and (21) are formed to identify the time points (as computed by using constraints (18) and (19)) with which the operations coincide.

Constraints (22) and (23) are used to determine the total number of workers required in the operations for each time point computed by using constraints (18) and (19).

Constraints (24) and (25) are formed to identify the time points (as computed by using constraints (18) and (19)) with which the setups coincide.

Constraints (26) and (27) are used to determine the total number of workers required for the setups for each time point computed by using constraints (18) and (19).

Constraint (28) is used to determine the total number of workers for each time point computed by using constraints (18) and (19).

Constraint (29) is used to determine the number of workers for the sequence-dependent setup time.

Constraint (30) shows the processing time of operations in the cells and machines.

Constraint (31) and (32) show the variation in the cycle time depending on the number of workers required for the operations in the cells. Long cycle times require small numbers of workers, while short cycle times require large numbers of workers (Nicholas and Soni, 2005).

Constraint (33) enforces the binary and non-negative restrictions on the variables.

## NSGA-II and FastPGA

Two different MOEAs are used within the scope of this study. One of these algorithms is the NSGA-II, which is widely used in the literature, while the other is the FastPGA, which is a comparatively recent algorithm. The effectiveness of the NSGA-II algorithm has been previously demonstrated in many large-sized problems. Studies of Eskandari and Geiger (2008) have shown that the NSGA-II and FastPGA algorithms, which are performed with well-known test problems, provide similar results. In addition, the FastPGA algorithm provides better results than the NSGA-II algorithm regarding the fast convergence, which indicates that the FastPGA is a promising algorithm. Eskandari and Geiger (2008) also emphasized the necessity of comparing the NSGA-II and FastPGA algorithms with other problems in order to assess the performance of FastPGA. In this study, the NSGA-II algorithm is compared to the FastPGA algorithm by using a MOP.

The FastPGA algorithm introduced a new ranking and fitness assignment strategy that use information which are related to Pareto dominance among individuals and niching relations. In addition, this algorithm also introduced a population regulation operator that dynamically changes the population size. Such an algorithm design allows solving computationally expensive MOPs more effectively (Tan et al., 2001; Leong and Yen, 2008). (For more information on the NSGA-II and FastPGA algorithms, please refer to the studies of Deb et al., 2002 and Eskandari and Geiger, 2008)

The algorithms of NSGA-II and FastPGA can be expressed as follows:

## NSGA-II

Input: P (a set of initial solutions randomly generated and fixed to feasible).
N_iteration (termination condition).
Output: Non-dominated solutions of P.
(1) Initialization. Set t=0, sort P based on non-domination and assign fitness to each solution equal to non-domination rank.
(2) Generate offspring population Q by P in binary tournament selection, crossover and polynomial mutation.
(3) Combine P and Q into new population R and sort R based on non-domination rank and get new NP (if it is needed use crowding distance to get new NP).
(4) t = t+1.
P = NP.
If termination condition (N_iteration) is satisfied then.
Go to step (5);
Else:
Go to step (2);
(5) Output solution P.

FastPGA

Input: P (a set of initial solution randomly generated and fixed to feasible).
N_iteration (termination condition).
Output: Non-dominated solutions of P.
(1) Initialization. Set t=0, sort P and assign fitness to each solution based on the new ranking and fitness assignment strategy.
(2) Generate offspring population O by P in binary tournament selection, crossover and polynomial mutation.
(3) Combine P and O into new population CP and sort CP based on the new ranking and fitness assignment strategy using their fitness values.
(4) Regulate the population size based on the new population regulation operator and generate new population NP from the composite population CP.
(5) t = t+1.
P = NP.
If termination condition (N_iteration) is satisfied then:
Go to step(5);
Else:
Go to step (2);
(6) Output solution P.

## Chromosome Representation

The following chromosome structure proposed by Yılmaz and Durmusoglu (2017) is adopted in this chapter. Serial scheduling scheme (SSS) is applied to obtain feasible schedules.

It is important to take into account both the encoding-decoding and the chromosome structures that are used for representing the solution of the problems. The chromosome used in this study consists of five lines. The first line ($b_{i,j,k}$) represents the sequencing of the batches within the cells and machines. The second line ($wa_{i,k}$) represents the elapsed time before the start of operations of first batches within the cells and machines. The third line ($wa1_{i,k}$) represents the elapsed times before the start of setups. The fourth line ($wa2_{i,k}$) represents the elapsed times following the end of setups. The fifth line ($x_{i,k}$) shows the number of workers for the cells and machines. Another important point in this context is that the second, the third and the fourth lines in the chromosome structure are dependent on the first and last lines. The decisions are made based on the first and last lines. The other lines show the elapsed times which are caused by decision variables. A sample chromosome structure is shown in Figure 4.

## Selection, Crossover, and Mutation Operators

The binary tournament selection approach (Beyer and Deb, 2001), which is commonly used in the literature, is applied to both algorithms in this study.

The crossover operator is formed by adapting the crossover operator and is used for both algorithms. This operator consists of two stages: In the first stage, the cells and machines are divided into two different groups for two parents, while in the second stage, one of these groups is selected and the operation sequencing is swapped according to a certain probability and two children are produced.

*Figure 4. Chromosome representation (Yılmaz and Durmusoglu, 2017)*

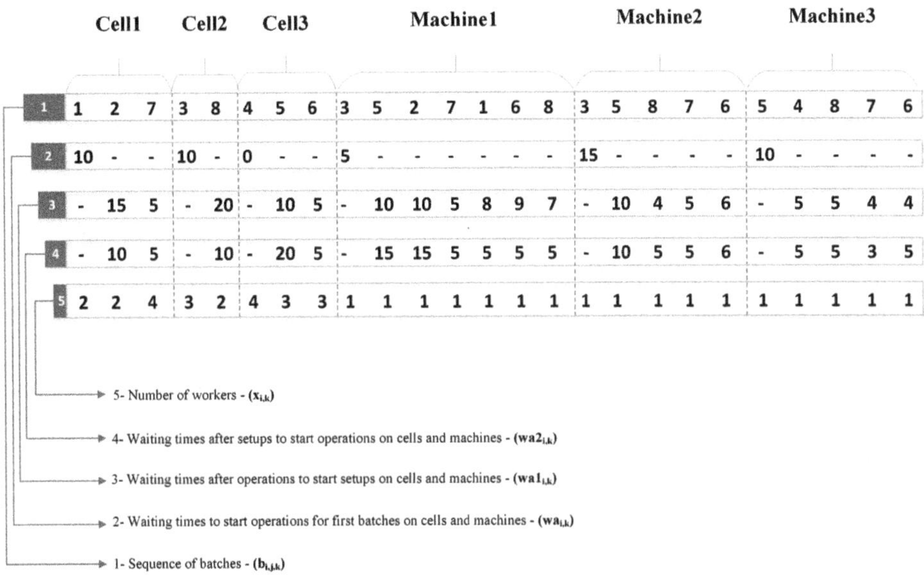


Since the chromosomes in this study are formed through real-coding, polynomial mutation (Hamdan, 2012) is used for both algorithms.

## EXPERIMENTAL STUDY

In this study, a comparison is made between the performance of the NSGA-II and the FastPGA algorithms. The coding of both algorithms was performed using the Matlab R2015a software with an Intel Core i7, 16 GB RAM, and 2.4 GHz computer.

### Experimental Test Data

In this study, five different hypothetical cases are formed to compare algorithms regarding performances. The five different cases are formed in order to create small, middle and large size problems by varying the number of cells, machines, and batches.

The first case is formed using four batches, three cells, and a machine; the second case is formed using eight batches, three cells, and a machine. The third case is formed using eight batches, three cells, and four machines. The fourth case is formed using sixteen batches, three cells, and four machines. Finally, the fifth case is formed using sixteen batches, four cells, and six machines.

Table 1 shows the five different cases. The numbers in the parentheses indicate the case number, while the numbers outside the parentheses indicate the order of operations for the batches. If a batch is not assigned to a cell or machine, the order of operation is indicated as 0. For example, first four batches, first three cells, and first machine are used for Case 1. The first operation of the first batch is assigned to the first cell; and since this is a Case 1 assignment, the intersection of the first batch and first cell is 1(1) on Table 1. Similarly, the second operation of the first batch is assigned to the first machine, and the intersection of the first batch and first machine is 2(1) on Table 1. Looking at the column of the fourth cell in Table 1, it is possible to see only 5 in the parentheses. This situation indicates that the fourth cell is not included in the first four cases and that it is only included into Case 5.

Since it would not be possible to show the sequence-dependent setup times in all cells and machines for each case, only the sequence-dependent setup times and the sequence-dependent number of workers of the first cell in the fifth case are shown in Table 3. The values on the left side of table indicate the setup times in minutes, while the immediately adjacent values indicate the sequence-dependent number of workers for setups. The setup times for the other cells and machines, along with the number of workers, are selected from similar values.

Table 4 provides the time elapsed between two finished batches (cycle times) for the cells and machines in Case 1. The values on the upper part of Table 4 indicate the lower bounds for the cycle times, while the values on the lower part of Table 4 indicate the upper bounds of the cycle times. It can be noted that the cycle times for the machines are all constant.

Table 5 describes the final order and batch size for Case 3.

## Algorithm Parameter Settings

When determining the parameter values of the algorithms, we utilized values that are determined in previous studies (Deb et al., 2002; Eskandari and Geiger, 2008). Experiments are performed to determine the population size and crossover probability, and best values obtained through these experiments are used within this study. The parameter values are shown in Table 2.

## Performance Metrics

In general, there are two main goals when solving MOPs, which are the convergence and the diversity (detailed information on both goals are provided within this study in the previous sections). In addition to these goals, another important requirement is the fast convergence, which is particularly important for algorithms used for computationally expensive MOPs. Numerous different performance metrics have been developed in the literature (Deb et al., 2002; Knowles and Corne, 2000; Erbas et al., 2006; Eskandari and Geiger, 2008; Durillo et al., 2010). Many of these performance metrics have been developed for cases in which the Pareto front is known. However, we utilized two different metrics that are generally used for determining convergence and diversity in case the optimum Pareto front is not known (Xu et al., 2015; Rabiee et al., 2012). Since the algorithms are similar, a different approach is introduced in this study to enable the comparison of these two algorithms regarding the fast convergence performances. This approach is based on the following process: if the best value of the average flow time objective does not change during the pre-specified number of iterations, the search process is concluded. The iteration number on which the search process concludes provides information about the algorithm's fast convergence performance. A search process that ends at a lower iteration value will indicate a solution with better fast convergence performance; this is because the algorithm converges more rapidly. The interpretation of the information regarding the fast convergence performance of the algorithms is provided in the following section.

The equations below describe the metric regarding convergence and diversity.

*Table 1. Order of operations for cases*

| Batches | Cell1 | Cell2 | Cell3 | Cell4 | Machine1 | Machine2 | Machine3 | Machine4 | Machine5 | Machine6 |
|---|---|---|---|---|---|---|---|---|---|---|
| 1 | 1(1)-1(2)-1(3)-1(4)-1(5) | 0(1)-0(2)-0(3)-0(4)-0(5) | 0(1)-0(2)-0(3)-0(4)-0(5) | 0(5) | 2(1)-2(2)-2(3)-2(4)-0(5) | 0(3)-0(4)-0(5)-0(5) | 0(3)-0(4)-2(5) | 3(3)-3(4)-0(5) | 0(5) | 0(5) |
| 2 | 0(1)-1(2)-0(3)-1(4)-0(5) | 1(1)-0(2)-1(3)-0(4)-0(5) | 0(1)-0(2)-0(3)-0(4)-0(5) | 1(5) | 2(1)-2(2)-0(3)-2(4)-2(5) | 2(3)-0(4)-0(5) | 0(3)-0(4)-0(5) | 3(3)-3(4)-3(5) | 0(5) | 0(5) |
| 3 | 0(1)-0(2)-0(3)-0(4)-0(5) | 1(1)-1(2)-1(3)-1(4)-1(5) | 0(1)-0(2)-0(3)-0(4)-0(5) | 0(5) | 2(1)-2(2)-2(3)-0(4)-2(5) | 3(3)-2(4)-0(5) | 0(3)-0(4)-3(5) | 4(3)-3(4)-0(5) | 4(5) | 0(5) |
| 4 | 0(1)-0(2)-0(3)-0(4)-0(5) | 0(1)-1(2)-0(3)-1(4)-0(5) | 1(1)-0(2)-1(3)-0(4)-1(5) | 0(5) | 2(1)-2(2)-0(3)-0(4)-2(5) | 0(3)-2(4)-0(5) | 2(3)-0(4)-3(5) | 3(3)-4(4)-4(5) | 0(5) | 0(5) |
| 5 | 0(2)-0(3)-0(4)-0(5) | 1(2)-0(3)-1(4)-0(5) | 0(2)-1(3)-0(4)-0(5) | 1(5) | 2(2)-2(3)-2(4)-2(5) | 3(3)-3(4)-0(5) | 4(3)-0(4)-3(5) | 5(3)-4(4)-4(5) | 0(5) | 0(5) |
| 6 | 0(2)-0(3)-0(4)-0(5) | 1(2)-0(3)-1(4)-0(5) | 0(2)-0(3)-0(4)-1(5) | 0(5) | 2(2)-1(3)-2(4)-2(5) | 2(3)-3(4)-3(5) | 3(3)-0(4)-0(5) | 4(3)-4(4)-0(5) | 4(5) | 5(5) |
| 7 | 0(2)-1(3)-0(4)-0(5) | 0(2)-0(3)-0(4)-1(5) | 1(2)-0(3)-1(4)-0(5) | 0(5) | 2(2)-2(3)-0(4)-0(5) | 3(3)-0(4)-0(5) | 4(3)-2(4)-0(5) | 5(3)-3(4)-2(5) | 3(5) | 0(5) |
| 8 | 0(2)-0(3)-0(4)-0(5) | 0(2)-1(3)-0(4)-0(5) | 1(2)-0(3)-1(4)-1(5) | 0(5) | 2(2)-2(3)-0(4)-0(5) | 3(3)-0(4)-2(5) | 4(3)-2(4)-3(5) | 5(3)-3(4)-0(5) | 4(5) | 0(5) |
| 9 | 0(4)-0(5) | 0(4)-0(5) | 1(4)-1(5) | 0(5) | 2(4)-0(5) | 3(4)-2(5) | 4(4)-3(5) | 5(4)-4(5) | 5(5) | 0(5) |
| 10 | 0(4)-1(5) | 0(4)-0(5) | 1(4)-0(5) | 0(5) | 2(4)-2(5) | 3(4)-3(5) | 4(4)-4(5) | 5(4)-0(5) | 0(5) | 5(5) |
| 11 | 0(4)-0(5) | 0(4)-0(5) | 0(4)-0(5) | 1(5) | 1(4)-2(5) | 2(4)-3(5) | 3(4)-4(5) | 4(4)-0(5) | 0(5) | 0(5) |
| 12 | 0(4)-0(5) | 0(4)-1(5) | 0(4)-0(5) | 0(5) | 1(4)-2(5) | 2(4)-3(5) | 3(4)-4(5) | 4(4)-0(5) | 0(5) | 0(5) |
| 13 | 1(4)-1(5) | 0(4)-0(5) | 0(4)-0(5) | 0(5) | 2(4)-2(5) | 3(4)-3(5) | 4(4)-4(5) | 5(4)-5(5) | 0(5) | 0(5) |
| 14 | 1(4)-1(5) | 0(4)-0(5) | 0(4)-0(5) | 0(5) | 2(4)-2(5) | 3(4)-3(5) | 4(4)-4(5) | 5(4)-5(5) | 0(5) | 0(5) |
| 15 | 0(4)-0(5) | 1(4)-1(5) | 0(4)-0(5) | 0(5) | 2(4)-2(5) | 3(4)-3(5) | 4(4)-0(5) | 5(4)-0(5) | 0(5) | 4(5) |
| 16 | 0(4)-0(5) | 1(4)-1(5) | 0(4)-0(5) | 0(5) | 2(4)-2(5) | 3(4)-3(5) | 4(4)-4(5) | 5(4)-0(5) | 0(5) | 0(5) |

*Table 2. Parameter settings for FastPGA and NSGA-II*

| Algorithm Parameter | FastPGA | NSGA-II |
|---|---|---|
| Initial Population Size | 300 | 300 |
| Maximum Population Size | 300 | |
| Crossover Probability | 0.8 | 0.7 |
| Mutation Probability | 1/n (n is number of variables) | 1/n |
| Mutation Type | Swap | Swap |
| Selection Scheme | Binary Tournament Selection | Binary Tournament Selection |

Equation (34) describes diversification performance metric. The "$\max f_{1i}$" and "$\min f_{1i}$" values within the equation describe the largest and smallest values on the Pareto front for the first objective function. The same applies for the two other objectives. A higher value for this metric is indicative of a better diversification performance for the solution.

*Table 3. Sequence-dependent setup times and number of workers for the operations of batches assigned to the first cell*

| | 1 | 10 | 13 | 14 |
|---|---|---|---|---|
| 1 | - | 20-2 | 30-2 | 20-2 |
| 10 | 20-2 | - | 10-1 | 20-1 |
| 13 | 30-2 | 10-1 | - | 20 |
| 14 | 20-2 | 20-1 | 20-2 | - |

*Table 4. Cycle times for batches for cells and functional area (machines)*

| Lower Bound of Cycle Time | Cell1 | Cell2 | Cell3 | Machine1 |
|---|---|---|---|---|
| 1 | 1.4 | 1 | 3 | 2 |
| 2 | 1.2 | 1 | 3.2 | 1.5 |
| 3 | 1 | 1.4 | 3.6 | 4 |
| 4 | 1.8 | 1.5 | 4 | 3 |
| **Upper Bound of Cycle Time** | **Cell1** | **Cell2** | **Cell3** | **Machine1** |
| 1 | 14 | 15 | 13 | 2 |
| 2 | 14.2 | 15 | 15 | 1.5 |
| 3 | 14.3 | 14 | 15.1 | 4 |
| 4 | 14.4 | 14.4 | 15.5 | 3 |

*Table 5. Final orders and batch sizes for the third case*

| | 1 | 2 | 3 | 4 | 5 | 6 | 7 | 8 |
|---|---|---|---|---|---|---|---|---|
| Batch Size | 100 | 50 | 75 | 100 | 50 | 100 | 75 | 50 |
| Final Order | 100 | 50 | 75 | 100 | 50 | 100 | 75 | 50 |

$$DM = \sqrt{\left(\max f_{1i} - \min f_{1i}\right)^2 + \left(\max f_{2i} - \min f_{2i}\right)^2 + \left(\max f_{3i} - \min f_{3i}\right)^2} \quad (34)$$

Equation (35) describes the convergence performance metric. The Mean Ideal Distance (*MID*) indicates the distance between the ideal point (0, 0, 0) and the solutions on the best Pareto front. Within this equation, "*n*" represents the number of non-dominated solutions. In this equation, the "$c_i$" values $\left(c_i = \sqrt{\left(f_{1i}^{\ 2} + f_{2i}^{\ 2} + f_{3i}^{\ 2}\right)}\right)$ are obtained through the use of all objective functions. A smaller value for this metric is indicative of a better convergence performance for the solution.

$$MID = \frac{\sum_{c=1}^{n} c_i}{n} \tag{35}$$

For all hypothetical cases designed within the frame of this study, both MOEAs are randomly run. Based on these runs, the mean and standard deviation of the metrics provided by equations 34 and 35 (*DM* and *MID*) and the fast convergence metric are determined. Using these values, the 95% confidence intervals of all three metrics are computed, and the MOEAs (NSGA-II and FastPGA) are compared with one another. The $\bar{x} \pm t_{\alpha/2,n-1} * s/\sqrt{n}$ equation is used to compute the confidence interval. In this equation, the sample size is set at 30, while the significance level α is set at 0.05.

In this study, the algorithms are not compared in terms of CPU time. This is because, regardless of the approach used for expensive MOPs, the actual CPU time tends to dominate at higher levels (Eskandari and Geiger 2008).

## Computational Results

The FastPGA and the NSGA-II algorithms are compared to each other in this section.

Table 6 shows the mean value, standard deviation, 95% confidence interval (*CI*) of the performance metrics as determined based on the random runs. For each case 30 different instances are generated randomly based on the values given in tables using uniform distribution. The results are interpreted based on these values.

Due to the absence of any overlap between the confidence intervals of the *DM* and *MID* metrics (except for the *MID* metric of case 1), it can be stated that the FastPGA algorithm outperforms the NSGA-II algorithm. In case 1, the confidence intervals of the algorithms regarding *MID* metrics overlap with one another. This indicated that the convergence performance of the algorithms is similar; this is further illustrated in Figure 5. Since the *MID* metrics of the algorithms overlap, it is possible to compare these two algorithms with respect to fast convergence performance as well. Comparing fast convergence would not be suitable in case where there is no overlap between the *MID* metrics. Comparison with respect to fast convergence demonstrates that the FastPGA algorithm provides better results. In addition, when compare with respect to *DM* performance, FastPGA outperforms the NSGA-II algorithm for all cases.

In addition to these results, it can also be stated that the FastPGA algorithm provided better and more effective results regarding the large-sized problems.

# CONCLUSION

It is a well-known in literature that multi-objective evolutionary algorithms are effective to solve MOPs. Two evolutionary algorithms, which are commonly used in the literature, are implemented in this study. The study not only provides information about these two algorithms but also compares their performance on a set of problems. For comparison purposes, we used a multi-objective scheduling problem in HMS, which is an expensive MOP.

*Figure 5. Confidence interval of performance metrics for Case 1*

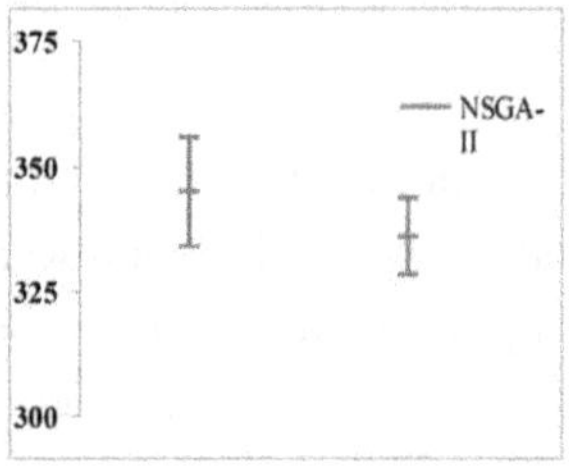


(a) Mean Ideal Distance (MID) – Case 1

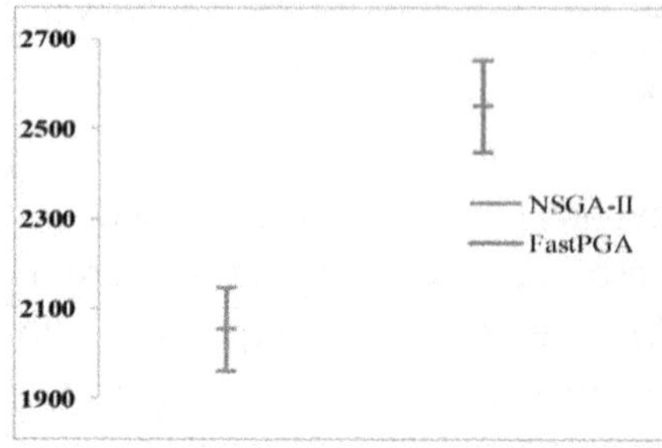


(b) Diversification Metric (DM) – Case 1

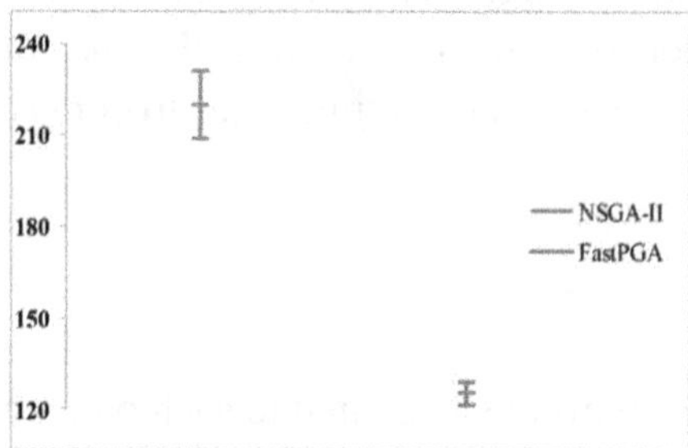


(c) Fast Convergence – Case 1

*Table 6. Means, standard deviations, and 95% confidence intervals of the performance metrics*

| Case | Number of Cells-Mach.-Prod. | Algorithm | Diversification Metric | | | Mean Individual Distance | | | Fast Convergence | | |
|---|---|---|---|---|---|---|---|---|---|---|---|
| | | | Avg. | Std. Dev. | %95 C.I. | Avg. | Std. Dev. | %95 C.I. | Avg. | Std. Dev. | %95 C.I. |
| 1 | 3-1-4 | FastPGA | 2550.1 | 273.6 | [2447.9 – 2652.2] | 336.1 | 20.6 | [328.4 – 343.7] | 94.05 | 9.4 | [90.54 – 97.6] |
| | | NSGA-II | 2053.4 | 250.3 | [1959.9 – 2146.9] | 345 | 29.2 | [334.1 – 355.9] | 165.02 | 19.8 | [157.6 – 172.4] |
| 2 | 3-1-8 | FastPGA | 7444.4 | 890.4 | [7111.9 – 7776.9] | 1193.4 | 100.4 | [1155.9 – 1230.9] | 173.8 | 19.1 | [166.6 – 180.9] |
| | | NSGA-II | 6779.7 | 852.5 | [6461.4 – 7097.9] | 1786.6 | 160.5 | [1726.7 – 1846.5] | 209.3 | 25.4 | [199.8 – 218.7] |
| 3 | 3-4-8 | FastPGA | 53482.8 | 7004.9 | [50867.4 – 56098.1] | 1792 | 148.3 | [1736.5 – 1847.4] | 277.8 | 32.6 | [265.6 – 289.9] |
| | | NSGA-II | 37288.5 | 4611.3 | [35566.9 – 39010.2] | 9149.5 | 834.1 | [8838.1 – 9460.9] | 266.7 | 34 | [254 – 279.3] |
| 4 | 3-4-16 | FastPGA | 544885.9 | 78193.9 | [515691.1 – 574080.7] | 7098.6 | 686.8 | [6842.2 - 7355] | 399.8 | 49.3 | [381.3 – 418.2] |
| | | NSGA-II | 330939.6 | 45258.9 | [314041.5 – 347837.7] | 22740.4 | 2155.8 | [21936 - 23545] | 387.6 | 52.1 | [368.1 - 407] |
| 5 | 4-6-16 | FastPGA | 719868.4 | 113794.1 | [677381.8 - 762355] | 8059.5 | 812.3 | [7756.2 – 8362.8] | 501.3 | 62.8 | [477.8 – 524.7] |
| | | NSGA-II | 612522.4 | 86847.3 | [580096.7 – 644948.1] | 64326.2 | 6324.4 | [61965 - 66688] | 460.8 | 64 | [436.9 – 484.7] |

Three different objectives are used for comparison purposes. One of these objectives is related to scheduling period while other two are related to the number of workers. When the results of the original NSGA-II and FastPGA algorithm for the five cases are compared, it is observed that the FastPGA algorithm shows better performance especially for the large-sized problems. This is mainly due to the adaptive population sizing used by the FastPGA algorithm. In this sense, it is concluded that FastPGA algorithm provides better solutions for the expensive MOPs regarding the performance metrics. In addition, comparison between two algorithms with the same problem instances demonstrated that the adaptive population sizing strategy is highly effective for solving large-sized problems.

Since the MOP addressed in the study is a new type of problem, different evolutionary algorithms can be applied in solving this problem. The problem can be adapted to other manufacturing environments through the mathematical model. Last but not least, different problem instances can be generated and used for comparison purposes for other metaheuristics.

## REFERENCES

Aglan, C., & Durmusoglu, M. B. (2015). Lot-splitting approach of a hybrid manufacturing system under CONWIP production control: A mathematical model. *International Journal of Production Research*, *53*(5), 1561–1583. doi:10.1080/00207543.2014.957873

Akturk, M. S. (2011). Joint cell loading and scheduling approach to cellular manufacturing systems. *International Journal of Production Research*, *49*(21), 6321–6341. doi:10.1080/00207543.2010.532165

Beyer, H. G., & Deb, K. (2001). On self-adaptive features in real-parameter evolutionary algorithms. *IEEE Transactions on Evolutionary Computation*, *5*(3), 250–270. doi:10.1109/4235.930314

Cámara, M., de Toro, F., & Ortega, J. (2013). An analysis of multiobjective evolutionary algorithms for optimization problems with time constraints. *Applied Artificial Intelligence*, *27*(9), 851–879. doi:10.1080/08839514.2013.835237

Coello, C. A. C., Pulido, G. T., & Lechuga, M. S. (2004). Handling multiple objectives with particle swarm optimization. *IEEE Transactions on Evolutionary Computation*, *8*(3), 256–279. doi:10.1109/TEVC.2004.826067

Corne, D. W., Jerram, N. R., Knowles, J. D., & Oates, M. J. (2001, July). PESA-II: Region-based selection in evolutionary multiobjective optimization. In *Proceedings of the 3rd Annual Conference on Genetic and Evolutionary Computation* (pp. 283-290). Morgan Kaufmann Publishers Inc.

Das, S., & Panigrahi, B. K. (2009). Multi-objective evolutionary algorithms. In Encyclopedia of artificial intelligence (pp. 1145-1151). Hershey, PA: IGI Global. doi:10.4018/978-1-59904-849-9.ch167

Das, S. R., & Canel, C. (2005). An algorithm for scheduling batches of parts in a multi-cell flexible manufacturing system. *International Journal of Production Economics*, *97*(3), 247–262. doi:10.1016/j.ijpe.2004.07.006

Deb, K. (2001). *Multi-Objective Optimization Using Evolutionary Algorithms*. New York, NY: John Wiley& Sons. Inc.

Deb, K., Agrawal, S., Pratap, A., & Meyarivan, T. (2002). A fast elitist non-dominated sorting genetic algorithm for multi-objective optimization: NSGA-II. In *Proceedings of the International Conference on Parallel Problem Solving From Nature* (pp. 849-858). Springer Berlin Heidelberg.

Durillo, J. J., Nebro, A. J., Luna, F., Coello Coello, C. A., & Alba, E. (2010). Convergence speed in multi-objective metaheuristics: Efficiency criteria and empirical study. *International Journal for Numerical Methods in Engineering*, *84*(11), 1344–1375. doi:10.1002/nme.2944

Durmusoglu, M. B., & Kaya, G. (2012). Lean Thinking Based Investment Planning at Design Stage of Cellular/Hybrid Manufacturing Systems. In *Industrial Engineering: Concepts, Methodologies, Tools, and Applications: Concepts, Methodologies, Tools, and Applications* (Vol. 1, p. 409).

Durmusoglu, M. B., & Satoglu, S. I. (2011). Axiomatic design of hybrid manufacturing systems in erratic demand conditions. *International Journal of Production Research*, *49*(17), 5231–5261. doi:10.1080/00207543.2010.510487

Erbas, C., Cerav-Erbas, S., & Pimentel, A. D. (2006). Multiobjective optimization and evolutionary algorithms for the application mapping problem in multiprocessor system-on-chip design. *IEEE Transactions on Evolutionary Computation*, *10*(3), 358–374. doi:10.1109/TEVC.2005.860766

Eskandari, H., Geiger, C., & Lamont, G. (2007). FastPGA: A dynamic population sizing approach for solving expensive multiobjective optimization problems. In Evolutionary Multi-Criterion Optimization (pp. 141-155). Springer Berlin/Heidelberg. doi:10.1007/978-3-540-70928-2_14

Eskandari, H., & Geiger, C. D. (2008). A fast Pareto genetic algorithm approach for solving expensive multiobjective optimization problems. *Journal of Heuristics*, *14*(3), 203–241. doi:10.1007/s10732-007-9037-z

Eskandari, H., & Geiger, C. D. (2009). Evolutionary multiobjective optimization in noisy problem environments. *Journal of Heuristics*, *15*(6), 559–595. doi:10.1007/s10732-008-9077-z

Feyzioglu, O., & Pierreval, H. (2009). Hybrid organization of functional departments and manufacturing cells in the presence of imprecise data. *International Journal of Production Research*, *47*(2), 343–368. doi:10.1080/00207540802425898

Fonseca, C. M., & Fleming, P. J. (1993, June). Genetic Algorithms for Multiobjective Optimization: FormulationDiscussion and Generalization. Icga, 93(July), 416-423.

Goldberg, D. E. (1989). *Genetie Algorithms in Seareh.* Optirnization and Machine Leaming.

Hamdan, M. (2010). On the disruption-level of polynomial mutation for evolutionary multi-objective optimisation algorithms. *Computing and Informatics*, *29*(5), 783–800.

Han, Y. Y., Gong, D. W., Sun, X. Y., & Pan, Q. K. (2014). An improved NSGA-II algorithm for multi-objective lot-streaming flow shop scheduling problem. *International Journal of Production Research*, *52*(8), 2211–2231. doi:10.1080/00207543.2013.848492

Harhalakis, G., Lu, T., Minis, I., & Nagi, R. (1996). A practical method for design of hybrid-type production facilities. *International Journal of Production Research*, *34*(4), 897–918. doi:10.1080/00207549608904942

Horn, J., Nafpliotis, N., & Goldberg, D. E. (1994, June). A niched Pareto genetic algorithm for multiobjective optimization. In *Proceedings of the First IEEE Conference on Evolutionary Computation* (pp. 82-87). Ieee. doi:10.1109/ICEC.1994.350037

Ioannou, G. (2006). Time-phased creation of hybrid manufacturing systems. *International Journal of Production Economics*, *102*(2), 183–198. doi:10.1016/j.ijpe.2005.03.005

Iranmanesh, H., Skandari, M. R., & Allahverdiloo, M. (2008). Finding Pareto optimal front for the multi mode time, cost quality trade-off in project scheduling. *Issues, 59*.

Kanthababu, M. (2013). Multi-Objective Optimization of Manufacturing Processes Using Evolutionary Algorithms. In *Industrial Engineering: Concepts, Methodologies, Tools, and Applications*. doi:10.4018/978-1-4666-1945-6.ch021

Knowles, J. D., & Corne, D. W. (2000). Approximating the nondominated front using the Pareto archived evolution strategy. *Evolutionary Computation*, *8*(2), 149–172. doi:10.1162/106365600568167 PMID:10843519

Konak, A., Coit, D. W., & Smith, A. E. (2006). Multi-objective optimization using genetic algorithms: A tutorial. *Reliability Engineering & System Safety*, *91*(9), 992–1007. doi:10.1016/j.ress.2005.11.018

Leong, W. F., & Yen, G. G. (2008). PSO-based multiobjective optimization with dynamic population size and adaptive local archives. *Systems, Man, and Cybernetics, Part B: Cybernetics. IEEE Transactions on*, *38*(5), 1270–1293.

Li, Y., Li, X., & Gupta, J. N. (2015). Solving the multi-objective flowline manufacturing cell scheduling problem by hybrid harmony search. *Expert Systems with Applications*, *42*(3), 1409–1417. doi:10.1016/j.eswa.2014.09.007

Lin, S. W., Ying, K. C., Lu, C. C., & Gupta, J. N. (2011). Applying multi-start simulated annealing to schedule a flowline manufacturing cell with sequence dependent family setup times. *International Journal of Production Economics*, *130*(2), 246–254. doi:10.1016/j.ijpe.2011.01.004

Luke, S. (2013). *Essentials of Metaheuristics* (Vol. II). Fairfax: Lulu.com.

Murthy, C. V., & Srinivasan, G. (1995). Fractional cell formation in group technology. *The International Journal of Production Research, 33*(5), 1323-1337.

Murugan, P., Kannan, S., & Baskar, S. (2009). NSGA-II algorithm for multi-objective generation expansion planning problem. *Electric Power Systems Research*, *79*(4), 622–628. doi:10.1016/j.epsr.2008.09.011

Nicholas, J., & Soni, A. (2005). *The portal to lean production: Principles and practices for doing more with less*. CRC Press.

Pires, D. F., Antunes, C. H., & Martins, A. G. (2012). NSGA-II with local search for a multi-objective reactive power compensation problem. *International Journal of Electrical Power & Energy Systems*, *43*(1), 313–324. doi:10.1016/j.ijepes.2012.05.024

Rabiee, M., Zandieh, M., & Ramezani, P. (2012). Bi-objective partial flexible job shop scheduling problem: NSGA-II, NRGA, MOGA and PAES approaches. *International Journal of Production Research*, *50*(24), 7327–7342. doi:10.1080/00207543.2011.648280

Ramesh, S., Kannan, S., & Baskar, S. (2012). Application of modified NSGA-II algorithm to multi-objective reactive power planning. *Applied Soft Computing*, *12*(2), 741–753. doi:10.1016/j.asoc.2011.09.015

Satoglu, S. I., Durmusoglu, M. B., & Ertay, T. (2010). A mathematical model and a heuristic approach for design of the hybrid manufacturing systems to facilitate one-piece flow. *International Journal of Production Research*, *48*(17), 5195–5220. doi:10.1080/00207540903089544

Satoglu, S. I., & Suresh, N. C. (2009). A goal-programming approach for design of hybrid cellular manufacturing systems in dual resource constrained environments. *Computers & Industrial Engineering*, *56*(2), 560–575. doi:10.1016/j.cie.2008.06.009

Schaller, J. E., Gupta, J. N., & Vakharia, A. J. (2000). Scheduling a flowline manufacturing cell with sequence dependent family setup times. *European Journal of Operational Research*, *125*(2), 324–339. doi:10.1016/S0377-2217(99)00387-2

Solimanpur, M., & Elmi, A. (2013). A tabu search approach for cell scheduling problem with makespan criterion. *International Journal of Production Economics*, *141*(2), 639–645. doi:10.1016/j.ijpe.2012.10.001

Talbi, E. G. (2009). *Metaheuristics: From design to implementation*. Hoboken, N.J: John Wiley & Sons. doi:10.1002/9780470496916

Tan, K. C., Lee, T. H., & Khor, E. F. (2001). Evolutionary algorithms with dynamic population size and local exploration for multiobjective optimization. *IEEE Transactions on Evolutionary Computation*, *5*(6), 565–588. doi:10.1109/4235.974840

Tang, L., & Wang, X. (2013). A hybrid multiobjective evolutionary algorithm for multiobjective optimization problems. *IEEE Transactions on Evolutionary Computation*, *17*(1), 20–45. doi:10.1109/TEVC.2012.2185702

Tavakkoli-Moghaddam, R., Gholipour-Kanani, Y., & Cheraghalizadeh, R. (2008). A genetic algorithm and memetic algorithm to sequencing and scheduling of cellular manufacturing systems. *International Journal of Management Science and Engineering Management*, *3*(2), 119–130.

Tavakkoli-Moghaddam, R., Javadian, N., Khorrami, A., & Gholipour-Kanani, Y. (2010). Design of a scatter search method for a novel multi-criteria group scheduling problem in a cellular manufacturing system. *Expert Systems with Applications*, *37*(3), 2661–2669. doi:10.1016/j.eswa.2009.08.012

Venkataramanaiah, S. (2008). Scheduling in cellular manufacturing systems: An heuristic approach. *International Journal of Production Research*, *46*(2), 429–449. doi:10.1080/00207540601138577

Venkataramanaiah, S., & Krishnaiah, K. (2002). Hybrid heuristic for design of cellular manufacturing systems. *Production Planning and Control*, *13*(3), 274–283. doi:10.1080/09537280110073978

Viguier, F., & Pierreval, H. (2004). An approach to the design of a hybrid organization of workshops into functional layout and group technology cells. *International Journal of Computer Integrated Manufacturing*, *17*(2), 108–116. doi:10.1080/09511920310001593092

Wang, X., Tang, J., & Yung, K. L. (2010). A scatter search approach with dispatching rules for a joint decision of cell formation and parts scheduling in batches. *International Journal of Production Research*, *48*(12), 3513–3534. doi:10.1080/00207540902922828

Xu, Z., Ming, X. G., Zheng, M., Li, M., He, L., & Song, W. (2015). Cross-trained workers scheduling for field service using improved NSGA-II. *International Journal of Production Research*, *53*(4), 1255–1272. doi:10.1080/00207543.2014.955923

Yilmaz, O. F., Cevikcan, E., & Durmusoglu, M. B. (2016). Scheduling batches in multi hybrid cell manufacturing system considering worker resources: A case study from pipeline industry. *Advances in Production Engineering & Management*, *11*(3), 192–206. doi:10.14743/apem2016.3.220

Yilmaz, O. F., & Durmusoglu, M.B. (in press). A Multi-objective scheduling of hybrid manufacturing systems with walking workers.

Yilmaz, O. F., & Erbiyik, H. (2016). Effective Applications of Optimization Methods in the Manufacturing Environment in Turkey. In Comparative Economics and Regional Development in Turkey (pp. 319–335). Hershey, PA: IGI Global. doi:10.4018/978-1-4666-8729-5.ch015

Yu, Y., Tang, J., Sun, W., Yin, Y., & Kaku, I. (2013). Combining local search into non-dominated sorting for multi-objective line-cell conversion problem. *International Journal of Computer Integrated Manufacturing*, *26*(4), 316–326. doi:10.1080/0951192X.2012.717717

Zhou, A., Qu, B. Y., Li, H., Zhao, S. Z., Suganthan, P. N., & Zhang, Q. (2011). Multiobjective evolutionary algorithms: A survey of the state of the art. *Swarm and Evolutionary Computation*, *1*(1), 32–49. doi:10.1016/j.swevo.2011.03.001

Zitzler, E., Laumanns, M., & Thiele, L. (2002). SPEA2: Improving the strength Pareto evolutionary algorithm.

## KEY TERMS AND DEFINITIONS

**Batch Scheduling in Hybrid Manufacturing:** A type of operations research problem involving sequencing and scheduling of batches on cells and machines in functional area.

**Evolutionary Algorithm:** An evolutionary algorithm is inspired by biological evolution and uses crossover, mutation, selection mechanisms.

**Hybrid Manufacturing System:** A manufacturing system that contains both cells and a functional area. It provides flexibility and allows the production of many kinds of products.

**Multi-Objective Evolutionary Algorithm:** A multi-objective evolutionary algorithm uses Pareto front and deals with multiple objective functions simultaneously.

**Multi-Objective Optimization:** A topic of multiple-criteria decision-making dealing with mathematical optimization problems which consists of more than one objective to be optimized at the same time.

**Pareto Front:** The set that corresponds to Pareto set and is composed of all Pareto optimal decision vectors is called Pareto front.

**Pareto Optimal Solution:** A solution is called Pareto optimal, if none of the objective functions can be improved without degrading some of the other objective values.

# Section 2
# Supply Chain and Inventory Management

*This section reveals the principles of Supply Chain and Inventory Management. The first chapter deals with the pricing, the lotsizing and the shipment for a two-echelon supply chain. Metaheuristic algorithms are used in the first chapter to solve the problem for two-echelon supply chain. The second chapter considers the key supply chain risks which could cause abnormalities and occur from rapid changes in customer demand, unpredictable price fluctuations, defect variations and delivery delays and provides the correction of these problems automatically. The third chapter focuses the components which help to constitute a supply chain strategy and classify the supply chain strategies described in the literature.*

# Chapter 9
# Differential Return on Investment Optimization:
## Pricing, Lotsizing, and Shipment Considerations in a Two-Echelon Supply Chain

**Reza Ghasemy Yaghin**
*Amirkabir University of Technology, Iran*

**Hadi Mosadegh**
*Amirkabir University of Technology, Iran*

**S. M. T. Fatemi Ghomi**
*Amirkabir University of Technology, Iran*

## ABSTRACT

*A two-echelon supply chain is studied that involves a retailer who faces demand from two or more market segments and enable to set different prices and marketing expenditures and a supplier who desires to find optimal number of shipments through an integrated system. A new mixed-integer non-linear fractional programming (MINLFP) model is developed. In order to solve the resultant MINLFP model, the constrained non-linear programming model is reformulated as an unconstrained one using penalty terms. Two meta-heuristics, namely simulated annealing (SA) and imperialist competitive algorithm (ICA), are applied to solve the relaxed unconstrained model. Numerical results show that ICA can reach better solutions in comparison with SA. However, SA has the ability of providing more robust solutions which are converged to a good solution. The chapter concludes with superiority of SA.*

DOI: 10.4018/978-1-5225-2944-6.ch009

## INTRODUCTION AND BACKGROUND

An up-to-date review by Chen and Simchi-Levi (2012) reveals there is a growing literature in presenting and analyzing optimization models integrating pricing and lotsizing policies. In today's global markets, the revenue management (RM) models are becoming a powerful instrument, where a retail industry desires to provide different levels of marketing mix (named four P's: price, product, promotion and place) to different market segmentations (i.e. channels). Since the pioneering review research of Kleijn and Dekker (1998), the concept of inventories' price differentiation has been one of the most pervasive activities in both the marketing and operations academic literature and practice.

One of the underlying principles of RM is to divide a single market into multiple sub-markets/segments and then set different prices in each sub-market. Price differentiation is a powerful way for sellers to improve their profitability (Phillips, 2005). Sen and Zhang (1999) considered the newsboy problem with multiple demand classes, where demands were realized sequentially and demand dependency was modeled through the diversion. Zhang and Bell (2007) extended the newsvendor problem with backlogged demand to the case where the single product can be sold to different demand classes at different prices. Zhang et al. (2010) evaluated the simultaneous determination of price and inventory replenishment in a two-segment market with a fence. All of these research papers focus on profit aspects of the retailer/manufacturer without any other criterion.

Ghasemy Yaghin et al. (2013) presented a joint pricing and lot-sizing model with multiple demand classes to set different prices and marketing expenditure in each sub-market. Traditionally, numerous papers have employed the profit maximization or cost minimization as their objective in designing and analyzing inventory models. Many researchers also optimized the inventory systems under return on investment (ROI) maximization. As Lenskold (2003) mentions, it is completely reasonable, and highly beneficial, to expect a return on investment for each incremental marketing dollar spent. An inventory model using the criterion of ROI maximization is proposed by Schroeder and Krishnan (1976). Also, Rosenberg (1991) compares and contrasts profit maximization versus return on inventory investment with respect to logarithmic concave demand functions. Otake et al. (1999) proposed an ROI maximization model with the lot size and setup cost reduction investment as the strategic joint decision variables. Otake and Min (2001) constructed and analyzed inventory and investment in quality improvement policies under ROI maximization.

Li et al. (2008) constructed and analyzed inventory and capital investment in setup and quality under ROI maximization. Wee et al. (2009) proposed a joint replenishment model under profit and ROI maximization. Ghasemy Yaghin et al. (2013) developed a return on inventory investment (ROII) maximization model in inventory-marketing problems under uncertainty to manage some marketing mix. Literature review on integrated models involving inventory-related decision, starting with Goyal (1976) and Banerjee (1986), reflects various mathematical and heuristic techniques developed to implement specific strategies. Good state-of-the-art review papers on integrated decision making on inventory management known as joint economic lot size (JELS) models and methodologies can be found in Ben-Daya (2008) and Glock (2012). Recently, Ghasemy Yaghin et al. (2014) have improved the marketing aspects of existing JPLM models in an integrated two-echelon supply chain in which a very interesting demand function is used to formulate the customer behaviour in a more realistic way while maximizing the system's total profit. Hammami and Frein (2014) have presented an optimization model for the design of global supply chains where the emphasis is made on transfer pricing for both tangible and intangible elements. Qin (2014) has considered the pricing and lot-sizing problem for products with quality and physical quantity deteriorating simultaneously. In a two-echelon supply chain, Yildirmaz et al. (2009) have incorporated transportation costs into the lot-sizing and pricing problems, and present an approximate algorithm to solve the resulting optimization problem. Wang et al. (2015) have studied a linear supply chain consisting of a supplier and a retailer selling the product in a market, in which the demand for the product is decreasing in the price set by the retailer.

Despite abundant research in recent decades, the literature of integrated marketing and operational planning models is yet scarce. Generally, supply chain planners do desire to take logistics operations into account and focus on minimizing total costs without accounting for demand and marketing managements. Literature review demonstrates that there are considerable amount of papers integrating retailing and supplying decisions (e.g. Goyal, 1976; Banerjee, 1986 and Glock, 2012). But, there are very limited papers integrating production and retailing stages with marketing. Additionally, there are some research works integrating pricing policies and meeting customer demand (e.g. Sajadieh and Jokar, 2009, and Ho, 2011). These studies on inventory planning take only pricing into account in order to determine joint order lot size without considering any market segmentation problems. Accordingly, our main motivation in this chapter is to take those policies into account which can integrate differential pricing decisions with lotsizing and shipment policies.

Similar to this chapter, Naeij and Shavandi (2010) studied inventory and pricing policy in a two-echelon supply chain with pricing, replenishment cycle and investment decisions. However, there are some differences in problem definition and assumptions. For example, they considered the pricing problem as a whole selling price, while this chapter primarily focuses on different prices at different markets. In addition, there is a significant difference in solution approaches. Naeij and Shavandi (2010) investigate different scenarios which are modeled as Stackelberg game, but this study provides a mixed-integer non-linear fractional programming model and applies two meta-heuristic algorithms to deal with the model. In addition, another major difference lies in the objective concept, i.e., maximizing return on inventory investment that has not been considered for this problem yet.

This chapter aims to propose a market-segmented supply chain model that can be applied by every supply chain dealing with shipping, multiple demand classes, pricing and inventory decisions and trying to perfect science of demand management. In addition, due to the lack of exact and efficient algorithms suitable for the developed MINLFP model, two well-known meta-heuristics are applied and examined through two test problems. As a single solution meta-heuristic, simulated annealing (SA) can find robust solutions which are converged to a good solution, i.e., the variance of SA outputs is relatively small. On the other hand, imperialist competitive algorithm (ICA), as a population-based meta-heuristic, although has the ability of finding better solutions compared to SA, it may end up with a very undesirable output. In fact, ICA's outputs have very variations that make it less favorable for more applications. Despite longer execution of SA vs. ICA, this chapter concludes with superiority of SA for other larger problems.

## MAIN FOCUS OF THE CHAPTER

To the best of our knowledge, there is no research work integrating differential pricing and lotsizing in a two-echelon supply chain based on ROII criterion to meet revenue management goals especially with considering shipment decisions of supplier. To fill this gap, the authors develop a novel integrated ordering, shipping and differential pricing model in a two-echelon supply chain under return on inventory investment maximization called differential ROII pricing. On the other hand, it can be considered as an integrated operations-marketing model developed for a two-echelon supply chain with multiple demand classes under ROII maximization.

The main contributions of this chapter can be summarized as follows:

- First, it introduces a comprehensive and practical market-based model for jointly making some major decisions in two-echelon supply chains involving a supplier and a retailer who has multiple distribution channels.
- Second, our model integrates the marketing-inventory and price discrimination decisions into a single model under ROII maximization of the chain as a financial performance criterion.
- Third, a third party can buy the product at a low price and resell it at a high price. Price discrimination is subject to arbitrage whenever a product can be purchased in a low-price submarket and delivered (transported) cheaply to be resold at a higher price elsewhere. For this reason, retailers determine prices for submarkets to avoid resale from low-price submarkets that would cannibalize sales in higher-price submarkets (Phillips, 2005). To overcome this deficiency, the researchers study and formulate differential pricing to avoid arbitrage and resale between submarkets. In fact, price discrimination creates a strong motivation for arbitrageurs to find a way to purchase the product at the low price and resell it to high willingness-to-pay customers below the market price, keeping the difference for themselves.
- Fourth, many firms use marketing expenditures to increase sales volume, encourage customers to switch to their firm. Therefore, by applying differential marketing expenditure it is attempted to synthesize marketing-operation interaction in order to influence consumer demand more effectively in each market segment.

## NOTATIONS AND PROBLEM DESCRIPTION

### Assumptions

Consider a supply chain for a product which consists of a single supplier and single buyer operating in a monopoly market. The objective of the whole supply chain is optimizing return on inventory investment by making optimal decisions regarding the lot-sizing, differential marketing expenditures, setting prices in all market segments and number of shipments of supplier. The retailer can sell the product to $n$ distribution channels which have potentially different demand functions. Therefore, the total ROII of supply chain is to be maximized through the following issues.

- **Marketing Plan:** The differential prices and marketing expenditures for each channel.
- **Retailing Plan:** The lotsizing quantity.
- **Shipment Plan:** The number of shipments from the supplier to the retailer.

This chapter considers the case of a single-supplier and single-retailer of a single product. The product is not perishable and shortage is not allowed. The planning horizon and the replenishment rate are assumed to be infinite. The demand for each market segment is a function of its price as well as marketing expenditure and is applied by a general function of price and marketing expenditure that is not specific. Market segments are not completely independent and an arbitrageur may find a way to purchase the

*Figure 1. Underlying structure of considered supply chain system*

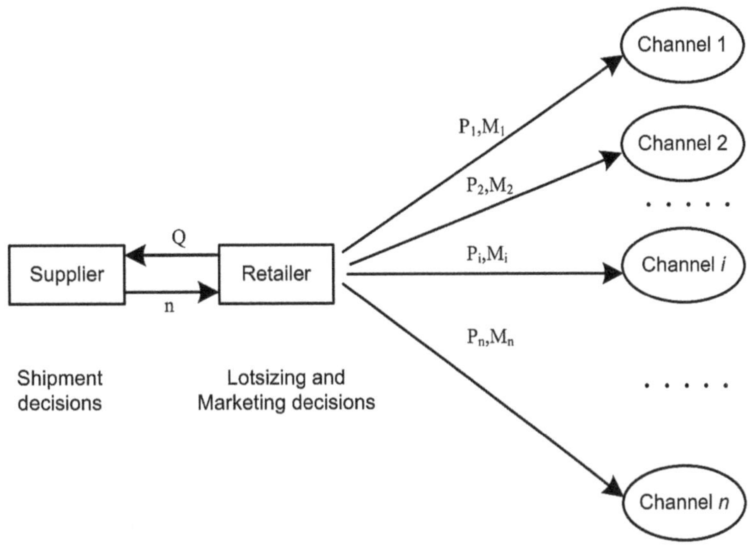


product at the low price and resell it to high willingness-to-pay customers below the market price. Finally, the supplier follows equal-sized shipments.

Figure 1 reveals the underlying structure of considered supply chain with distribution channels.

## Decision Variables and Input Parameters

$P_i$: Selling price in the ith market, $i = 1,2,...,n$ (decision variable)
$T$ : Duration of inventory cycle / cycle time (decision variable)
$M_i$: Marketing expenditure per unit in the ith market, $i = 1,2,..,n$ (\$/unit) (decision variable)
$n$ : Number of shipments (decision variable)
$A_R$: Retailer's ordering cost per order
$A_S$ : Supplier's ordering cost per order
$c_R$ : The retailer unit purchasing price (determined by the supplier)
$c_S$: The supplier unit purchasing price
$D_i(P_i)$: Demand rate for the retailer in ith market, $i = 1,2,..,n$ (units/period)
$K_i(M_i)$: Marketing function for the retailer in ith market $i = 1,2,..,n$
$h_R$ : Holding cost per unit for the retailer
$h_S$ : Holding cost per unit for the supplier
$i$: The cost of having one dollar of the item tied up in inventory for a unit time interval
$a_i$: Parameter of the price sensitive demand in ith market $i = 1,2,..,n$
$b_i$ : Parameter of the price sensitive demand in ith market $i = 1,2,..,n$
$\lambda_i$: Parameter of the marketing expenditure sensitive demand in ith market $i = 1,2,..,n$
$tc_{ij}$ : The cost which an arbitrageur transports the product from channel i to channel j $i < j$

$M^{\max}$ : The maximum allowed total marketing cost
$AP^R$ : The average profit of retailer
$AI^R$ : The average inventory investment of retailer
$ROII^R$ : The return on inventory investment of retailer
$AP^S$ : The average profit of supplier
$ROII^S$ : The return on inventory investment of supplier
$ROII^T$ : Total system ROII

As a result, the problem has $2n + 2$ decision variables.

## Model Formulation

The two-echelon supply chain's model is to set differential prices, differential marketing expenditures, lot size and number of shipments in an inventory-marketing optimization manner via total ROII (the ratio of the profit over the average investment) maximization. For inventory cycle of time span $[0, T]$, the following terms are derived:

The average profit of retailer is given by

$$AP^R\left(P, M, T\right) = \sum_{i=1}^{n} P_i D_i(P_i) K_i(M_i) - \sum_{i=1}^{n} M_i D_i(P_i) K_i(M_i) - A^R/T - 1/2\, ic^S \sum_{i=1}^{n} D_i(P_i) K_i(M_i) \quad (1)$$

where $P, M$ stand for the vector of prices and marketing expenditures. The average inventory investment of retailer, is calculated by

$$AI^R(P, M, T) = \sum_{i=1}^{n} M_i D_i(P_i) K_i(M_i) + \frac{1}{2} c^S \sum_{i=1}^{n} D_i(P_i) K_i(M_i) \quad (2)$$

Then, the return on inventory investment of retailer is

$$ROII^R(P, M, T) = AP^R(P, M, T) \,/\, AI^R(P, M, T) \quad (3)$$

and

$$ROII^R(P, M, T) = \left(\sum_{i=1}^{n} P_i D_i(P_i) K_i(M_i) - \sum_{i=1}^{n} M_i D_i(P_i) K_i(M_i) - A^R / T - \frac{1}{2} ic_h^S \sum_{i=1}^{n} D_i(P_i) K_i(M_i)\right) \Big/ \left(\sum_{i=1}^{n} M_i D_i(P_i) K_i(M_i) + \frac{1}{2} c^S \sum_{i=1}^{n} D_i(P_i) K_i(M_i)\right) \quad (4)$$

The average profit of supplier is

$$AP^S(n) = c^S \sum_{i=1}^{m} D_i(P_i)K_i(M_i) - A^S / n - \frac{1}{2} ic^V (n-1) \sum_{i=1}^{m} D_i(P_i)K_i(M_i) - c^V \sum_{i=1}^{m} D_i(P_i)K_i(M_i) \quad (5)$$

Then, the return on inventory investment of supplier is given by

$$ROII^S(n) = \frac{\left( c^S \sum_{i=1}^{m} D_i(P_i)K_i(M_i) - A^S / n - \frac{1}{2} ic^V (n-1) \sum_{i=1}^{m} D_i(P_i)K_i(M_i) - c^V \sum_{i=1}^{m} D_i(P_i)K_i(M_i) \right)}{\left( \frac{1}{2} c^V (n-1) \sum_{i=1}^{m} D_i(P_i)K_i(M_i) \right)} \quad (6)$$

Suppose that retailer and supplier decide to cooperate and agree to follow the jointly optimal integrated policy. The total system ROII under joint optimization is going to be maximized.

$$\max\ ROII^T(\mathrm{P}, \mathrm{M}, T, n) = \left( \begin{array}{l} \sum_{i=1}^{n} P_i D_i(P_i)K_i(M_i) - \sum_{i=1}^{n} M_i D_i(P_i)K_i(M_i) - A^R / T - \frac{1}{2} ic_h^S \sum_{i=1}^{n} D_i(P_i)K_i(M_i) \\ + c^S \sum_{i=1}^{m} D_i(P_i)K_i(M_i) - A^S / n - \frac{1}{2} ic^V (n-1) \sum_{i=1}^{m} D_i(P_i)K_i(M_i) \end{array} \right) \Bigg/ \left( \sum_{i=1}^{n} M_i D_i(P_i)K_i(M_i) + \frac{1}{2} c^S \sum_{i=1}^{n} D_i(P_i)K_i(M_i) + \frac{1}{2} c^V (n-1) \sum_{i=1}^{m} D_i(P_i)K_i(M_i) \right) \quad (7)$$

Subject to:

$$\sum_{i=1}^{n} M_i D_i(P_i)K_i(M_i) \leq M^{\max} \quad (8)$$

$$P_i - P_j \leq tc_{ij} \quad \forall i < j \quad (9)$$

$$P_i \leq P_j \quad \forall i < j \quad (10)$$

$$D_i(P_i) \geq 0 \quad \forall i \quad (11)$$

$$P_i \geq 0, M_i \geq 0 \quad \forall i \tag{12}$$

$$T > 0 \tag{13}$$

*n integer*

It should be mentioned that demand of each market is determined via a linear function of selling price in that market with two parameters, *a* and *b* as follows.

$$D_i(P_i) = a_i - b_i P_i \tag{14}$$

The marketing function is formulated as equation (15).

$$K_i(M_i) = M_i^{\lambda_i} \tag{15}$$

## SOLUTION PROCEDURES

The formulation given in (7)-(13) is a highly constrained mixed-integer non-linear fractional programming model (MINLFP). This characteristic causes the model to be hard enough to be solved by an exact method. If a problem includes non-convex functions, there are, in general, no direct optimization methods available, which would guarantee global optimality (Porn et al. 1999). Accordingly, a heuristic search algorithm is required to solve the model. Indeed, as well as combinatorial optimization problems, meta-heuristics are remarkable tools for optimization problems with continuous search spaces. But, some algorithms are only capable of searching combinatorial spaces, and not suitable for the problem of this chapter.

Generally, meta-heuristics are divided into single-solution and population algorithms (Gendreau and Potvin, 2005). In single-solution meta-heuristics only one solution is considered, while population meta-heuristics work with a multiplicity of solutions. In this chapter, a Simulated Annealing (SA) as well as an Imperialist Competitive Algorithm (ICA), as a single-solution and a population meta-heuristic respectively, are employed to deal with the problem. This section briefly describes the algorithms. The results are provided in the next section.

### Simulated Annealing

The Simulated Annealing (SA) algorithm is a well-known neighborhood search-based algorithm (Damodaran and Vélez-Gallego 2012). The SA algorithm derives its acceptance mechanism from the annealing process in order to escape from local optima. In high temperatures, the algorithm may replace the current solution, $x_1$ by its worse neighbor, $x_2$, through calculating the acceptance probability, given by $\exp\left\{\left(f(x_2) - f(x_1)\right)/\left(c.T\right)\right\}$, where c is a coefficient of the temperature (*T*). The value of *T* varies from a relatively large value to a small value close to zero. These values are controlled by a cooling schedule that specifies the initial and incremental temperature values at each stage of the algorithm (Tavakkoli-Moghaddam et al. 2006).

The SA algorithm is used to solve the following mixed integer non-linear fractional programming model:

$$Min\, f\left(X,Z\right) \tag{16}$$

Subject to:

$$h_j\left(X,Z\right) \le 0, j = 1,..,m \tag{17}$$

$$g_j\left(X,Z\right) = 0, j = 1,..,n \tag{18}$$

$$X^L \le X \le X^U \tag{19}$$

$Z$ *integer*

where $X$ is the vector of continuous decision variables and $Z$ represents the vector of integer variables. Eqs. (17) and (18) are *m* inequality constraints and *n* equality constraints, respectively. In addition, $X^L$ and $X^U$ are lower and upper bounds of the decision variables.

## Initialization

The SA algorithm starts from an initial random solution. During the search process the algorithm generates, based on some mechanism, a new solution in the neighborhood of the current solution (Damodaran and Vélez-Gallego, 2012). Providing a feasible start value to initialize the search process is important and has a significant influence on the algorithm output. In this chapter, the initial solution is a vector, elements of which determine the prices of channels, marketing expenditures, lot size and number of shipments.

## Searching Mechanism

Due to the continuous nature of most the decision variables, application of the normal distribution function seems suitable for generating new neighborhoods. For this purpose, the current solution vector X is considered as the mean of the distribution function and the standard deviation is obtained with multiplying an arbitrary coefficient of variance, i.e., cv by the mean vector, i.e. X. After generating new X, the new integer variable is gained by rounding its corresponding continuous value.

## Temperature Setting and the Cooling Pattern

Settings related to the temperature of the algorithm include an initial temperature, final temperature, cooling schedule and the number of observed temperatures. Both initial and final temperatures need to be selected carefully. The initial temperature should be high enough so that all states of the search space have an equal chance to be accepted in the first stages of the algorithm. However, the final temperature should be low enough to not permit worse solutions to replace better ones in the final stages. Apparently, increasing the number of observed temperatures would improve the solution quality. On the other hand, it would increase the running time of the algorithm (Seyed-Alagheband et al., 2011). The idea of accepting non-improving solutions, with certain probability, helps the algorithm to avoid being trapped at a local optimum. This chapter incorporates two types of cooling schedule given by Eqs. (20) and (21).

$$T_k = T_{k-1} - 1 \tag{20}$$

$$T_k = \theta T_{k-1} \tag{21}$$

where $T_k$ is the temperature at stage k, $\theta$ is the temperature reduction rate.

## Evaluating Fitness Function and Treating Constraints

The performance of each solution is evaluated with the fitness function. The constrained non-linear programming models can be modified into equivalent unconstrained ones using penalty terms. In this chapter, there is no equality constraint. In general, a suitable penalty function must incur a positive penalty for infeasible points and no penalty for feasible points (Bazaraa et al. 1993). Therefore, the equivalent models can be generally presented as follows.

$$\varphi\left(X,Z\right) = f\left(X,Z\right) + \mu\left[\sum_{j=1}^{m}\left\{max\left\{0,h_j\left(X\right)\right\}\right\}^2 + \left\{max\left\{0,X-X^U\right\}\right\}^2 + \left\{max\left\{0,X^L-X\right\}\right\}^2\right] \tag{22}$$

where $\mu$ is a positive large number, i.e., penalty number? The authors refer to the function $\varphi\left(X,Z\right)$ as the augmented Lagrangian function in which objective function is augmented with a penalty term. If a constraint is satisfied, then maximum $max\left\{0,.\right\} = 0$ and no penalty are incurred. On the other hand, if a constraint is not satisfied, then maximum $max\left\{0,.\right\} > 0$, and the penalty term is realized. In fact, the technique transforms a constrained problem into a sequence of unconstrained problems. The constraints are placed into the objective function via a penalty parameter in a way that penalizes any violation of the constraints. The pseudo code of the proposed SA is outlined as follows.

Generate a solution $\left(X,Z\right)$ and set $T = T_0$
Save $\left(X,Z\right)$ as the best solution

*While T*>1
*For each* $n \in N$

1. *Produce a new solution* $\left(X,Z\right)^{New}$ *by Normal distribution function and modify it* according to the lower and upper bound values
2. *If* $\mathrm{ROII}^{(X,Z)^{New}} \geq \mathrm{ROII}^{(X,Z)}$ *then* accept $\left(X,Z\right)^{New}$
3. *Else if* $\exp\left\{\left(\varphi\left(X^{new},Z^{new}\right)-\varphi\left(X,Z\right)\right)/\left(c.T\right)\right\}$ *is greater than the* ***acceptance probability***

*then* $\left(X,Z\right)=\left(X,Z\right)^{New}$

4. Update the best solution

End For
Decrease ***T*** by the cooling scheme
End While

## Imperialist Competitive Algorithm

As a population-based meta-heuristic, ICA (Atashpaz-Gargariand & Lucas, 2007), was originally developed to deal with continuous search space optimization problems. Recently, many applications of ICA for combinatorial optimization problems has been developed indicating the increasing attention to this algorithm (Hosseini and Al Khaled, 2014). ICA is initialized with a number of random solutions, namely countries. There exist a few countries with high value of performance measure which are called imperialists. Each imperialist possesses some other weaker countries named colonies of that empire. During the algorithm execution colonies are moved toward their imperialists, while empires compete to possess the weakest colony of the weakest empire. This procedure is continued until criteria measures are met or all colonies are possessed by one empire.

### Initialization

Each country is a vector of decision variables which almost all of them are continuous. A country is conceptually an exact match of what is called solution in SA algorithm. Creating a feasible initial country, each variable is uniformly generated in its domain. Putting together all random countries constitutes the population. Then, each country is evaluated based on the objective function of the problem, i.e., $\varphi$, formulated in Eq. (22). $N_{imp}$ countries of top ones (with the highest value of $\varphi$) are adopted as imperialists. In order to divide remained $N_{col}$ countries among the empires, for each imperialist, a normalized value of objective function, say $C_n$, is determined using the following equation.

$$C_n = \varphi_n - \min_i\{\varphi_i\} \tag{23}$$

Using the normalized value of each imperialist, the normalized power is defined by Eq. (24).

$$p_n = \left| \frac{C_n}{\sum_{i=1}^{N_{imp}} C_i} \right| \qquad (24)$$

By the use of normalized power, the initial number of colonies which belong to each imperialist is determined via equation (25).

$$NC_n = round\{p_n . N_{col}\} \qquad (25)$$

Based on initial number of each imperialist, $N_{col}$ countries are randomly assigned to the empires.

## Assimilation

At each iteration, colonies are moved toward their imperialists. Let $d_{ij}$ be the distance between the $j^{th}$ colony and the $i^{th}$ imperialist. The movement amount, $x_{ij}$, is a random uniform number calculated via the following equation, where $\beta$ is a real positive number, usually greater than 1.

$$x_{ij} \sim U(0, \beta d_{ij}) \qquad (26)$$

It is obvious that the position of all colonies is changed to a new position during the assimilation process. Then, there is a possibility of reaching a position better than former positions or even better than the imperialist. In the latter case, labels of colony and imperialist are exchanged between the colony and its former imperialist.

## Imperialistic Competition

Empires compete with each other to reach the possession of more powers. This power is attainable through taking more colonies. Hence, the weakest colony of the weakest empire is indicated as the candidate of possession by the most likelihood imperialist. For this purpose, total power of the $n^{th}$ empire is determined using the following equation.

$$TC_n = \varphi_i + \xi \frac{\sum_{i=1}^{NC_n} \varphi_i}{NC_n} \qquad (27)$$

In Eq. (27), $\xi$ is a parameter that is normally set to a positive decimal number, e.g., 0.1. As the objective of this chapter is maximization of return on inventory investment, the empire with highest $TC_n$ has the greater possibility of taking possession of the weakest colony. Similar to the initialization step, the normalized total power as well as the possession probability of the $n^{th}$ empire is obtained by equations (28) and (29) respectively.

$$NTC_n = TC_n - \min_{i=1}^{N_{imp}}\{TC_i\} \tag{28}$$

$$pr_n = \left| \frac{NTC_n}{\sum_{i=1}^{N_{imp}} NTC_i} \right| \tag{29}$$

Let P be a vector containing *pr* values obtained via equation (29). Consider R as a uniformly distributed random vector having the same size of P. By subtracting R from P a new vector is created whose maximum value indicates the index of empire with the right of taking possession of the weakest colony.

### General Framework

The structure of assimilation process of ICA is basically similar to Particle Swarm Optimization (PSO) algorithm. However, the empires competition feature plays a key role in convergence of ICA and improving its efficiency versus PSO. The general procedure of ICA is provided as follows.

**Step 1:** Initialization.
**Step 2:** Assimilation: moving colonies toward their imperialists.
Exchanging position of colony and its empire if the colony has better performance.
**Step 3:** Competition: assigning the weakest colony of the weakest empire to the empire with the right of possessing that colony.
Eliminating the empire with no colony.
**Step 4:** Stopping criteria: if it is met then stop, otherwise repeat steps 2 and 3.

## NUMERICAL EXPERIMENTS

### Test Problems

In order to examine SA and ICA, six test problems have been generated including 3-, 5- and 10-market segments. Tables 1 to 6 provide input parameters of each test problem.

### Tuning the SA via the Taguchi Design

Due to the random nature of SA, the output results may differ for different executions. The aim is to determine an appropriate set of parameters for which the output results converge to the optimal solution. Experimentally, there are five factors with high degrees of influence on the output results, some of them obtained from Mosadegh et al. (2012). As a result, orthogonal array $L_{16}(4^5)$ is fitted to our experiments. Table 7 shows levels of each factor in detail. It should be mentioned that the cooling pattern as well as the number of iterations at any temperature are simultaneously embedded in factor D, where 1 and 2

*Table 1. Input parameters of 3-market segment problem (set3_1)*

| Markets | 1 | 2 | 3 |
|---|---|---|---|
| $(a_i, b_i)$ | (38000, 20) | (50000, 45) | (63000, 65) |
| $\lambda_i$ | 0.61 | 0.5 | 0.43 |
| $tc_{ij}$ | - | 63 | 40 |
| | - | - | 28 |
| $A_b$=9000 | $C_b$=166 | $M^{max}$=70500 | |
| $A_v$=7600 | $C_v$=285 | $i_h$=0.18 | |

*Table 2. Input parameters of 3-market segment problem (Set3_2)*

| Markets | 1 | 2 | 3 |
|---|---|---|---|
| $(a_i, b_i)$ | (48000, 25) | (60000, 65) | (74000, 71) |
| $\lambda_i$ | 0.41 | 0.33 | 0.22 |
| $tc_{ij}$ | - | 63 | 40 |
| | - | - | 28 |
| $A_b$=9000 | Cb=290 | Mmax=85000 | |
| $A_v$=7600 | Cv=195 | ih=0.18 | |

*Table 3. Input parameters of 5-market segment problem (Set5_1)*

| Markets | 1 | 2 | 3 | 4 | 5 |
|---|---|---|---|---|---|
| $(a_i, b_i)$ | (38000, 20) | (50000, 45) | (63000, 65) | (65000, 72) | (70000, 80) |
| $\lambda_i$ | 0.61 | 0.5 | 0.43 | 0.35 | 0.3 |
| $tc_{ij}$ | - | 63 | 40 | 44 | 45 |
| | - | - | 28 | 40 | 41 |
| | - | - | - | 35 | 36 |
| | - | - | - | - | 29 |
| $A_b$=9000 | | $C_b$=166 | | $M^{max}$=179000 | |
| $A_v$=7600 | | $C_v$=285 | | $i_h$=0.18 | |

stand for the cooling type by either Eq. (20) or Eq. (21) and the adjacent number indicates the number of iterations per temperature. The aim is to find levels of factors at which the S/N ratio, i.e., the robustness measure, is maximized. For optimization problems with maximization objective, equivalently the larger the better, the S/N ratio is computed as S/N = -10 log (1 / objective function)$^2$.

To perform the executions, test problems Set3_1 and Set5_1 have been employed. By the use of MINITAB 16.1.1, the S/N ratios are computed and depicted in Figure 2. The factors are set at levels with high S/N ratio values. As an exception, there is no significant difference between D(2) and D(4), where the researchers opt D(4) instead of D(2) since it saves more running times. Table 8 gives the tuned parameters of the proposed SA.

*Table 4. Input parameters of 5-market segment problem (Set5_2)*

| Markets | 1 | 2 | 3 | 4 | 5 |
|---|---|---|---|---|---|
| $(a_i, b_i)$ | (31000, 18) | (44500, 39) | (56000, 58) | (62000, 65) | (75000, 75) |
| $\lambda_i$ | 0.7 | 0.58 | 0.47 | 0.38 | 0.5 |
| $tc_{ij}$ | - | 60 | 38 | 28 | 40 |
| | - | - | 27 | 50 | 48 |
| | - | - | - | 35 | 23 |
| | - | - | - | - | 28 |
| $A_b$=9000 | | $C_b$=262 | | $M^{max}$=160000 | |
| $A_v$=7600 | | $C_v$=179 | | $i_h$=0.18 | |

*Table 5. Input parameters of 10-market segment problem (Set10_1)*

| Markets | 1 | 2 | 3 | 4 | 5 | 6 | 7 | 8 | 9 | 10 |
|---|---|---|---|---|---|---|---|---|---|---|
| $(a_i, b_i)$ | (33000, 21) | (45400, 34) | (55000, 55) | (63000, 72) | (38000, 61) | (43000, 75) | (53000, 45) | (75000, 63) | (65000, 82) | (70000, 85) |
| $\lambda_i$ | 0.7 | 0.59 | 0.47 | 0.3 | 0.41 | 0.28 | 0.39 | 0.54 | 0.65 | 0.42 |
| $tc_{ij}$ | - | 45 | 22 | 25 | 36 | 61 | 55 | 27 | 14 | 12 |
| | - | - | 10 | 15 | 25 | 53 | 65 | 25 | 58 | 55 |
| | - | - | - | 35 | 45 | 52 | 22 | 53 | 60 | 57 |
| | - | - | - | - | 30 | 24 | 17 | 30 | 73 | 13 |
| | - | - | - | - | - | 42 | 17 | 70 | 32 | 28 |
| | - | - | - | - | - | - | 12 | 31 | 36 | 26 |
| | - | - | - | - | - | - | - | 26 | 11 | 19 |
| | - | - | - | - | - | - | - | - | 22 | 76 |
| | - | - | - | - | - | - | - | - | - | 59 |
| $A_b$=15000 | | | | $C_b$=250 | | | $M^{max}$=700000 | | | |
| $A_v$=8500 | | | | $C_v$=185 | | | $i_h$=0.18 | | | |

## Tuning the ICA

The parameters of ICA are adjusted based on Atashpaz Gargari (2014) and Wang et al. (2015), dealing with six well-known benchmarks.

## Empirical Results

Both SA and ICA are examined thorough six test problems with three, five and ten market segments. Parameters of each test problem is provided in Tables 1 to 6 The algorithms have been coded in MATLAB 8.1 and executed on a laptop computer equipped with a 2.4 GHz CPU and 2 GB RAM. Among five random executions of each test problem via SA and ICA, the best obtained solutions are reported

*Table 6. Input parameters of 10-market segment problem (Set10_2)*

| Markets | 1 | 2 | 3 | 4 | 5 | 6 | 7 | 8 | 9 | 10 |
|---|---|---|---|---|---|---|---|---|---|---|
| $(a_i, b_i)$ | (33000, 21) | (45400, 34) | (55000, 55) | (63000, 72) | (38000, 61) | (43000, 75) | (53000, 45) | (75000, 63) | (65000, 82) | (70000, 85) |
| $\lambda_i$ | 0.5 | 0.45 | 0.38 | 0.3 | 0.8 | 0.62 | 0.39 | 0.41 | 0.56 | 0.37 |
| $tc_{ij}$ | - | 24 | 25 | 65 | 69 | 42 | 35 | 51 | 36 | 28 |
| | - | - | 35 | 44 | 46 | 22 | 38 | 62 | 71 | 55 |
| | - | - | - | 30 | 33 | 26 | 73 | 45 | 26 | 44 |
| | - | - | - | - | 28 | 51 | 66 | 55 | 62 | 51 |
| | - | - | - | - | - | 34 | 28 | 33 | 24 | 46 |
| | - | - | - | - | - | - | 27 | 32 | 61 | 45 |
| | - | - | - | - | - | - | - | 46 | 55 | 65 |
| | - | - | - | - | - | - | - | - | 52 | 36 |
| | - | - | - | - | - | - | - | - | - | 77 |
| $A_b$=11500 | | | | | $C_b$=286 | | | $M^{max}$=780000 | | |
| $A_v$=12000 | | | | | $C_v$=200 | | | $i_h$=0.18 | | |

*Table 7. factors' levels of L16(45)*

| A: Coefficient of C | B: Initial Temperature | C: Acceptance Probability | D: Cooling Scheme | E: Coefficient of Variance |
|---|---|---|---|---|
| A(1): 1 | B(1): 1000 | C(1): 0.50 | D(1): (I, 10) | E(1): 0.005 |
| A(2): 100 | B(2): 10000 | C(2): 0.85 | D(2): (I, 20) | E(2): 0.010 |
| A(3): 1000 | B(3): 100000 | C(3): 0.95 | D(3): (II, 10) | E(3): 0.050 |
| A(4): 100000 | B(4): 500000 | C(4): 0.98 | D(4): (II, 20) | E(4): 0.100 |

*Table 8. Final parameter setting*

| A: Coefficient of C | B: Initial Temperature | C: Acceptance Probability | D: Cooling Scheme | E: Coefficient of Variance |
|---|---|---|---|---|
| A(1): 1 | B(4): 500000 | C(4): 0.98 | D(4): (II, 20) | E(1): 0.005 |

in Table 9. According to this table, ICA has better performance in terms of objective function dealing with both 3-, 5- and 10-market segment problems in comparison with SA. Figure 3 depicts convergence of SA and ICA during their executions. Based on these results, it seems that ICA should be chosen to be utilized for more real case applications. However, another analysis about algorithm robustness reveals that although ICA could find better solutions, it has higher volatility in its outputs, compared to SA.

Table 10 shows a brief statistics of five runs of each test problem as well as average running times. It indicates that SA performs more robust than ICA. Figure 4 confirms this statement as well. ICA may find solutions better than SA, but it needs more executions to increase this possibility. On the other hand,

*Figure 2. S/N ratio values*

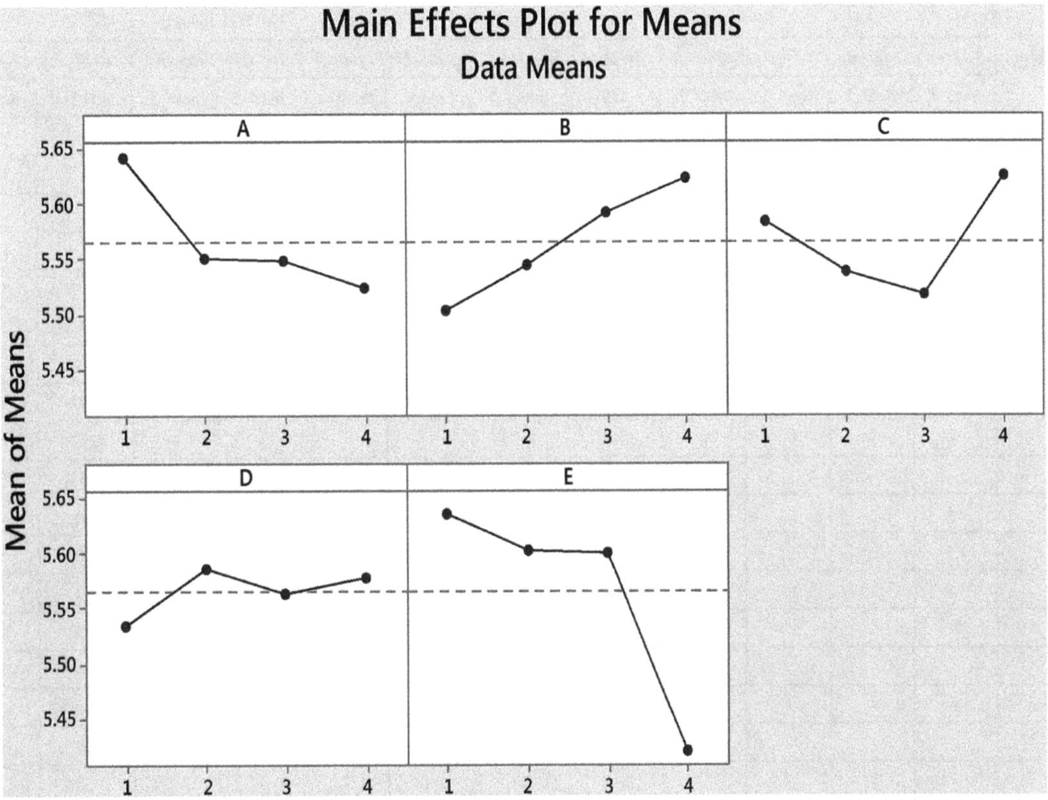


*Figure 3. Convergence diagram of SA (left) and ICA (right). The blue plot shows the obtained best objective at each iteration. The red plot (only for ICA) reveals average objective values of all countries at each iteration*

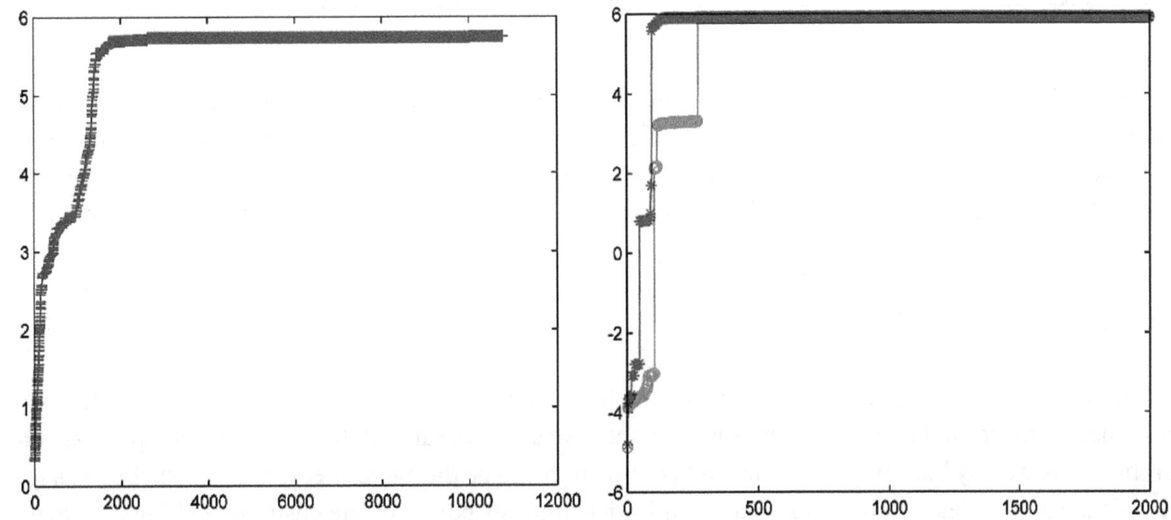


**For a more accurate representation see the electronic version.*

*Table 9. Best obtained prices, marketing expenditures, shipment and lotsizes*

| Variables | SA Results | | | | | | ICA Results | | | | | |
|---|---|---|---|---|---|---|---|---|---|---|---|---|
| | *3-Market Segments* | | *5-Market Segments* | | *10-Market Segments* | | *3-Market Segments* | | *5-Market Segments* | | *10-Market Segments* | |
| | Set3_1 | Set3_2 | Set5_1 | Set5_2 | Set10_1 | Set10_2 | Set3_1 | Set3_2 | Set5_1 | Set5_2 | Set10_1 | Set10_2 |
| $P_1$ | 989.5 | 1003.9 | 988.7 | 975.4 | 530.2 | 567.3 | 990.0 | 1005.0 | 990.0 | 978.0 | 841.7 | 752.8 |
| $P_2$ | 977.1 | 952.8 | 975.3 | 974.8 | 527.9 | 560.2 | 978.0 | 966.8 | 978.0 | 977.0 | 841.7 | 750.5 |
| $P_3$ | 949.7 | 949.6 | 949.0 | 949.2 | 525.8 | 558.9 | 950.0 | 950.0 | 950.0 | 950.0 | 841.7 | 748.9 |
| $P_4$ | - | - | 948.1 | 948.6 | 525.1 | 555.3 | - | - | 950.0 | 950.0 | 841.7 | 748.7 |
| $P_5$ | - | - | 944.2 | 948.6 | 524.2 | 553.0 | - | - | 950.0 | 950.0 | 829.8 | 747.9 |
| $P_6$ | - | - | - | - | 523.6 | 549.7 | - | - | - | - | 829.8 | 728.8 |
| $P_7$ | - | - | - | - | 523.5 | 547.6 | - | - | - | - | 829.8 | 727.9 |
| $P_8$ | - | - | - | - | 521.9 | 546.4 | - | - | - | - | 829.8 | 727.7 |
| $P_9$ | - | - | - | - | 519.7 | 543.4 | - | - | - | - | 829.8 | 726.9 |
| $P_{10}$ | - | - | - | - | 519.5 | 542.4 | - | - | - | - | 829.7 | 726.3 |
| $M_1$ | 1.1 | 2.5 | 1.1 | 1.5 | 1.3 | 2.3 | 1.0 | 1.0 | 1.0 | 1.0 | 3.0 | 3.0 |
| $M_2$ | 1.3 | 2.6 | 1.1 | 2.1 | 1.9 | 1.6 | 1.0 | 3.0 | 1.0 | 1.0 | 3.0 | 3.0 |
| $M_3$ | 1.9 | 0.0 | 1.0 | 0.2 | 1.7 | 1.5 | 1.0 | 1.0 | 1.0 | 3.0 | 1.0 | 3.0 |
| $M_4$ | - | - | 1.1 | 1.6 | 0.3 | 2.5 | - | - | 3.0 | 3.0 | 3.0 | 1.0 |
| $M_5$ | - | - | 1.6 | 1.5 | 0.3 | 1.1 | - | - | 3.0 | 3.0 | 3.0 | 1.0 |
| $M_6$ | - | - | - | - | 2.2 | 1.5 | - | - | - | - | 3.0 | 1.0 |
| $M_7$ | - | - | - | - | 1.0 | 0.1 | - | - | - | - | 3.0 | 1.0 |
| $M_8$ | - | - | - | - | 1.5 | 2.4 | - | - | - | - | 1.0 | 3.0 |
| $M_9$ | - | - | - | - | 1.6 | 1.2 | - | - | - | - | 1.0 | 3.0 |
| $M_{10}$ | - | - | - | - | 1.6 | 2.4 | - | - | - | - | 3.0 | 1.0 |
| T | 0.6 | 0.6 | 0.6 | 0.6 | 1.5 | 0.6 | 0.6 | 0.6 | 0.6 | 0.6 | 0.6 | 0.6 |
| n | 2.0 | 2.0 | 2.0 | 2.0 | 2.0 | 2.0 | 2.0 | 2.0 | 2.0 | 2.0 | 2.0 | 2.0 |
| Objective function | 5.6564 | 5.2744 | 5.8176 | 5.7125 | 0.8847 | 2.2 | 5.6769 | 5.2950 | 6.0671 | 5.7063 | 4.7534 | 3.4347 |

SA is capable of finding good solutions that are guaranteed with respect to their objective functions. In addition, SA takes more execution time than ICA. However, robustness of SA is more preferable than quickness and luckiness of ICA in finding better solutions, because one execution of SA would be sufficient to ensure a good solution, but ICA has to be executed at least five times to verify its solutions.

## FUTURE RESEARCH DIRECTIONS

For future researches, the authors recommend the followings areas: Some parameters of the model may be either fuzzy or random variable. In this case, the researchers deal with fuzzy/stochastic optimization problems. A worthy but complex extension of our work could be the game-theoretic version of this chapter, if the supplier and retailer coordinate in a leader-follower policy or the customer exhibits strategic behavior especially in a multi- period planning horizon. In addition, it is appropriate to involve the other

*Table 10. Comparison of SA and ICA in terms of objective value and execution time*

| Performance | | | Worst Solution | Average Solution | Best Solution | Range = Best – Worst | Average Run Time (s) |
|---|---|---|---|---|---|---|---|
| 3-Market Segment | Set3_1 | *SA* | 5.6423 | 5.6507 | 5.6564 | 0.0142 | 27 |
| | | *ICA* | 5.6250 | 5.6636 | 5.6769 | 0.0520 | 24 |
| | Set3_2 | *SA* | 5.2124 | 5.2400 | 5.2744 | 0.0620 | 21 |
| | | *ICA* | 5.2730 | 5.2906 | 5.2950 | 0.0221 | 15 |
| 5-Market Segment | Set5_1 | *SA* | 5.7273 | 5.7805 | 5.8176 | 0.0903 | 45 |
| | | *ICA* | 0.7647 | 4.1684 | 6.0671 | 5.3023 | 24 |
| | Set5_2 | *SA* | 5.6702 | 5.6914 | 5.7125 | 0.0423 | 21 |
| | | *ICA* | 0.2013 | 1.9305 | 5.7063 | 5.5051 | 15 |
| 10-Market Segment | Set10_1 | *SA* | 0.4477 | 0.6259 | 0.8847 | 0.4369 | 22 |
| | | *ICA* | 0.4303 | 1.6897 | 4.7534 | 4.3231 | 17 |
| | Set10_2 | *SA* | 1.1438 | 1.8676 | 2.1915 | 1.0477 | 22 |
| | | *ICA* | 0.0652 | 1.2856 | 3.4347 | 3.3695 | 17 |

*Figure 4. Comparison of output volatility between SA and ICA solving Set3_1 and Set5_1*

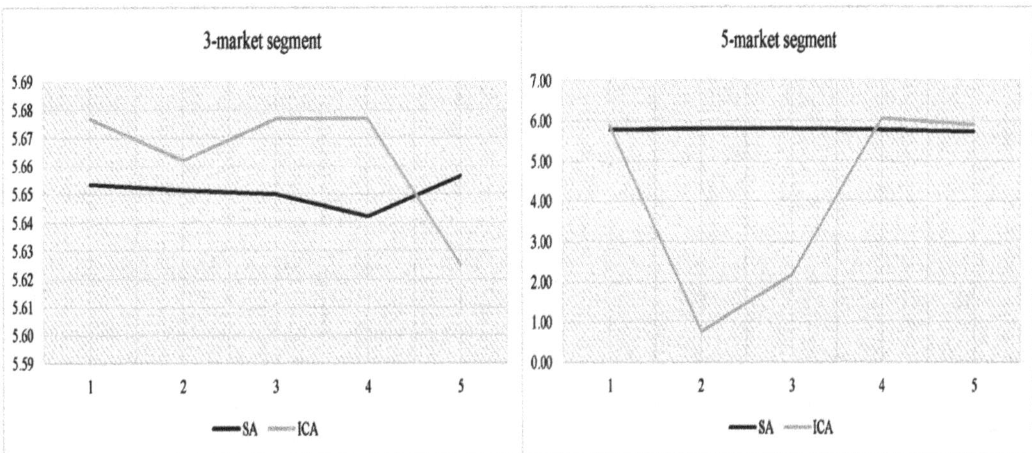


popular objectives in the literature and develop a multi-objective model to deal with the problem. As an application, the underlying problem can be illustrated in a simple purchasing scenario in food industries where a retailer orders a particular quantity of a kind of food to a vendor, who in turn supplies the food upon receipt of the order and ships the entire lot to the retailer facing with the price-sensitive demand.

## CONCLUSION

In order to make the joint pricing and lotsizing models more applicable to real-world production and inventory control problems, in this chapter, the authors extended this model by assuming several market segments in which return on inventory investment was involved as a criterion in a two-echelon supply chain, the total marketing cost had limited budget and there were some constraints to take arbitrage into account. The proposed model of this chapter was applied to a two-level supply chain consisting of a single supplier and a single retailer that operates under return on inventory investment of multiple channels. Under these conditions, the researchers formulated the problem as a mixed-integer non-linear fractional programming model and proposed two meta-heuristic algorithms, i.e., simulated annealing and imperialist competitive algorithm, to solve it. Empirical results showed that variation of ICA's outputs is relatively high, but it may find better solutions in comparison with SA. Since simulated annealing provides more robust solutions, it is recommended to utilize SA for real cases or other larger problems, although its running time is more than ICA.

## REFERENCES


Atashpaz Gargari, E. (n. d.). Retrieved September 1st, 2014, from http://atashpaz.com/

Atashpaz-Gargari, E., & Lucas, C. (2007). Imperialist competitive algorithm: An algorithm for optimization inspired by imperialistic competition. In *Proceedings of the IEEE Congress on Evolutionary Computation.*

Banerjee, A. (1986). A joint economic-lot-size model for purchaser and vendor. *Decision Sciences*, *17*(3), 292–311. doi:10.1111/j.1540-5915.1986.tb00228.x

Bazaraa, M. S., Sherali, H. D., & Shetty, C. M. (1993). *Nonlinear Programming: Theory and Algorithms.* New York: John Wiley.

Ben-Daya, M., Darwish, M., & Ertogral, K. (2008). The joint economic lot sizing problem: Review and extensions. *European Journal of Operational Research*, *185*(2), 726–742. doi:10.1016/j.ejor.2006.12.026Chen, X., & Simchi-Levi, D. (2012). *Pricing and Inventory Management. Handbook of Pricing* (R. Philips & O. Ozalp, Eds.). Oxford University Press.

Damodaran, P., & Vélez-Gallego, M. (2012). A simulated annealing algorithm to minimize makespan of parallel batch processing machines with unequal job ready times. *Expert Systems with Applications*, *39*(1), 1451–1458. doi:10.1016/j.eswa.2011.08.029

Gendreau, M., & Potvin, J.-Y. (2005). Metaheuristics in Combinatorial Optimization. *Annals of Operations Research*, *140*(1), 189–213. doi:10.1007/s10479-005-3971-7

Ghasemy Yaghin, R., Fatemi Ghomi, S. M. T., & Torabi, S. A. (2013). A possibilistic multiple objective pricing and lot-sizing model with multiple demand classes. *Fuzzy Sets and Systems*, *231*(16), 26–44. doi:10.1016/j.fss.2012.11.012

Ghasemy Yaghin, R., Fatemi Ghomi, S. M. T., & Torabi, S. A. (2014). Enhanced joint pricing and lotsizing problem in a two-echelon supply chain with logit demand function. *International Journal of Production Research*, *52*(17), 4967–4983. doi:10.1080/00207543.2014.885665

Glock, C. H. (2012). The joint economic lot size problem: A review. *International Journal of Production Economics*, *135*(2), 671–686. doi:10.1016/j.ijpe.2011.10.026

Goyal, S. K. (1976). An integrated inventory model for a single supplier–single customer problem. *International Journal of Production Research*, *15*(1), 107–111. doi:10.1080/00207547708943107

Hammami, R., & Frein, Y. (2014). Integration of the profit-split transfer pricing method in the design of global supply chains with a focus on offshoring context. *Computers & Industrial Engineering*, *76*, 243–252. doi:10.1016/j.cie.2014.07.021

Ho, C. (2011). The optimal integrated inventory policy with price-and-credit-linked demand under two-level trade credit. *Computers & Industrial Engineering*, *60*(1), 117–126. doi:10.1016/j.cie.2010.10.009

Hosseini, S., & Al Khaled, A. (2014). A survey on the Imperialist Competitive Algorithm metaheuristic: Implementation in engineering domain and directions for future research. *Applied Soft Computing*, *24*, 1078–1094. doi:10.1016/j.asoc.2014.08.024

Kleijn, M. J., & Dekker, R. (1998). *An overview of inventory systems with several demand classes. Econometric Institute Report 9838/A*. Rotterdam, The Netherlands: Erasmus University.

Lenskold, J. (2003). *Marketing ROI: The Path to Campaign, Customer, and Corporate Profitability*. San Diego, CA, U.S.A.: McGraw-Hill.

Li, J., Min, K. J., Otake, T., & Voorhis, T. M. (2008). Inventory and investment in setup and quality operations under return on investment maximization. *European Journal of Operational Research*, *185*(2), 593–605. doi:10.1016/j.ejor.2006.11.045

Mosadegh, H., Zandieh, M., & Fatemi Ghomi, S. M. T. (2012). Simultaneous solving of balancing and sequencing problems with station-dependent assembly times for mixed-model assembly lines. *Applied Soft Computing*, *12*(4), 1359–1370. doi:10.1016/j.asoc.2011.11.027

Naeij, J., & Shavandi, H. (2010). An optimal lot sizing and pricing planning in two-echelon supply chain. *International Journal of Industrial Engineering Computations*, *1*(1), 11–32. doi:10.5267/j.ijiec.2010.01.002

Otake, T., & Min, K. J. (2001). Inventory and investment in quality improvement under return on investment maximization. *Computers & Operations Research*, *28*(10), 113–124. doi:10.1016/S0305-0548(00)00022-8

Otake, T., Min, K. J., & Chen, C. (1991). Inventory and investment in setup operations under return on investment maximization. *Computers & Operations Research*, *26*(9), 883–899. doi:10.1016/S0305-0548(98)00095-1

Phillips, R. L. (2005). *Pricing and Revenue Optimization*. Stanford, CA: Stanford University Press.

Porn, R., Harjunkoski, I., & Westerlund, T. (1999). Convexification of different classes of non-convex MINLP problems. *Computers & Chemical Engineering*, *23*(3), 439–448. doi:10.1016/S0098-1354(98)00305-6

Qin, Y., Wang, J., & Wei, C. (2014). Joint pricing and inventory control for fresh produce and foods with quality and physical quantity deteriorating simultaneously. *International Journal of Production Economics*, *152*, 42–48. doi:10.1016/j.ijpe.2014.01.005

Rosenberg, D. (1991). Optimal price-inventory decisions: Profit vs. ROII. *IIE Transactions*, *23*(1), 17–22. doi:10.1080/07408179108963837

Sajadieh, M. S., & Jokar, M. R. A. (2009). Optimizing shipment, ordering and pricing policies in a two stage supply chain with price-sensitive demand. *Transportation Research Part E, Logistics and Transportation Review*, *45*(4), 564–571. doi:10.1016/j.tre.2008.12.002

Schroeder, R. G., & Krishnan, R. (1976). Return on investment as a criterion for inventory model. *Decision Sciences*, *7*(4), 697–704. doi:10.1111/j.1540-5915.1976.tb00713.x

Sen, A., & Zhang, A. (1999). The newsboy problem with multiple demand classes. *IIE Transactions*, *31*(5), 431–444. doi:10.1080/07408179908969846

Seyed-Alagheband, S. A., Fatemi Ghomi, S. M. T., & Zandieh, M. (2011). A simulated annealing algorithm for balancing the assembly line type II problem with sequence-dependent setup times between tasks. *International Journal of Production Research*, *49*(3), 805–825. doi:10.1080/00207540903471486

Tavakkoli-Moghaddam, R., Safaei, N., & Gholipour, Y. (2006). A hybrid simulated annealing for capacitated vehicle routing problems with the independent route length. *Applied Mathematics and Computation*, *17*(2), 445–454. doi:10.1016/j.amc.2005.09.040

Wang, C., Huang, R., & Wei, Q. (2015). Integrated pricing and lot-sizing decision in a two-echelon supply chain witha finite production rate. *International Journal of Production Economics*, *161*, 44–53. doi:10.1016/j.ijpe.2014.11.011

Wee, H., Lo, C., & Hsu, P. (2009). A multi-objective joint replenishment inventory model of deteriorated items in a fuzzy environment. *European Journal of Operational Research*, *197*(2), 620–631. doi:10.1016/j.ejor.2006.08.067

Yıldırmaz, C., Karabat, S., & Sayın, S. (2009). Pricing and lot-sizing decisions in a two-echelon system with transportation costs. *OR-Spektrum*, *31*(3), 629–650. doi:10.1007/s00291-008-0156-1

Zhang, M., Bell, P., Cai, G., & Chen, X. (2010). Optimal fences and joint price and inventory decisions in distinct markets with demand leakage. *European Journal of Operational Research*, *204*(3), 589–596. doi:10.1016/j.ejor.2009.11.032

Zhang, M., & Bell, P. C. (2007). The effect of market segmentation with demand leakage between market segments on a firms price and inventory decisions. *European Journal of Operational Research*, *182*(2), 738–754. doi:10.1016/j.ejor.2006.09.034

## ADDITIONAL READING

Ghasemy Yaghin, R., Torabi, S. A., & Fatemi Ghomi, S. M. T. (2012). Integrated markdown pricing and aggregate production planning in a two echelon supply chain: A Hybrid Fuzzy Multiple Objective Approach. *Applied Mathematical Modelling*, *36*(12), 6011–6030. doi:10.1016/j.apm.2012.01.029

Lenskold, J. (2003). *Marketing ROI: The Path to Campaign, Customer, and Corporate Profitability*. San Diego, CA, U.S.A.: McGraw-Hill.

Phillips, R. L. (2005). *Pricing and Revenue Optimization, CA*. Stanford: Stanford University Press.

## KEY TERMS AND DEFINITIONS

**Inventory:** Materials and components that businesses hold in stock.

**Marketing:** Marketing's primary focus is to identify and satisfy customers in a way that helps build a solid and, hopefully, sustained relationship that encourages customers to continue doing business with the marketer.

**Mathematical Optimization:** The selection of a best element (with regard to some criterion) from some set of available alternatives based on mathematical formulation.

**Pricing:** The method a company uses to set the price its product. Pricing is one of the four aspects of marketing. The other three parts of the marketing mix are product management, promotion, and distribution.

**Return on Inventory Investment:** The ratio of profit to investment, and is a widely utilized economic performance measure dealing with finished goods inventories.

# Chapter 10
# A Comprehensive Risk Management Tool Based on Multi-Agents and System Dynamics for Traditional and E-Commerce Supply Chain

**Sultan Ceren Oner**
*Istanbul Technical University, Turkey*

**Mahir Oner**
*Istanbul Technical University, Turkey*

## ABSTRACT

*Supply chain management paradigms are becoming increasingly common management perspectives all over the world due to violent global competition of trade organizations and rapid changes in technology. In recent years, thanks to the communication improvements, customers have become more conscious about purchasing goods or services. Furthermore, organizations have to be customer oriented and more flexible against the dynamism of supply chain environment which increases uncertainties in supply chain parameters. Although a considerable amount of risk factors appearing in supply chain operations, this study concentrates on detecting key supply chain risks which could cause abnormalities and occur from rapid changes in customer demand, unpredictable price fluctuations, defect variations and delivery delays and provides the correction of these problems automatically. Thus, a system dynamics model is established for determining risks. This combined approach would be helpful for integrated supply chain risk management.*

DOI: 10.4018/978-1-5225-2944-6.ch010

## INTRODUCTION

In recent years, thanks to the communication improvements, customers have become more conscious about purchasing goods or services. Because of increasing knowledge about products and services, customer requirements are changing rapidly and these changings needs are forcing companies to be speedier to satisfy customer orders with more qualified products at acceptable prices. Furthermore, organizations have to be customer oriented and more flexible against the dynamism of supply chain environment which increases uncertainties in supply chain parameters. Flexibility and customer oriented attitude help organizations to have a chance about making more profit but also bring pressure to take risks such as inventory shortages, decreasing demand, late shipments, quality losses etc. Within this context, supply chain risk management applications are sought after dramatic losses appeared. For example, Toyota recalled automobiles owing to quality and safety problems which led to loss of millions of dollars and reputability. In addition to Toyota, Boeing had intensity problems with their production schedule and it was declared that aircrafts would be delivered with about fifteen-month tardiness. Thus, the main problem in a supply chain appears to be the changes in supply chain behavior with different elements. In addition to the conflicting aims of supply chain elements, supply chains may also be influenced from the of government restrictions such as the purchasing quota, sales limitations etc. (Nagurney et al., 2005). External factors such as tax regulations, high inflation rates, final customer demand etc., cause additional uncertainty, thereby increasing the risk potential in supply chains. In this sense, organizations are in need of cohesive management approaches, like supply chain risk management, against supply chain variances considering the requirements of a supply chain. To consider all of the external and internal factors, supply chain management must reflect the interactions among diversified supply chain elements (Ellegaard, 2008).

First supply chain risk management applications were implemented in financial issues that include cash flows and late payments. From 1995 to 1999, most favorable topics in supply chain risk management were lean production and supplier selection. However, especially from 2000 to 2004, analysts realized that risk was not only in financial issues but also in other supply chain components such as manufacturing and transportation. Besides that, supply chain risk management studies gave rise up to now and recently, due to increasing uncertainties, environmental information management has become a significant issue in risk management (Tang and Musa, 2010).

As realized from the literature review, significant number of models related to supply chain management generally focused on different optimization methods and decision-making tools such as linear programming, stochastic modeling, ANP, SOM and deterministic modeling (Pham et al., 2012, Liang et al., 2012; Azaron et al., 2007). Although sufficient information is provided in most of these models they could not be implemented due to lack of interpretation of the results and excessive computational times (Hanafizadeh et al., 2009). Furthermore, they are also critisized for not considering interactions between the risk elements (Ritchie and Brindley, 2007; Nagurney et al., 2005, Chatzidimitriou et al., 2008). Due to the emerging and dynamic virtual relations between supply chain members, models should acquire sudden changes and cope with uncertainties and provide continuous monitoring. From this perspective, fuzzy based models that formulate supply chains could cope with the difficulties about instant variations and continuous monitoring of the entire system. (Ngai and Wat, 2004) In addition to the static structure, interactions and variations could not be included in the mathematical modeling. Therefore, models based on multi criteria decision making are more suitable for determining causal relationships and control parameters that effects the system behavior.

According to Giannakis and Louis (2011) different simulation paradigms could help modeling complex systems. They also mentioned that there is an increasing interest in combined modeling with agent based simulation. On the other hand, from literature reviews, models based on both system dynamics and agent based modeling are generally considered to be the sub components of supply chain risk management. Although they described supply chain sub systems in detail, their proposed subsystem may contradict other sub systems when analysts want to implement these approaches in their supply chain. Because of this reason, an integrated point of view which could be able to handle the whole supply chain issues is needed for effective supply chain risk control. This paper aims to evaluate supply chain management, reflecting both external and internal risk factors, by handling major risk factors with an integrated point of view. In order to include external and internal supply chain risk parameters, a decision-making tool which is combined with multi agents and system dynamics model is purposed. First, a system dynamics model is established for determining risks and secondly an agent based model is applied for making corrections. Using this hybrid model, risk management models could estimate whole supply chain problems with different aspects.

## BACKGROUND

As mentioned before, studies applied agent based modeling or systems dynamics are very rare in supply chain risk management literature. The reason could be that agent based modeling and supply chain risk management are emerging topics for authors. Most of the earlier studies are generally focused on sub areas of supply chain risk management.

In literature, supply chain system dynamics studies are intensified on supply chain risk management sub systems such as manufacturing, distribution, supplier improvement, quality control and capacity planning. Özbayrak et al. (2007) described supply chain manufacturing process with system dynamics and made scenario analysis according to the demand changes, non-reliable manufacturer, non-reliable supplier and production times. Tesfamariam and Lindberg (2005) evaluated quality control theme in manufacturing with systems dynamics modeling. Georgiadis et al. (2005) analyzed distributor and retailer relationship with systems dynamics for capacity planning in food supply chain. Ovalle and Marquez (2003) purposed an integrated systems dynamics perspective that includes information, material and financial flow in supply chain and due to this characteristic of their research, this study separated from other system dynamics applications appeared in previous studies.

Agent based simulation provides most attractive modeling tools due to its visual expressions and its ability to make individual decisions with respect to experiments (Giannakis and Louis, 2011). In addition to learning experiments, agents interact with each other and modify their behaviors according to the environmental changes. Barbati et al. (2011) indicated increasing number of studies with agent based simulation in supply chain management functions such as transportation, logistics and supply chain planning. Supply chain risk management applications using agent based modeling approach are increasing in recent years with different perspectives Giannakis and Louis 's (2011) agent based study about demand and supply discrepancy could be given as a salient example in literature. Wang et al. (2011) explained B2B and B2C negotiation process included problems and negotiation criteria. They also represented purchasing process, customer behaviors and negotiations with intelligent agents. Hanafizadeh and Sherkat (2009) proposed genetic learner model for distribution systems in supply chain as well.

All these studies indicate us insufficient number and variety of risk management studies that consider different aspects of risks appearing in chain operations from a holistic and integrated point of view, rather using just a few of the risk factors in the model. The proposed model is combined with agent based and system dynamics approaches for the aim of reflecting internal and external risk elements and changes in supply chain, simultaneously.

## SUPPLY CHAIN RISK MANAGEMENT AND RISK FACTORS

Supply chain risk management system consists of five stages: Specifying risk factors, selecting most appropriate parameters to measure risk factors, gathering data for measurements, determining risk scores with risk models and evaluating risk results (Ngai and Wat,2004). In the first step, risk factors should be defined with respect to most problematic issues in the supply chain. For instance, Kull and Talluri (2008) specified risk factors as delivery delay, unexpected costs, inadequate quality level, flexibility and reliability. Besides that, Sodhi and Tang (2009) mentioned problems which cause disruptions in the supply chain are financial problems, unsatisfied demand and working with over stock or out of stock. After determining supply risks, proper variables should be selected for risk measurements. Hanafizadeh and Sherkat (2009) suggested delivery times, shipment capacity, inventory level and accuracy of shipment amount as risk variables. For Ovalle and Marquez (2002), critical parameters are capacity utilization, demand, quality level, product or service price, lead time and delivery delays. Gathering data is another step for risk calculation. Generally, companies utilize ERP solutions such as Oracle, SAP, and CRM etc. But these tools should be integrated with risk models for working with real time data. Subsequently, risk calculation model should be established for risk analysis. In literature, variety of models were proposed but mathematical modeling and stochastic approaches were most practiced ones for risk management models (Azaron et al., 2007). Ivanov et al. (2012) utilized control theory and emphasized that dynamic risk management tools should be unified with control theory models for the adaptation of continuously changing environment. As mentioned in literature review section, Giannakis and Louis's (2010) agent based simulation study could also be given as a good risk calculation tool for supply chain.

After risk scores are calculated, risk management analysts should assess the results. If current risk scores are above the expected values, making improvements like applying quality management tools such as six sigma, Kaizen or manufacturing strategies such as lean production, JIT or performance based management approaches like BPR are needed for preventing and eliminating risk effects or any other unexpected changes in critical parameters.

The developed model in this study purposes to combine various risk factors listed in literature with final chain risk. The model includes delays risk, demand fluctuation risk, quality risk and price fluctuation risk as total stage risk. As defined in the previous papers, delays risk means not to satisfy customer needs on time. Demand fluctuation risk is present if the customer needs cannot be forecasted in advance. Quality risk indicates that supplier could not send the finished goods in keeping with customer requirements. Finally, price fluctuation risk is the difference between expected price and actual price. As distinct from former studies, this model involves service level as a main indicator of total stage risk.

## SYSTEM DYNAMICS APPROACH IN SIMULATION AND ROLE IN SUPPLY CHAIN RISK MANAGEMENT

System dynamics approach was introduced by Forrester (1961) for explaining causal relations between variables and determining effects of these relations. According to effects, decision process will be linked to actions. In other words, key variables that influence system behavior could be connected with final indicators which provide essential information about system performance. With information feedbacks and chancing conditions, model could be able to evaluate for different kinds of problems (Forrester, 1961). Simulation is a necessity for system dynamic approach to represent system agglomerations, nonlinearity, time delays and feedback mechanism. Generally, most of modeling techniques could not be able to capture causal relations which determine actual root of the problems (Sterman, 2000).

Causal loop and stock flow diagrams are the most significant tools for implementing system dynamics approach. By using causal loop diagrams, main reasons of dynamics could be easily identified, interactions of the feedbacks responsible for problems could be foreseen and finally, dependent and independent variables could be defined. Together with stock and flow diagram, system agglomerations and delays will appear as a "memory" for simulation in stock variables. In addition to stock variables, flow variables depend on time and provide accumulations or empty accumulations for stocks. Apart from these variables, auxiliary variables consist of functions for stock variables (Forrester, 1961; Sterman, 2000).

The papers dealing with supply chain risk management used mathematical modeling (Nagurney et al., 2005), stochastic modeling (Azaron et al., 2007; Sodhi and Tang, 2009), C-means clustering, fuzzy logic (Ngai and Wat, 2005), AHP and ANP (Kull and Talluri, 2008), agent based simulation and case based reasoning-agent based simulation hybrid models (Giannakis and Louis, 2011). On the other hand, system dynamics approach could not be used directly in supply chain risk management. This approach confronted as scenario analysis in capacity planning of supply chain (Georgiadis et al., 2005), demand evaluation (Canetta et al., 2010), manufacturing supply chain modeling (Özbayrak et al., 2007), supply chain collaboration (Marquez and Ovalle, 2003). Additionally, former studies clearly emphasized that not only modeling inventory, manufacturing etc. but also representing interactivity among dynamically changing different variables are important (Özbayrak et al., 2007). In addition, Forrester (1961) and Campuzano and Mula (2011) mentioned the basis of the supply chain control systems and gave hints about how risk management could be implemented in supply chains.

In supply chain risk management, simulation models should reflect actual structure of the supply chain to control the performance variables. However, supply chain risk model formulation is hard to understand, reserves lots of variables which may lead neglects about major points of the basic structure. Because of the inadequacy in modeling dynamic demand and various units of control parameters, discrete event simulation becomes useful (Forrester, 1961). Because of the main focus of this paper is to present entire supply chain risk management system operations and complex interactions between variables, system dynamics is one of the most appropriate method for modeling supply chain risk management with agent based simulation. By using system dynamics, causal relations between risk factors and parameters could be visually described and while formulating the model, mathematical equations could be represented without skipping any detail. In addition to visualization, the model can be revised for various alternative conditions. The more detailed models could be brought out for complex systems due to its system dynamics flexibility (Tesfamariam and Lindberg, 2004). These capabilities make risk evaluation process easier and computational time would be shortened. System could also reorganize disruptive events automatically by utilizing feedbacks (Georgiadis et al., 2005). Associated with predefined performance criteria, questions as to how supply chain risk system behavior occurs in specific conditions could be foreseen for strategic decisions as well as sensitivity analysis and harmonic working of supply chain could be screened (Campuzano and Mula, 2011). From information, money, material, order and labor flow, critical control factors and uncertainties of the model components could be pursued (Forrester, 1961).

Although advantages of the system dynamics are clear, there are also undesirable issues in modeling process. For example, if model boundary could not be specified correctly, results could change considerably. Additionally, developer should decide where and when the customer should test the model, because in some cases, the model validation could not be provided in certain areas. Furthermore, time period and unit mismatch are significant for model parameters. (Sterman, 2000).

## A COMBINED DECISION-MAKING MODEL BASED ON MULTI AGENTS AND SYSTEM DYNAMICS FOR SUPPLY CHAIN RISK MANAGEMENT

The proposed model mainly focuses on the development of risk management in supply chain critical parts that were determined by Forrester (1961) and Campuzano and Mula (2011). Forrester (1961) analyzed four supply chain elements: manufacturer, distributor, retailer and wholesaler. While Forrester, Campuzano and Mula (2011) investigated only manufacturer and retailer levels, this paper is interested in both supply and demand side: supplier, manufacturer, retailer and final customer. In this respect, all critical parts of supply chain are included (Öner and Oztayşi, 2017).

Supply chain risk score is determined as the summation of supplier, manufacturer and retailer side risk scores which imply chain risk relays on the overall retailer risk score, the retailer risk depends on manufacturer total risk and manufacturer risk is directly influenced from supplier risk. Using this approach, risk management modeling could be achieved more effectively and better solutions can be obtained. Supply chain risk management subcomponents that are extracted from literature review are presented as as price fluctuation risk, demand fluctuation risk, delays risk and quality risk and include all the necessary variables used in the model and these variables are represented in determining the finalized risk score.

According to the subcomponents, total risk score includes four major risk factors named as price fluctuation risk, demand fluctuation risk, delays risk and quality risk. The price fluctuation risk for time t for the supplier stage is given by:

$$\begin{aligned} Supplier\,price\,fluctuation\,risk\left(t\right) &= Supplier\,price\,fluctuation\,risk\left(t-1\right) \\ &+\int_{t-1}^{t}\left(Supplier\,raw\,material\,price\right).dt \end{aligned} \tag{1}$$

Supplier raw material price can be defined as a function of tax reduction, inflation rate and exchange rate where coefficients a, b, c and d are determined by asking experts about the importance of these four parameters with Analytic Hierarch Process (AHP) method. Similar to the Equation (1), it is accumulated by supplier raw material price and it is affected by previous price fluctuations.

$$\text{Supplier raw material price} = \begin{pmatrix} \text{a} \times \text{tax reduction} \\ +\text{b} \times \text{inflation rate} \\ +\text{c} \times \text{exchange rate} \end{pmatrix} \times \text{supplier cost} \tag{2}$$

where

$$\sum \text{a} + \text{b} + \text{c} + \text{d} = 1$$

Supplier price fluctuation risk process outcome is supplier time varying price that indicates mass order discounts and specifies supplier total risk.

$$\text{Supplier time varying price} = \left( \frac{\begin{array}{l} \text{Suppliers expected price} \\ + \text{ Supplier price fluctuation risk} \end{array}}{\textit{time period}_t} \right) \tag{3}$$

In manufacturer tier, price fluctuation risk could be defined as supplier price fluctuation risk:

$$\begin{aligned} \textit{Manufacturer price fluctuation risk}(t) &= \textit{Manufacturer price fluctuation risk}(t-1) \\ &+ \int_{t-1}^{t} (\textit{Manufacturers price})\, dt \end{aligned} \tag{4}$$

Manufacturer's price is affected by the gap between manufacturer actual price and expected price from manufacturer. Manufacturer's time varying price is shown in Equation 5 where *w* implies a constant to reflect changes.

$$\text{Manufacturer's price} = \frac{\text{Manufacturer's actual price} + \text{w} \times \begin{pmatrix} \text{Manufacturer's actual price} \\ -\text{Manufacturers expected price} \end{pmatrix}}{\text{time period}_t} \tag{5}$$

Finally, retailer price fluctuation risk and retailer's price are given in Equations (6) and (7). Retailer's price only depends on inventory, transportation and unit order costs which constitute retailer total costs and expresses difference between manufacturer and supplier.

$$\begin{aligned} \text{Retailer price fluctuation risk}(\text{t}) &= \text{Retailer price fluctuation risk (t-1)} \\ &+ \int_{\text{t-1}}^{\text{t}} (\text{Retailer unit price})\, \text{dt} \end{aligned} \tag{6}$$

$$Retailer's\,price = \frac{Retailer\,costs}{time\,period_t} \tag{7}$$

In addition to price fluctuation risk, second risk factor is identified for measuring "Butterfly effect", demand fluctuation risk. Basically, demand fluctuation risk represents difference between average sales and average order quantity. It shows how much a company can carry out its sales against its incoming orders. Accordingly, demand fluctuation risk is also determined from the gap between expected demand which has the same meaning with average order quantity, and actual demand which is another term of actual sales. Demand fluctuation risk can be formulated for each supply chain tiers as follows:

$$\begin{aligned}\text{Demand fluctuation risk}(t) &= \text{Demand fluctuation risk}(t-1) \\ &+ \int_{t-1}^{t} \begin{pmatrix}\text{Average order quantity} \\ -\text{average sales ratio}\end{pmatrix} dt\end{aligned} \tag{8}$$

For supplier level, average order quantity and average sales ratio can be defined as:

$$\text{Supplier average order quantity} = \frac{(\text{Suppliers actual sales-Supplier risk})}{time\,period_t} \tag{9}$$

$$\text{Supplier average sales ratio} = \frac{\begin{matrix}\text{Raw material stock taken from supplier} \\ \text{-Suppliers backlogged order quantity}\end{matrix}}{time\,period_t} \tag{10}$$

Supplier risk is a critical key indicator about future sales because manufacturers select their suppliers according to this parameter. In Equation (9), subtraction of supplier's risk and demand fluctuation from actual sales specifies net sales that could appear in the future sales. Besides, supplier average sales ratio can be calculated from manufacturer raw material stock taken from supplier. Subtraction of backlogged orders from raw material stock defines net sales under the name of average sales ratio.

In manufacturer level, manufacturer average order quantity and average sales ratio are similar to supplier level but there is a difference: manufacturer average order quantity is also influenced by retailer quality risk which is the key performance indicator for retailer to state the order amount. However, average sales ratio is calculated similar to supplier level.

$$\text{Manufacturer average order quantity} = \frac{\begin{pmatrix}\text{Manufacturer actual sales} \\ \text{-manufacturer risk-retailer quality risk}\end{pmatrix}}{time\,period_t} \tag{11}$$

$$\text{Manufacturer average sales rate} = \frac{\begin{array}{l}\text{Manufacturer actual sales}\\ \text{-manufacturer backlogged order quantity}\end{array}}{\mathit{time\ period}_t} \tag{12}$$

Finally, retailer demand fluctuation risk is calculated using average sales and retailer service level. Service level is a critical issue that identifies how much final customer's orders are satisfied on time. Furthermore, service level is considered instead of average order quantity, because final customer demand is determined directly from service level in the model. Besides that, service level calculated in Equation (15) returns demand fluctuation risk and causes a loop.

$$\begin{aligned}\text{Retailer demand fluctuation risk}\left(t\right) &= \text{Retailer demand fluctuation risk}\left(t-1\right)\\ &+\int_{t-1}^{t}\begin{pmatrix}\text{Retailer average sales}\\ \text{−Retailer service level}\end{pmatrix}dt\end{aligned} \tag{13}$$

$$\text{Retailer average sales} = \frac{\begin{array}{l}\text{Final customer actual sales}\\ \text{-Retailer delayed goods quantity}\end{array}}{\mathit{time\ period}_t} \tag{14}$$

$$\text{Retailer service level} = \frac{\begin{array}{l}\text{Retailer actual sales}\\ \text{−Retailer backlogged order quantity}\\ \text{−Retailer demand fluctuation risk}\end{array}}{\mathit{time\ period}_t} \tag{15}$$

Third risk character considered in this paper is delays risk. Actually, delays could be calculated within system dynamics simulation model. However, in risk management, that directly causes malfunctions in the system. While implementing the systems dynamics model, historical data is needed for determining delays. Supplier and manufacturer delays risk can be stated as follows:

$$\begin{aligned}\begin{array}{l}\text{Supplier \& Manufacturer}\\ \quad\text{delays risk}\left(t\right)\end{array} &= \mathit{Supplier\ \&\ Manufacturer\ delays\ risk}\left(t-1\right)\\ &+\int_{t-1}^{t}\begin{pmatrix}\text{Manufacturer \& retailer delivery time}\\ \times\text{ supplier \& manufacturer delay caused by inventory}\end{pmatrix}dt\end{aligned} \tag{16}$$

As one can see in Equation (16), delays risk is calculated with using "delivery time" and "delay quantity caused by inventory" variables. On the other hand, in retailer stage, delays risk also depends on the divergence of service level and shipment rate which constraints additional transfers that could be shipped to the customer.

$$\begin{aligned} Retailer\,delays\,risk(t) &= Retailer\,delays\,risk(t-1) \\ &+\int_{t-1}^{t} \begin{pmatrix} Final\,customer\,delivery\,time \\ \times Retailer\,delay\,caused\,by\,inventory\,level \end{pmatrix} \\ &\times (retailer\,service\,level - retailer\,shipment\,rate).dt \end{aligned} \tag{17}$$

Delivery time can be determined from shipments ready for transfer and capacity constraint. That means although company has enough goods which can be transferred, transfer or customer inventory level may limit the shipment rate to the customer. As in Equation (19), negative or out of stock inventory levels may induce delays caused by inventory level.

$$\begin{matrix}\text{Manufacturer \& Retailer \&} \\ \text{Final customer delivery time}\end{matrix} = \frac{\begin{matrix}\text{Supplier \& Manufacturer \& Retailer ready to be shipped product rate} \\ -\text{ Supplier \& Manufacturer \& Retailer capacity}\end{matrix}}{\text{Supplier \& Manufacturer \& Retailer delays risk}} \tag{18}$$

$$\begin{matrix}\text{Supplier \& Manufacturer \& Retailer} \\ \text{delay caused by inventory level}\end{matrix} = \frac{\text{Negative inventory level}}{\text{Supplier \& Manufacturer \& Retailer delay risk}} \tag{19}$$

The last risk indicator mentioned in this paper is quality risk which is one of the major causes of the troubles within the operations of the system. On the other hand, quality risk can be considered as a qualitative variable rather than quantitative variable. From Equation (20), while quality risk can be directly measured by observing defect rate for supplier and manufacturer level, retailer's quality risk also could be affected from manufacturer defects and customer returns.

$$Quality\,risk(t) = Quality\,risk(t-1) + \int_{t-1}^{t} (1 - Nondefective\,goods\,rate).dt \tag{20}$$

$$\begin{aligned} \text{Retailer quality risk}(t) &= \text{Retailer quality risk}(t-1) \\ &+\int_{t-1}^{t} \left[1 - \begin{pmatrix} \text{Nondefective goods from final customer} \\ +\text{Nondefective goods from manufacturer} \end{pmatrix}\right].dt \end{aligned} \tag{21}$$

Non-defective goods rate could be identified by eliminating defective goods from inventory. Besides, non-defective goods coming from final customer directly influenced from manufacturer faults is assumed to be constant in this paper.

$$\text{Nondefective goods rate} = \frac{\left(\begin{matrix}\text{Supplier \& Manufacturer} \\ \text{\& Retailer inventory}\end{matrix} - \text{Quality risk}\right)}{time\,period_t} \tag{22}$$

$$\text{Nondefective goods from final customer} = k \times \text{Nondefective goods from manufacturer} \tag{23}$$

where $k$ is a constant.

In conclusion, the summation of all these risk factors comprises total stage risk which is effectuated directly by demand fluctuation risk, delays risk and indirectly from quality risk which constitutes actual sales in the model. In addition to these factors, time varying price fluctuations is considered with a risk multiplier which promotes time varying price as a risk indicator.

$$\text{Total stage risk}\left(t\right) = \text{Total stage risk}\left(t-1\right) + \int_{t-1}^{t} \left(\text{Service level} + \left(\begin{matrix}\text{risk multiplier} \\ \times \text{ time varying price}\end{matrix}\right)\right) dt \tag{24}$$

Service level is a measurement of how much a company responds to customer orders on time. In accordance with this definition, service level could be calculated by using Equations (25) and (26). Apart from these equations, demand fluctuations should be taken into account in retailer level due to presence of the reflections of final customer demand variations to the entire supply chain.

$$\begin{matrix}\text{Supplier} \\ \text{service} \\ \text{level}\end{matrix} = \frac{\left(\begin{matrix}\text{Supplier actual} \\ \text{sales}\end{matrix}\right) - \left(\begin{matrix}\text{Supplier backlogged} \\ \text{order quantity}\end{matrix}\right) - \left(\begin{matrix}\text{supplier demand} \\ \text{fluctuation risk}\end{matrix}\right)}{\text{Supplier delays risk}} \tag{25}$$

$$\begin{matrix}\text{Manufacturer} \\ \text{service level}\end{matrix} = \frac{\begin{matrix}\text{Manufacturer actual} \\ \text{sales}\end{matrix} - \begin{matrix}\text{Manufacturer backlogged} \\ \text{order quantity}\end{matrix}}{\text{Manufacturer delays risk}} \tag{26}$$

$$\text{Retailer service level} = \frac{\text{Retailer actual sales} - \text{Retailer backlogged order quantity} - \text{Retailer demand fluctuation risk}}{\text{Retailer delays risk}} \quad (27)$$

Risk multipliers are given in Equations (28), (29) and (30).

$$\text{Supplier risk multiplier} = \frac{1}{\text{Supplier risk level}} \quad (28)$$

$$\text{Manufacturer risk multiplier} = \frac{1}{\text{Manufacturer risk level}} \quad (29)$$

$$\text{Retailer risk multiplier} = \frac{1}{\text{Retailer risk level}} \quad (30)$$

Using the equations that are briefly explained above, supplier and manufacturer side risk determination model is driven in Powersim package.

A problem arises from system dynamics model particularly in calculating demand and price risks: how could the model follow the environmental changes in economic variations such as demand- supply balance and who controls whether risk parameters are under control or not? In response to these questions, system dynamics model should be composed of multi agents.

In agent based model, all supply chain members are connected with each other and system dynamics model. A risk evaluation agent checks whether economic changes constitute critical situations for each supply chain members. If supply chain activities are under disruption risks, risk evaluation agent makes decisions to prevent further damages caused by environmental changes.

In risk detection process, internal risk changes are compared with other supply chain members in order to monitor sudden variations. First, environmental indicators and supply chain internal situation are evaluated and if any risk factor is above the standards which are determined by supply chain members, risk evaluation agent decides what could be done for the adjustment of the system to normal risk values. Therefore, risk evaluation agent conducts experiments which are stored in ERP systems. Risk evaluation agent selects the most appropriate approach by checking system conditions and environmental changes appeared in the past supply chain operations.

By making corrections to abnormalities in supply chain, risk could be taken under control. This demonstrates that model could eliminate the impact of sudden changes in the supply chain.

The most important advantage of this study is its ability of integration of different kinds of risk factors including all supply chain members. Thus, a supply chain analyst could see the entire flow of the key indicators and directs the whole system in a correct way. Additionally, combined model could collect decisions in a database in order to reuse them in further situations. The model focused on four risk factors but it would be expanded for considering different risk aspects such as environmental damages risk, supplier reliance risk etc.

E-commerce based supply chains are more consolidated than traditional supply chains in terms of chain length and information sharing effectiveness. Thus, supplier side in e- commerce based supply chain behaves similar as presented in manufacturer side in traditional supply chain risk management. Although price determination process, expected price determination, quality improvement applications, delay risk, backlogged order identification and inventory control are similar as discussed in traditional supply chain risk management, order picking is totally different from the processed faced in traditional supply chain. The reason of this difference is the diversification of the products and customer requirements.

## FINDINGS AND DISCUSSION

In order to observe the effects of critical control parameters such as targeted lateness effect, inventory level control period, supplier share, distributor risk effect, supplier cost elasticity, distributor price elasticity on risk scores in diversified levels of supply chain members, one-at-a-time sensitivity analysis is applied for e-commerce supply chain risk management. First, a sensitivity analysis is conducted for observing the effects of the changes in targeted lateness (weeks) on distributor delay risk. As seen from Figure 1 and Figure 2 if lateness period has very few changes, in terms of hours, the effect would not affect delay risk but if period extends as days, delay risk will decrease. The reason in this situation is more confidence in delivery of the products. On the other hand, if lateness period constricts, potential customer rate will decrease due to the impatience of the customers in delivery process.

As seen in Figure 3 and Figure 4, inventory level control period is the other critical parameter to be monitored in risk management processes. If inventory level control period increases, distributor tends to keep large number of products and these circumstances should cause higher holding costs and escalation of the product prices. Figure 5 and Figure 6 presents the related conditions.

*Figure 1. Distributor delay risk(week)*

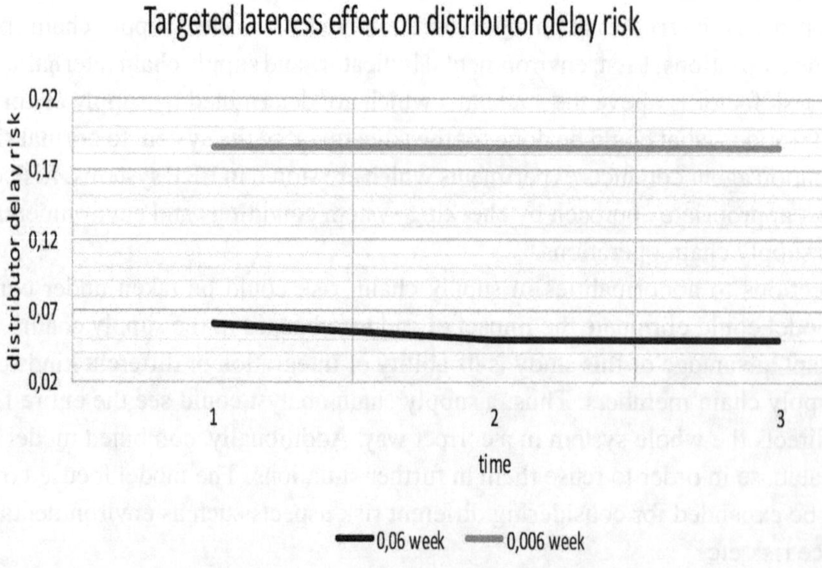

*Figure 2. Potential customer rate (week)*

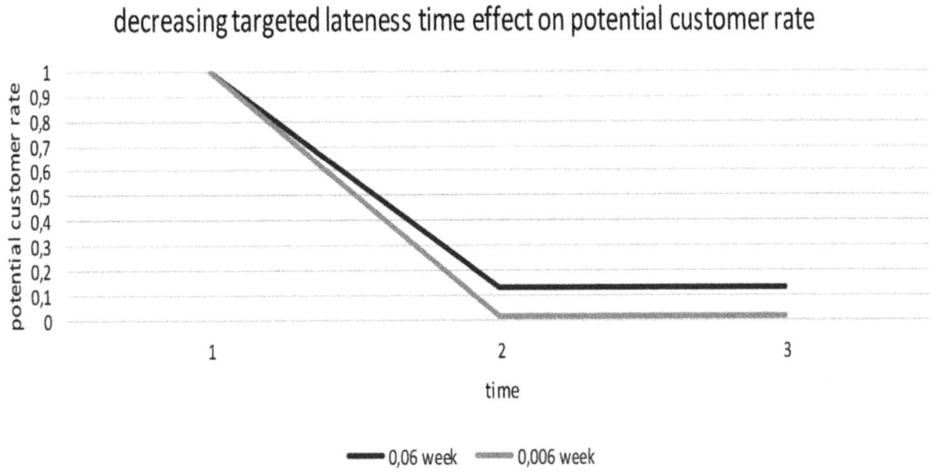


*Figure 3. Inventory level (units)*

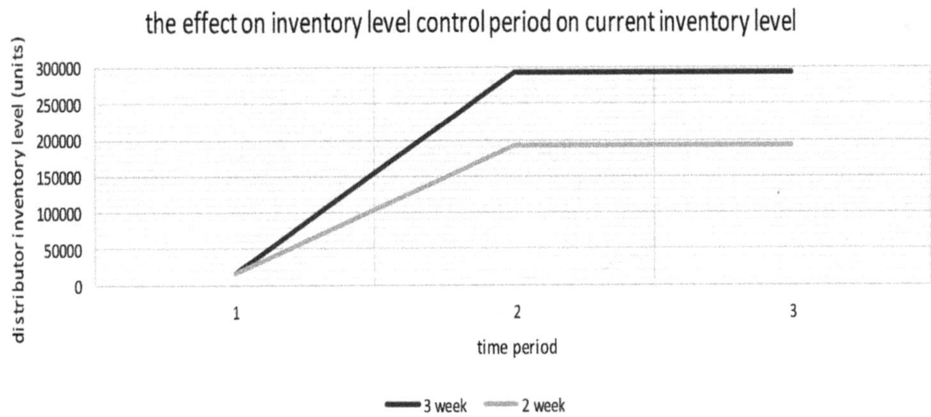


*Figure 4. Distributor price (units)*

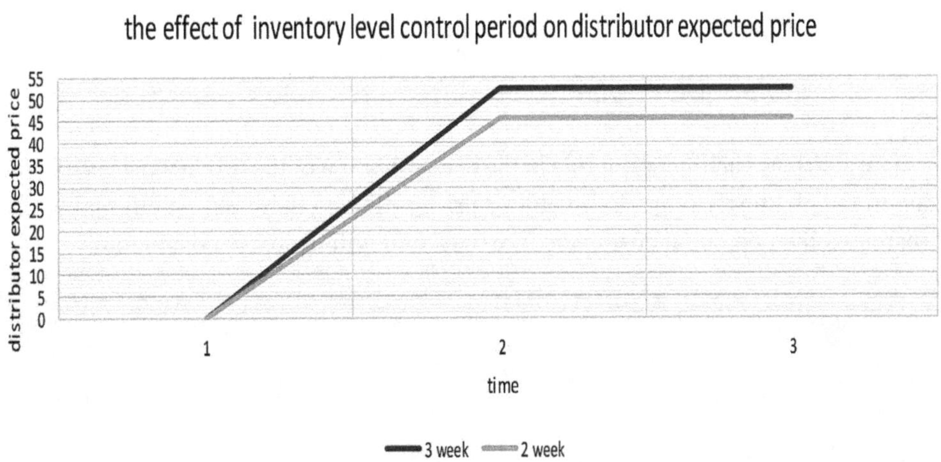

*Figure 5. Distributor price ($)*

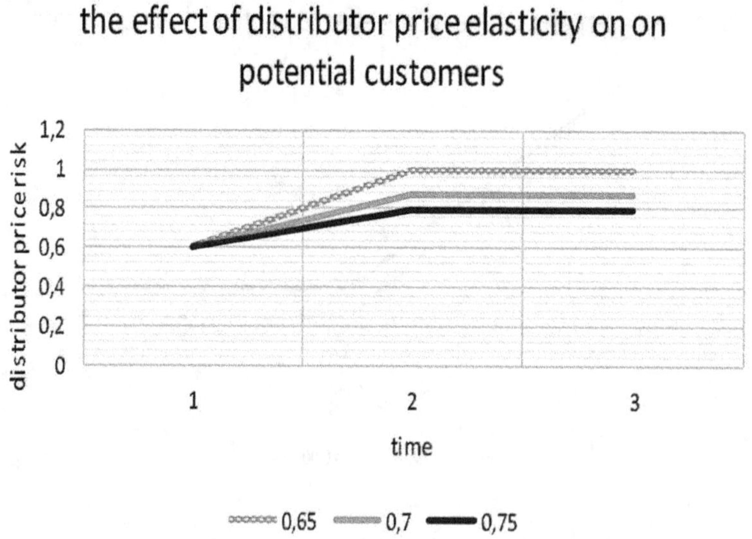


*Figure 6. Number of potential customer*

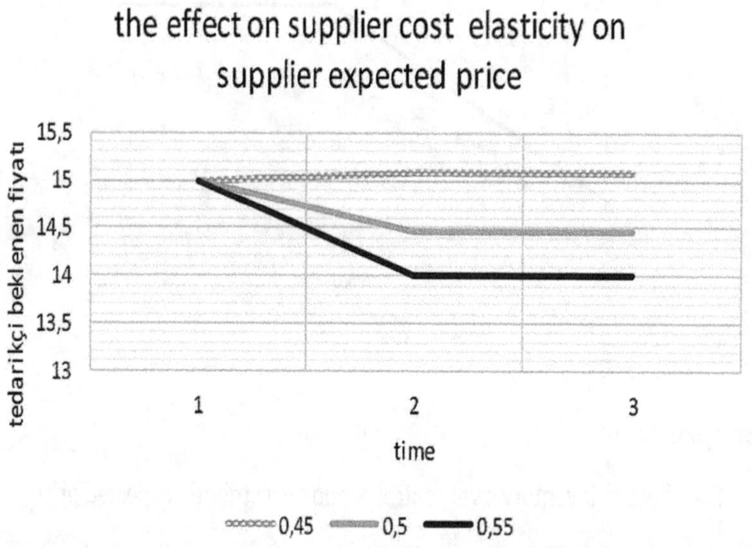


Supplier share also plays an important role in the determination of the product prices. If supplier share increases, due to the "butterfly effect", distributor price will also increase. Additionally, if distributor risks score increases then potential customers in per second will decrease considerably.

Distributor price elasticity and supplier cost elasticity are other crucial parameters to be evaluated for the pricing of goods. From Figure 7 and Figure 8, if distributor price elasticity decreases, which imply the reactiveness of price fluctuations, will increase and adaptability to price changes will not be diminishing, naturally, potential customers will increase due to the trust on price stability. On the other hand, if cost elasticity decreases, supplier would have few impacts on product price determination and this status encourages the collaborative pricing strategies. Thus, expected prices will increase leading to decreasing number of purchases and potential customers. In these circumstances, quota should be specified for expected prices or government guarantee should be provided in order to organize the purchases in the related sectors.

*Figure 7. Potential customer*

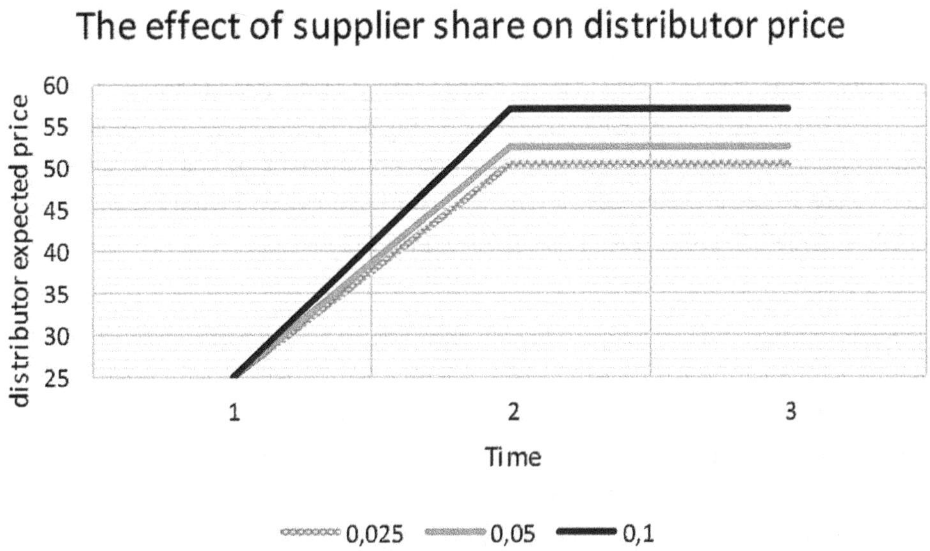


*Figure 8. Supplier expected price ($)*

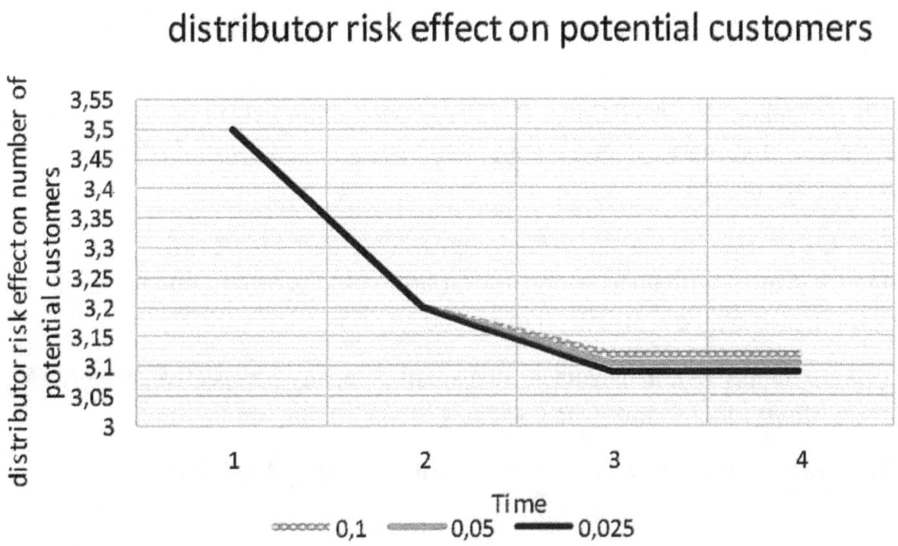

## CONCLUSION

This study deals with supply chain risk parameters' calculation such as quality, delay, price variation and demand fluctuation and evaluation of risk scores according to the environmental changes. A combined decision-making model based on multi agents and system dynamics is proposed for determining risk scores and making adjustments according to internal and external changes. A one at a time sensitivity analysis is adopted for testing the validity of the proposed methodology. According to the sensitivity analysis, distributor delay risk, distributor price and inventory level are the most sensitive parameters. Additionally, distributor price elasticity and supplier cost elasticity are other crucial parameters to be evaluated in pricing of goods. From the sensitivity analysis, one could conclude that expected price will increases which could lead to decreasing number of purchases and potential customers.

The most important contribution of this study is there was a need for the integration of different kinds of the risk factors including all supply chain members. Thus, supply chain analysts could see the entire flow of the key indicators and manages the entire system in a correct way. Additionally, combined model stores decisions in a database in order to reuse of them in further situations. The model focused on four risk factors but it easily be expanded by considering different risk factors such as environmental damages risk, supplier reliance risk etc. Thus, this combined model will help the adaptation of supply chain activities to environmental changes and internal risk factors. Conducted experiments have shown that the model will reduce the total risk effect in the supply chain.

## REFERENCES


Azaron, A., Brown, K. N., Tarim, S. A., & Modarres, M. (2007). A multi-objective stochastic programming approach for supply chain design considering risk. *International Journal of Production Economics, 116*(1), 129–138. doi:10.1016/j.ijpe.2008.08.002

Barbati, M., Bruno, G., & Genovese, A. (2011). Applications of agent-based models for optimization problems: A literature review. *Expert Systems with Applications, 39*(5), 6020–6028. doi:10.1016/j.eswa.2011.12.015

Campuzano, F., & Mula, J. (2011). *Supply Chain Simulation*. London: Springer-Verlag. doi:10.1007/978-0-85729-719-8

Canetta, L., Cheikhrouhou, N., & Glardon, R. (2010). Modelling hybrid demand (e- commerce "+" traditional) evolution: A scenario planning approach. *International Journal of Production Economics*. doi:10.1016/j.ijpe.2010.06.003

Chatzidimitriou, C., Symeonidis, A. L., Kontogounis, I., & Pericles, A. M. (2008). Agent Mertacor: A robust design for dealing with uncertainty and variation in SCM environments. *Expert Systems with Applications, 35*(3), 591–603. doi:10.1016/j.eswa.2007.07.050

Ellegard, C. (2008). Supply risk management in a small company perspective, *Supply Chain Management. International Journal (Toronto, Ont.), 13*(6), 425–434.

Forrester, J. W. (1961). *Industrial Dynamics*. Massachusetts: The M.I.T. Press.

Georgiadis, P., Vlachos, D., & Iakovou, E. (2005). A system dynamics modeling framework for the strategic supply chain management of food chains. *Journal of Food Engineering*, *70*(3), 351–364. doi:10.1016/j.jfoodeng.2004.06.030

Georgiadis, P., Vlachos, D., & Iakovou, E. (2005). A system dynamics modeling framework for the strategic supply chain management of food chains. *Journal of Food Engineering*, *70*(3), 351–364. doi:10.1016/j.jfoodeng.2004.06.030

Giannakis, M., & Louis, M. (2010). A multi-agent based framework for supply chain risk management. *Journal of Purchasing and Supply Management*, *17*(1), 23–31. doi:10.1016/j.pursup.2010.05.001

Hanafizadeh, P., & Sherkat, M. H. (2009). Designing fuzzy-genetic learner model based on multi-agent systems in supply chain management. *Expert Systems with Applications*, *36*(6), 10120–10134. doi:10.1016/j.eswa.2009.01.008

Ivanov, D., Dolgui, A., & Sokolov, B. (2011). Applicability of optimal control theory to adaptive supply chain planning and scheduling. *Annual Reviews in Control*, *36*(1), 73–84. doi:10.1016/j.arcontrol.2012.03.006

Kull, T. J., & Talluri, S. (2008). A Supply Risk Reduction Model Using Integrated Multi Criteria Decision Making. *IEEE Transactions on Engineering Management*, *55*(3), 409–419. doi:10.1109/TEM.2008.922627

Liang, W., Huang, C., Tseng, T., Lin, Y., & Tseng, J. (2012). The evaluation of intelligent agent performance: An example of B2C e-commerce negotiation. *Computer Standards & Interfaces*, *34*(5), 439–446. doi:10.1016/j.csi.2012.02.003

Nagurney, A., Cruz, J., Dong, J., & Zhang, D. (2005). Supply chain networks, electronic commerce, and supply side and demand side risk. *European Journal of Operational Research*, *164*(1), 120–142. doi:10.1016/j.ejor.2003.11.007

Ngai, E. W. T., & Wat, F. K. T. (2004). Fuzzy decision support system for risk analysis in e-commerce development. *Decision Support Systems*, *40*(2), 235–255. doi:10.1016/j.dss.2003.12.002

Öner, C., & Oztayşi, B. (2017). Sustainable Supply Chains and Risk Management for E-Commerce Companies Using Fuzzy Inference System. In Kahraman, Cengiz, Ucal Sari, İrem (Eds.), Intelligence Systems in Environmental Management: Theory and Applications. Springer.

Ovalle, O. R., & Marquez, A. C. (2002). The effectiveness of using e-collaboration tools in the supply chain: An assessment study with system dynamics. *Journal of Purchasing and Supply Management*, *9*(4), 151–163. doi:10.1016/S1478-4092(03)00005-0

Özbayrak, M., Papadopoulou, T. C., & Akgün, M. (2007). Systems dynamics modeling of a manufacturing supply chain system. *Simulation Modelling Practice and Theory*, *15*(10), 1338–1355. doi:10.1016/j.simpat.2007.09.007

Pham, H. V., Cooper, E. W., Cao, T., & Kamei, K. (2012). Hybrid Kansei-SOM model using risk management and company assessment for stock trading. *Inform. Sci.* doi:10.1016/j.ins.2011.11.036

Ritchie, B., & Brindley, C. (2007). An emergent framework for supply chain risk management and performance measurement. *The Journal of the Operational Research Society*, *58*(11), 1398–1411. doi:10.1057/palgrave.jors.2602412

Sodhi, M. S., & Tang, C. S. (2009). Modeling supply- chain planning under demand uncertainty using stochastic programming: A survey motivated by asset–liability management. *International Journal of Production Economics*, *121*(2), 728–738. doi:10.1016/j.ijpe.2009.02.009

Sterman, J. D. (2000). *Business Dynamics: Systems Thinking and Modeling for a Complex World*. Mc Graw Hill.

Tang, O., & Musa, S. N. (2010). Designing fuzzy- genetic learner model based on multi-agent systems in supply chain management. *International Journal of Production Economics*, *133*, 25–34. doi:10.1016/j.ijpe.2010.06.013

Tesfamariam, D., & Lindberg, B. (2005). Aggregate analysis of manufacturing systems using system dynamics and ANP. *Computers & Industrial Engineering*, *49*(1), 98–117. doi:10.1016/j.cie.2005.05.001

Wang, G., T.N. Wong, C. Yu. (2011). A computational model for multi-agent E-commerce negotiations with adaptive negotiation behaviors, *J. Comput. Sci*. doi:10.1016/j.jocs.2011.10.003

# Chapter 11
# Optimal Strategy Selection in a Supply Chain

**Ömer Faruk Gürcan**
*Istanbul Technical University, Turkey*

**Ahmet Erdoğan**
*Yıldız Technical University, Turkey*

## ABSTRACT

*Uncertainties and unpredictability in the market force companies to develop strategies which enable them to perform better than their competitors. Developing proper strategies for a supply chain is crucial. Strategies are affected by the nature of the firm's products or services, customer preferences, operations, process design of the firm, etc. Companies should form adaptive supply chain strategies which enable them to be resilient and flexible enough in the flow of materials, products, information, and money along the supply chain. There are many studies about supply chain management and supplier selection in the literature. However, the number of studies about the selection of the right supply chain strategy are very limited. This study presents the components which help to constitute a supply chain strategy and classify the supply chain strategies described in the literature. Lastly, it offers a strategy and criteria matrix which can be used as a road map for selecting the most appropriate supply chain strategy by firms.*

## INTRODUCTION

All organizations operating in service or manufacturing industries are members of a supply chain. Firm and industry characteristics affect the design and management of supply chains. The nature of the firm's products or services, customer preferences, the operations and process design of the firm determine the structure of the supply chain. Any supply chain should be strategically planned to gain competitive advantage in the market (Magutu et al., 2015).

The supply chain can be defined as a group of organizations included in the upstream and downstream flow of products, services, finances, and information from a source to a final customer. Supply chain management describes the strategic re-organization of processes among networks of companies which are included in the chain (Sharifi et al., 2013). "Supply chain management is the integration of

DOI: 10.4018/978-1-5225-2944-6.ch011

key business processes from end user through original suppliers that provides products, services, and information that add value for customers and other stakeholders" (Lambert et al., 1998, p.1).

Supply Chain Strategy (SCS) as an emerging research area of supply chain management, requires a study from both the academia and the practitioners of supply chain management (de la O & Matis, 2014). These strategies are crucial to the success of a firm's product and market growth strategy when today's great demand uncertainty, higher risk, increasing competitive intensity and complex business environment are present (Sharifi et al., 2013; Roh et al., 2014).

Failures in supply chain management are still common in today's industries. Not sorting different products with appropriate SCSs is one of the failure reasons (Li & O'Brien, 2001). SCSs must be dynamically matched customer needs and problems to maximize competitiveness in the market. There is no single SCS that is applicable to all product types in industry (Aitken et al., 2003). Therefore, developing a proper SCS by taking into consideration various criteria will be helpful for companies' sustainability and overall success.

This study identifies and explains the components which affect the SCS of a firm and classify the SCSs described in the literature. Five components and seven kinds of SCS models are described. These strategies are grouped under three main headings: efficiency oriented, responsiveness oriented, and hybrid SCSs. In the last section of the study, SCS selection criteria and sub criteria are defined. These criteria, are intended to guide companies in strategy selection.

## COMPONENTS AFFECTING SUPPLY CHAIN STRATEGY

Supply chains include the flow of information, product and money. These concepts are basis of organizations in terms of cost, market power, service level, competitiveness, etc. Today organizations should offer low cost, high customer service level, fast delivery and flexible solutions by using high technologies. In reality, accomplishing all of them is difficult. Thus, companies need to apply specific SCSs according to their priorities. For example, Ryanair focuses on cost and offers cheap service. One of the key decisions for supply chain managers is to choose the right strategies (Waters, 2003). Market characteristics, type of product or services, internal capabilities and available external resources are critical components in forming SCSs (Sharifi et al., 2013).

Before designing a supply chain, demand structure of the product should be considered. The first step is to determine the structure of product demand for an effective supply chain. In this process, the product life cycle, demand predictability, product variety, market and service standards for the replenishment period are considered. According to literature review, product type, industry framework, managerial focus, and internal processes are basic components that affect the selection of strategy. Each of these components are discussed in the next section.

### Product Type

Functional and innovative products are identified by Fisher (1997). Long product life cycle, low-profit margin, low variety and long lead times are the main characteristics of functional products while short product life cycle, high-profit margin, high variety and short lead times are the main characteristics of innovative products. These product types require different SCSs. Other researchers expanded considerations about products; product uniqueness, product complexity or relevance of both supply and demand uncertainty were studied (Caniato et al. 2011; Birhanu et al., 2014).

Functional products satisfy basic needs. Customers can buy these products in a wide range of retailers, such as bakeshops and gas stations. Such products have life cycle longer than two years, fewer than 20 variants in the product line or family, contribution margins are under 20%. Innovativeness help companies to avoid low margins and give customers an additional reason to buy companies' offerings. Such as Starbucks Coffee Company strategy (Fisher, 1997; Birhanu et al., 2014).

A company's product type is called as innovative if the product has a life cycle of up to a year, more than 30 variants and margins higher than 20% (Birhanu et al., 2014). Table 1 shows the comparison between functional and innovative products.

Functional products tend to have more mature and stable supply process. For example, the annual demand for electricity and other utility products in an area is stable and predictable, but the supply of hydroelectric power, which depends on rainfall in an area, can be unstable. Likewise, innovative products can have a stable supply process. Fashion apparel products have short selling seasons and demand is highly unpredictable but the supply process very stable (Lee, 2002).

## Industry Framework

Industry framework includes the interaction of suppliers, customers, technological developments, and economic factors in any sector. This framework has four basic drivers which affect SCS selection (Perez, 2013).

- Demand variation is a main driver of production efficiency and product cost.
- Market mediation cost which is related to the imbalance of demand and supply. Because of functional products have predictable demand, their market mediation cost is low. But innovative products have higher market mediation costs.
- Product lifecycle which is continuously getting shorter, affects the predictability of demand and market mediation costs.

*Table 1. Comparison between functional and innovative products (Lee, 2002, p. 106)*

| Functional | Innovative |
|---|---|
| Low demand uncertainties | High demand uncertainties |
| More predictable demand | Difficult to forecast |
| Stable demand | Variable demand |
| Long product life | Short selling season |
| Low inventory cost | High inventory cost |
| Low profit margins | High profit margins |
| Low product variety | High product variety |
| High volume per SKU | Low volumes per SKU |
| Low stock out cost | High stock out cost |
| Low obsolescence | High obsolescence |

- The ratio between relevance of the cost of assets and the total cost is important. When the relevance of the cost of assets to total cost is low, companies can value responsiveness oriented strategy.

## Company's Competitive Positioning

A company's competitive positioning is related to which strategies does a company apply to gain a competitive advantage regarding product or service strategies. A company should determine the factors which are thought as important in competitiveness for selection of proper strategy. Different strategies may be appropriate for companies of various sectors. For instance, a laptop manufacturer has different SCSs than a soap manufacturer. Even if two manufacturers who sell same products, have different SCSs (Gorener, 2013). To be best in Performance/Cost, offering the best price, sustained portfolio renewal regarding fashion or technology trend can be given as example strategies to gain competitive advantage (Perez, 2013).

## Managerial Focus

Managerial focus aims to define main focuses of a company about process design, customer service, product design, capacity management and supplier management.

## Process Design

Critical components for an accurate process designs are listed below (Perez, 2013):

- Collaborative relationship to build synergy; to increase the level of service and establish cooperation programs with key customers to reduce costs.
- Efficiency; maximizing the rate of asset utilization to ensure low production costs.
- Continuous portfolio renewal.
- Product configurability for customer's specific requirements.
- Agile response to changes in demand.
- Agile response and process flexibility to adapt to specific requirements of customers.

## Customer Service

Critical components of customer service in terms of production, planning, management and distribution processes are listed below (Perez, 2013):

- To adapt to customer's specific needs for each unique order
- To provide short delivery time
- To provide the accuracy of the order
- Focus on delivering innovative and / or renewable products on an ongoing basis
- To provide perfect order fulfillment
- Information sharing for continuous improvement

## Product Design

Critical components for product design are listed below (Perez, 2013):

- Product design for reducing the manufacturing time per unit
- Product design for fast changeover
- Low cost at standard performance
- Modular design for multiple configurations
- Providing value-added service that supports the product

## Capacity Management

Critical components for capacity management are listed below (Perez, 2013):

- High rate of asset utilization
- Programming products regularly in the appropriate order
- Additional capacity in production
- Having flexible asset and capacity to meet specific customer needs

## Supplier Management

Critical components for supplier management are listed below (Perez, 2013):

- The product portfolio offered to the market requires the newest and most innovative raw materials and components from suppliers.
- The selection of the lowest-cost supplier
- The ability of suppliers to respond to demand fluctuations
- To establish strong relationships with suppliers in order to minimize costs
- Lead times of suppliers
- Process flexibility of suppliers

## Internal Processes

Internal processes enable to companies to make connection and combination within the supply chain activities which are included categories of source, make, and deliver. Asset utilization and the location of the decoupling point are the most important factors. The decoupling point is the process in the value chain where a product takes on unique characteristics or specifications according to customer requirements (Perez, 2013). If an organization's competitive position is based on low cost, high efficiency on asset utilization is required. In this case, decoupling point stays at the end of the manufacturing process. Processes push each other before decoupling point. So making workload leveling becomes easy by the forecast and production cycle tends to be long to increase production efficiency.

After the decoupling point, processes are pull, so asset utilization is about medium level, the workload is driven by demand and is therefore highly variable.

*Figure 1.*

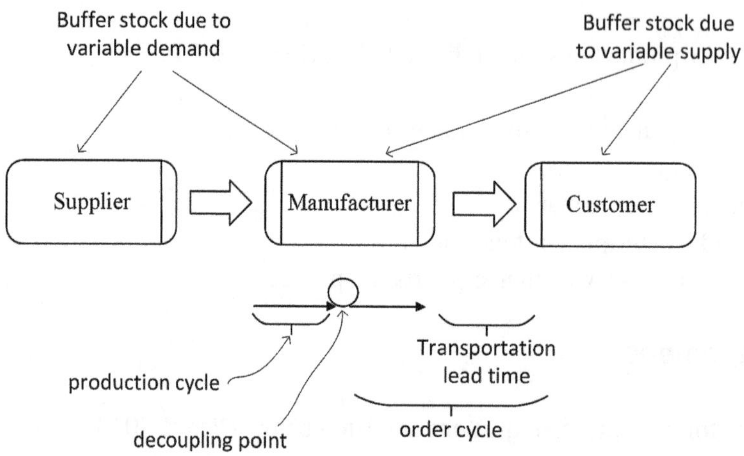


A supply chain process is shown in Figure 1 (Adopted from (Perez, 2013)). Before the decoupling point, processes are push so that the workload level can be defined smoothly with a forecast, the production cycle is disposed to be long which increase production efficiency, and the asset-utilization rate is high. After the decoupling point, processes turn into pull, so asset utilization is low level, the workload is determined by demand, so variation is high, and the production cycle is disposed to be shorter (Perez, 2013).

## SUPPLY CHAIN STRATEGIES

Strategic decisions are very risky and crucial decisions that involve the whole organization, have a lot of resource needs and have long-lasting impacts. SCS of an organization involves all processes from procurement of raw materials to delivering products to customers and if needed strategic decisions, policies, plans and cultural relations related to management of reverse supply chain activities (Gorener, 2013). SCSs help to show the companies' competitiveness and the position in the market against their competitors (Birhanu et al., 2014).

The SCS should correspond to corporate strategy to be successful. When a firm builds supply chain practices upon an SCS well, the firm's and its supply chain partners' business performance and thus their competitiveness increase (Roh et al., 2014). SCSs in the literature are listed in Table 2.

When SCSs' features, similarities and dissimilarities are considered, strategies can be grouped into three categories which are efficiency oriented, response oriented and hybrid. These strategies are shown in Table 3.

### Efficiency Oriented Supply Chain Strategies

Lean/Efficient, Fast and Continuous Flow SCSs are efficiency oriented strategies. The basic differences between efficiency oriented and market-responsive SCSs were listed in Table 4.

*Table 2. Supply chain strategies in literature*

| Researchers | SCS | Examples |
|---|---|---|
| Fisher (1997); Lee (2002); Agarwal et al. (2006); Gorener (2013); Perez (2013); Roh et al. (2014); Morita et al. (2014); Chibba (2007) | Lean/Efficient | Food, petroleum, steel, cement production companies |
| Perez (2013); Chibba (2007) | Fast | retailers that sell trendy apparel |
| Perez (2013) | Continuous- flow | manufacturers of intermediate products; short-shelf-life production companies |
| Fisher (1997); Lee (2002); Roh et al. (2014) | Responsive | ready-made clothing sector |
| Agarwal et al. (2006); Waters (2003); Gorener (2013); Perez (2013); Roh et al. (2014); Chibba (2007) | Agile | manufacturers of intermediary goods to satisfy customer's specific needs |
| Agarwal et al. (2006); Waters (2003); Gorener (2013); Roh et al. (2014); Chibba (2007) | Leagile/Hybrid | white goods manufacturers |
| Perez (2013) | Custom-configured | Personalized products manufacturers, some fast food restaurants, paper manufacturing industry |
| Perez (2013); Topoyan (2011) | Flexible | manufacturer of spare parts for industrial customers |

*Table 3. Supply chain strategy groups*

| Efficiency Oriented | Response Oriented | Hybrid |
|---|---|---|
| Lean/Efficient | Agile | Leagile |
| Fast | Custom Configured | |
| Continuous Flow | Flexible | |

*Table 4. Basic differences between SCSs (Fisher, 1997, p. 108)*

| | Efficiency Oriented SC | Market-Responsive SC |
|---|---|---|
| Primary purpose | supply predictable demand efficiently at the lowest possible cost | respond quickly to unpredictable demand in order to minimize stock outs, forced markdowns, and obsolete inventory |
| Manufacturing focus | maintain high average utilization rate | deploy excess buffer capacity |
| Inventory strategy | generate high turns and minimize inventory throughout the chain | deploy significant buffer stocks of parts or finished products |
| Lead-time focus | shorten lead time as long as it doesn't increase cost | invest aggressively in ways to reduce lead time |
| Approach to choosing suppliers | select primarily for cost and quality | select primarily for speed, flexibility, and quality |
| Product-design strategy | maximize performance and minimize cost | use modular design to postpone product differentiation for as long as possible |

## Lean/Efficient Supply Chain Strategy

Early work on lean operations was done in the automobile industry by Toyota. A lean strategy looks for ways of eliminating waste. Material movements which don't add value, unnecessary waiting, complicated designed processes, poor quality to satisfy external and internal customers, wrong production level or capacity and holding too many stocks are most likely occurring wastes (Waters, 2003).

The principles of lean manufacturing have been applied in the manufacturing area for several decades. The notion of lean supply chain emerged from the principles of lean manufacturing being applied in supply chains. A lean supply chain is directly related to upstream and downstream flows of products, services, finances, and information which work collaboratively to reduce cost and waste by pulling what is required to meet individual customer needs efficiently. In a lean supply chain, partners make coordinated effort to eliminate waste across the supply chain (Vitasek et al., 2005).

It is not possible for firms to eliminate all of the logistics cost so that firms try to minimize cost. Logistics costs should be minimized by offering acceptable service level. Lean strategy uses a collection of operational techniques focused on productive use of resources (Sanchez & Nagi, 2001). The resources can be employees, facility, time, stock or equipment. The lean strategy aims short lead times, low stock levels and in parallel with lower costs by decreasing waste to provide efficient material flow.

A lean SCS tries to constitute a value stream from the suppliers to the end customers to eliminate all kinds of buffering cost in the supply chain and make sure that a stable schedule in production in order to improve process efficiency and then maintain the competitive advantage through economies of scale in marketplace which is a stable and predictable (Roh et al., 2014).

In the case of lean-focused applications, it is firstly necessary to describe in detail the activities carried out from the supply stage of a product to the delivery to the customer. After this flow is expressed, activities that do not add value are identified and removed. Processes that add value are operated as requested by the customer. The results are analyzed and necessary improvements are carried out (Gorener, 2013).

Lean/efficient supply chain is oriented on low cost or high relevance of asset utilization to the total cost. Efficiency is a must through the chain. The efficient SCSs best suited to industries where market competition is intense. Competition is based almost solely on price. This strategy uses a model based on a "make to forecast" decoupling point which was shown in figure 1. Production is scheduled based on sales expectations. Therefore, competitive positioning depends on best price offering and perfect order fulfillment (Perez, 2013). It is aimed to satisfy the demand with minimum cost by forecasting accurately then make to stock production.

Managers should focus on end-to-end efficiency in processes. Managers can accomplish this with two main actions. Firstly, they should provide high rates of asset utilization coupled with high overall equipment efficiency (OEE) which help to decrease cost. Secondly, they should provide high levels of forecast accuracy which guarantees product availability. There are factors which should be carried out to be successful using this strategy (Perez, 2013). These factors are listed below:

- By ensuring efficiency, non-value-added activities should be eliminated, scale economics and optimization techniques should be deployed (Lee, 2002).
- There should be extra capacity in outbound logistics to absorb demand peaks successfully (Perez, 2013).

- When transportation cost is high compared to total cost, a minimum order-size policy of a full truckload is recommended. Another alternative is a fixed order-cycle policy which allows the company to combine certain customers' orders in the same truck (Perez, 2013).
- Information linkages should be constituted to provide the most efficient, accurate and cost-effective transmission of information across the supply chain. Internet play a significant role (Lee, 2002).
- Collaborative programs should be developed with customers whose buying behavior follows a regular, predictable pattern. Supplier and customer share supply and demand forecasts and schedules which reduces demand variability in this programs. Aim is building a continuous–replenishment model and then to convert the supply chain model from efficient to continuous-flow gradually. So then higher levels of customer loyalty will be achieved (Perez, 2013).

## Fast Supply Chain Strategy

The fast supply chain applications are the most appropriate for trendy products manufacturing companies with short lifecycles. A company's ability to keep product portfolios update in accordance with the latest trends is crucial from the customer's perspective. Manufacturers should sell their products at an affordable price in parallel to their effort on continuously developing new products. Hence, trying to reduce market mediation costs becomes the main driver of competitiveness (Perez, 2013).

Encouraging continuous portfolio renewal by management is supported by three basic capabilities. These are shortening the time from idea to market, increasing end to end efficiency to enable affordable cost for customers and increasing forecast accuracy which helps to reduce market mediation cost (Perez, 2013).

Continuous portfolio renewal and extensive product portfolio lead many stock keeping unit (SKU) with low sales volume. In this case, the ability to produce small lots and purchasing small quantity products gain importance. Fast product development and manufacturability can be achieved through standardization of raw materials and limiting their variety, modular processes, and sharing of raw materials within some SKUs. Companies can cope up with high levels of seasonal demand with a pool of suppliers which provides additional capacity as needed (Perez, 2013).

## Continuous Flow Supply Chain Strategy

The main consideration of the continuous flow supply chain model is supply and demand stability. This model usually serves to mature supply chains where customer demand profile has a little variation (Perez, 2013).

This supply chain model applies built to stock (BTC) strategy where product is produced prior to demand. BTC strategy offers fastest response time to the customers. Customer's order is met either from a retail shelf or a finished goods stock place. Manufacturers supply predetermined configurations of products, so customers have limited selectivity. So customers buy offered product features against actually desired product features (Reeve & Srinivasan, 2005).

Stock levels are predefined and replenished based on a specified reorder point for inventory. Companies should use a prescheduled order cycle. Receiving orders from a group of customers the same day every month is a typical example of this strategy. High levels of inventory are required if there is high variance in SKUs in order to avoid unexpected changes in production schedule. Collaborative relations can be developed for demand variability. Competitive advantage is gained through offering a continuous replenishment system to provide high service levels and low inventory levels at customer's facilities (Perez, 2013).

## Responsiveness Oriented Supply Chain Strategies

Unpredictable demand can lead to excessive inventory levels. Because of product life cycles getting shorter, the inventory cost of innovative products may be significant. Innovative companies should follow responsive SCSs. Furthermore, companies should apply responsive SCSs when high demand uncertainty and stable supply process exist in the market. Changing and diverse needs of customers require companies to be responsive and flexible. To satisfy these requirements, companies use build to order strategy and mass customization processes. Mass customization processes are designed as flexible. Order accuracy is crucial in customization success. Internet technology can be very helpful through the effective information flow required during the customization process (Lee, 2002).

## Agile Supply Chain Strategy

Agility means being able to recombine processes, operations and business relations efficiently in continuous changing environment (Hormozi, 2001).

The agile SCS aims to offer a higher level of service quality by responding quickly to different or changing demands of customers. Agile SCS keeps a close check on customers and regulate logistics operations to customer demands. Lean strategy tries to minimize costs and sees customer service as a constraint, but agile strategy tries to maximize customer service and sees costs as a constraint. If a company is in a market where more variety and customization are demanded, a flexible SCS application is necessary (Waters, 2003).

When a manufacturer satisfies unique product specification needs of customers, the agile strategy is very helpful. This kind of firms operates in markets with unpredictable demand. Make to orders strategy is applied; manufacturers produce after receiving the customer's purchase order. Competitive success is gained with the ability to satisfy unpredictable demand such as shortening lead time. Ability is proportioned with a ratio between excess capacity and the average rate of asset usage. Without excess capacity, to be agile very difficult (Perez, 2013).

With an agile SCS, firms try to develop a flexible and reconfigurable network with partners. Competencies and market knowledge are shared in this network. Thus, firms can respond to rapid market changes. Furthermore, agile SCS focuses on new product development and streamlining information flows across the chain (Roh et al., 2014).

Mason-Jones et al. (2000) compared agile and lean strategies regarding market qualifiers and market winners. The developed matrix is shown in Table 5. According to the matrix, both agility and leanness demand high levels of product quality and short lead time. The difference between these two strategies are in cost and service level categories. While service is the critical factor for agility, in contrast, cost is critical in leanness.

For this supply chain model to be successful:

*Table 5. Comparison of agile and lean SCSs (Mason-Jones et al., 2000, p. 55)*

| | **Market Qualifiers** | **Market Winners** |
|---|---|---|
| Agile supply | Quality | Service Level |
| | Cost | |
| | Lead Time | |
| Lean Supply | Quality | Cost |
| | Lead Time | |
| | Service Level | |

- Customers should be able to access organization easily. Collaborative relationships with key customers are important so that required capacity changes can be anticipated. In order to reduce lead time, modular design is required where a group of products share some key components.
- Pricing should be adjustable based on demand variation. Customers with high demand variation should pay higher prices while customers with low-variance demand should be protected by lower prices.
- When extra capacity decreases to low levels, the company should invest in additional assets. This is required to maintain agile ability (Perez, 2013).

This strategy is applied especially by manufacturers of intermediary goods where customers have specific product needs and care about short lead times (Perez, 2013). High-end personal computer and semiconductor industries can be given as examples of practitioners of this strategy (Roh et al., 2014).

## Custom Configured Supply Chain Strategy

Competitive positioning is gained through offering a unique configuration of the finished product or service according to the consumer's needs. Modular production is an important process which can be attributed to some factors, including the potential for increased flexibility, increased speed to market, reduced cost and the ability to configure new product variations in short times (Perez, 2013; Doran & Giannakis, 2011).

According to custom configured SCS, a product is assembled to demand using standard modules. Customer demands are introduced before assembly and then the order is pushed to the customer. The customer is offered rich end-item choices. The automobile industry can be given as an example. Manufacturers aim to offer a wide range of options/color combination to customers. When customers demand some of specific combinations, they must wait until the automobile is assembled to specifications. The critical point for manufacturers and dealer is to decrease lead time from assembly to final delivery (Reeve & Srinivasan, 2005).

The difference between agile and custom-configured strategy is that product can be customized to meet virtually any customer requirement in agile strategy while the product is configurable within a limited combination of product specifications in custom configured strategy. Product configuration is done during an assembly process; where some of the parts are assembled according to an individual customer's requirements. However, product configuration can be done in other types of processes, such as mixing, packaging etc. The assembly of personalized products, such as computers and cars can be given as examples (Perez, 2013).

## Flexible Supply Chain Strategy

Flexibility is defined as an ability to change or react with little penalty in time, effort, cost or performance (Upton, 1994). There are many sources of uncertainty in organizations. These uncertainties arise from organization environment to inner organizational processes (Giachettia et al., 2003; Sawhney, 2006).

Literature about supply chain flexibility focuses mainly on different types of flexibility available to the supply chain members and how the industry reacts towards them. The rigid system produces products or services which do not satisfy customer's requirements fully and in turn affecting the quality negatively. Whereas, the flexible system adapts to changing customer's requirements so perceived, thus increasing the quality (Kumar & Deshmukh, 2006). Comparison between Rigid and Flexible SCSs are shown in Table 6.

The flexible strategy is suitable for companies which operate in a market with unexpected demand. High demand peaks and long periods of low workload are challenges for these companies. Adaptability is an important characteristic for this strategy. It has the capability which enables to reconfigure internal processes to satisfy customers' specific needs. This strategy is mainly used by service companies because they meet unexpected situations. Customers value supplier's ability to offer solutions to their needs besides the speed of response. The price is not first concern for them (Perez, 2013).

Management should focus on extra capacity of critical resources, rapid-response capability, technical strengths in process and product engineering, and a process flow that is designed to be quickly reconfigurable. These components support suppliers to be flexible (Perez, 2013).

## Hybrid Supply Chain Strategy

Leagile supply chain strategy falls into hybrid category.

## Leagile Supply Chain Strategy

Agility and leanness have some common requirements. Both of them demand a high level of product quality or require short lead times. When demand is highly volatile and fast moving, minimizing lead

*Table 6. Comparison between Rigid and Flexible SCSs (Kumar & Deshmukh, 2006, p. 18)*

| Attributes | Rigid Supply Chain | Flexible Supply Chain |
|---|---|---|
| Customer Satisfaction | High | Very High |
| Unit cost of production | Relatively higher | Relatively lower |
| Initial investment | Relatively lower | Relatively higher |
| Time frame for recovery of investment | Relatively certain | Relatively uncertain |
| Potential for increasing the market share | Low | Relatively higher |
| Time frame for making changes | Short time duration | Long time duration |
| Costs for future changes | Very High | Low |

time becomes an important aspect of agile strategy. Some of the reasons for long lead times can be partly because of waste as defined in lean manufacturing systems. Leanness calls for the elimination of waste by reducing manufacturing lead times. Lean and agile strategies offer value to customers differently. Agility places importance on service level to customers while cost is crucial in lean strategy (Mason-Jones et al., 2000). The lean and agile terms are different but can be combined and applied in supply chains (Agarwal et al., 2006).

When some elements of both lean and agile supply strategies are combined, leagile SCS is formed. A leagile SCS uses make-to-stock/lean strategies for the products which require high volume, stable demand; make-to-order/agile strategies for the rest of product demand. So this strategy can provide flexible production capacity to satisfy unexpected customer requirements and use postponement strategies for the products which are produced to forecast, and then assembled and configured according to customer needs (Roh et al., 2014). Comparison of lean, agile and leagile SCSs are shown in Table 7.

## PROPER SUPPLY CHAIN STRATEGY SELECTION

Based on previously described strategies, evaluation criteria were established. According to these criteria, it is aimed to guide companies in strategy selection. Supply chain selection criteria are competition, manufacturing, product-service, supplier and order point. Sub criteria of these main criteria were also defined.

### Competition

It is about which strategies should be used to ensure competitive advantage in product or service strategy of an organization. To be best in performance/cost rate in the market, offering the best price, continu-

*Table 7. Attributes of agile, lean and leagile SCSs (Agarwal et al., 2006)*

| | SCSs | | |
|---|---|---|---|
| **Attributes** | **Agile** | **Lean** | **Leagile** |
| Customer demand | Volatile | Predictable | Both |
| Product variety | High | Low | Medium |
| Product life cycle | Short | Long | Short |
| Customer drivers | Lead time, availability | Cost | Service level |
| Profit margin | High | Low | Medium |
| Dominant costs | Marketability costs | Physical costs | Both |
| Information enrichment | Obligatory | Desirable | Essential |
| Typical products | Fashion goods | Commodities | Product as per customer demand |
| Lead time compression | Essential | Essential | Desirable |
| Rapid reconfiguration | Essential | Desirable | Essential |
| Quality | Essential | Essential | Essential |

ously portfolio renewal according to fashion or technology and product customizing are sub criteria in competition (Brandimarte & Zotteri, 2007; Perez, 2013).

## Manufacturing

It is about strategies used in manufacturing process along the organization's supply chain. Within this scope, sub criteria are described below.

- Asset utilization rate has critical importance in industrial sectors where business profits are highly correlated with active asset efficiency. Companies fitting this profile, should assure high utilization rates. When the relevance of the cost of assets is low, companies can choose strategies that focus on responsiveness.
- Lead time includes the time from ordering to end of production and delivery to customer.
- Providing order accuracy.
- Increasing fill rate.
- Additional capacity expresses presence of additional capacity in manufacturing to fulfill orders in peak periods of demand (Perez, 2013).

## Product/ Service

It is about how product/service policy is adopted for customers within the frame of organizational management focus. Five sub-criteria are listed below.

- Rapid product development.
- Low cost/ standard product.
- Modular design.
- Small lot design; when product portfolio is extensive and often changes, based on many product varieties consistent with low sales volume, it emphasizes small lot production ability.
- Supporting with additional services; it includes offering value added services which support the product.

## Supplier

It expresses choosing of suppliers according to characteristics of them within the scope of supplier management. Four sub-criteria are defined.

- Innovativeness: product portfolio requires newest and the most innovative raw material and components which are offered to the market.
- Low cost.
- Short delivery time.
- Process flexibility of suppliers which satisfies variable customer needs.

## Order Point

When receiving an order from customers, order point is decoupling point of supply chain or push-pull bounds of product or semi-product in supply chain. It is related to how the production will be planned after a customer order is received (Reeve & Srinivasan, 2005; Perez, 2013). In this context, five sub-criteria are defined.

- Forecasting: Final product is produced before customer order is received. Production is made according to sales forecasts (Perez, 2013).
- Build to stock means building the product before demand with a standard bill of materials (Reeve & Srinivasan, 2005).
- Configure to order: The product is assembled according to demand with standard modules or components. Desktop computers can be given as an example. Customers receive greater end-item choice but some of the immediacy of order fulfillment are given up (Reeve & Srinivasan, 2005).
- Build to order: Customer orders are introduced before fabrication or at the start of the production process. Products are customized generally to customer specifications (Reeve & Srinivasan, 2005).
- Engineer to order: The product is produced and assembled to order with unique parts and drawings. A thermo-chemical reactor can be an example. This kind of supply chain reply to exactly customized products (Reeve & Srinivasan, 2005). Studies in the literature about criteria are listed in Table 8.

According to Table 9, "+" signed cells show important criteria for each supply chain strategy. For example, Performance/Cost, The Best Price, Asset Utilization Rate, Fill Rate, Low Cost/Standard Product, Short Delivery Time, Low Cost, Forecasting and Build to Stock are important criteria of Lean/Efficient Supply Chain Strategy.

The matrix is prepared based on literature. In this table, there is no order of importance between criteria. Companies can assign weighting coefficient to each criterion, and by using multiple criteria decision-making techniques, they can choose proper supply chain strategy.

*Table 8. Supply chain strategy selection criteria*

| Criteria | Authors |
|---|---|
| **Competition** | (Perez, 2013; Fisher, 1997; Upton, 1994; Roh et al., 2014; Brandimarte & Zotteri, 2007; Reeve & Srinivasan, 2005) |
| **Manufacturing** | (Perez, 2013; Reeve & Srinivasan, 2005; Waters, 2003; Agarwal et al., 2006; Roh et al., 2014; Vitasek et al., 2005; Lee, 2002) |
| **Product/ Service** | (Fisher, 1997; Reeve & Srinivasan, 2005; Perez, 2013; Birhanu et al., 2014; Lee, 2002; Agarwal et al., 2006, Roh et al., 2014) |
| **Supplier** | (Perez, 2013; Fisher, 1997; Roh et al., 2014; Agarwal et al., 2006) |
| **Order Point** | (Perez, 2013; Naylor et al., 1999; Reeve & Srinivasan, 2005; Roh et al., 2014) |

Table 9. Supply chain strategy and criteria matrix

| | | Competition | | | | Manufacturing | | | | | Product/Service | | | | | Supplier | | | | Order Point | | | | |
|---|---|---|---|---|---|---|---|---|---|---|---|---|---|---|---|---|---|---|---|---|---|---|---|---|
| | Supply Chain Strategies | Performance/ Cost | The Best Price | Portfolio Renewal | Product Customizing | Asset Utilization Rate | Additional Capacity | Lead Time | Order Accuracy | Fill Rate | Rapid Product Development | Low Cost/ Standard Product | Moduler Design | Small Lot Design | Supporting with additional services | Innovativeness | Flexible | Short Delivery Time | Low Cost | Forecasting | Build to Stock | Configure to Stock | Build to Order | Engineer to Order |
| 1 | Lean/ Efficient | + | + | | | + | + | | | + | | + | | | | | | + | + | + | + | | | |
| 2 | Fast | | | + | | | + | | | | + | | | + | | | | | + | + | | | | |
| 3 | Continuous Flow | + | | | | | | | | + | | + | | | | | | + | + | | + | | | |
| 4 | Agile | | + | | + | | + | + | | | | | | | + | | + | | + | | | + | + | |
| 5 | Custom Configured | | | + | + | | + | | + | | | | + | + | + | + | + | | | | | + | | + |
| 6 | Flexible | | | | + | | + | | | | | | | | + | + | + | | | | | + | + | + |
| 7 | Leagile | + | + | | + | + | + | | + | | | + | | | + | | + | + | + | + | + | + | + | |

## CONCLUSION

Companies are facing with ever increasing global challenges. Uncertainty related to market factors such as low-cost competitors, fluctuating commodity prices, increasing customer expectations force companies continuously to re-evaluate and adjust their competitive strategies, supply chains, and manufacturing strategies and technologies (Alomar and Pasek, 2014). Supply chain strategy is regarded as an increasingly important subject in supply chain management literature. Recent studies have examined supply chain strategies in particular industries such as food, fashion, retail as well as the relationship between supply chain strategy and other organizational functions. Developing a proper supply chain strategy is critical to a firm's long-term competitive success. Supply chain strategies can not only help managers to integrate with suppliers and customers, but also enhance the business performance of the firm and its supply chain partners (Perez-Franco et al., 2016). Companies compete with each other through their supply chain strengths. Developing proper strategies in terms of characteristics of product/service, customer demand, market, internal processes etc. affect companies' overall success in supply chains.

This study begins with the components which forms supply chain strategy. Strategies in the literature are given and these 7 strategies are grouped under 3 categories which are efficiency oriented, response oriented and hybrid strategies. After attributes of these strategies are given, strategy selection criteria which are competition, manufacturing, product-service, supplier and order point are mentioned. In the last section, a supply chain strategy and criteria matrix is defined. According to this matrix, companies can select the most proper strategy.

## REFERENCES


Agarwal, A., Shankar, R., & Tiwari, M. K. (2006). Modeling the metrics of lean, agile and leagile supply chain: An ANP-based approach. *European Journal of Operational Research*, *173*(1), 211–225. doi:10.1016/j.ejor.2004.12.005

Aitken, J., Childerhouse, P., & Towill, D. (2003). The impact of product life cycle on supply chain strategy. *International Journal of Production Economics*, *85*(2), 127–140. doi:10.1016/S0925-5273(03)00105-1

Alomar, M., & Pasek, Z. J. (2014). Linking supply chain strategy and processes to performance improvement. *Procedia CIRP*, *17*, 628–634. doi:10.1016/j.procir.2014.01.144

Birhanu, D., Lanka, K., & Rao, A. N. (2014). A survey of classifications in supply chain strategies. *Procedia Engineering*, *97*, 2289–2297. doi:10.1016/j.proeng.2014.12.473

Brandimarte, P., & Zotteri, G. (2007). *Introduction to distribution logistics* (Vol. 21). John Wiley & Sons. doi:10.1002/9780470170052

Caniato, F., Caridi, M., Castelli, C., & Golini, R. (2011). Supply chain management in the luxury industry: A first classification of companies and their strategies. *International Journal of Production Economics*, *133*(2), 622–633. doi:10.1016/j.ijpe.2011.04.030

Chibba, A. (2007). *Measuring supply chain performance measures: prioritizing performance measures* [Doctoral dissertation]. Luleå Tekniska Universitet.

de la O, G., & Matis, T. (2014). Supply Chain Strategies for the International Interoceanic Mazatlan Houston Logistic Corridor. *Journal of Applied Research and Technology*, *12*(4), 666–673. doi:10.1016/S1665-6423(14)70083-5

Doran, D., & Giannakis, M. (2011). An examination of a modular supply chain: A construction sector perspective. *Supply Chain Management: An International Journal*, *16*(4), 260–270. doi:10.1108/13598541111139071

Fisher, M. (1997). What is the right supply chain for your product? *Harvard Business Review*, *75*(2), 62–68.

Giachetti, R. E., Martinez, L. D., Sáenz, O. A., & Chen, C. S. (2003). Analysis of the structural measures of flexibility and agility using a measurement theoretical framework. *International Journal of Production Economics*, *86*(1), 47–62. doi:10.1016/S0925-5273(03)00004-5

Gorener, A. (2013). Supply Chain Strategy Selection: An Application in Manufacturing Industry by Fuzzy VIKOR Method. *International Journal of Alanya Faculty of Business*, *5*(3), 47–62.

Hormozi, A. M. (2001). Agile manufacturing: The next logical step. *Benchmarking: An International Journal*, *8*(2), 132–143. doi:10.1108/14635770110389843

Kumar, P., & Deshmukh, S. G. (2006). A model for flexible supply chain through flexible manufacturing. *Global Journal of Flexible Systems Management*, *7*(3/4), 17.

Lambert, D. M., Cooper, M. C., & Pagh, J. D. (1998). Supply chain management: implementation issues and research opportunities. *The international journal of logistics management*, *9*(2), 1-20.

Lee, H. L. (2002). Aligning supply chain strategies with product uncertainties. *California Management Review*, *44*(3), 105–119. doi:10.2307/41166135

Li, D., & OBrien, C. (2001). A quantitative analysis of relationships between product types and supply chain strategies. *International Journal of Production Economics*, *73*(1), 29–39. doi:10.1016/S0925-5273(01)00094-9

Magutu, P. O., Aduda, J., & Nyaoga, R. B. (2015). Does Supply Chain Technology Moderate the Relationship between Supply Chain Strategies and Firm Performance? Evidence from Large-Scale Manufacturing Firms in Kenya. *International Strategic Management Review*, *3*(1), 43–65. doi:10.1016/j.ism.2015.07.002

Mason-Jones, R., Naylor, B., & Towill, D. R. (2000). Engineering the leagile supply chain. *International Journal of Agile Management Systems*, *2*(1), 54–61. doi:10.1108/14654650010312606

Morita, M., Machuca, J. A., Flynn, E. J., & de los Ríos, J. L. P. (2015). Aligning product characteristics and the supply chain process–A normative perspective. *International Journal of Production Economics*, *161*, 228–241. doi:10.1016/j.ijpe.2014.09.024

Naylor, J. B., Naim, M. M., & Berry, D. (1999). Leagility: Integrating the lean and agile manufacturing paradigms in the total supply chain. *International Journal of Production Economics*, *62*(1), 107–118. doi:10.1016/S0925-5273(98)00223-0

Perez, H. D. (2013). Supply chain strategies: Which one hits the mark. *CSCMP Supply Chain Quarterly*, 1.

Perez-Franco, R., Phadnis, S., Caplice, C., & Sheffi, Y. (2016). Rethinking supply chain strategy as a conceptual system. *International Journal of Production Economics*, *182*, 384–396. doi:10.1016/j.ijpe.2016.09.012

Reeve, J. M., & Srinivasan, M. M. (2005). Which supply chain design is right for you?.

Roh, J., Hong, P., & Min, H. (2014). Implementation of a responsive supply chain strategy in global complexity: The case of manufacturing firms. *International Journal of Production Economics*, *147*, 198–210. doi:10.1016/j.ijpe.2013.04.013

Sanchez, L. M., & Nagi, R. (2001). A review of agile manufacturing systems. *International Journal of Production Research*, *39*(16), 3561–3600. doi:10.1080/00207540110068790

Sawhney, R. (2006). Interplay between uncertainty and flexibility across the value-chain: Towards a transformation model of manufacturing flexibility. *Journal of Operations Management*, *24*(5), 476–493. doi:10.1016/j.jom.2005.11.008

Sharifi, H., Ismail, H. S., Qiu, J., & Tavani, S. N. (2013). Supply chain strategy and its impacts on product and market growth strategies: A case study of SMEs. *International Journal of Production Economics*, *145*(1), 397–408. doi:10.1016/j.ijpe.2013.05.005

Topoyan, M. (2011). An Integrated Measurement Model of Supply Chain Flexibility. *Ege Academic Review*, *11*(4).

Upton, D. M. (1994). The management of manufacturing flexibility. *California Management Review*, *36*(2), 72–89. doi:10.2307/41165745

Vitasek, K. L., Manrodt, K. B., & Abbott, J. (2005). What makes a lean supply chain? *Supply chain management review*, *9*(7), 39-45.

Waters, D. (2003). *Logistics: an introduction to supply chain management*. Editorial Aardvark.

## KEY TERMS AND DEFINITIONS

**Efficiency Oriented Supply Chain Strategies:** Lean/Efficient, fast and continuous flow supply chain strategies are efficiency oriented.

**Hybrid Supply Chain Strategy:** Leagile supply chain strategy is hybrid strategy.

**Logistics:** It is the general management of how resources are acquired, stored and transported to their final destination.

**Response Oriented Supply Chain Strategies:** Agile, custom configured and flexible supply chain strategies are response oriented.

**Supply Chain:** A supply chain is a system of organizations, people, activities, information, and resources involved in moving a product or service from supplier to customer.

**Supply Chain Management:** Supply chain management is the integration of key business processes from end user through original suppliers that provides products, services, and information that add value for customers and other stakeholders.

**Supply Chain Selection Criteria:** Competition, manufacturing, product-service, supplier and order point are supply chain selection criteria.

**Supply Chain Strategy:** By definition a forward-looking document - which anticipates changing customer needs and defines how the supply chain is going to evolve to meet those new requirements.

# Section 3

# Differential Return on Investment Optimization:

## Pricing, Lotsizing and Shipment Considerations in a Two-Echelon Supply Chain

*This section introduces most used techniques that increase the efficiency of manufacturing systems. These techniques include the mathematical models, the simulation methods, the optimal control theory, the lean manufacturing principles, and the multi-criteria decision making.*

# Chapter 12
# Mathematical Optimization Models for the Maintenance Policies in Production Systems

**Alperen Bal**
*Istanbul Technical University, Turkey*

**Sule Itir Satoglu**
*Istanbul Technical University, Turkey*

## ABSTRACT

*This chapter initially presents a brief information about production systems. At these systems, different types of maintenance policies are developed to cope with wear out failures. Mainly used maintenance policies can be classified as corrective, preventive, and condition-based maintenance. In the corrective maintenance, repair or replacement is applied whenever components of the machine breakdown. In the preventive maintenance approach maintenance activities are applied to the critical components on a periodic basis. On the other hand, maintenance activities are applied whenever critical reliability level is reached or exceeded. These types of maintenance policies are modeled using mathematical modeling techniques such as linear programming, goal programming, dynamic programming, and simulation. A review of current literature about the mathematical models, the simulation-based optimization studies examining these maintenance policies are categorized and explained. Besides, the solution methodologies are discussed. Finally, the opportunities for future research are presented.*

## INTRODUCTION

Manufacturing environment and production technologies evolved rapidly in the past few decades. Proper maintenance has been drawing more and more attention to sustain the manufacturing systems' reliability, maintainability and availability. However, the maintenance cost can even reach 15 to 70% of the total expenditures (Madu 2000, Mobley 2002, Wang, Chu, & Mao, 2008). Although the maintenance was considered as a cost factor in the past, companies are more conscious about the importance of the maintenance activities and they noticed that these can add value to their business. Different decision

DOI: 10.4018/978-1-5225-2944-6.ch012

tools can help companies to ensure a proper maintenance policy. In the literature, too many decision models exist. On the other hand, most of these models only remain in theory. Dekker (1996) focused on the maintenance models and the factors which prevent applications of them. The six reasons for the gap between theory and practice according to Dekker (1996) are that the maintenance models are difficult to understand; many papers are written only for math purposes, but the companies are not interested in publication; maintenance activities consist of many different aspects, optimization is not always necessary, and the models often focus on the wrong type of maintenance. Langberg (1988), Sherwin (2000), Li (2005) reviewed overall models for maintenance management. Wang (2002) provided a literature review, classified and compared maintenance policies taking into account both single-unit and multi-unit systems. Garg and Deshmukh (2006) reviewed the literature on maintenance management including the optimization models, maintenance techniques, scheduling, and information systems. Horenbeek et al. (2010) reviewed the literature on maintenance optimization models especially focusing on the optimization objectives used. Sharma and Yadava (2011), Ding and Kamaruddin (2015) reviewed the literature on maintenance management by considering the associated case studies and applications. Recently Alaswad and Xiang (2017) reviewed overall optimization models on Condition Based Maintenance (CBM) for stochastically deteriorating system. As one of the most important aspect of CBM, optimization criteria of CBM policies is also subjected to review in their research.

The specific objectives of this chapter are as follows:

1. To identify the characteristics of the production systems and the suitable maintenance operation strategies for them.
2. To consolidate the literature on maintenance management as much as possible with regard to both mathematical modelling and simulation based optimization.
3. To identify new trends in maintenance policies to be applied in the production systems.

## PRODUCTION SYSTEMS

Production systems transform raw materials or semi-products into end-products. This transformation process typically uses resources such as material, energy, labor to make a change. A value adding activity is conducted in this process. The production systems are categorized according to different criterions. Traditional production systems can be categorized in terms of production process, type of product, amount of product produced, and implemented stock policy. On the other hand, modern production systems can be classified such as Just-in-Time Production, Cellular Manufacturing, Computer Integrated Manufacturing, Flexible Manufacturing, and Additive Manufacturing. Since the manufacturing and maintenance activities cannot be performed simultaneously, cutting-edge production systems require well planned maintenance activities, in order to increase the availability of the machines and lines.

## MAINTENANCE MODELS

Maintenance is perceived as a value-adding activity, instead of an unavoidable cost item, (Ben-Daya and Duffuaa, 1995). Most of the time, maximizing reliability can bring maintenance cost up to an unacceptable level. So, to obtain the best performance, maintenance policies aim to maximize system reliability while the system maintenance cost are justified. Besides, other policies aim to minimize the system maintenance cost while the system reliability requirements are satisfied (Sharma et al., 2011).

In general maintenance optimization models cover the following aspects:

- Model description of a technical system and its function: Modeling of the deterioration of the system in time and its possible consequences for the system.
- A description of the available information about the system and the action that may be taken by the management.
- Objective function and an optimization technique, which help in finding the best balance between reliability level and cost of the maintenance.

In reliability theory, hazard function displays a bathtub shape (Klutke et al., 2003). A sample curve is shown in Figure 1. The bathtub curve has a great importance in reliability practice. The curve represents three distinct areas to demonstrate failure rates of equipment or machine. Early life period depicts the decreasing failure rate over time. In this period of time, the equipment may be exposed of many initial failures which may stem from assembly errors, poor design and poor joints. During the useful life period, a constant failure rate is observed. It should be noted that the reliability analysis is based on the random failures occurring during this period. Because the failure rate is low in this period and failures can happen unexpectedly, in other words, randomly. The wear-out period depicts the end-of-life period of the equipment and, there is an increasing failure rate during this period. The wear-out failures can be caused by factors such as aging of equipment, fatigue of material, poor lubrication, improper design. Effective maintenance policies can help to enhance useful life period and also to decrease the failure-rate of wear-out period.

*Figure 1. The Bathtub Curve (Klutke et al., 2003)*

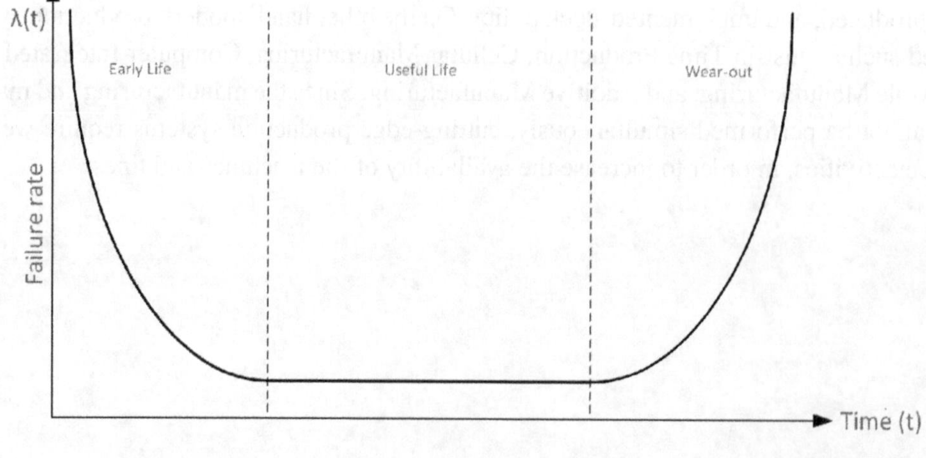

Many different methods and strategies are used for the implementation of maintenance management. The most common maintenance management strategies in the literature and also in practice are as follows:

## Corrective Maintenance (CM)

This type of maintenance is the first and the oldest type of maintenance policy applied in manufacturing industry (Mechefske 2001). A corrective maintenance is only implemented after the failure occurs. The aim of this policy is to repair or restore the system back to its operational condition. This procedure is strictly related with the experience of the workers or technicians and has no scientific study beneath it. It is generally assumed that the control policies and frequency of failures are dependent on the machine age. However, it has been pointed out in the papers of Boukas (1998) and Kenné (2003) that applying corrective maintenance which is not age-dependent can increase the availability of the production systems. Iyoob (2003) investigated the effectiveness of five different corrective maintenance models for equipment availability. These models are renewal, minimal, Kijima I, Kijima II, and quasi renewal. Ho and Quinino (2012) proposed an economic model that minimizes the cost per unit, taking into account the on-line process control and corrective maintenance. Wang et al. (2014) proposed a complete corrective maintenance procedure for an engineering equipment using failure mode, effects and criticality analysis. Corrective maintenance policy can be preferred when profit margins are large (Sharma 2005). On the other hand, this production policy may result in unexpected failures. These unexpected failures can lead to massive production losses and serious damage, especially in high-tech manufacturing enterprises.

## Preventive Maintenance (PM)

The production systems have become more complex with the increasing size of the companies and developing technology. This situation could let some disastrous results when failure of a small component occurs, implying the loss of large amount of money. Therefore, maintenance is used to prevent the occurrence of a possible failure and to sustain the system in proper operating conditions. Consequently, preventive maintenance policy was introduced which is repairing or replacing system components at each of the pre-defined time instants (Gertsbakh, 2013). Since the basic preventive maintenance (PM) policies proposed by Barlow and Hunter (1960), many PM studies have been conducted.

Mechefske (2001) stated that most of the systems maintain its remaining useful life when preventive maintenance policy is applied. On the other hand, it is difficult to define the optimal time interval for maintenance when the historical data is insufficient and this can lead to unnecessary maintenance (Wang 2007). In Figure 2, the corrective and preventive maintenance policies are compared and illustrated. In a corrective maintenance policy, maintenance is only performed when the equipment fails. However, according to preventive maintenance policy, maintenance is carried out in certain periods. A preventive maintenance is imperfect because it restores a system to a state between as-good-as-new and as-bad-as-old (Castro, 2009). Kenné and Nkeungoue (2008) proposed a stochastic control model that allows simultaneous planning of corrective and preventive maintenance activities. They studied the problem of production planning for a flexible production system by considering corrective and preventive maintenance. Xu and Hu (2008) considered both corrective maintenance and preventive maintenance to investigate optimal steady availability. Thus, they investigated the optimum time required to apply preventive maintenance theoretically and also with numerical examples. Ahmad and Kamaruddin (2012) reviewed the current state of time-based maintenance and condition-based maintenance in industrial applications and summarized the latest state-of-the-art condition monitoring techniques. Castro (2015) proposed a model of a complex cold standby system by analyzing the preventive maintenance with regard to cost and effectiveness, in an algorithmic form. The overall availability of the production systems can be significantly increased by the use of the corrective and preventive maintenance in the same model. Sidibe et al. (2017) proposed a preventive maintenance optimization taking into account used systems such as second hand products. These used systems are modeled with an uncertain parameter to begin the second life cycle. Imperfect preventive maintenance was extensively studied in the literature. Interested readers should see (Ben-Daya 2000), (Nakawaga 2005), (Osaki 2002) and (Pham 2003). In addition, preventive maintenance has been studied in many studies integrated with scheduling problems (Graves & Lee, 1999; Kubzin & Strusevich, 2006; Ruiz et al., 2007; Budai et al., 2008; Ma et al., 2010; Berrichi et al., 2010; Berrichi and Yalaoui, 2013; Ciu et al., 2014).

## Condition-Based Maintenance (CBM)

Condition-based maintenance is a rapidly growing maintenance strategy in research area and practical life. It recommends maintenance decisions based on the collected information through condition monitoring. In this type of maintenance policy, the decision of maintaining the system is taken dynamically based upon the observed condition of the system. Therefore, CBM attempts to avoid unnecessary maintenance activities. The operational condition of the system is tracked of by means of advanced sensor technologies. This allows to easily and clearly detect an abnormal situation in a manufacturing system. However, CBM is not always the best maintenance policy, especially with regard to cost effectiveness (Arunraj, 2010). The need for complex and expensive monitoring technologies makes the CBM impractical for a number of applications (Jonge et al., 2015).

Martin (1994) summarized condition monitoring and fault diagnosis techniques of the machine tools and its brief development history. Jardine et al. (2006) reviewed the CBM systems with emphasis on models, algorithms and technologies. Peng et al. (2010) reviewed various techniques and algorithms on CBM and divided prognostic models into four categories. Namely physical model, knowledge-based model, data-driven model, and combination model. Prajapati (2012) provided a brief overview of the CBM from the viewpoint of history, recent developments, applications, and research challenges. Bousdekis et al. (2015a) reviewed the literature in the area of decision making for CBM and identified the possibilities for implementation of proactive Decision Support System (DSS) by considering the real-time sensor data. Bousdekis et al. (2015b) also provided a literature review for prognostic-based decision support methods for CBM.

*Figure 2. The Comparison of Corrective and Preventive Maintenance*

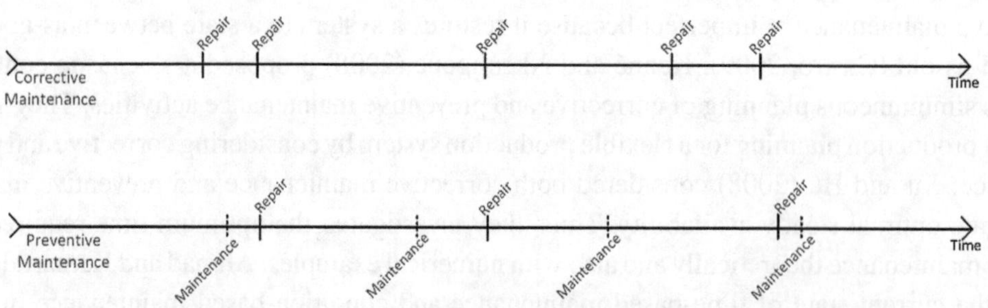

Besides, Marseguera et al. (2002) focused on the problem of determining the optimal on-condition maintenance strategy considering the optimum degradation level of components beyond which maintenance has to be performed. The optimal degradation level is determined using a Genetic Algorithm (GA). The problem intends to minimize the two typical objectives of profit and availability. Besides, Monte Carlo simulation is used for the predictive model describing the evolution of the degrading system. Do et al. (2015) proposed a proactive CBM policy with both perfect and imperfect maintenance for a single component deteriorating system. First, they investigated the impact of imperfect maintenance actions, and then they proposed an adaptive maintenance policy which can select perfect or imperfect maintenance actions. Castro et al. (2015) proposed a maintenance model by analyzing a condition-based maintenance for a system subject to internal degradation and external shocks. They developed a CBM with periodic inspection times for this competing failure model. Caballé et al. (2015) considered a CBM policy for a system subject to multiple degradation and sudden shocks in their paper. Multiple degradation is modelled by a gamma process and the initiation of degradation process modelled by a Non-homogenous Poisson process. The sudden shocks are modelled by using a Doubly Stochastic Poisson Process in their study. Li et al. (2016) proposed a new adaptive maintenance policy for grouping maintenance actions and they investigated the impacts of both economic and stochastic dependencies. Levy copulas dependence is used for modelling the stochastic dependencies. Kaizer et al. (2017) considered a system with multi-components and the joint optimization of condition-based maintenance and condition-based spares. The model is formulated using Markov Decision Process and the long-run average cost per time unit is aimed to be minimized.

Maintenance strategies are compared in terms of different aspects by Jin et al. (2016). In the face of this evaluation, their table is adapted to our study as shown in Table 1.

*Table 1. The Classification of the Maintenance Strategies*

| Maintenance Strategy | Corrective Maintenance | Preventive Maintenance | Condition-Based Maintenance |
|---|---|---|---|
| Maintenance interval | Fail and fix | Time based, usage based | Condition based, improve & sustain |
| Object | Component | Component, function | Component, function, system |
| Planning & Scheduling | Planning on the fly | Planning & scheduling based on optimal PM interval | Proactive planning & scheduling |
| Failure severity and frequency | Low severity, low frequency | Low to medium severity, high frequency | High impact, high frequency |
| Human factor (inspection & decision-making) | High | Intermediate | Low (false alarm) |
| Cost effectiveness | Labor intensive, labor and material | Costly due to over maintenance or ineffective & inefficient PM | Cost-effective, substantially save failures & extend the life of equipment |
| Requirement for technology readiness | Low | Low | High |

## MATHEMATICAL MODELING-BASED OPTIMIZATION OF THE MAINTENANCE POLICIES

Mathematical modelling is frequently used in deterministic maintenance optimization. According to Almeida and Bohoris (1995) the general mathematical model steps can be divided into four main steps including the problem definition, necessary information collection, preference elicitation, and modelling the state of nature optimization. Many different performance measures are used as objective function in mathematical modeling. Alaswad and Xiang (2017) classified these optimization criterions into three groups. Namely cost minimization, reliability or availability maximization, and multi-objective. In offered models, it is generally aimed to minimize maintenance cost. However, in some cases it can be difficult to obtain maintenance cost. While it is difficult to obtain cost parameters, it can be easier and more measurable to calculate such criteria as uptime / downtime. For this reason, the availability, reliability, or safety criterions are used as easily measurable performance parameters.

Various parameters are optimized in different studies, but these parameters can conflict with each other in some cases. For example, when cost minimization is targeted in a multi-component system, the reliability / availability of the system can be low at a level that cannot be accepted (Pham and Wang, 1996 and Wang and Pham, 2006). Considering these possible differences especially in practice, it is necessary to minimize the cost of the system and to maximize the reliability / availability criteria in order to obtain a good maintenance policy. It is aimed to combine conflicting objectives in multi-objective optimization problems in the most harmonious way. Ferreira et al. (2009) developed a multi-objective decision model that takes into account the expected cost of maintenance policy and the duration of downtimes while simultaneously determining inspection interval. Zio and Viadana (2011) also aimed to optimize the maintenance intervals and, in doing so, considered the conflicting objective functions. These objective functions are determined as mean availability, cost of inspections/repairs, and exposure time of maintenance operators. Wang and Liu (2015) investigated a multi-objective scheduling problem with two kinds of resources (machines and molds) by adding flexible preventive maintenance work on these resources. Dalfard and Mohammadi (2012) have developed a multi-objective optimization model consisting of three objectives that take into account of a flexible job shop scheduling (FJSP) problem with maintenance.

Besides, to solve the multi-objective optimization problems, there are some other studies using Mixed-Integer Programming (MIP) in the literature. To find an integrated lot-sizing and preventive maintenance strategy of a production system Aghezzaf (2007) offered a mixed-integer programming model. Besides, Aghezzaf and Najid (2008) proposed a mixed-integer non-linear program for integrating production planning and a cyclic preventive maintenance policy. They also proposed a Mixed-Integer Linear Programming (MILP) model when a non-cyclic preventive maintenance policy is applied.

The proposed model of Fitouhi and Nourelfath (2012) determines simultaneously noncyclical preventive maintenance actions and the optimal production plan. Nouri et al. (2013) proposed a MILP model for determining sequence of jobs and implementation time of the maintenance activities. Lu et al. (2013) used MILP to formulate a joint model for integrating run-based preventive maintenance into the capacitated lot sizing problem. Besides, Zhao and Wang (2014) offered a MINLP (Mixed Integer Non-Linear Programming) model to better integrate production planning and maintenance. In the model, preventive and corrective maintenance are considered.

Recently, Hnainen (2016) proposed a mixed-integer programming model for planning production and PM simultaneously for a single machine capacitated lot-sizing. The main contribution is the integration of maintenance scheduling on the MILP model to minimize the sum of the total production and maintenance cost. Aghezzaf et al. (2016) investigated the integration of production planning and imperfect preventive maintenance using MINLP. A fix-and-optimize procedure is offered for the large instances.

## SIMULATION OPTIMIZATION OF THE MAINTENANCE POLICIES

Simulation is a powerful tool to reflect the complexity found in industrial systems while other modelling approaches for maintenance rely on oversimplified assumptions (Alrabgi and Tiwari, 2016). Andijani and Duffuaa (2002) reviewed the literature and evaluated the use of simulation in maintenance. They observed that most papers clearly define their objectives, simulation languages, and model performance measures. However, they criticize these studies due to unclear factors like verification, validation, experimental design and output analysis.

Marseguera and Zio (2000) and Marseguera et al. (2002) built the models by Monte Carlo simulation. In both of these studies, Monte Carlo simulation and a Genetic Algorithm are integrated. Barata et al. (2002) used Monte Carlo simulation to develop a maintenance model for a continuously monitored degrading system. The simulation model was initially built for a non-repairable single component and then was generalized to multi-component systems. Marseguera and Zio (2000) proposed an approach to the optimal maintenance and repair strategies of an industrial plant taking into account the reliability and economic constraints. Marseguera et al. (2002) focused on condition-based maintenance strategy to determine the thresholds of the component degradation. Wang et al. (2008) modelled the deterioration of units based-on discrete time Markov chains and proposed a condition-based replacement and provisioning policy for the deteriorating systems.

Bal and Satoglu (2013) tested different maintenance strategies using discrete event simulation and they investigated economic feasibility of the RFID based maintenance management system via Monte-Carlo simulation method. Satoglu and Ustundag (2012) performed a Monte-Carlo simulation analysis to estimate the cost savings provided by RFID-enhanced maintenance, by considering different possible levels of cost savings. Alrabghi and Tiwari, (2015) provided an overview of the state of the art simulation-based optimization of maintenance. Later, the same authors proposed an optimization technique that integrates discrete event simulation with some optimization algorithms such as Simulated Annealing (SA), Hill Climb and Random Solutions to optimize the maintenance strategies while taking account of production dynamics and spare parts management (Alrabghi and Tiwari, 2016).

Recently, Cheng et al. (2016) proposed a new approach for analyzing the machining accuracy reliability of machine tool based on the Markov Chain simulation. They defined the machining accuracy reliability as the ability of a machine tool to perform at its specified machining accuracy under the stated conditions for a given period of time. Besides, the key error causes of the machining accuracy reliability were identified via sensitivity analysis. After obtaining the results, a chosen machine tool was selected to experimentally validate the effectiveness of the proposed method.

Authors believe that although, simulation is a powerful technique for modeling the maintenance models under dynamic conditions, it is not an optimization technique. Therefore, it must be integrated with other optimization techniques such as meta-heuristics or with the statistical techniques such as response surface methodology.

## SOLUTION METHODOLOGIES FOR THE OPTIMIZATION MODELS

Finding the optimal solution for mathematical optimization problems is rather difficult for the solvers. These types of problems are very complex. Therefore, heuristic and meta-heuristic algorithms are used to obtain a good solution. Different types of meta-heuristics are used for various maintenance problems such as Particle swarm optimization (Wang et al., 2011; Pereira, 2010; Loganathan, 2016), Genetic Algorithm (Levitin and Lisnianski, 2000; Martorell et al., 2005; Ruiz et al., 2007; Lin and Wang, 2012), Artificial Bee Colony Algorithm (Yeh & Hsieh, 2011), Ant Colony Optimization (Samrout, 2005; Berrichi, 2010), and Simulated Annealing (Dahal and Chakpitak, 2007; Doostparast et al., 2015). Jin et al. (2008) proposed a multi-objective genetic algorithm where the objectives were minimizing the maintenance cost, makespan, total weighted completion time of jobs, total weighted tardiness and machine unavailability. Ba et al. (2017) used Genetic Algorithm to optimize combined opportunistic and preventive maintenance strategies. Two critical opportunities are considered in this study: non-homogeneous opportunity arrivals and stochastic opportunity duration. Dalfard and Mohammadi (2012) proposed two meta-heuristic algorithms, a genetic algorithm and a simulated annealing to solve a multi-objective maintenance problem. Dong (2013) used a branch and bound algorithm to find exact solution of a parallel machine scheduling problem with flexible maintenance activities to minimize the total cost. Sarker et al. (2013) developed a hybrid evolutionary algorithm for a job shop scheduling problem with maintenance activities with the objective of minimizing the makespan.

In addition, self-improved heuristic methods are also used for a solution methodology besides the meta-heuristics. Aghezzaf and Najid (2009) used a Lagrangian-based heuristic procedure for the solution of the proposed MINLP model for integrating production planning and preventive maintenance in the production systems. Lu et al. (2013) used a Lagrangian-based heuristic to solve an integrated model of the capacitated lot-sizing problem including preventive maintenance activities. Yalaoui et al. (2014) improved Aghezzaf and Najid's mixed-integer linear model and they adapted a relaxed heuristic for the lot sizing problem of their case. Zhao and Wang (2014) used linear relaxation and presented an iterative approach to find a solution for the non-linear model that they proposed.

## CONCLUSION AND FUTURE STUDIES

Many published papers show that maintenance management is very important both in the academic field and in the industry. In this literature review study, the papers related to the subject are tried to be handled as much as possible. On the other hand, the papers considered not to be directly related to the topic of this study are not included.

First, a brief information about the production systems is given and maintenance policies are divided into the main classes. Later, the optimization methods reviewed have been examined in terms of mathematical modeling and simulation studies. Discrete-event simulation is the most frequently used technique to model the maintenance systems. Besides, the Genetic Algorithm is the most frequently used meta-heuristic as a modern optimization algorithm in the literature to solve the maintenance optimization models (Alrabghi and Tiwari, 2015). In the literature, many studies cannot be implemented in real-case systems due to the over-simplified assumptions, such as perfect maintenance/inspection, and a single-unit system. Therefore, more studies that comply with the real maintenance conditions are required. Besides, maintenance modeling and optimization of the complex manufacturing systems should be performed in future studies to better assist the practitioners.

In addition, there are a limited number of studies that employ discrete-event simulation and response surface methodology for simulation-optimization purposes. Therefore, for parameter optimization of the maintenance models in dynamic conditions, these two techniques should be integrated to each other in future studies.

Modelling the maintenance policies in accordance with the production and spare parts management is an emerging issue in this field. Therefore, more studies are expected to be made that optimize the maintenance policies and the spare parts inventories simultaneously. Besides, modelling simultaneous maintenance strategies for each asset with different characteristics in a manufacturing system is an interesting issue in optimization of the maintenance strategies. Lack of this kind of studies still remains as a research gap.

In addition, in parallel with the fourth industry revolution, predictive data analytics methods for failure estimation integrated with the condition-based monitoring should be performed in near future. So, in the field of maintenance optimization, more condition-based monitoring and failure prevention studies must be carried out.

## REFERENCES

Aghezzaf, E. H., Jamali, M. A., & Ait-Kadi, D. (2007). An integrated production and preventive maintenance planning model. *European Journal of Operational Research*, *181*(2), 679–685. doi:10.1016/j.ejor.2006.06.032

Aghezzaf, E. H., Khatab, A., & Le Tam, P. (2016). Optimizing production and imperfect preventive maintenance planning′ s integration in failure-prone manufacturing systems. *Reliability Engineering & System Safety*, *145*, 190–198. doi:10.1016/j.ress.2015.09.017

Aghezzaf, E. H., & Najid, N. M. (2008). Integrated production planning and preventive maintenance in deteriorating production systems. *Information Sciences*, *178*(17), 3382–3392. doi:10.1016/j.ins.2008.05.007

Ahmad, R., & Kamaruddin, S. (2012). An overview of time-based and condition-based maintenance in industrial application. *Computers & Industrial Engineering*, *63*(1), 135–149. doi:10.1016/j.cie.2012.02.002

Alaswad, S., & Xiang, Y. (2017). A review on condition-based maintenance optimization models for stochastically deteriorating system. *Reliability Engineering & System Safety*, *157*, 54–63. doi:10.1016/j.ress.2016.08.009

Almeida, A. T., & Bohoris, G. A. (1995). Decision theory in maintenance decision making. *Journal of Quality in Maintenance Engineering*, *1*(1), 39–45. doi:10.1108/13552519510083138

Alrabghi, A., & Tiwari, A. (2015). State of the art in simulation-based optimisation for maintenance systems. *Computers & Industrial Engineering*, *82*, 167–182. doi:10.1016/j.cie.2014.12.022

Alrabghi, A., & Tiwari, A. (2016). A novel approach for modelling complex maintenance systems using discrete event simulation. *Reliability Engineering & System Safety*, *154*, 160–170. doi:10.1016/j.ress.2016.06.003

Andijani, A., & Duffuaa, S. (2002). Critical evaluation of simulation studies in maintenance. *Production Planning and Control*, *13*(4), 336–341. doi:10.1080/09537280110077578

Arunraj, N. S., & Maiti, J. (2010). Risk-based maintenance policy selection using AHP and goal programming. *Safety Science*, *48*(2), 238–247. doi:10.1016/j.ssci.2009.09.005

Ba, H. T., Cholette, M. E., Borghesani, P., Zhou, Y., & Ma, L. (2016). Opportunistic Maintenance Considering Non-Homogenous Opportunity Arrivals and Stochastic Opportunity Durations. *Reliability Engineering & System Safety*. doi: 10.1016/j.ress.2016.12.011

Bal, A., & Satoglu, S. I. (2013) Maintenance Management of Production Systems with Sensors and RFID: A Case Study. In *Proceedings of the Global Conference on Engineering Technology and Management*.

Barata, J., Soares, C. G., Marseguerra, M., & Zio, E. (2002). Simulation modelling of repairable multi-component deteriorating systems for on condition maintenance optimisation. *Reliability Engineering & System Safety*, *76*(3), 255–264. doi:10.1016/S0951-8320(02)00017-0

Barlow, R., & Hunter, L. (1960). Optimum preventive maintenance policies. *Operations Research*, *8*(1), 90–100. doi:10.1287/opre.8.1.90

Ben-Daya, M., & Duffuaa, S. O. (1995). Maintenance and quality: The missing link. *Journal of Quality in Maintenance Engineering*, *1*(1), 20–26. doi:10.1108/13552519510083110

Ben-Daya, M., & Duffuaa, S. O. (2000). *Overview of maintenance modelling areas* (pp. 3–35). Springer, US: Maintenance, Modeling and Optimization. doi:10.1007/978-1-4615-4329-9_1

Berrichi, A., & Yalaoui, F. (2013). Efficient bi-objective ant colony approach to minimize total tardiness and system unavailability for a parallel machine scheduling problem. *International Journal of Advanced Manufacturing Technology*, *68*(9-12), 2295–2310. doi:10.1007/s00170-013-4841-0

Berrichi, A., Yalaoui, F., Amodeo, L., & Mezghiche, M. (2010). Bi-objective ant colony optimization approach to optimize production and maintenance scheduling. *Computers & Operations Research*, *37*(9), 1584–1596. doi:10.1016/j.cor.2009.11.017

Boukas, E. K. (1998). Hedging point policy improvement. *Journal of Optimization Theory and Applications*, *97*(1), 47–70. doi:10.1023/A:1022670932536

Bousdekis, A., Magoutas, B., Apostolou, D., & Mentzas, G. (2015a). A proactive decision making framework for condition-based maintenance. *Industrial Management & Data Systems*, *115*(7), 1225–1250. doi:10.1108/IMDS-03-2015-0071

Bousdekis, A., Magoutas, B., Apostolou, D., & Mentzas, G. (2015b). Review, analysis and synthesis of prognostic-based decision support methods for condition based maintenance. *Journal of Intelligent Manufacturing*. doi: 10.1007/s10845-015-1179-5

Budai, G., Dekker, R., & Nicolai, R. P. (2008). Maintenance and production: a review of planning models. In *Complex system maintenance handbook* (pp. 321–344). Springer London. doi:10.1007/978-1-84800-011-7_13

Caballé, N. C., Castro, I. T., Pérez, C. J., & Lanza-Gutiérrez, J. M. (2015). A condition-based maintenance of a dependent degradation-threshold-shock model in a system with multiple degradation processes. *Reliability Engineering & System Safety, 134*, 98–109. doi:10.1016/j.ress.2014.09.024

Castro, I. T. (2009). A model of imperfect preventive maintenance with dependent failure modes. *European Journal of Operational Research, 196*(1), 217–224. doi:10.1016/j.ejor.2008.02.042

Castro, I. T., Caballé, N. C., & Pérez, C. J. (2015). A condition-based maintenance for a system subject to multiple degradation processes and external shocks. *International Journal of Systems Science, 46*(9), 1692–1704. doi:10.1080/00207721.2013.828796

Cheng, Q., Sun, B., & Zhao, Y. (2016). A Method to Analyse The Machining Accuracy Reliability Sensitivity Of Machine Tools Based On Fast Markov Chain Simulation *Eksploatacja I Niezawodnosc, 18*(4), 552.

Cui, W. W., Lu, Z., & Pan, E. (2014). Integrated production scheduling and maintenance policy for robustness in a single machine. *Computers & Operations Research, 47*, 81–91. doi:10.1016/j.cor.2014.02.006

Dahal, K. P., & Chakpitak, N. (2007). Generator maintenance scheduling in power systems using metaheuristic-based hybrid approaches. *Electric Power Systems Research, 77*(7), 771–779. doi:10.1016/j.epsr.2006.06.012

Dalfard, V. M., & Mohammadi, G. (2012). Two meta-heuristic algorithms for solving multi-objective flexible job-shop scheduling with parallel machine and maintenance constraints. *Computers & Mathematics with Applications (Oxford, England), 64*(6), 2111–2117. doi:10.1016/j.camwa.2012.04.007

de Jonge, B., Dijkstra, A. S., & Romeijnders, W. (2015). Cost benefits of postponing time-based maintenance under lifetime distribution uncertainty. *Reliability Engineering & System Safety, 140*, 15–21. doi:10.1016/j.ress.2015.03.027

Dekker, R. (1996). Applications of maintenance optimization models: A review and analysis. *Reliability Engineering & System Safety, 51*(3), 229–240. doi:10.1016/0951-8320(95)00076-3

Ding, S. H., & Kamaruddin, S. (2015). Maintenance policy optimization—literature review and directions. *International Journal of Advanced Manufacturing Technology, 76*(5-8), 1263–1283. doi:10.1007/s00170-014-6341-2

Do, P., Voisin, A., Levrat, E., & Iung, B. (2015). A proactive condition-based maintenance strategy with both perfect and imperfect maintenance actions. *Reliability Engineering & System Safety, 133*, 22–32. doi:10.1016/j.ress.2014.08.011

Dong, M. (2013). Parallel machine scheduling with limited controllable machine availability. *International Journal of Production Research, 51*(8), 2240–2252. doi:10.1080/00207543.2012.714002

Doostparast, M., Kolahan, F., & Doostparast, M. (2015). Optimisation of PM scheduling for multi-component systems–a simulated annealing approach. *International Journal of Systems Science, 46*(7), 1199–1207. doi:10.1080/00207721.2013.815822

Ferreira, R. J., de Almeida, A. T., & Cavalcante, C. A. (2009). A multi-criteria decision model to determine inspection intervals of condition monitoring based on delay time analysis. *Reliability Engineering & System Safety*, *94*(5), 905–912. doi:10.1016/j.ress.2008.10.001

Fitouhi, M. C., & Nourelfath, M. (2012). Integrating noncyclical preventive maintenance scheduling and production planning for a single machine. *International Journal of Production Economics*, *136*(2), 344–351. doi:10.1016/j.ijpe.2011.12.021

Garg, A., & Deshmukh, S. G. (2006). Maintenance management: Literature review and directions. *Journal of Quality in Maintenance Engineering*, *12*(3), 205–238. doi:10.1108/13552510610685075

Gertsbakh, I. (2013). *Reliability theory: with applications to preventive maintenance*. Springer.

Graves, G. H., & Lee, C. Y. (1999). Scheduling maintenance and semiresumable jobs on a single machine. *Naval Research Logistics*, *46*(7), 845–863. doi:10.1002/(SICI)1520-6750(199910)46:7<845::AID-NAV6>3.0.CO;2-#

Hnaien, F., Yalaoui, F., Mhadhbi, A., & Nourelfath, M. (2016). A mixed-integer programming model for integrated production and maintenance. *IFAC-PapersOnLine*, *49*(12), 556–561. doi:10.1016/j.ifacol.2016.07.694

Ho, L. L., & Quinino, R. C. (2012). Integrating on-line process control and imperfect corrective maintenance: An economical design. *European Journal of Operational Research*, *222*(2), 253–262. doi:10.1016/j.ejor.2012.04.019

Iyoob, I. M. (2003). Analysis of equipment availability under varying corrective maintenance models. In *Proceedings of the IIE annual conference* (p. 1). Institute of Industrial and Systems Engineers (IISE).

Jardine, A. K., Lin, D., & Banjevic, D. (2006). A review on machinery diagnostics and prognostics implementing condition-based maintenance. *Mechanical Systems and Signal Processing*, *20*(7), 1483–1510. doi:10.1016/j.ymssp.2005.09.012

Jin, X., Siegel, D., Weiss, B. A., Gamel, E., Wang, W., Lee, J., & Ni, J. (2016). The present status and future growth of maintenance in US manufacturing: results from a pilot survey. *Manufacturing review, 3*.

Keizer, M. C. O., Teunter, R. H., & Veldman, J. (2017). Joint condition-based maintenance and inventory optimization for systems with multiple components. *European Journal of Operational Research*, *257*(1), 209–222. doi:10.1016/j.ejor.2016.07.047

Kenné, J. P., Boukas, E. K., & Gharbi, A. (2003). Control of production and corrective maintenance rates in a multiple-machine, multiple-product manufacturing system. *Mathematical and Computer Modelling*, *38*(3-4), 351–365. doi:10.1016/S0895-7177(03)90093-8

Kenné, J. P., & Nkeungoue, L. J. (2008). Simultaneous control of production, preventive and corrective maintenance rates of a failure-prone manufacturing system. *Applied Numerical Mathematics*, *58*(2), 180–194. doi:10.1016/j.apnum.2006.11.010

Klutke, G. A., Kiessler, P. C., & Wortman, M. A. (2003). A critical look at the bathtub curve. *IEEE Transactions on Reliability*, *52*(1), 125–129. doi:10.1109/TR.2002.804492

Kubzin, M. A., & Strusevich, V. A. (2006). Planning machine maintenance in two-machine shop scheduling. *Operations Research*, *54*(4), 789–800. doi:10.1287/opre.1060.0301

Langberg, N. A. (1988). Comparisons of replacement policies. *Journal of Applied Probability*, *25*(04), 780–788. doi:10.1017/S0021900200041577

Levitin, G., & Lisnianski, A. (2000). Optimization of imperfect preventive maintenance for multi-state systems. *Reliability Engineering & System Safety*, *67*(2), 193–203. doi:10.1016/S0951-8320(99)00067-8

Li, H. (2005). Stochastic comparison of age-dependent block replacement policies. *Methodology and Computing in Applied Probability*, *7*(4), 473–488. doi:10.1007/s11009-005-5004-z

Li, H., Deloux, E., & Dieulle, L. (2016). A condition-based maintenance policy for multi-component systems with Lévy copulas dependence. *Reliability Engineering & System Safety*, *149*, 44–55. doi:10.1016/j.ress.2015.12.011

Li, R., & Ryan, J. K. (2011). A Bayesian Inventory Model Using Real-Time Condition Monitoring Information. *Production and Operations Management*, *20*(5), 754–771. doi:10.1111/j.1937-5956.2010.01200.x

Lin, T. W., & Wang, C. H. (2012). A hybrid genetic algorithm to minimize the periodic preventive maintenance cost in a series-parallel system. *Journal of Intelligent Manufacturing*, *23*(4), 1225–1236. doi:10.1007/s10845-010-0406-3

Loganathan, M. K., & Gandhi, O. P. (2016). Maintenance cost minimization of manufacturing systems using PSO under reliability constraint. *International Journal of System Assurance Engineering and Management*, *7*(1), 47–61. doi:10.1007/s13198-015-0374-2

Lu, Z., Zhang, Y., & Han, X. (2013). Integrating run-based preventive maintenance into the capacitated lot sizing problem with reliability constraint. *International Journal of Production Research*, *51*(5), 1379–1391. doi:10.1080/00207543.2012.693637

Ma, Y., Chu, C., & Zuo, C. (2010). A survey of scheduling with deterministic machine availability constraints. *Computers & Industrial Engineering*, *58*(2), 199–211. doi:10.1016/j.cie.2009.04.014

Madu, C. N. (2000). Competing through maintenance strategies. *International Journal of Quality & Reliability Management*, *17*(9), 937–949. doi:10.1108/02656710010378752

Marseguerra, M., & Zio, E. (2000). Optimizing maintenance and repair policies via a combination of genetic algorithms and Monte Carlo simulation. *Reliability Engineering & System Safety*, *68*(1), 69–83. doi:10.1016/S0951-8320(00)00007-7

Marseguerra, M., Zio, E., & Podofillini, L. (2002). Condition-based maintenance optimization by means of genetic algorithms and Monte Carlo simulation. *Reliability Engineering & System Safety*, *77*(2), 151–165. doi:10.1016/S0951-8320(02)00043-1

Martin, K. F. (1994). A review by discussion of condition monitoring and fault diagnosis in machine tools. *International Journal of Machine Tools & Manufacture*, *34*(4), 527–551. doi:10.1016/0890-6955(94)90083-3

Martorell, S., Villanueva, J. F., Carlos, S., Nebot, Y., Sánchez, A., Pitarch, J. L., & Serradell, V. (2005). RAMS+ C informed decision-making with application to multi-objective optimization of technical specifications and maintenance using genetic algorithms. *Reliability Engineering & System Safety*, *87*(1), 65–75. doi:10.1016/j.ress.2004.04.009

Mechefske, C. K., & Wang, Z. (2001). Using fuzzy linguistics to select optimum maintenance and condition monitoring strategies. *Mechanical Systems and Signal Processing*, *15*(6), 1129–1140. doi:10.1006/mssp.2000.1395

Mobley, R. K. (2002). *An introduction to predictive maintenance*. Butterworth-Heinemann.

Nakawaga, T. (2005). *Maintenance Theory of Reliability*. Springer-Verlag.

Osaki, S. (2002). *Stochastic Models in Reliability and Maintenance*. Berlin: Springer. doi:10.1007/978-3-540-24808-8

Peng, Y., Dong, M., & Zuo, M. J. (2010). Current status of machine prognostics in condition-based maintenance: A review. *International Journal of Advanced Manufacturing Technology*, *50*(1-4), 297–313. doi:10.1007/s00170-009-2482-0

Pereira, C. M., Lapa, C. M., Mol, A. C., & Da Luz, A. F. (2010). A particle swarm optimization (PSO) approach for non-periodic preventive maintenance scheduling programming. *Progress in Nuclear Energy*, *52*(8), 710–714. doi:10.1016/j.pnucene.2010.04.009

Pham, H. (2003). *Handbook of Reliability Engineering*. London: Springer. doi:10.1007/b97414

Pham, H., & Wang, H. (1996). Imperfect maintenance. *European Journal of Operational Research*, *94*(3), 425–438. doi:10.1016/S0377-2217(96)00099-9

Prajapati, A., Bechtel, J., & Ganesan, S. (2012). Condition based maintenance: A survey. *Journal of Quality in Maintenance Engineering*, *18*(4), 384–400. doi:10.1108/13552511211281552

Ruiz, R., García-Díaz, J. C., & Maroto, C. (2007). Considering scheduling and preventive maintenance in the flowshop sequencing problem. *Computers & Operations Research*, *34*(11), 3314–3330. doi:10.1016/j.cor.2005.12.007

Ruiz-Castro, J. E. (2015). A preventive maintenance policy for a standby system subject to internal failures and external shocks with loss of units. *International Journal of Systems Science*, *46*(9), 1600–1613. doi:10.1080/00207721.2013.827258

Samrout, M., Yalaoui, F., Châtelet, E., & Chebbo, N. (2005). New methods to minimize the preventive maintenance cost of series–parallel systems using ant colony optimization. *Reliability Engineering & System Safety*, *89*(3), 346–354. doi:10.1016/j.ress.2004.09.005

Sarker, R., Omar, M., Hasan, S. K., & Essam, D. (2013). Hybrid Evolutionary Algorithm for job scheduling under machine maintenance. *Applied Soft Computing*, *13*(3), 1440–1447. doi:10.1016/j.asoc.2012.04.032

Satoglu, S. I., & Ustundag, A. (2013). Value of RFID Enhanced Maintenance in Aerospace Industry. In *The value of RFID* (pp. 141–153). Springer London. doi:10.1007/978-1-4471-4345-1_11

Sharma, A., Yadava, G. S., & Deshmukh, S. G. (2011). A literature review and future perspectives on maintenance optimization. *Journal of Quality in Maintenance Engineering*, *17*(1), 5–25. doi:10.1108/13552511111116222

Sharma, R. K., Kumar, D., & Kumar, P. (2005). FLM to select suitable maintenance strategy in process industries using MISO model. *Journal of Quality in Maintenance Engineering*, *1*(11), 359–374. doi:10.1108/13552510510626981

Sherwin, D. (2000). A review of overall models for maintenance management. *Journal of Quality in Maintenance Engineering*, *6*(3), 138–164. doi:10.1108/13552510010341171

Sidibe, I. B., Khatab, A., Diallo, C., & Kassambara, A. (2017). Preventive maintenance optimization for a stochastically degrading system with a random initial age. *Reliability Engineering & System Safety*, *159*, 255–263. doi:10.1016/j.ress.2016.11.018

Vahedi Nouri, B., Fattahi, P., & Ramezanian, R. (2013). Hybrid firefly-simulated annealing algorithm for the flow shop problem with learning effects and flexible maintenance activities. *International Journal of Production Research*, *51*(12), 3501–3515. doi:10.1080/00207543.2012.750771

Van Horenbeek, A., Pintelon, L., & Muchiri, P. (2010). Maintenance optimization models and criteria. *International Journal of System Assurance Engineering and Management*, *1*(3), 189–200. doi:10.1007/s13198-011-0045-x

Wang, C. H., & Lin, T. W. (2011). Improved particle swarm optimization to minimize periodic preventive maintenance cost for series-parallel systems. *Expert Systems with Applications*, *38*(7), 8963–8969. doi:10.1016/j.eswa.2011.01.113

Wang, H. (2002). A survey of maintenance policies of deteriorating systems. *European Journal of Operational Research*, *139*(3), 469–489. doi:10.1016/S0377-2217(01)00197-7

Wang, H., & Pham, H. (2006). *Reliability and optimal maintenance*. Springer Science & Business Media.

Wang, L., Chu, J., & Mao, W. (2008). An optimum condition-based replacement and spare provisioning policy based on Markov chains. *Journal of Quality in Maintenance Engineering*, *14*(4), 387–401. doi:10.1108/13552510810909984

Wang, L., Chu, J., & Wu, J. (2007). Selection of optimum maintenance strategies based on a fuzzy analytical hierarchy process. *International Journal of Production Economics*, *107*(1), 151–163. doi:10.1016/j.ijpe.2006.08.005

Wang, S., & Liu, M. (2015). Multi-objective optimization of parallel machine scheduling integrated with multi-resources preventive maintenance planning. *Journal of Manufacturing Systems*, *37*, 182–192. doi:10.1016/j.jmsy.2015.07.002

Wang, Y., Deng, C., Wu, J., Wang, Y., & Xiong, Y. (2014). A corrective maintenance scheme for engineering equipment. *Engineering Failure Analysis*, *36*, 269–283. doi:10.1016/j.engfailanal.2013.10.006

Xu, H., & Hu, W. (2008). Availability optimisation of repairable system with preventive maintenance policy. *International Journal of Systems Science*, *39*(6), 655–664. doi:10.1080/00207720701872057

Yalaoui, A., Chaabi, K., & Yalaoui, F. (2014). Integrated production planning and preventive maintenance in deteriorating production systems. *Information Sciences*, *278*, 841–861. doi:10.1016/j.ins.2014.03.097

Yeh, W. C., & Hsieh, T. J. (2011). Solving reliability redundancy allocation problems using an artificial bee colony algorithm. *Computers & Operations Research*, *38*(11), 1465–1473. doi:10.1016/j.cor.2010.10.028

Zhao, S., Wang, L., & Zheng, Y. (2014). Integrating production planning and maintenance: An iterative method. *Industrial Management & Data Systems*, *114*(2), 162–182. doi:10.1108/IMDS-07-2013-0314

Zio, E., & Viadana, G. (2011). Optimization of the inspection intervals of a safety system in a nuclear power plant by Multi-Objective Differential Evolution (MODE). *Reliability Engineering & System Safety*, *96*(11), 1552–1563. doi:10.1016/j.ress.2011.06.010

## KEY TERMS AND DEFINITIONS

**Additive Manufacturing:** The production process which is usually performed by adding layer upon layer using joining materials to produce objects from the 3D model data, as opposed to subtractive manufacturing methodologies.

**Cellular Manufacturing:** It is a lean manufacturing method that consists of a specific group of employees, workstations or equipment used to produce similar types of products, and which removes the setup and unneeded costs between production processes to facilitate operations.

**Computer Integrated Manufacturing:** The use of computer controlled machines and automation systems in the production process of manufactured products. It constitutes from combination of separate applications, such as Computer Aided Design (CAD), Computer Aided Engineering (CAE), Computer Aided Manufacturing (CAM), Computer Aided Process Planning (CAPP), Computer Aided Quality Assurance (CAQ), Production Planning and Control (PPC), Enterprise Resource Planning (ERP).

**Condition Based Maintenance:** Maintenance technique in which the machine status is controlled by sensors and the machine failures are predicted.

**Corrective Maintenance:** Maintenance activities applied to identify and correct the cause of the failure in a failed system.

**Flexible Manufacturing:** A method for producing goods which can be easily adapted to changes in the product being manufactured whose machines are capable of producing parts and capable of processing varying production levels.

**Just-in-Time Manufacturing:** Production processes are made according to Just in Time inventory system, which can respond to customer requests more quickly without having large quantities of finished product or work-in-process inventory.

**Maintenance Management:** Assessment and planning of maintenance operations according to a scheduled plan considering administrative, financial and technical evaluation.

**Preventive Maintenance:** The systematic control, detection, correction and prevention of faults that are still in its infancy before turning into a major breakdown.

**Production System:** Manufacturing subsystem that covers all stages required for design, manufacture, distribution and service of a manufactured product.

# Chapter 13
# A Simulation-Optimization Approach for the Production of Components for a Pharmaceutical Company

**Nicolas Zufferey**
*University of Geneva, Switzerland*

**David Dal Molin**
*IPros, Switzerland*

**Rémy Glardon**
*IPros, Switzerland*

**Christos Tsagkalidis**
*IPros, Switzerland*

## ABSTRACT

*The considered problem (P) concerns the production of strains (also called jobs or batches), which are the used components in the final products that are bought by the consumers. (P) contains two components that have to be tackled sequentially: the inventory management problem (IMP) and the job scheduling problem (JSP). (IMP) is solved with a reorder-point policy, defined on the basis of critical demand coverage. To tackle (JSP), a descent local search (DLS) is used, based on swap moves. In other words, for a given job sequence, a series of modifications is performed on it in order to try to improve the solution, where each modification consists of exchanging the positions of two jobs. Because of random events (some jobs might be rejected if they do not meet predefined standards) and stochasticity (the duration of each job follows a normal distribution), simulation is required to evaluate any sequence of jobs that is a solution to (JSP). A simulation-optimization approach is therefore proposed to accurately tackle (JSP). This work is motivated by a real pharmaceutical company.*

DOI: 10.4018/978-1-5225-2944-6.ch013

## INTRODUCTION

The problem (P) considered in this chapter concerns the production of strains (also called jobs or batches hereafter), which are the used components in the final products that are bought by the consumers located at the very end of the supply chain. Problem (P) contains two components that have to be tackled sequentially. The first one is called the inventory management problem and it is denoted as (IMP), whereas the second one is called the job scheduling problem and it is denoted as (JSP).

(IMP) is solved with a reorder-point policy, defined on the basis of critical demand coverage. More precisely, it must assure a minimum amount of stock at all time, for each product type. To tackle (JSP), a descent local search (DLS) is used, based on swap moves. In other words, for a given job sequence, a series of modifications is performed on it in order to try to improve the solution, where each modification consists of exchanging the positions of two jobs. Because of random events (e.g., some jobs might be rejected if they do not meet the predefined standards) and stochasticity (e.g., the duration of each job follows a normal distribution), simulation is required to evaluate any sequence of jobs that is a solution to (JSP). A simulation-optimization approach is therefore proposed to accurately tackle (JSP). The reader is referred to (Silver & Zufferey, 2011) for more information on such type of solution methods.

This work is motivated by a real pharmaceutical company, denoted here as PHARMA. It cannot be named because of a non-disclosure agreement. For this reason, the accurate data and results will not be provided, as it is highly confidential. The presented information will however allow the reader to fully capture all the features of the considered problem (P), as well as the associated methods used to solve (IMP) as well as (JSP). To highlight the success of the proposed overall approach, it is important to mention that the resulting decision-making tool is planned to be used in PHARMA for the inventory management and the production planning of the concerned strains.

The reminding part of this chapter is organized as follows. The next section describes the optimization elements that are necessary to better understand the proposed solution methods. It will be discussed that quality is not the only criteria for measuring the efficiency of a solution method. Next, (IMP) is formally presented, along with the proposed inventory management policy, based on the well-known reorder-point policy. (JSP) is described afterwards (in a production environment corresponding to a flow shop), for which an optimization model and an optimization method are successively designed. Finally, a discussion and a conclusion are provided at the end.

## OPTIMIZATION ELEMENTS

As presented in (Zufferey, 2015), let $f$ be an objective function that has to be minimized (e.g., a production cost, a shortage function, a sum of lateness and/or earliness penalties) over a solution space (i.e., the set of all the possible solutions to a problem). A solution $s$ is optimal for $f$ if there is no better solution than it; that is, there is no solution $s'$ such that $f(s') < f(s)$. An *exact* method guarantees the optimality of the provided solution. However, for a large number of applications and most real-life optimization problems (as for the studied problem (P)), such methods need a prohibitive amount of time to find an optimal solution, because such problems are *NP-hard* (Garey & Johnson, 1979). For these difficult problems, one should prefer to quickly find a satisfying solution, which is the goal of heuristic and metaheuristic solution methods. The reader is referred to (Blum & Roli, 2003; Gendreau & Potvin, 2010) for accurate information on several metaheuristics, and to (Zufferey, 2012; Hertz & Widmer, 2003) for general guidelines on how to adapt them efficiently to various types of problems. There also exists a family of algorithms that efficiently integrate metaheuristics and exact methods in a single approach. For a survey on such methods, the reader is referred to (Dumitrescu & Stuezle, 2003) and to (Puchinger & Raidl, 2005).

On the one hand, a heuristic can be defined as a streamline solution method able to provide a satisfying solution to a problem, in a reasonable amount of computing time, but without any guarantee on optimality. On the other hand, and informally, a metaheuristic is a more refined but usually more efficient heuristic. The following formal definition is proposed in (Osman & Laporte, 1996). "A metaheuristic is formally defined as an iterative generation process which guides a subordinate heuristic by combining intelligently different concepts for exploring and exploiting the search space, learning strategies are used to structure information in order to find efficiently near-optimal solutions."

There exist three main families of (meta)heuristic methods. First, in a constructive algorithm, a solution is built step by step from scratch. For instance, at each step of the greedy algorithm, the best feasible (according to the constraints that have to be respected) element is added to the solution in construction. Second, in a local search method, a solution is iteratively modified (as discussed in the next paragraph) to hopefully generate better solutions. At the end of the search process, the best encountered solution is provided to the user. Third, in an evolutionary metaheuristic, a population of solutions is managed (e.g., genetic algorithms, ant algorithms, adaptive memory algorithms, scatter search). In each generation of an evolutionary algorithm, two complementary phases are used: a cooperation phase (for instance, the history of the search can be used to generate new solutions) and a self-adaptation phase (for instance, a mutation operator or a local search can be employed to modify or improve a solution).

A local search algorithm starts with an initial solution (generated at random or with the use of a constructive heuristic) and tries to improve it iteratively. At each iteration, a modification, called a move, of the current solution $s$ is performed in order to generate a neighbor solution $s'$. Let $N(s)$ denote the set of all the neighbor solutions of $s$. The definition of a move, that is the definition of the neighborhood structure $N$, depends on the considered problem. A move will have more chance to be successful if it is able to take advantage of the problem structure (which will be the case in the method proposed below for (JSP)). Popular local search methods are the descent local search (DLS), simulated annealing, tabu search, guided local search, and variable neighborhood search. In DLS, the best move is performed at each iteration and the process stops when a local optimum is found (i.e., there is no solution better than the incumbent in its neighborhood). DLS can be restarted as long as a time limit $T$ is not reached. In this case, at the end of the search process, the best solution found within time $T$ is returned to the decision-maker. Such a restarting approach will be used to tackle (JSP). Indeed, from a practical point of view, the allowed computing time $T$ can be seen as a limited resource for solving a problem.

## INVENTORY MANAGEMENT PROBLEM (IMP)

In most inventory management problems, two types of decision have to be taken at the manufacturer level: *when* and *how much* to order to suppliers. PHARMA faces the following system in the considered plant. (1) The production unit that produces the jobs (there are $K$ different type of strains to produce). (2) The assembly unit that assembles a certain number of strains into a product. (3) The strain inventory that must be refilled to assure availability of strains for assembly at all time. As validated by PHARMA, a stable demand without seasonality nor trend can be assumed.

The overall problem (i.e., involving both (IMP) and (JSP)) consists of creating a short-term batch list (BL) for production, which must assure the replenishment of the strain inventory on a short-term horizon (typically three weeks). BL is the input of the scheduling algorithm that will find a good schedule for a planning horizon of typically one week.

The following terminology and notation are used. For each notation, its nature (e.g., given data, observation, formula, decision variable) is indicated at the end, between brackets. The reader is referred to (Silver et al., 1998; Frazelle, 2015) for more information about the main features and approaches of inventory management.

$F_k$: Estimated demand rate of strain $k$ for assembly (i.e., the stock outgoing flow), in units per week (given data).
$S_k(t)$: Stock level of strain $k$ at week $t$ (observation).
*SSC*: Safety stock coverage, in weeks (decision variable).
$Ss_k = SSC \cdot F_k$: Safety stock for strain $k$, without any unit (formula).
*ERT*: Expected replenishment time (i.e., the lead time) of any strain batch, in weeks (given data).
*CDC*: Critical demand coverage, which is equal to $ERT + SSC$, in weeks (formula).
$Sc_k$: Reorder point for strain $k$, which is equal to $CDC \cdot F_k$, without any unit (formula).
$\beta_k$: Strain specific coverage coefficient, without any unit (given parameter).
$Y_k$: Yield of strain $k$, which is the number of produced units per batch, without any unit (given data).
*K*: Number of strains (given data).
*W*: Number of weeks for the BL horizon (given data).
*Z*: Priority factor. It is a multiplication factor used to make the priority better understandable for the production planning. In the detailed example below, the value $Z = 10$ is chosen.

For each strain $k$, a production order *PO(k)* is planned when the stock level reaches the reorder point. The reorder point is defined on the basis of a *CDC*: it must assure a minimum amount of stock at all time (safety stock concept). The *CDC* is the same for all strains. A specific coverage coefficient $\beta_k$ can be introduced to allow for some differentiations among the strain categories.

By each algorithm execution (typically every week), one or several *PO* are planned if the stock level is lower than the reorder point. A priority is associated with each *PO*. It is proportional to the indicator provided in Equation (1).

$$[Sc_k - S_k(t)] / (Sc_k - Ss_k) \quad (1)$$

The priority is positive if the stock level is below the reorder point (numerator), and this priority is reinforced if there is only a small gap between the reorder-point level and the safety-stock level (denominator). If the ratio $[Sc_k - S_k(t)] / Y_k$ exceeds 1 (i.e., if the delivery of a single batch is not sufficient to replenish the inventory above *Sc*), a second *PO* is then added to the batch list with a priority defined as in Equation (2), where $i$ is the PO number.

$$P_k^i = Z \times \frac{Sc_k - [S_k(t) + (i-1) \cdot Y_k]}{Sc_k - Ss_k} \quad (2)$$

For the first batch (i.e., $i = 1$), the above priority reduces to the product of the priority factor $Z$ and the indicator given in Equation (1). This process is extended to any required number of $PO$ (for PHARMA, the number of required $PO$ is likely to never exceed 2, and it will most often be 0 or 1). If the stock level is higher than the reorder point during the first week of the BL horizon, the process is extended to the second week by taking into account the projected stock level, which is $S_k(2) = S_k(1) - F_k$. This process is repeated for the whole BL horizon (i.e., $W$ times).

For (IMP), it was preferred to work with priorities instead of with earliness and tardiness penalties. For the latter context, the reader is referred to (Thevenin et al., 2016b), where a single-machine (but deterministic) environment is studied in an order acceptance and scheduling context. This typically occurs if the production capacity of the company is overloaded, which is the case of PHARMA. The problem consists therefore in deciding which jobs to perform, and according to which sequence.

In the context of PHARMA, it is assumed that the following information is known: $W$, $K$, $CDC$, $SSC$, the $\beta_k$'s, the $Y_k$'s, and the current (on-hand) stock levels $S_k$'s. It is further supposed that the forecast computation algorithm allowing the computation of the estimated demand rate $F_k$ is known. The algorithm contains three main loops:

- Loop from $k = 1$ to $K$ strains
- Loop from $w = 1$ to $W$ weeks (BL horizon)
- Loop from $i = 1$ to the maximum number of $PO$'s ($i$ should rarely be larger than

The priority associated with a $PO$ in the BL depends on the one hand on the level of the current (i.e., $w = 1$) or the projected (i.e., $w > 1$) stock, and on the other hand on the anticipation in the BL horizon. The priority $P(w)$ increases linearly from $P(w) = 0$ if $S_k \geq Sc_k$ to $P(w) = 10$ if $S_k = Ss_k - (w\text{-}1)\cdot F_k$. More precisely, the following situations occur.

- For the first week (i.e., $w = 1$), the linear variation of priority $P(1)$ is defined by the two points $P(1) = 0$ for $S_k = Sc_k$ and $P(1) = 10$ (selected value for parameter $Z$) for $S_k = Ss_k$
- For the second week (i.e., $w = 2$), the linear variation of priority $P(2)$ is defined by the two points $P(2) = 0$ for $S_k = Sc_k$ and $P(2) = 10$ for $S_k = Ss_k - F_k$
- For the third week (i.e., $w = 3$), the linear variation of priority $P(3)$ is defined by the two points $P(3) = 0$ for $S_k = Sc_k$ and $P(3) = 10$ for $S_k = Ss_k - 2\cdot F_k$

The BL algorithm is depicted in Algorithm 1, where the priority formula is a generalization of Equation (2), as it concerns more than one week and more than one batch. The last test checks if another batch is needed to put the inventory level above the reorder point.

## Algorithm1: Batch List Algorithm

For $k = 1$ to $K$, do,

1. Compute $F_k$, $Sc_k$, $Ss_k$, and set $w = 1$.
2. If $[S_k(w) / Sc_k] \geq 1$, do
   a. Set $w = w + 1$.
   b. If $w \leq W$, set $S_k(w) = S_k(w\text{-}1) - F_k$ and go to step (2).

c. If $w > W$, STOP (i.e., move to the next $k$ in the above "For" loop).

3. If $[S_k(w) / Sc_k] < 1$, do
   a. Set $i = 1$.
   b. Compute $P_k^i = Z\dfrac{[Sc_k - (S_k(w) + (i-1)Y_k)]}{Sc_k - (Ss_k - (w-1)F_k)}$ and add $PO_k^i(P_k^i)$ to the batch list BL.
   c. If $[S_k(w) + i\,Y_k / Sc_k] < 1$, set $i = i + 1$ and go to step (3.b).
   d. If $[S_k(w) + i\,Y_k / Sc_k] \geq 1$, STOP (i.e., move to the next $k$ in the above "For" loop).

The BL algorithm is now illustrated for strain 3 on a simple example which involves various strains (the same type of strain can occur several times). Consider the following initial priorities: priority (strain 1) = 1, priority (strain 2) = 2.5, priority (strain 3) = 2, priority (strain 3) = 1, priority (strain 4) = 5, priority (strain 5) = 13.3, priority (strain 5) = 6.7, priority (strain 6) = 2.5. Consider also the following input data: $F_3 = 40$ units per week, $S_3 = 480$, $CDC = 12$ weeks, $\beta_3 = 1$, $Y_3 = 20$ units per batch, $SSC =$ 8 weeks, $Sc_3 = CDC \cdot F_3 = 12{\cdot}40 = 480$, $Ss_3 = SSC \cdot F_3 = 8{\cdot}40 = 320$, and $Z = 10$. The computational sequence is detailed below.

- Set $w = 1$.
- **Check:** $[\,S_3(1) / Sc_3\,] = (480 / 480) < 1 \rightarrow$ no.
- Set $w = 2$.
- Set $S_3(2) = S_3(1) - F_3 = 480 - 40 = 440$.
- **Check:** $[\,S_3(2) / Sc_3\,] = (440 / 480) < 1 \rightarrow$ yes.
- Set $i = 1$.

Compute:

$$P_3^i = Z\frac{[Sc_3 - (S_3(2) + (i-1)Y_3)]}{Sc_3 - (Ss_3 - (w-1)F_3)}$$

$$P_3^1 = 10\frac{[480 - (440 + (1-1)20)]}{480 - (320 - (2-1)40)} = 10\frac{480 - 440}{480 - 280} = 2$$

add $PO_3^1$(priority = 2) to the batch list BL.

- Check: $(S_3(2) + 1 \cdot Y_3) \cdot Sc_3 = (440 + 1 \cdot 20) / 480 < 1 \rightarrow$ yes.
- Set $i = i + 1 = 2$.

Compute:

$$P_3^2 = Z\frac{\left[Sc_3 - (S_3(2) + (i-1)Y_3)\right]}{Sc_3 - (Ss_3 - (w-1)F_3)}$$

$$P_3^2 = 10\frac{\left[480 - (440 + (2-1)20)\right]}{480 - (320 - (2-1)40)} = 10\frac{480 - 460}{480 - 280} = 1$$

add $PO_3^2$ (priority = 1) to the batch list BL.

- **Check:** $(S_3(2) + 2 \cdot Y_3) \cdot Sc_3 = (440 + 2 \cdot 20) / 480 < 1 \rightarrow$ no $\rightarrow$ next $k$.

The output priorities computed as above are used as an input to the solution methods for (JSP). Three zones are defined for the scheduling problem.

- The bottom zone (i.e., $0 \leq P < 7$) is associated with a free choice of the batch, without any inventory penalty. This zone corresponds to a green zone.
- The intermediate zone (i.e., $7 \leq P < 10$) corresponds to the critical zone. In this case, the batch should be prioritized, and the inventory penalty is equal to the priority. This zone corresponds to an orange zone.
- The top zone (i.e., $P \geq 10$) involves jobs that must be scheduled in the first week. It corresponds to urgent situations. This zone corresponds to a red zone.

## Job Scheduling Problem (JSP)

The overall project concerns the production of components, strains of finished products. This work only focuses on the first production phase (i.e., the strains). The second production phase consists of assembling the components into finished products. Therefore, if a job (i.e., a component) is not available on time, it delays the assembly production phase. To avoid such delays, PHARMA uses a safety stock corresponding to at least three months of component requirement.

A set of $n$ jobs has to be scheduled on $m$ machines that are put in a linear production line. In other words, the production environment corresponds to a flow shop. The reader is referred to (Pinedo, 2016) for detailed information on the job scheduling literature. $K$ different products are considered. For each machine $i$, it is known in advance if a quality control is required or not.

Because a job $j$ consists in building a bacteriologic component, as soon as it is started on machine 1, it cannot wait more than 30 minutes between any pair of consecutive machines (until it is completed on the last machine $m$). This is called the waiting time constraint. The associated idle time between machine 1 and machine 2 is a sensitive parameter of the problem. This idle time issue is fixed below when discussing the parameter $r$, which is a tolerated percentage risk of failures. Next, there are two possible cases for a job $j$ that is finished on machine $i$: (1) if the next machine $i+1$ is available, job $j$ is immediately started on it; (2) if the next machine $i+1$ is not available, job $j$ waits until it becomes available. But if the waiting time exceeds 30 minutes, then $j$ is lost. However, between some specific pairs of machine, it is possible to put jobs (belonging to specific types) in inventory or in a cooler for a larger time period (typically up to roughly 100 hours). This option is helpful to store some jobs during the non-working hours (e.g., weekends, nights, some holidays). It can be seen as a way to interrupt a job and then continue it later. This is called preemption in the job-scheduling literature (Thevenin et al., 2016). A preemption could be interesting in situations for which negligible (when compared to the total production time of a batch) setup times are possible. However, in general, preemptions have to be avoided and minimized (Liu & Cheng, 2002). Indeed, the use of preemptions leads to an augmentation of the throughput time of a job, which is defined as the difference between its end time and its start time. Moreover, a partially-filled job has to stay on the production floor until it is completed, and it can therefore not be sent to any downstream level (e.g., distributors, shops, consumers) of the supply chain or of the production line. As a consequence, preempted jobs will increase the work-in-progress inventory costs. In the considered problem (P) and as in the study of (Shachnai et al., 2002), a maximum number of preemptions is imposed (it is likely to never exceed one in the context of PHARMA).

Let $p_{ij}$ be the processing time (i.e., the duration) of job $j$ on machine $i$, which follows a truncated normal distribution $N(\mu_{ij}, \sigma_{ij})$. All the jobs of a fixed type of product have the same size (i.e., it involves the same number of units). Therefore, for a fixed type of product $k$ belonging to the set $\{1, 2, \ldots, K\}$, all the corresponding $p_{ij}$'s are the same. The processing times range typically from 1 to 20 hours.

Let HR denote the considered human resource (i.e., the operators). The limitations on HR are on the availability of the workers and not on their quantity. In PHARMA, HR is daily available in interval $D$ = [6:00, 23:00], which corresponds to two shifts. However, it is possible to make operators work one hour before $D$ and two and a half hours after $D$. In such a case, it is penalized with an extra-hour cost. Therefore, a working day actually corresponds to the time interval [05:00, 01:30]. On each machine, each job $j$ needs a HR intervention at its beginning (it counts typically for 25% of the HR intervention time) and at its end (typically 75% of the HR intervention time). During each week, there is a strict limitation on the costs that can be spent for the extra-hours. Such a limitation is also important from a managerial point of view. Such variable production costs (i.e., depending on the time slot) are also considered in (Bloechliger & Zufferey, 2013) in a deterministic production environment. In their study, the assignment of a job to a time slot results in an assignment cost that depends on both the job and the time period. Local search procedures are also recommended in their work.

Multi-objective scheduling problems consider often the minimization of the makespan, with setup costs and setup times constraints. The reader is referred to (Allahverdi et al., 2009) for a survey on scheduling techniques that account for setup issues. In problem (P), the setup times correspond to cleaning, decontamination and sterilization times. Given a solution to (JSP) (i.e., a production sequence), after the completion of job $j$ on machine $i$, a setup time has to be allocated for these side-operations before machine $i$ is again available for the next job $j'$. This is also true if $j'$ is of the same type as $j$, as PHARMA is dealing with bacteriologic components. The setup time depends on the machines and not on the products. Moreover, they are constant and deterministic (i.e., their values are perfectly known in advance, and they typically do not exceed 8 hours).

There are two types of failures: the first one can be considered as an unlucky job, whereas the second one is due to the generation of a bad production sequence (in other words, it concerns the optimization approach). These two types of failures are detailed below. If a job $j$ is rejected, in order to avoid tardiness issues, it has to be rescheduled (typically within the same week, or at the latest within the next week). The reader is referred to (Thevenin et al., 2015) for a reference on scheduling with rejection and tardiness penalties. It was shown that local search algorithms are efficient methods for such problems. First, a failure rate of $q_i$% is encountered for each machine i (i.e., $q_i$% of the jobs are rejected), independently of the production sequence. The production planner has thus no impact on such a failure type. The failure rate is different for each $i$, and is typically varying from 0% to 2% depending on the process. Second, if a job $j$ waits more than 30 minutes between two machines, it has to be rejected as it loses its chemical properties. Note that such a chemical-property constraint is also known in contexts where blood samples have to be first collected (at different locations involving doctors) and then delivered to the labs in charge of their analysis (Zufferey et al., 2016). The production planner has an impact on such a failure type. Let $r$ be the tolerated percentage risk of such failures. Parameter $r$ is fixed by the management as a kind of service level (as defined in supply chain management). Typical values are below 2%. For each pair $(j, j')$ of jobs, from any $r$ value, one can deduce the waiting time $w_{jj'}$ to wait before triggering job $j$ when job $j'$ is completed on the first machine. In such a case, jobs $j$ and $j'$ will violate the waiting time constraints (between machines 1 and $m$) with a probability equal to $r$.

## OPTIMIZATION MODEL FOR (JSP)

The input of the optimization model is listed below.

- A prioritized batch list $J$ (i.e., a priority is assigned to each job $j$ of $J$, and the same product can appear several times in $J$) associated with the considered planning horizon (three weeks).
- The $p_{ij}$'s distributions (processing times).
- The organizational structure of the plant (i.e., typical duration of a working day, applied policy for the extra-hours, allow or not night shifts, determined number of working days per week).
- The risk level $r$, which has an impact on the simulation (as some initially expected jobs will actually be rejected), but absolutely not impact on the optimization.

The planning horizon consists of three weeks, but the system will be simulated for 44 weeks for quality evaluation (which corresponds to one full year of production). At the end of each week, the batch list BL is updated (i.e., the successfully performed jobs are removed from the list), as well as their corresponding priorities. The next three weeks are then rescheduled, and so on. Therefore, any output solution is a schedule over three weeks, and a rolling horizon approach is employed (Silver & Zufferey, 2005) in order to cover a full year.

The following constraints have to be satisfied.

- A machine can only process one job at a time. In other words, overlaps are not allowed. If it happens between the jobs $j'$ that precedes $j$, then job $j'$ is simply rejected. This contrasts with some studies (e.g., Bloechliger & Zufferey, 2013), where the execution of some jobs are allowed overlapping, and if it occurs, it is strongly penalized in a component of the objective function. Indeed, additional costs would have to be encountered for specific time periods if additional resources (e.g., staff, machines, a cooling unit) are employed to face the situation.
- It is not possible to use HR outside of the allowed time intervals. More precisely, there are some green time-zones without additional HR costs (regular hours), orange time-zones with additional HR costs (extra-hours), and red time-zones during which it is forbidden to work (e.g., middle of the night).
- Do not leave a job wait more than 30 minutes between any pair of machines (waiting time constraint).

- Do not exceed the allowed budget on the HR penalties. This threshold is a parameter tuned by the production planner. An upper bound on the HR additional cost is associated with each week.

As presented in (Respen et al., 2016), production systems involve various important aspects such as cost, time, and available (human and material) resource. Multi-objective planning problems are of growing practical relevance. As mentioned in (Loukil et al., 2005), multi-objective scheduling problems can be tackled using five different approaches: (1) lexicographic (see below); (2) utility; (3) goal programming; (4) simultaneous; (5) interactive. No approach is dominating the others as each one has its own advantages and drawbacks. Some approaches require more parameters whereas others are unable to generate efficient solutions under certain conditions. The reader is referred to the survey of (Ehrgott & Gandibleux, 2000) on multi-objective combinatorial optimization, including theoretical results as well as exact and heuristic methods, and to the survey of (Jones et al., 2002) on multi-objective metaheuristics.

The lexicographic multi-objective optimization approach proposed below for (JSP) is a common practice, as reported in (Solnon et al., 2008) and (Prats et al., 2010). This approach is very convenient when a ranking can be established among the objectives, as it is the case for (JSP). The following functions have to be minimized, following a lexicographic optimization (i.e., $f_i$ is infinitely more important than $f_{i+1}$, or in other words, no augmentation on an objective function component can be compensated with reductions of lower level objectives).

$f_0$: $U$ – number of produced batched, where $U$ is an upper bound on the number of jobs that can be performed during any week (U is typically equal to 10 for PHARMA).

$f_1$: Sum of the red (i.e., the inventory levels below the safety stock) inventory penalties of the non-performed jobs at the end of week 1.

$f_2$: Makespan of the total batch list (roughly corresponding to three weeks of production).

$f_3$: Sum of the orange (tolerance range, typically two weeks of coverage above the safety stock) inventory penalties of the non-performed jobs at the end of week 1.

Note that in the specific considered case, two solutions *s* and *s'* are makespan-different only if the difference of their associated makespans is at least four hours. In other words, makespan($s$) $\neq$ makespan($s'$) if $|f_2(s) - f_2(s')| \geq 4$ hours. Indeed, PHARMA is likely to not trigger any action below this threshold.

## OPTIMIZATION METHOD FOR (JSP)

As already presented above, the priority $P(t)$ to produce product $t$ is determined on the basis of the coverage, which is computed as Coverage($t$) = (inventory level of $t$) / (demand rate of $t$). A greedy constructive heuristic is first proposed. It starts from an empty solution. At each step, the job $j \in J'$ (defined below) with the largest attractiveness $A(j)$ (also defined below) is chosen as the next job to perform. A new step is performed as long as the week is not fully loaded with jobs.

The following procedure is used to define the set $J'$.

- Compute $R$ as the subset of $J$ (given batch list BL) with red inventory penalties. The jobs of $R$ are urgent as they directly contribute to the objective function component $f_1$.

- Compute $O$ as the subset of $J$ with orange inventory penalties. Such products have a kind of urgency, as they will belong to the red zone in the next time period.
- Compute $Q$ as the subset of $J$ with the largest $p$ priorities, where $p$ is a parameter (after preliminary experiments, it was tuned to three in the case study).
- Set $J' = R$. If $J'$ is empty, then set $J' = O$. If $J'$ is still empty, then set $J' = Q$.

The attractiveness $A(j)$ associated with the execution of job $j$ is based on its impact on the HR costs (for the considered week). More precisely, it is inversely proportional to the percentage of the consumed budget envelope if job $j$ is the next scheduled job of the sequence.

The optimization steps are summarized in Algorithm 2. It is stopped as soon as a time limit $T$ is reached, where $T$ is a parameter chosen by the decision-maker. At the end of the search process, the best encountered solution is returned to the user. Usually, $T$ is chosen such that the improvement potential of the method is poor if it is allowed to continue (this technique to fix $T$ is well-known in the metaheuristics community). Preliminary experiments have to be performed in order to set $T$ (it was tuned to five hours in the case study).

## Algorithm 2: Optimization Method for (JSP)

- **Construction:** Build a week sequence with the greedy heuristic (left-right sequencing according to time).
- **Improvement:** Perform some switch moves within a local search framework (using the DLS algorithm, stopping at the first local minimum). Both internal moves (i.e., within the week sequence) and external moves (i.e., switch an internal non-red job with an external job) are used.
- **Extension:** Add a new job $j \in J$ to the week sequence (using a best insertion fashion according to the lexicographic optimization). If the resulting makespan exceeds the week, then restart the method (from scratch) from the above construction step.

As the duration of each job is stochastic and because of the possible quality control failures, the only way to evaluate a solution (i.e., a production sequence) is to use simulation (based on a discrete event generator). For this reason, and because of the allowance of idle times, no optimal timing algorithm can be used. More precisely, in a deterministic context, from any given sequence of jobs, a timing algorithm (Pinedo, 2016) can be used to determine the starting time of each job, which then allows the accurate measure of the quality of the solution. The used computer is a Portable computer Thinkpad Edge (Processor: inter® Core™ i5-3230M CPU @2.60GHZ, installed (RAM) 8GB (7.84 usable), System 64-bit operating system, Windows 10). The simulation software enables to use a max of 4096 Mb of RAM. In the studied case, it has been observed that simulating three weeks requires roughly two minutes. In order to have robust and reliable results, 50 replications of the simulation are necessary. As the simulation duration is significant, it cannot be used too often in the optimization method. Indeed, there are typically hundreds of candidate neighbor solutions to evaluate when moving from the current solution $s$ to the selected neighbor solution $s'$. For this reason, an efficient evaluation strategy is now proposed for each iteration of the DLS procedure.

Let $f(s)$ be the expected value of solution $s$, and let $F_x(s)$ be the simulated value (with $x$ replications) of solutions s. Remember that $x$ replications are required to have a robust and accurate evaluation of any solution. In DLS, while an improvement is encountered, the steps described in Algorithm 3 are performed in order to generate a neighbor solution *s'* from the current solution *s*.

### Algorithm 3: Evaluation of Neighbor Solutions in DLS

- Generate and evaluate (with function *f*) all the possible switch moves (i.e., internal and external). Any resulting neighbor solution that violates a constraint is not considered for further investigations. Such non-feasible solutions are simply discarded from the set of candidate neighbor solutions. If the latter is empty, DLS stops.
- Put the $q_1$ (parameter tuned to 20) best (according to *f*) solutions in the set *B*.
- Partially simulate (i.e., with parameter *x* tuned to 20) the solutions of the set *B*, and put the best (according to $F_{20}$) $q_2$ (parameter tuned to 5) solutions in the set *B'*.
- Fully simulate (i.e., use $x = 50$) the solutions of the set *B'*, and let *s'* denote the best simulated one (according to $F_{50}$).
- If $F_{50}(s') < F_{50}(s)$, set $s = s'$ (i.e., *s'* becomes the new current solution, as there is an improvement). Otherwise, stop (as *s'* is not strictly better than the current solution *s*).

## DISCUSSION AND CONCLUSION

In this chapter, a production problem is studied in the context of a pharmaceutical company denoted as PHARMA. Two optimization problems are investigated. On the one hand, a reorder-point policy is proposed for the inventory management problem (IMP). On the other hand, a simulation-optimization solution method is designed for the job-scheduling problem (JSP). It relies on two procedures: a constructive greedy heuristic able to build a job sequence covering a week, and a descent local search, based on swap moves, to try to improve a given job sequence according to various objectives (considered in a lexicographic order).

The following values characterize the instances faced by PHARMA: $K = 21$ product types, $m = 10$ machines, $n = 250$ jobs (per year), all the $\beta_k$'s are equal to 1 (i.e., no product type is more important than another). Because of the non-disclosure agreement, no detailed result can be presented. However, as a main impact, it is important to mention that the proposed approaches for both (IMP) and (JSP) are planned to be used in the company, instead of their current decision-making tool, as the resulting solutions are satisfying. The interpretation of "satisfying" can, for instance, be based on the following ingredients: (1) history (e.g., the proposed solutions outperform the previous ones that were used by the involved company); (2) concurrence (e.g., the provided solutions have a better value than the ones employed by the competitors); (3) direction (e.g., the top management of the firm is happy with the presented solutions). Note that the quality (according to the considered objective function) of the obtained results is not the only criterion that has to be used to evaluate a solution method. Indeed, the speed (i.e., the computation time to get the provided solutions), the robustness (i.e., the sensitivity to variation in data quality and in problem characteristics), the ease of adaptation (from a practical implementation standpoint), the intensification ability (i.e., the capability to focus the search process on promising regions of the solution space), and the diversification ability (i.e., the capability to generate solutions with structures that significantly differ from the already explored solutions) are important criteria as well. Unsurprisingly, it is a real challenge to design a solution method that has a good overall score for each of the above criteria. In other words, tradeoffs have to be accepted among the above measures.

Two avenues of research can be mentioned as future works. First, other categories of metaheuristics (e.g., various types of ant algorithms, as proposed in (Zufferey, 2012b)) can be developed for both (IMP) and (JSP). Second, dedicated filtering techniques (e.g., Hertz et al., 2005) can be used to efficiently reduce the search space (i.e., poor solutions are removed once and for all from the solutions space), which is particularly relevant if the evaluation of a single solution requires a significant amount of time, as it is the case for (JSP).

## REFERENCES


Allahverdi, A., Ng, C. T., Cheng, T. C. E., & Kovalyov, M. (2009). A survey of scheduling problems with setup times or costs. *European Journal of Operational Research*, *187*(3), 985–1032. doi:10.1016/j.ejor.2006.06.060

Bloechliger, I., & Zufferey, N. (2013). Multi-Coloring and Project-Scheduling with Incompatibility and Assignment Costs. *Annals of Operations Research*, *211*(1), 83–101. doi:10.1007/s10479-013-1397-1

Blum, C., & Roli, A. (2003). Metaheuristics in combinatorial optimization: Overview and conceptual comparison. *ACM Computing Surveys*, *35*(3), 268–308. doi:10.1145/937503.937505

Dumitrescu, I., & Stuetzle, T. (2003). Combinations of local search and exact algorithms. *Lecture Notes in Computer Science*, *2611*, 211–223. doi:10.1007/3-540-36605-9_20

Ehrgott, M., & Gandibleux, X. (2000). A survey and annotated bibliography of multiobjective combinatorial optimization. *OR-Spektrum*, *22*(4), 425–460. doi:10.1007/s002910000046

Frazelle, E. H. (2015). *Inventory Strategy: Maximizing Financial, Service and Operations Performance with Inventory Strategy*. Mc Graw-Hill.

Garey, M., & Johnson, D. S. (1979). *Computer and Intractability: A Guide to the Theory of NP-Completeness*. San Francisco: Freeman.

Gendreau, M., & Potvin, J.-Y. (2010). *Handbook of Metaheuristics. International Series in Operations Research & Management Science, 146*. Springer.

Hertz, A., Schindl, D., & Zufferey, N. (2005). Lower Bounding and Tabu Search Procedures for the Frequency Assignment Problem with Polarization Constraints. *4OR*, *3*(2), 139-161.

Hertz, A., & Widmer, M. (2003). Guidelines for the use of meta-heuristics in combinatorial optimization. *European Journal of Operational Research*, *151*(2), 247–252. doi:10.1016/S0377-2217(02)00823-8

Jones, D. F., Mirrazavi, S. K., & Tamiz, M. (2002). Multi-objective meta-heuristics: An overview of the current state-of-the-art. *European Journal of Operational Research*, *137*(1), 1–9. doi:10.1016/S0377-2217(01)00123-0

Liu, Z., & Cheng, T. C. E. (2002). Scheduling with job release dates, delivery times and preemption penalties. *Information Processing Letters*, *82*(2), 107–111. doi:10.1016/S0020-0190(01)00251-4

Loukil, T., Teghem, J., & Tuyttens, D. (2005). Solving multiobjective production scheduling problems using metaheuristics. *European Journal of Operational Research*, *161*(1), 42–61. doi:10.1016/j.ejor.2003.08.029

Osman, I. H., & Laporte, G. (1996). Metaheuristics: A bibliography. *Annals of Operations Research*, *63*(5), 513–623. doi:10.1007/BF02125421

Pinedo, M. (2016). *Scheduling: Theory, Algorithms and Systems*. New York: Springer. doi:10.1007/978-3-319-26580-3

Prats, X., Puig, V., Quevedo, J., & Nejjari, F. (2010). Lexicographic optimisation for optimal departure aircraft trajectories. *Aerospace Science and Technology*, *14*(1), 26–37. doi:10.1016/j.ast.2009.11.003

Puchinger, J., & Raidl, G. R. (2005). Combining metaheuristics and exact algorithms in combinatorial optimization: A survey and classification. *Lecture Notes in Computer Science*, *3562*, 41–53. doi:10.1007/11499305_5

Respen, J., Zufferey, N., & Amaldi, E. (2016). Metaheuristics for a Job Scheduling Problem with Smoothing Costs Relevant for the Car Industry. *Networks*, *67*(3), 246–261. doi:10.1002/net.21656

Shachnai, H. T., & Woeginger, G. (2002). Minimizing makespan and preemption costs on a system of uniform machines. Springer Berlin / Heidelberg. doi:10.1007/3-540-45749-6_74

Silver, E. A., Pyke, D. F., & Peterson, R. (1998). *Inventory management and production planning and scheduling*. New York: Wiley.

Silver, E. A., & Zufferey, N. (2005). Inventory Control of Raw Materials under Stochastic and Seasonal Lead Times. *International Journal of Production Research*, *43*(24), 5161–5179. doi:10.1080/00207540500219866

Silver, E. A., & Zufferey, N. (2011). Inventory Control of an Item with a Probabilistic Replenishment Lead Time and a Known Supplier Shutdown Period. *International Journal of Production Research*, *49*(4), 923–947. doi:10.1080/00207540903449888

Solnon, C., Cung, V.-D., Nguyen, A., & Artigues, C. (2008). The car sequencing problem: Overview of state-of-the-art methods and industrial case-study of the ROADEF2005 challenge problem. *European Journal of Operational Research*, *191*(3), 912–927. doi:10.1016/j.ejor.2007.04.033

Thevenin, S., Zufferey, N., & Potvin, J.-Y. (2017). Makespan minimization for a parallel machine scheduling problem with preemption and job incompatibility. *International Journal of Production Research*, *55*(6), 1588–1606. doi:10.1080/00207543.2016.1181285

Thevenin, S., Zufferey, N., & Widmer, M. (2015). Metaheuristics for a Scheduling Problem with Rejection and Tardiness Penalties. *Journal of Scheduling*, *18*(1), 89–105. doi:10.1007/s10951-014-0395-8

Thevenin, S., Zufferey, N., & Widmer, M. (2016b). Order acceptance and scheduling with earliness and tardiness penalties. *Journal of Heuristics*, *22*(6), 849–890. doi:10.1007/s10732-016-9321-x

Zufferey, N. (2012). Metaheuristics: Some Principles for an Efficient Design. *Computer Technology and Applications*, *3*(6), 446–462.

Zufferey, N. (2012b). Optimization by Ant Algorithms: Possible Roles for an Individual Ant. *Optimization Letters*, *6*(5), 963–973. doi:10.1007/s11590-011-0327-x

Zufferey, N. (2015), October 22-24. Multiple Neighborhoods in Tabu Search: Successful Applications in Operations Management. In*Proceedings of the 4th M-Sphere Conference M-SPHERE 2015*, Dubrovnik, Croatia.

Zufferey, N., Cho, B. Y., & Glardon, R. (2016, February 23-25). Dynamic Multi-Trip Vehicle Routing with Unusual Time-Windows for the Pick-Up of Blood Samples and Delivery of Medical Material. In *Proceedings of the 5th International Conference on Operations Research and Enterprise Systems ICORES 2016*, Rome, Italy. doi:10.5220/0005733303660372

## KEY TERMS AND DEFINITIONS

**Exact Method:** Method able to find an optimal solution to an optimization problem. Such method are not appropriate for a NP-hard problem, except if its size (e.g., number of decision variables) is small.

**Inventory Management:** Policy used in a company to manage its stocks.

**Job Scheduling:** Problem for which various jobs have to be sequenced, for instance, on a production resource (typically a machine or a production line).

**Local Search Algorithm:** Meta/heuristic that starts from an initial solution, and then tries to improve is iteratively, by performing a sequence of modifications.

**Meta/Heuristic:** Method able to generate a satisfying solution to an optimization problem, without any guarantee on optimality, but in a reasonable amount of time. Such methods are appropriate for NP-hard problems.

**Np-Hard Problem:** Informally, optimization problem for which an optimal solution cannot be found in a reasonable amount of time on real/realistic data.

**Reorder-Point Policy:** Specific inventory management approach, where an order is triggered if the available stock reaches a specific level, called the reorder point.

**Simulation:** Computer tool used to reproduce a sequence of events, which are not all deterministic (i.e., some of these events are stochastic).

**Stochasticity:** Refers to the non-deterministic elements of a problem.

# Chapter 14
# The Use of Optimal Control Theory as a Benchmarking Tool in Production–Inventory Systems

**Paulo Nocera Alves Jr.**
*University of São Paulo (USP), Brazil*

**Elmer Pablo Tito Cari**
*University of São Paulo (USP), Brazil*

## ABSTRACT

*This chapter addresses some issues related to some Optimal Control Theory (OCT) problems (for example, impossible analytical solution because of an unsolvable integral, or punctual parameters that were unrealistic). It is proposed the use of OCT as a benchmarking tool to analyze inventory control systems to enhance parameters. In addition, the application of methods and heuristics in solving these problems is also described. These methods are discussed and applied in calculating the production and inventory functions using data of accounting variables of USA and Brazil companies, available in the Economatica software data base. Eventually, the results are compared and some recommendations about the advantages and disadvantages of each method are accomplished.*

## INTRODUCTION

### Motivation

Effective inventory management is critical for a company's success. In general, a business spends more than a quarter of its budget just in inventory costs, and those that manage their inventory ineffectively can spend considerably more, incurring unnecessary costs, lower profits, and even increased the risk of bankruptcy.

DOI: 10.4018/978-1-5225-2944-6.ch014

Inventory control models can be used to support decision making, for example, determining the order or production quantities, the inventory level to be held, etc. A diversity of production and inventory control models have been developed by other researchers, using several different techniques, as shown in Bag, Chakraborty, and Roy (2009), Hsieh and Dye (2013), Maddah, Yassine, Salameh, and Chatila (2014), Moori, Marcondes, and Avila (2002), Varyani, Jalilvand-Nejad, and Fattahi (2014), Sana (2010), Segic, Marinovic, and Potokar (2007), Thornhill and Naim (2006), Wee and Widyadana (2013), among others.

Optimal control theory (OCT) has been widely applied in inventory control. For example, Ignaciuk and Bartoszewicz (2012) studied the optimal inventory control of products that can be considered perishable in a periodic review system. Keblis and Feng (2012) studied optimal inventory control in a make-to-assembly system, but with make-to-stock components. Oliva (2005) considered the transportation costs to make the decision on how much to buy to stock. Rosa (2010) proposed a model with dynamic networks using stochastic dynamic programming (used in stochastic optimal control) to obtain an optimal policy for inventory control at places with demand to be satisfied and to get the cost. Many other authors, such as, Ceryan, Duenyas, and Koren (2012), Federgruen and Zheng (1992), Hwang and Koh (1992), Lahmar and Kulkarni (2006), Lototskii and Mandel (1979), Sana (2010), Sethi and Thompson (2006) and Song (2009) have also studied this topics.

In addition, in Zhu (2013) is studied a model for joint decision under uncertainty of demand and supply. Nowadays, with the risk of supply interruption, many companies hold a safety stock, affecting the selling price and the price-sensitive demand. A model that combines price and inventory was studied by Zhu (2013) and a concern is shown by the fact that the reliable supply had more significant impact on profit than the demand variability. Thus, companies should not use all raw material inventories because it would increase the quantity and the necessity to resupply, affecting negatively on earnings in case of interruption. Therefore, at least a minimum inventory or safety stock is required.

One way to solve this problem is using Optimal Control Theory (OCT) in an inventory control model taking into account the safety stock to avoit the stockout problem. Some of the common formulation of an optimal control inventory can be seen at Holt, Modigliani, Muth, and Simon (1960) and Sethi and Thompson (1980). Some other models can be found in Sethi and Thompson (2006).

Few OCT problems can be solved analytically or with an analytical solution from Variational Calculus (Chilan& Conway, 2015). Many of them will lead to unsolvable integrals or it will lead to a good solution, but with an unworkable computational time.

Since there are some methodologies issues in solution of OCT and parameter estimation problems, in this chapter is addressed some topics related about them. Moreover, the use of OCT solution is performed in an entire sector as benchmarking tool to enhance punctual parameters that were unrealistic.

## Contribution

This research proposes the use of optimal control theory as a benchmarking tool to analyze inventory control systems. The results are compared with other similar companies (e.g. companies from the same sector) to confront the parameters and solutions of each company, including results of two models. In addition, an easy way to solve optimal inventory control problems is recommended, and it is applied to companies of a sector using financial reports data, so the companies could compare themselves to benchmarking companies instead to just the optimal solution (depending on parameters). The problem is modeled as OCT and its solution is accomplished using metaheuristics, such as genetic algorithm, discretization, and numerical methods, such as finite difference.

OCT was applied using data from industrial machine companies in Brazil and the United States, from the fourth quarter of 2010 to the first quarter of 2016.This period was chosen because in Brazil from 2010 became mandatory to adopt the accounting method described in international accounting standards (International Financial Reporting Standards - IFRS) to make financial statements. The selection of data of those countries was due to information availability at Economática.

## Organization of the Research

The following section provides a theoretical background on OCT and other inventory control systems. Section 3 (Methodology) addresses methodology; Section 4 (Empirical Research) reports results; and Section 5 (Conclusion) presents the authors' conclusions.

# THEORY AND TOOLS USED IN THIS RESEARCH

The literature on inventory control involves from simple concepts to complex mathematical formulations, and it is related to various topics, such as inventory management, demand forecasting and optimal control theory.

## OCT and Production-Inventory Systems

The OCT, as other techniques of Operations Research, consists of models with an objective function to be optimized (subject to constraints), but the objective function in control optimization is a functional integral, where the constraint may be a variation (derivative) with control and state variables, which make the problem a complex optimization.

Many methods exist to solve such problems. They include genetic algorithm, (Michalewicz, Janikow,&Krawczyk, 1992), direct search (Ghosh, Das, Chowdhury,&Giri, 2011), ant colony optimization (Borzabadi&Mehne, 2009), and stochastic dual dynamic programming algorithm (Rotting &Gjelsvik, 1992).

In this work, some metaheuristics are applied in a model proposed by Sethi and Thompson (1980) and the results are compared in terms of accuracy and computation time.

In production-inventory systems, OCT may be applied to generate dynamically optimal production and inventory functions, by solving an objective function "J". It is written as minimizing the costs associated with production and inventory at time t, subject to the condition of the differential equation, representing the variation of the inventory (difference between production P(t) and demand S(t)). The model of Sethi and Thompson (1980) with a penalty function as the objective function is as follows:

Where T is the horizon time, P(t) is the production function, I(t) is the inventory function, $dI/dt$ is the inventory variation, S(t) is the demand function, I(0) is the initial inventory level, $\hat{I}$ is the "goal" of the inventory function (e.g. safety stock based on the standard deviation of the demand), $\hat{P}$ is the "goal" of the production function (e.g. the average demand forecast), $P_{min}$ is the minimal production (it could be 0), $P_{max}$ is the maximal production (if there is a limited capacity), Ce is the inventory costs, and Cp is the production costs. J is the objective function and it can be interpreted as follows:

T is the time horizon and the problem can be set as infinite horizon ($T \to +\infty$), $Ce(I - \hat{I})^2 / 2$ imposes higher h storage costs when the inventory level I is below the safety stock $\hat{I}$ (for example: cost of lack of stock, cost for not achieving demand, or penalty for increasing the risk of not meeting it, otherwise the best alternative would be 0 of inventory level) or above (for example: cost for excess of inventory). $Cp(P - \hat{P})^2 / 2$ impose that it will have higher production costs when the production rate is below the production target (for example: cost for not achieving the production target, or also cost for not achieving the expected demand, otherwise the best alternative would be 0 production) or above (eg cost for excess of production that leads to excess of inventory).

Some terms could be added or changed in the modeling, e.g. it could be considered maximum production limit and an expansion cost of the installed capacity, a discount factor $e^{-\lambda t}$, inventory with deteriorating items, stochastic components, among other factors.

The sum-of-costs model, without penalty and with discrete-time is as follows:

$$\max \quad J = \sum_{0}^{T} -\left[\mathrm{Ce}\left(\mathrm{I}\left(\mathrm{t}\right)\right) + \mathrm{Cp}\left(\mathrm{P}\left(\mathrm{t}\right)\right)\right] \qquad \text{s.t.}: \begin{cases} \mathrm{I}_t - \mathrm{I}_{t-1} = \mathrm{P}_t - \mathrm{S}_t \\ \mathrm{I}_{t=0} = \mathrm{I}_0 \end{cases}$$

It can be interpreted similarly as the penalty model, but the objective function is simply the "sum of costs". This discrete-time model can be easily solved by linear programming.

Considering a production-inventory system with production, inventory, and demand variables, Sethi and Thompson (2006) show some examples and solutions, through discrete-time modeling or using analytical solution for some continuous-time cases, for example, demand is well-known described as a polynomial function, so the problem is solvable.

## How the Models Work

Considering a company with constant demand *S*, and no initial inventory ($I_0$=0), that produces *P* products. The lead time is constant (unitary) and without delay. Since the demand is constant, the forecasted demand is the average demand, the safety stock is 0 (because there is no standard deviation) and the production goal is the average demand.

The initial inventory is equal to the safety stock, so the minimum inventory cost should be keeping the inventory level equal to 0. According to the inventory variation constraint, if there is no variation (always equal to 0), so the production should be equal to the demand. The production goal is equal to the forecasted demand, so the costs should be minimum at the optimal solution I = 0 and P = S in both models for all periods, since it is all constant.

Figure 1 is a plot of the optimal solution of this constant example with S = 30, I0 = 0, Ce = Cp = 1, for both models, and $\hat{I} = \hat{\mathrm{P}} = 0$ for the penalty model (both models lead to the same results).

The solutions of the models differ according to the change of the demand and the initial inventory. The "sum of the costs" model always tends to lead the inventory level to 0, unless it is considering a Pmax limit. But considering a variable demand, in the penalty model the safety stock is not 0, so it tends to lead the inventory level to the safety stock level. Depending on the relation of the inventory and production

*Figure 1. Optimal solution of the constant example*

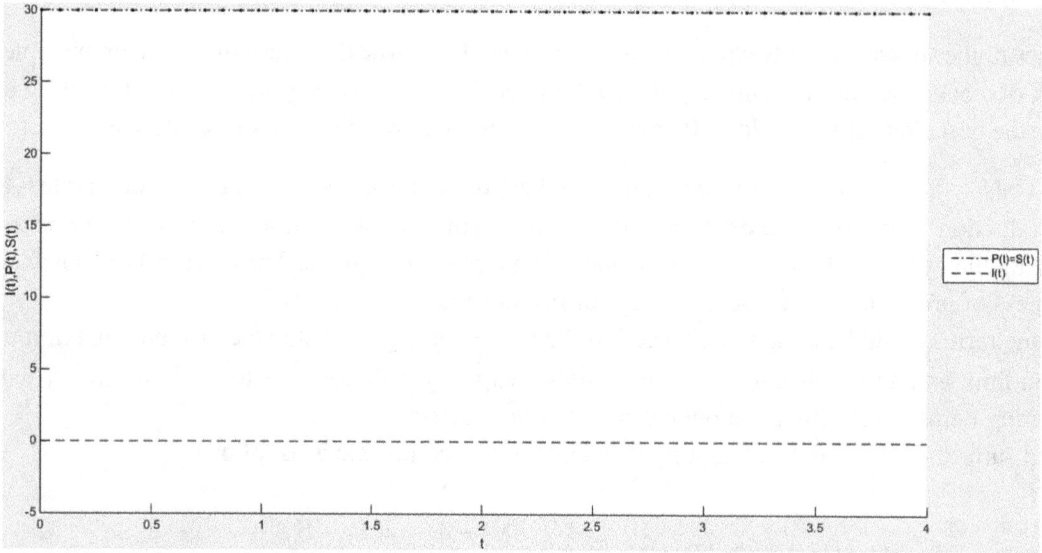


costs and goals, it could be necessary to incur in risk of inventory shortage, overstock, or not achieving the production goal, so the model obtains the optimal solution that minimizes all considered costs.

Figure 2 is a plot of the optimal solution of this variable example with S1 = 30, S2 = 15, S3 = 30, S4 = 39, S5 = 30, S6 = 15, S7 = 30, I0 = 30, $\hat{I}$ = 9, $\hat{P}$ = 27, and Ce = Cp = 1. If necessary, use $\hat{I}$ = 9 and $\hat{P}$ = 27 as an initial guess for the optimal solution of I and P of the seven periods.

OCT can be used in general in sectors the same way it can be applied to one company. The next section shows some methods to deal with demand forecasting.

## Demand Management and Forecasting

The demand is an important variable to the inventory control and the main process of demand management is the forecasting.

According to Correa, Gianesi, and Caon (2009), the ability to make the demand forecast with some accuracy is very important and the company must maintain a historical database of sales, obtain and understand information explaining its variation (internal factors, discount sales, external factors, market, climate, economy, etc.), and make use of mathematical models to explain their behavior and estimate the future demand.

According to Correa, Gianesi, and Caon (2009), planning are affected from the forecasting errors, because it is impossible to get a 100% correct forecast. One source of these uncertainties is the unstable market, which also affect competitors. Other source of error is in demand forecasting method used by the company that will make a difference to their performance, but not to competitors.

Godinho Filho and Fernandes (2010) and Correa, Gianesi, and Caon (2009) describe the forecasting process, approaches and forecasting models, which can be influenced by the term (short, medium or long term) and should be done in different ways.

*Figure 2. Optimal solution of the variable example (penalty model)*

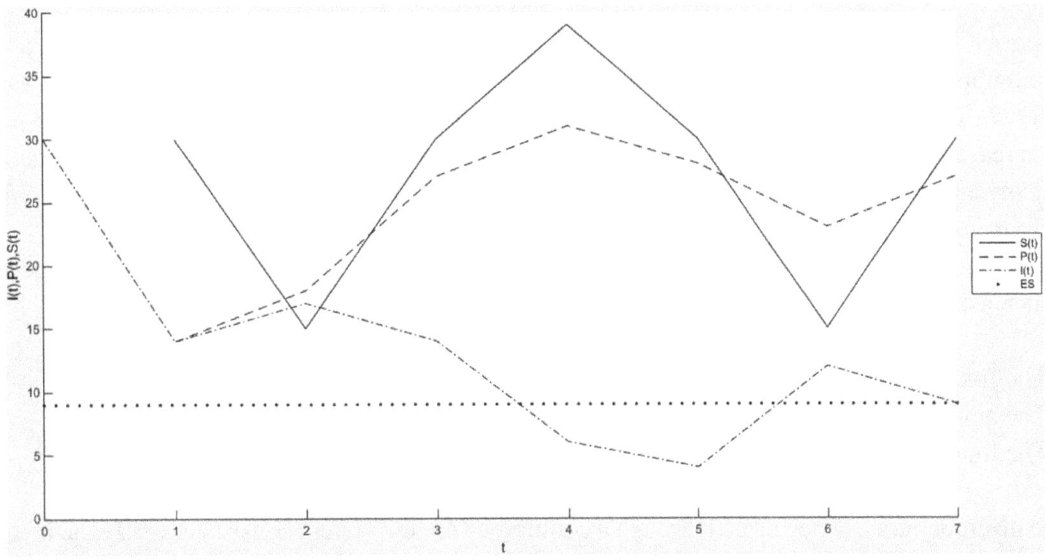


The forecasting process may be composed of five steps: identify the purpose of forecasting, selecting a forecasting approach (qualitative, causal or temporal series), selecting methods of forecasting and estimating the parameters, preparing the forecasting, monitoring, interpreting and updating the forecasting.

The most common for short-term forecasting models are: moving average, weighted moving average and exponential smoothing (considering the behavior trend, with or without seasonality). Causal models are used to medium term, and a qualitative approach and expert opinion for long-term.

Other types of forecasting models are regression analysis, which can be linear, curvilinear or multiple. According to Sethi and Thompson (2006), it is common to assume that the demand behavior as a polynomial time-varying. Since the analyzed data seems to be linear per quarter, but curvilinear per year, due to the actual behavior of the studied demand (with 3 concavities per year), and it was the lowest order able to fit the data with high coefficient of determination, it could be made four linear regressions (one for each quarterly data) and one curvilinear regression to obtain a function of demand (polynomial of order 4) for one year of forecasting.

Demand forecast using curvilinear regression: Polynomial function of order 4:

$$S(t) = a_1 + a_2 t + a_3 t^2 + a_4 t^3 + a_5 t^4$$

Where: S(t) is the demand (dependent variable), t is the time period (independent variable) and a1, a2, a3, a4 and a5 are the coefficients of the curve.

According to Beutel and Minner (2011), the inaccuracy of the demand forecasting can lead to stockout, then you should hold a safety stock to achieve a certain service level.

## Inventory Management

"Inventories are items held to be used later by internal or external customers, in other words, a buffer between supply and demand" (Filho & Fernandes, 2010, p. 163) and it is presented whenever exit a difference between the supply and demand rates. When the supply rate is larger than the demand the inventory level increase; but when it is lower, the inventory level decrease (Slack; Chambers; & Johnston, 2008).

"An inventory is justified if you have to lose money when you get rid of it" (Hobbs, 1976, p. 15), but it is also desirable to try to minimize the inventory levels and its costs.

According to Fioriolli (2002), it is important to take into consideration some issues to manage inventory, such as:

- The frequency measurement of inventories;
- The amount that must be ordered;
- The time to be ordered.

The inventory control performed through a continuous review system consists in verify the inventory level when there is a material output. If the inventory level reaches the point of resupply, it is resupplied a certain quantity of material (Correa, Gianesi, & Caon, 2009).

The periodic review is based on checking periodically the inventory levels and determine the quantity to be resupplied to the inventory to reach a predetermined maximum level (Correa, Gianesi, & Caon, 2009).

According to Godinho Filho and Fernandes (2010), it is convenient to use the periodic review system for inexpensive items, since there is no significant impact on the holding cost.

## Costs

According to Ching (1999) and Pozo (2004), the most important function of inventory control is the management of inventory levels, and the main types of costs are:

- Ordering cost: price of order processing, including the transportation and conference.
- Holding cost: combination of the costs involved in the maintenance of materials in inventory (opportunity, operation, occupation and obsolescence costs).
- Shortage cost: incurred costs when there is a stockout, lost sales, penalty, truancy, among others.
- Total cost: sum of all costs.

According to Araújo and Neto (2010), in financial reports there are other important types of costs:

- Costs of the period production cost: price of the production in the current period.

$$CPP = Cp * P$$

where CPP is the cost of period production; Cp is the unit cost of the production, P is the production.

- Cost of good sold: cost attributable to the production of the goods sold.

$$CPV = Cd * S$$

where CPV is the cost of good sold; Cd is the unit cost of the good sold, S is the good sold (equal to the demand).

And inventory costs could be divided into final and initial:

$$CEI = Ci * I_i$$

$$CEF = Cf * I_f,$$

where: CEI is the cost of inventorymaintaining at the onset of the period equal to the close of the previous period (termed initial inventory); CEF is the cost of maintaining inventory at the close of the period (termed final inventory); Ci is the unit cost of the initial inventory, Cf is the unit cost of the final inventory, $I_i$ is the initial inventory level and $I_f$ is the final inventory level.

It is possible to estimate the unit cost of inventories ($Cf$ and $Ci$) based on the cost of average inventory of the period $C\left(I_f + I_i\right)/2$ (GodinhoFilho, &Fernandes, 2010), and its sum is equivalent to the final and initial quantities of inventories, incurring average costs. Remark that the final inventory $I_{ft}$ of a period t is equal to the initial inventory $I_{i(t+1)}$ for the next period t+1, i.e. $I_{ft} = I_{i(t+1)}$:

$$\sum_{t=1}^{T} Cd_t \frac{\left(I_{ft} + I_{it}\right)}{2} = Cd_1 \frac{I_{it}}{2} + \sum_{t=1}^{T-1} I_{ft} \frac{\left(Cd_t + Cd_{t+1}\right)}{2} + Cd_T \frac{I_{fT}}{2}$$

Considering that the costs are proportional to the average Cd to estimate Cf and Ci for intermediate periods (t = 2, …, T-1):

$$Cf_t = \frac{\left(Cd_t + Cd_{t+1}\right)}{2} \text{ and } Ci_t = \frac{\left(Cd_{t-1} + Cd_t\right)}{2}$$

## Other Financial Measures and Relations

Similarly, as relations among P, S, and I in the inventory variation equation $\mathrm{P}_t = \mathrm{S}_t + \mathrm{I}_t - \mathrm{I}_{t-1}$ of OCT model constraint, the costs have a similar relation, described in Araújo and Neto (2010):

CPP = CPV + CEF - CEI

where CPP is the cost of period production; CPV is the cost of good sold; CEF is the cost of maintaining inventory at the end of the period, and CEI is the cost of maintaining inventory in the initial period (equal to the final of the last period).

In this chapter, it was used relative cost (CR) to generate $1.00 of revenue (R = 1) as a financial measure, because it is not possible to estimate unit costs and quantities only from total costs.

$$CR = \frac{CPV}{revenue} = \frac{Cd * S}{R * S} = \frac{Cd}{R}$$

where CR is the relative cost, CPV is the cost of good sold, Cd is the unit cost of the good sold, S is demand, R is the unit revenue.

For R = 1, CR = Cd

## Safety Stock

According to Correa, Gianesi and Caon (2009), the safety stock can be calculated more generally using a probabilistic approach, as follows:

$$\hat{I} = F_{Safety} * \sigma_{Slt}$$

where $\hat{I}$ is the safety stock (inventory goal); $F_{Safety}$ is the safety factor (depending on service level, based on the normal distribution); and $\sigma_{Dlt}$ is the standard deviation of demand during lead time, which one can be modeled in several different ways.

In this chapter, $F_{Safety} = 1$ and $\sigma_{Slt} = \sigma_{Demand}$ are used for those variables. It means that the service level based on the normal distribution is aproximately 84,14%, and the lead time is unitary and constant. So it is possible to compare the companies with those with the same parameters.

# METHODOLOGY

The method of this research has a quantitative approach and it is used modeling to describe the behavior of a system from mathematical models (Miguel, 2010).

## Issues, Controversies, Problems

Some OCT problems related to unsolvable integrals could be solved by numerical methods and approximations. In addition, those problems also have stochastic versions, making it more difficult to solve the models accurately.

For example, solving a simple OCT problem lead to a complex or impossible analytical solution, but using a simplified function (discretized model), solved by linear or nonlinear programming could lead to a simple discrete solution in a few seconds.

The modeling could be an issue too, because some formulations could lead to unrealistic or impossible solutions. For example, using the linear inventory control model without safety stock leads to a solution of "zero inventory level". In addition, a very high initial inventory level led to "negative production" if the problem is not properly modeled.

A nonlinear model needs an initial guess to find the optimal solution which could be solved using the goals as a starting point or by a hybrid method (for example, using a metaheuristic to find an initial guess and another method to find the optimal solution).

Another issue is due to the estimation of some parameters, if the demand forecasting and the safety stock is not properly calculated or do not represent the actual parameters; therefore, it will lead to unrealistic solutions. For example, if a company calculates the optimal inventory level based on an unrealistic high safety stock, it will obtain an optimal inventory control solution that keeps the company overstocked. However, looking other companie even with similar demands, it could identify a benchmarking company to compare its results and parameters.

## Data Sample and Research Variables

The study data sample was taken considering registered quarterly data from companies of the industrial machines sector in the Economática software (mainly from the United States and some from Brazil). The Economática software is one of the widest and reliable financial database of the continent. Brazilian and North-American companies represent the great majority of companies registered at Economática.

The data was from the fourth quarter of 2010 (to calculate the average cost of inventory and use the initial inventory in the next quarter) until the first quarter of 2016. This period was chosen because, in 2010 became mandatory in Brazil to adopt the accounting method described in international accounting standards (International Financial Reporting Standards - IFRS) to make financial statements, and most companies registered at Economática have followed this method since 2010. In the end, 36 companies were selected for the study.

The accounting reports contain total costs then it is not possible to directly obtain unit costs and quantities. As the behavior of OCT model depends only on the relation of the production and storage costs, and not the actual value itself, so, for example, whether the costs of production and storage were respectively \$10.00 and \$20.00 to generate revenue of \$60.00, or \$100.00 and \$200.00 to generate \$600.00. What matters is that a cost would be half the other, and revenue would be double the sum of the costs, so it was used as a financial measure the relative cost (CR) to generate \$1.00 of revenue (R = 1). To obtain this cost and other necessary values, unit costs and relative quantities were estimated from the data reported in accounting reports.

Available data are revenue, cost of goods sold (CPV) and inventories (CE), which can be divided into costs of final inventory (CEF) and initial inventory (CEI). It was used the formula of the costs relation described in Araújo and Neto (2010), exposed in section "Costs", calculatingCPP in the industrial sector (it also can be used in the case of commerce, but as cost of period purchase):

$$CPP = CPV + CEF\text{-}CEI$$

Then it is needed to calculate Cd by the CR formula Section of Other Financial Measures and Relations) with R=1:

*Table 1. Description of functions, variables, and parameters*

| Functions, Variables, and Parameters: | Description: | Functions, Variables, and Parameters: | Description: |
|---|---|---|---|
| $J$ | Objective function (cost function); | $Cp$ | Unitary production cost; |
| $t$ | Time; | $Cf$ | Unit. (final) inventory cost; |
| $S(t)$ | Demand (exogenous); | $Ci$ | Unit. (initial) inventory cost; |
| $P(t)$ | Production (control); | $Cd$ | Sold production cost; |
| $I(t)$ | Inventory (state); | $CPP$ | Cost of period production (or purchase); |
| $I_0$ | Initial inventory (inventory of period 0 at time 1); | $CEF$ | Cost of final inventory; |
| $\hat{I}$ | Safety stock (inventory goal) | $CEI$ | Cost of initial inventory; |
| $\hat{P}$ | Production goal | $CPV$ | Cost of goods sold. |
| $R$ | Revenue | $CR$ | Relative cost; |

$$Cd = CR = \frac{CPV}{revenue}$$

and Cf and Ci:

$$Cf_t = \frac{\left(Cd_t + Cd_{t+1}\right)}{2} \text{ and } Ci_t = \frac{\left(Cd_{t-1} + Cd_t\right)}{2}$$

When necessary $Cd_0$ and $Cd_{T+1}$ costs can be estimated respectively as $Cd_1$ and $Cd_T$, or delete the first and last period of analysis, then it is possible to calculate them instead of estimating, but horizon analysis is reduced by 2 periods. In the case of this article, $Cd_{T+1}$ as Cd are considered of the first quarter of 2016 and $Cd_0$ as Cd of the fourth quarter of 2010. After calculating the storage costs, it is possible to obtain the values of the intermediate variables (final and initial inventories):

$$I_f = CEF/Cf \text{ and } I_i = CEI/Ci$$

The values of demand (proportional to revenue) and production:

$$S = CPV/Cd \text{ and } P = S + I_f - I_i$$

The average unitary production cost can be calculated as:

$$Cp = CPP/P$$

## Methods for Solving OCT Problems

In this chapter, the discretization of the continuous-time model is made replacing the integral function by Riemann sum $\int f(t)dt = \sum f(t)\Delta t$, with $\Delta t = 1$, and the differential by difference equation $\frac{df(t)}{dt} = \frac{\Delta f(t)}{\Lambda t}$, with $\Delta t = 1$ and $\Delta f(t) = f(t) - f(t-1)$.

This discrete-time model is solved by nonlinear programming. This model is simpler and faster to solve than the continuous model. The variational difference constraint in the proposed model equivalent to the one used in the models of Sethi and Thompson (1980) and Holt et al. (1960), describes the variation of inventory by a difference at time t and the integral cost function by the sum of inventory and production costs.

Another solution is to try using metaheuristics such as genetic algorithm or a numerical method, such as finite difference to solve the problem. It can also be solved analytically by the Maximum Principle of Pontryagin, creating an Hamiltonian and solving it using Lagrange's method of variation of parameters, or using Laplace transform to make the problem simple. It is important to consider that, in this case of the optimal inventory control, the Demand function is known, but a variable coefficient, making the problem inhomogeneous and difficult to solve.

Before Solving the OCT problem by finite difference, it is important to transform the problem into a system of differential equations, creating the Hamiltonian and the system from the Maximum Principles of Pontryagin.

The Hamiltonian (associating a joint variable $\lambda(t)$ to the constraint and add it to the objective function):

$$H = \lambda(P - S) - \frac{h}{2}(I - \hat{I})^2 - \frac{c}{2}(P - \hat{P})^2$$

From the Maximum Principles of Pontryagin:

$$\frac{\partial H}{\partial P} = 0$$

$$\frac{\partial \lambda}{\partial t} = -\frac{\partial H}{\partial I}$$

$$\lambda(\mathrm{T}) = 0$$

then:

$$P = \hat{P} + \frac{\lambda}{c}$$

$$\frac{\partial \lambda}{\partial t} = h(I - \hat{I})$$

Finally:

$$\frac{\partial \mathrm{I}}{\partial \mathrm{t}} = \frac{\lambda}{c} + \hat{\mathrm{P}} - \mathrm{S}(\mathrm{t})$$

$$\frac{\partial \lambda}{\partial t} = h(I - \hat{I})$$

Or a second order differential equation:

$$\frac{\partial^2 I}{\partial t^2} - \frac{h}{c} I = -\frac{h}{c}\hat{I} - \frac{\partial S}{\partial t}$$

From this it is possible to derivate the Neumann boundary condition:

$$\frac{\partial \mathrm{I}(\mathrm{T})}{\partial \mathrm{t}} = \hat{\mathrm{P}} - \mathrm{S}(\mathrm{T})$$

Using this condition together with the initial value, it is possible to apply the finite difference to solve a two-point boundary value problem. The tridiagonal matrix used to solve this problem was adapted fromLeVeque (2007) and it is as follows:

$$\begin{bmatrix} -(2+k) & 1 & & & & \\ 1 & -(2+k) & 1 & & & \\ & 1 & -(2+k) & 1 & & \\ & & \ddots & \ddots & \ddots & \\ & & & 1 & -(2+k) & 1 \\ & & & & -\Delta t & \Delta t \end{bmatrix} \begin{bmatrix} I_1 \\ I_2 \\ I_3 \\ \vdots \\ I_n \\ I_{n+1} \end{bmatrix} = \begin{bmatrix} (-\dot{S}_1 - \hat{I} \cdot Ce / Cp) \cdot \Delta t^2 - I_0 \\ (-\dot{S}_2 - \hat{I} \cdot Ce / Cp) \cdot \Delta t^2 \\ (-\dot{S}_3 - \hat{I} \cdot Ce / Cp) \cdot \Delta t^2 \\ \vdots \\ (-\dot{S}_{T-1} - \hat{I} \cdot Ce / Cp) \cdot \Delta t^2 \\ (-S(T) + \hat{P}) \cdot \Delta t^2 \end{bmatrix}$$

where $k = \Delta t^2 \cdot Ce / Cp$, $\dot{S}_t = \frac{S(t+1) - S(t)}{\Delta t}$ is the demand variation, $\Delta t = 1$, Ce is the unitary cost of the inventory,Cp is the unitary production cost, S(T) is the demand in the last period T, $\hat{I}$ is the safety stock, $\hat{P}$ is the production goal, and $I_t$ is the inventory level at period t.

The next section shows some methods to solve OCT problems.

## RECOMMENDATIONS AND SOLUTIONS OF OCT PROBLEM USED IN THIS RESEARCH

Some recommendations to solve OCT problems are to transform the problem into a discrete-time OCT problem, so that if the problem keeps the nonlinear form, at least it is easier to solve sums and differences instead of integrals and differentials.

Depending on the method, it could be difficult to estimate a good initial value necessary to start the optimization. Even a metaheuristic could need an unworkable amount of time to solve the problem. So using finite difference is recommended to obtain a good solution, at least to have an idea of the approximated solution to solve it optimally.

## EMPIRICAL RESEARCH

### Application

An empirical test was conducted applying the OCT "sum of costs" and penalty models to data from industrial machine companies. Different methods are used to solve the above-mentioned models, using data of accounting variables of companies from USA and Brazil. The data were obtained from the software database Economatica.

The continuous models were possible to solve analytically (since the demand had a polynomial form, so it was possible to solve the integrals of the Hamiltonian method in the model). One of the issues of continuous-time models is that many manufacturing systems have integer variables (such as number of stored products). Due to the complexity of the model, it is difficult to set some constraints, and the time used to solve the problem. So it is recommended to use other methods to solve it.

The models were discretized as described in the solutions and recommendations section. Then it was solved as linear programming (in the case of "sum of costs") or nonlinear optimization (in the case of the penalty model). Both models were solved using genetic algorithm too.

### Analysis of the Results

Analyzing the results of the "sum of costs" model, it was possible to see that the model tends to assign null inventory level, so the relevant result is the production function (results for CPP on Table 2 of Appendix A). And the penalty model tends to assing values near to the average demand (results for CPP on Table 3 of Appendix B) and the safety stock (results for CEF on Table 4 of Appendix C).

Since the discretized models are simpler than the continuous models, the results using genetic algorithm and the nonlinear optimization without a good initial guess of the solution takes an unworkable time to be obtained. Thus, it is recommended to use discrete-time model and solve it as a linear programming or nonlinear optimization with linear constraints. To run the algorithm took only 23 seconds, including data exportation to a sheet, when solved by nonlinear programming. Using finite difference approximation, it takes only 4 seconds, but with 8.84% of error (in average).

Observing the results, it is possible to see that 13 of all 36 companies achieved a minimum cost as small as the optimal costs calculated with the penalty model. And 6 of these 13 companies were among the 10 companies that achieved a minimum cost as small (less than 4.27% of difference) as the optimal costs of the company that achieved the minimum costs of these 36 calculated with the sum-of-costs model. This results probably means that the demand of these 6 companies are more stable, so they can have low safety stock and be optimal within the two models.

The remaining four companies did not rank among the best companies in the sum-of-costs, possible means that they are working above the safety stock (at least 2 of these 4 companies), so they have high storage costs. The other 2 of these 4 companies did not achieve the minimum for less than 0,7% and they are among the 3 companies after the 13 best companies using the penalty model; thus, a total of 15 companies were considered as a benchmarking for the other 21 of all 36 companies.

On the other side, when companies are among the best companies only in the penalty model, it probably means that they are working a little above of the safety stock. There are 2 Brazilian companies in that situation. Among the worst results using the penalty model, 10 out of 13 companies were among the 12 worst results using the other model. It possible means that they are working overstocked. There are more 2 Brazilian companies in that situation. The results indicate that, in general, the USA companies have a better inventory control.

The average costs reduction of 36 companies studied could be 29.55% of actual costs. Using the 15 companies with best results in this study as a benchmarking for the other 21 companies, it is possible to conclude that the average reduction of inventory costs of these 21 companies studied could be 45% of actual costs. Many companies hold inventories, even with a constant demand.

## CONCLUSION

In this chapter, empirical evidence showed that even some metaheuristics as genetic algorithm need unworkable computational time to solve optimal control theory problems. The best alternatives were to simplify the model through discretization or work the model to be approximated by finite difference.

The "sum of costs" model tends to obtain null inventory values, but the penalty model could be subjective due to the "goals" (achieving the safety stock and average demand or forecasted demand). Thus, it is recommended to compare both results and analyze it together with the actual data and possible benchmarks.

The practical implications of using OCT as a benchmarking tool is that a company could use a similar company as a benchmark to improve its management, instead of relying only on its own forecasts and estimations. Using the estimations based on financial report data is an easy way to make that comparison, because it is easier to found data from these databases (any company could do it).

Another practical implication is a reduction in the costs. The average costs reduction of 36 companies studied could be 29.55% of actual costs. Using the 15 companies with best results in this study as a benchmarking for the other 21 companies, it is possible to conclude that the average reduction of inventory costs of these 21 companies studied could be 45% of actual costs. Many companies hold inventories, even with a constant demand.

Finally, the result showed that in general the USA companies from industrial machine sector have a good inventory control and the Brazilian companies almost always tends to overstock, even in small amounts, so it is suggested to make case studies to verify why some companies do not have efficient inventory control systems.

## Ongoing Research

The critical views regarding the manner in which OCT problems are solved indicate valid points. Sometimes it is unworkable to solve these problems analytically or even with some metaheuristics. Other alternative methods to those presently used to solve the OCT problem should be undertaken to optimize its calculation. And considering the managerial aspects of the companies, it is suggested to make case studies to see why Brazilian companies are always working above (even if it is a little) the safety stock necessary to achieve the forecasted demand and some nonforecasted deviation. Another suggestion is to use other benchmark tools to analyze the inventory control systems of the companies or even to use OCT to make analysis of relative efficiency as data envelopment analysis.

## REFERENCES

Araújo, A.M.P., & Neto, A.A. (2010). *Aprendendo a contabilidade: Como entender o processo contábil, como interpretar as deonstrações contábeis, como aplicar a contabilidade em negócios*. RibeirãoPreto: Inside Books.

Bag, S., Chakraborty, D., & Roy, A. R. (2009). A production inventory model with fuzzy random demand and with flexibility and reliability considerations. *Comput. Ind. Eng. J., 56*(1), 411–416. doi:10.1016/j.cie.2008.07.001

Bertesekas, D. P. (2005). *Dynamic programming and optimal control*. Belmont: Athena Scientific.

Beutel, A. L., & Minner, S. (2011). Safety stock planning under casual demand forecasting. *International Journal of Production Economics, 140*(2), 637–645. doi:10.1016/j.ijpe.2011.04.017

Borzabadi, A. H., & Mehne, H. H. (2009). Ant colony optimization for optimal control problems. *Journal of Information and Computing Science, 4*(4), 259–264.

Ceryan, O., Duenyas, I., & Koren, Y. (2012). Optimal control of an assembly system with demand for the end-product and intermediate components. *IIE Transactions, 44*(5), 386–403. doi:10.1080/0740817X.2011.609525

Chilan, C. M., & Conway, B. A. (2015). A Reachable Set Analysis Method for Generating Near-Optimal Trajectories of Constrained Multiphase Systems. *Journal of Optimization Theory and Applications, 167*(1), 161–194. doi:10.1007/s10957-014-0651-2

Ching, H. Y. (1999). *Gestão de estoques na cadeia de logística integrada: Supply chain*. São Paulo: Atlas.

Correa, H. L., Gianesi, I. G. N., & Caon, M. (2009). *Planejamento, programação e controle da produção. 5*. São Paulo: Atlas.

Federgruen, A., & Zheng, Y. S. (1992). Efficient algorithm for computing an optimal (r, Q) policy in continuous review stochastic inventory systems. *INFORMS Operations Research*, *40*(4), 808–813. doi:10.1287/opre.40.4.808

Fiorioli, J. C. (2002). *Modelagem estocástica de sistemas hierárquicos de estoques*. Porto Alegre: UFRGS.

Ghosh, A., Das, S., Chowdhury, A., & Giri, R. (2011). An ecologically inspired direct search method for solving optimal control problems with Bézier parameterization. *Engineering Applications of Artificial Intelligence*, *24*(7), 1195–1203. doi:10.1016/j.engappai.2011.04.005

Godinho Filho, M., & Fernandes, F. C. F. (2010). *Planejamento e controle da produção: dos fundamentos ao essencial*. São Paulo: Atlas.

Hobbs, J. A. (1976). *Control Over Inventory and Production*. São Paulo: McGraw-Hill.

Holt, C. C., Modigliani, F., Muth, J. F., & Simon, H. A. (1960). *Planning Production, Inventories, and Work Force*. Prentice-Hall.

Hsieh, T. P., & Dye, C. Y. (2013). A production-inventory model incorporating the effect of preservation technology investment when demand is fluctuating with time. *Journal of Computational and Applied Mathematics*, *239*, 25–36. doi:10.1016/j.cam.2012.09.016

Hwang, H., & Koh, S. G. (1992). Optimal control of work-in-process with dual limit switches in two-stage transfer line. *European Journal of Operational Research*, *63*(1), 66–75. doi:10.1016/0377-2217(92)90055-E

Ignaciuk, P., & Bartoszewicz, A. (2012). Linear-Quadratic optimal control of periodic-review perishable inventory systems. *IEEE Transactions on Automatic Control*, *20*, 1400–1407.

Keblis, M. F., & Feng, Y. (2012). Optimal pricing and production control in na assembly system with a general stockout cost. *IEEE Transactions on Automatic Control*, *57*(7), 1821–1826. doi:10.1109/TAC.2011.2178335

Lahmar, M., & Kulkarni, N. R. (2006). Optimal control policies for production/inventorysystems with exchange transactions. *Proceedings of the IIE Annual Conference and Exhibition*.

LeVeque, R. J. (2007). *Finite Difference Methods for Ordinary and Partial Differential Equations: Steady-State and Time-Dependent Problems*. Philadelphia: Society for Industrial and Applied Mathematics. doi:10.1137/1.9780898717839

Lototskii, V. A., & Mandel, A. S. (1979). Models and methods of multiproduct inventory control. *Automation and Remote Control*, *40*(6), 888–896.

Maddah, B., Yassine, A. A., Salameh, M. K., & Chatila, L. (2014). Reserve stock models: Deterioration and preventive replenishment. *European Journal of Operational Research*, *232*(1), 64–71. doi:10.1016/j.ejor.2013.06.043

Michalewicz, Z., Janikow, C. Z., & Krawczyk, J. B. (1992). A modified genetic algorithm for optimal control problems. *Computers & Mathematics with Applications (Oxford, England)*, *23*(12), 83–94. doi:10.1016/0898-1221(92)90094-X

Miguel, P. A. C. (2010). *Metodologia de pesquisa em engenharia de produção e gestão de operações*. Rio de Janeiro: Elsevier.

Moori, R. G., Marcondes, R. C., & Ávila, R. T. (2002). A análise de agrupamentos como instrumento de apoio à melhoria de qualidade dos serviços aos cliente. *J. Admin. Contemp*, *6*(1), 63–84. doi:10.1590/S1415-65552002000100005

Oliva, G. M. (2005). *Planejamento conjunto e colaborativo da cadeia de suprimentos: Modelo de controle ótimo multiobjetivo com custos de transporte*. Porto Alegre: UFRGS.

Pozo, H. (2004). *Administração de recursos materiais e patrimoniais: Uma abordagem logística*. São Paulo: Atlas.

Rosa, H. (2010). *Um modelo conjunto de localização e operação de estoque em redes dinâmicas*. Florianópolis: UFSC.

Rotting, T. A., & Gjelsvik, A. (1992). Stochastic dual dynamic programming for seasonal scheduling in the norwegian power system. *IEEE Transactions on Power Systems*, *7*(1), 273–279. doi:10.1109/59.141714

Sana, S. S. (2010). A production-inventory model in an imperfect production process. *European Journal of Operational Research*, *200*(2), 451–464. doi:10.1016/j.ejor.2009.01.041

Segic, M. R., Marinovic, M., & Potokar, M. (2007). Modification of production-inventory control model with quadratic and linear costs. *ISOR*, *9*, 369–376.

Sethi, S. P., & Thompson, G. L. (1980). Turnpike Horizons for Production Planning. *Management Science*, *26*(3), 229–241. doi:10.1287/mnsc.26.3.229

Sethi, S. P., & Thompson, G. L. (2006). *Optimal control theory: applications to management science and economics*. New York: Springer.

Slack, N., Chambers, S., & Johnston, R. (2010). *Operations management*. Pearson education.

Song, D. P. (2009). Stability and optimization of a production inventory system under prioritized base-stock control. *IMA J. Manag. Math*, *20*(1), 59–79. doi:10.1093/imaman/dpn010

Thornhill, N. F., & Naim, M. M. (2006). An exploratory study to identify rogue seasonality in a steel companys supply network using spectral principal component analysis. *European Journal of Operational Research*, *172*(1), 146–162. doi:10.1016/j.ejor.2004.09.044

Varyani, A., Jalilvand-Nejad, A., & Fattahi, P. (2013). Determining the optimum production quantity in three-echelon production system with stochastic demand. *International Journal of Advanced Manufacturing Technology*, *72*(1-4), 119–133. doi:10.1007/s00170-014-5621-1

Wee, H. M., & Widyadana, G. A. (2013). A production model for deteriorating items with stochastic preventive maintenance time and rework process with FIFO rule. *Omega*, *41*(6), 941–954. doi:10.1016/j.omega.2012.12.001

Zhu, S. X. (2013). Dynamic replenishment, production, and pricing decisions, in the face of supply disruption and random price-sensitive demand. *International Journal of Production Economics*, *146*(2), 612–619. doi:10.1016/j.ijpe.2013.08.009

## KEY TERMS AND DEFINITIONS

**CEF:** Cost of final inventory.

**CEI:** Cost of initial inventory.

**Control Variable:** Controllable variable to be optimized.

**CPP:** Cost of period production (or purchase)

**CPV:** Cost of goods sold.

**CR:** Relative cost to produce $1,00 of revenue.

**Exogenous Variable:** Is a known variable that it is out of the system and do not depend on the variable of the system.

**OCT:** Operation research technique that consists of models with an objective function to be optimized, subject to constraints, but optimizing a function based on a control variable (and a state variable that varies with the control).

**State Variable:** Variable that varies with the control and it is optimal when the control variable is optimal.

# APPENDIX A: CPP* USING THE "SUM OF COSTS" MODEL

*Table 2. Companies and optimal costs of production per period (CPP* in millions of Dolars), using the "sum of costs" model*

| | Period | | | | | | | | | | | | | | | | | | | | |
|---|---|---|---|---|---|---|---|---|---|---|---|---|---|---|---|---|---|---|---|---|---|
| **Company** | **1** | **2** | **3** | **4** | **5** | **6** | **7** | **8** | **9** | **10** | **11** | **12** | **13** | **14** | **15** | **16** | **17** | **18** | **19** | **20** | **21** |
| Agco | 208 | 3310 | 5004 | 6981 | 1764 | 3856 | 5676 | 7838 | 1855 | 4201 | 6136 | 8405 | 1813 | 3944 | 5691 | 7674 | 1355 | 2966 | 4348 | 5916 | 1250 |
| Applied Materials | 3 | 3226 | 4834 | 6248 | 1450 | 2895 | 4346 | 5423 | 979 | 2128 | 3318 | 4507 | 1279 | 2629 | 3914 | 5252 | 1416 | 2821 | 4301 | 5714 | 1346 |
| Baker Hughes | 906 | 7212 | 11129 | 15315 | 4318 | 8541 | 12696 | 17406 | 4371 | 8919 | 13662 | 18542 | 4677 | 9440 | 14554 | 20057 | 4678 | 7905 | 11344 | 14621 | 2782 |
| Bardella | 10 | 177 | 222 | 282 | 52 | 88 | 129 | 175 | 40 | 85 | 124 | 157 | 34 | 84 | 119 | 151 | 32 | 62 | 77 | 100 | 15 |
| Briggs & Stratton | 659 | 1719 | 336 | 700 | 1272 | 1692 | 262 | 610 | 1118 | 1525 | 272 | 597 | 1101 | 1507 | 236 | 583 | 1081 | 1517 | 235 | 548 | 1032 |
| Brooks Automation | 125 | 381 | 464 | 78 | 171 | 266 | 349 | 71 | 147 | 226 | 302 | 78 | 158 | 235 | 318 | 85 | 176 | 269 | 363 | 79 | 168 |
| Brunswick | 222 | 1570 | 2249 | 2872 | 732 | 1523 | 2180 | 2770 | 728 | 1528 | 2187 | 2870 | 699 | 1525 | 2096 | 2809 | 725 | 1538 | 2256 | 2998 | 792 |
| Colfax Corp | 47 | 227 | 337 | 456 | 629 | 1347 | 2045 | 2757 | 651 | 1391 | 2087 | 2902 | 727 | 1537 | 2330 | 3144 | 616 | 1316 | 1989 | 2714 | 595 |
| Cummins | 923 | 6331 | 9779 | 13440 | 3253 | 6522 | 9609 | 12859 | 2979 | 6322 | 9488 | 12918 | 3290 | 6897 | 10544 | 14360 | 3501 | 7184 | 10611 | 14190 | 3258 |
| Deere & Co | 1028 | 10200 | 16015 | 21938 | 4547 | 11394 | 18214 | 24993 | 4972 | 12475 | 19338 | 25717 | 5230 | 12075 | 18704 | 24891 | 4487 | 10089 | 15484 | 20207 | 3892 |
| Donaldson | 473 | 1081 | 1480 | 394 | 774 | 1192 | 1622 | 394 | 786 | 1183 | 1587 | 384 | 766 | 1165 | 1596 | 389 | 780 | 1156 | 1565 | 362 | 706 |
| Dover | 497 | 2552 | 3784 | 4901 | 1286 | 2620 | 3980 | 4997 | 1264 | 2633 | 4011 | 5386 | 1143 | 2402 | 3695 | 4794 | 1099 | 2188 | 3307 | 4393 | 1037 |
| Eaton | 1117 | 5541 | 8443 | 11252 | 2744 | 5567 | 8326 | 11460 | 3727 | 7591 | 11488 | 15381 | 3863 | 7869 | 11787 | 15634 | 3579 | 7268 | 10864 | 14285 | 3288 |
| Flowserve | 0 | 1172 | 2153 | 2999 | 720 | 1516 | 2287 | 3163 | 718 | 1540 | 2347 | 3257 | 683 | 1486 | 2268 | 3183 | 703 | 1471 | 2180 | 3080 | 640 |
| FMC Technologies | 276 | 1794 | 2777 | 3975 | 1109 | 2281 | 3381 | 4839 | 1311 | 2655 | 4004 | 5558 | 1392 | 2905 | 4385 | 6002 | 1297 | 2578 | 3758 | 4907 | 954 |
| Graco | 2 | 196 | 298 | 395 | 104 | 233 | 347 | 460 | 117 | 247 | 373 | 497 | 131 | 276 | 414 | 558 | 146 | 298 | 448 | 602 | 143 |
| Idex | 52 | 521 | 815 | 1097 | 284 | 578 | 863 | 1147 | 279 | 578 | 857 | 1147 | 297 | 606 | 904 | 1197 | 274 | 560 | 841 | 1117 | 280 |
| Illinois Tool Works | 1281 | 5762 | 8719 | 11508 | 2899 | 5871 | 8731 | 11439 | 2481 | 5095 | 6359 | 8556 | 2153 | 4369 | 6557 | 8665 | 1961 | 3992 | 5946 | 7881 | 1888 |
| Inds Romi | 0 | 40 | 185 | 256 | 72 | 101 | 161 | 234 | 48 | 93 | 152 | 201 | 48 | 98 | 141 | 185 | 34 | 60 | 77 | 121 | 30 |
| Ingersoll-Rand | 844 | 4950 | 8038 | 10498 | 2241 | 4875 | 7341 | 9767 | 2187 | 4849 | 7387 | 8709 | 1952 | 4375 | 6717 | 9008 | 2092 | 4553 | 6953 | 9312 | 2054 |
| ITT Industries | 1120 | 4160 | 6266 | 1429 | 405 | 797 | 1184 | 1545 | 415 | 828 | 1260 | 1697 | 460 | 906 | 1344 | 1785 | 387 | 805 | 1214 | 1680 | 415 |
| Joy Global | 0 | 1089 | 2013 | 2921 | 784 | 1790 | 2710 | 3789 | 775 | 1679 | 2568 | 3434 | 633 | 1248 | 1876 | 2690 | 535 | 1089 | 1682 | 2412 | 521 |
| Kennametal | 664 | 1512 | 407 | 825 | 1271 | 1758 | 439 | 859 | 1300 | 1745 | 428 | 908 | 1417 | 1939 | 483 | 957 | 1393 | 1856 | 419 | 783 | 1123 |
| Lam Research | 994 | 1757 | 416 | 749 | 1139 | 1599 | 597 | 1114 | 1614 | 2174 | 565 | 1208 | 1906 | 2595 | 646 | 1345 | 2137 | 2961 | 868 | 1684 | 2422 |
| Lennox Intl | 235 | 1208 | 1904 | 2472 | 517 | 1203 | 1714 | 2228 | 502 | 1158 | 1794 | 2344 | 528 | 1218 | 1874 | 2473 | 522 | 1219 | 1911 | 2525 | 536 |
| Manitowo Co | 0 | 1274 | 1995 | 2815 | 648 | 1407 | 2131 | 2994 | 671 | 1447 | 2207 | 3020 | 616 | 1367 | 2113 | 2910 | 572 | 1232 | 1893 | 2624 | 375 |

*continued on following page*

Table 2. Continued

| | Period | | | | | | | | | | | | | | | | | | | | |
|---|---|---|---|---|---|---|---|---|---|---|---|---|---|---|---|---|---|---|---|---|---|
| Company | 1 | 2 | 3 | 4 | 5 | 6 | 7 | 8 | 9 | 10 | 11 | 12 | 13 | 14 | 15 | 16 | 17 | 18 | 19 | 20 | 21 |
| Metalfrio | 29 | 226 | 262 | 338 | 79 | 156 | 229 | 300 | 62 | 156 | 237 | 301 | 80 | 181 | 241 | 291 | 62 | 148 | 169 | 225 | 67 |
| Middleby | 4 | 236 | 367 | 514 | 143 | 299 | 457 | 637 | 208 | 432 | 650 | 878 | 229 | 486 | 729 | 997 | 249 | 511 | 786 | 1124 | 321 |
| Nordson | 0 | 213 | 352 | 484 | 106 | 235 | 393 | 593 | 153 | 316 | 494 | 682 | 166 | 343 | 547 | 760 | 171 | 350 | 566 | 779 | 178 |
| Oil States Intl | 73 | 1188 | 1854 | 2589 | 793 | 1624 | 2433 | 3292 | 793 | 1520 | 1162 | 1664 | 426 | 587 | 885 | 1213 | 246 | 435 | 621 | 792 | 136 |
| Pentair | 136 | 1165 | 1784 | 2378 | 572 | 1207 | 1821 | 3178 | 1219 | 2515 | 3723 | 5001 | 1138 | 2384 | 3392 | 4580 | 970 | 2061 | 3076 | 4268 | 1039 |
| SPX Corp | 286 | 1854 | 2848 | 3939 | 874 | 1784 | 2696 | 3725 | 830 | 1698 | 2438 | 3360 | 772 | 1607 | 2424 | 3361 | 688 | 1500 | 1107 | 1417 | 294 |
| Timken | 92 | 1902 | 2878 | 3783 | 996 | 1983 | 2826 | 3637 | 824 | 1639 | 2454 | 3243 | 790 | 1053 | 1639 | 2183 | 525 | 1043 | 1557 | 2084 | 508 |
| Toro | 53 | 667 | 999 | 1247 | 277 | 733 | 1059 | 1280 | 276 | 735 | 1065 | 1315 | 282 | 765 | 1129 | 1400 | 307 | 852 | 1242 | 1548 | 299 |
| Weg | 0 | 984 | 1402 | 1943 | 539 | 1004 | 1551 | 2094 | 496 | 966 | 1484 | 1964 | 540 | 1118 | 1582 | 2024 | 478 | 1026 | 1260 | 1796 | 492 |

# APPENDIX B: CPP* USING THE PENALTY MODEL

*Table 3. Companies and optimal costs of production per period (CPP* in millions of Dolars), using the penalty model*

| | Period | | | | | | | | | | | | | | | | | | | | |
|---|---|---|---|---|---|---|---|---|---|---|---|---|---|---|---|---|---|---|---|---|---|
| **Company** | **1** | **2** | **3** | **4** | **5** | **6** | **7** | **8** | **9** | **10** | **11** | **12** | **13** | **14** | **15** | **16** | **17** | **18** | **19** | **20** | **21** |
| Agco | 1789 | 3390 | 4948 | 6954 | 2246 | 3858 | 5654 | 7528 | 2172 | 4246 | 6266 | 8204 | 2245 | 3933 | 5547 | 7093 | 1445 | 2950 | 4228 | 5631 | 1547 |
| Applied Materials | 1650 | 3371 | 4882 | 6009 | 1474 | 2755 | 4133 | 5298 | 997 | 2196 | 3365 | 4573 | 1419 | 2682 | 3907 | 5155 | 1474 | 2898 | 4324 | 4915 | 1343 |
| Baker Hughes | 3708 | 7349 | 11260 | 15433 | 4686 | 8687 | 12728 | 17258 | 4425 | 8875 | 13789 | 18477 | 4831 | 9545 | 14647 | 19670 | 4111 | 7649 | 11087 | 14157 | 2530 |
| Bardella | 94 | 182 | 204 | 269 | 52 | 84 | 130 | 171 | 36 | 80 | 123 | 160 | 33 | 80 | 112 | 153 | 28 | 66 | 65 | 104 | 21 |
| Briggs & Stratton | 1124 | 1673 | 396 | 781 | 1191 | 1639 | 330 | 652 | 1050 | 1459 | 330 | 613 | 1027 | 1487 | 369 | 621 | 988 | 1384 | 332 | 588 | 947 |
| Brooks Automation | 252 | 348 | 479 | 75 | 173 | 267 | 342 | 71 | 140 | 223 | 310 | 79 | 155 | 244 | 306 | 85 | 173 | 275 | 340 | 79 | 169 |
| Brunswick | 773 | 1548 | 2247 | 2876 | 764 | 1505 | 2227 | 2769 | 762 | 1474 | 2216 | 2892 | 785 | 1513 | 2096 | 2785 | 754 | 1535 | 2295 | 2636 | 841 |
| Colfax Corp | 124 | 229 | 338 | 430 | 1130 | 1324 | 2031 | 2736 | 667 | 1355 | 2073 | 2894 | 756 | 1636 | 2284 | 3066 | 630 | 1306 | 1990 | 2611 | 630 |
| Cummins | 3128 | 6414 | 9799 | 13305 | 3515 | 6715 | 9581 | 12477 | 3131 | 6425 | 9532 | 12786 | 3489 | 7063 | 10631 | 14393 | 3584 | 7247 | 10682 | 12806 | 3287 |
| Deere & Co | 5210 | 10710 | 15994 | 21602 | 5883 | 11845 | 17922 | 24310 | 6088 | 12427 | 18756 | 25008 | 5815 | 12362 | 18268 | 23547 | 4738 | 10212 | 15168 | 19641 | 4273 |
| Donaldson | 704 | 1109 | 1494 | 396 | 769 | 1188 | 1610 | 407 | 777 | 1164 | 1584 | 393 | 763 | 1179 | 1595 | 413 | 772 | 1169 | 1537 | 381 | 695 |
| Dover | 1341 | 2588 | 3753 | 4853 | 1360 | 2663 | 3986 | 4947 | 1296 | 2647 | 4044 | 5367 | 1062 | 2450 | 3708 | 4734 | 1087 | 2175 | 3274 | 4380 | 1080 |
| Eaton | 2785 | 5643 | 8447 | 11193 | 2832 | 5546 | 8361 | 11996 | 3780 | 7616 | 11486 | 15348 | 4008 | 7885 | 11763 | 15576 | 3589 | 7283 | 10821 | 13379 | 3359 |
| Flowserve | 784 | 1475 | 2137 | 2822 | 849 | 1521 | 2296 | 3027 | 792 | 1570 | 2348 | 3114 | 752 | 1524 | 2226 | 3023 | 806 | 1487 | 2171 | 2694 | 732 |
| FMC Technologies | 893 | 1875 | 2804 | 3956 | 1180 | 2377 | 3465 | 4835 | 1338 | 2693 | 4031 | 5500 | 1407 | 2953 | 4411 | 5975 | 1285 | 2537 | 3658 | 4754 | 922 |
| Graco | 104 | 206 | 295 | 389 | 107 | 245 | 343 | 463 | 127 | 249 | 371 | 498 | 144 | 276 | 415 | 530 | 167 | 311 | 451 | 500 | 153 |
| Idex | 267 | 584 | 802 | 1082 | 286 | 578 | 858 | 1136 | 287 | 570 | 862 | 1145 | 313 | 615 | 901 | 1185 | 282 | 577 | 832 | 1004 | 318 |
| Illinois Tool Works | 3103 | 5854 | 8594 | 11439 | 3019 | 5847 | 8694 | 11280 | 2431 | 5097 | 6199 | 8493 | 2209 | 4384 | 6519 | 8588 | 1977 | 3998 | 5909 | 7821 | 1944 |
| Inds Romi | 48 | 134 | 156 | 233 | 91 | 97 | 149 | 209 | 50 | 79 | 152 | 194 | 61 | 107 | 118 | 156 | 19 | 68 | 64 | 107 | 36 |
| Ingersoll-Rand | 2427 | 4971 | 7955 | 10263 | 2400 | 4936 | 7335 | 9605 | 2350 | 4840 | 7388 | 8394 | 2137 | 4435 | 6752 | 8923 | 2341 | 4603 | 6904 | 8938 | 2249 |
| ITT Industries | 2069 | 4189 | 6317 | 707 | 418 | 790 | 1194 | 1583 | 428 | 823 | 1249 | 1716 | 434 | 927 | 1349 | 1778 | 375 | 811 | 1216 | 1670 | 422 |
| Joy Global | 656 | 1374 | 2317 | 2745 | 997 | 1781 | 2724 | 3655 | 797 | 1609 | 2445 | 3280 | 600 | 1250 | 1962 | 2564 | 588 | 1116 | 1667 | 2157 | 393 |
| Kennametal | 1131 | 1513 | 447 | 829 | 1327 | 1669 | 460 | 850 | 1258 | 1723 | 475 | 1003 | 1429 | 1868 | 483 | 906 | 1362 | 1784 | 378 | 715 | 1136 |
| Lam Research | 1349 | 1781 | 396 | 723 | 1141 | 1838 | 508 | 1081 | 1638 | 2210 | 639 | 1258 | 1964 | 2535 | 722 | 1440 | 2141 | 2675 | 851 | 1640 | 2047 |

*continued on following page*

*Table 3. Continued*

| | Period | | | | | | | | | | | | | | | | | | | | |
|---|---|---|---|---|---|---|---|---|---|---|---|---|---|---|---|---|---|---|---|---|---|
| **Company** | **1** | **2** | **3** | **4** | **5** | **6** | **7** | **8** | **9** | **10** | **11** | **12** | **13** | **14** | **15** | **16** | **17** | **18** | **19** | **20** | **21** |
| Lennox Intl | 690 | 1219 | 1838 | 2412 | 584 | 1229 | 1682 | 2204 | 596 | 1170 | 1757 | 2287 | 639 | 1250 | 1863 | 2426 | 593 | 1240 | 1861 | 2294 | 631 |
| Manitowo Co | 676 | 1330 | 2071 | 2666 | 762 | 1442 | 2185 | 2834 | 772 | 1466 | 2245 | 2892 | 726 | 1370 | 2083 | 2736 | 620 | 1281 | 1865 | 2479 | 246 |
| Metalfrio | 114 | 218 | 246 | 329 | 91 | 150 | 230 | 297 | 89 | 144 | 239 | 307 | 94 | 182 | 221 | 283 | 69 | 142 | 146 | 222 | 85 |
| Middleby | 121 | 242 | 369 | 513 | 155 | 293 | 472 | 642 | 240 | 451 | 658 | 889 | 270 | 488 | 730 | 962 | 270 | 528 | 844 | 886 | 333 |
| Nordson | 110 | 237 | 354 | 489 | 113 | 239 | 413 | 577 | 164 | 320 | 488 | 691 | 167 | 352 | 552 | 732 | 172 | 376 | 553 | 669 | 174 |
| Oil States Intl | 605 | 1252 | 1867 | 2650 | 870 | 1621 | 2503 | 3187 | 771 | 1594 | 809 | 1673 | 421 | 557 | 886 | 1191 | 243 | 435 | 607 | 772 | 131 |
| Pentair | 548 | 1237 | 1757 | 2373 | 603 | 1191 | 2714 | 3147 | 1197 | 2538 | 3712 | 4939 | 1175 | 2392 | 3337 | 4448 | 1011 | 2083 | 3168 | 3668 | 1064 |
| SPX Corp | 900 | 1901 | 2876 | 3951 | 772 | 1772 | 2715 | 3652 | 863 | 1743 | 2390 | 3292 | 816 | 1614 | 2439 | 3292 | 713 | 1475 | 701 | 1386 | 318 |
| Timken | 1010 | 1911 | 2899 | 3816 | 1038 | 1926 | 2804 | 3555 | 787 | 1625 | 2487 | 3203 | 833 | 870 | 1630 | 2145 | 515 | 1037 | 1576 | 2026 | 511 |
| Toro | 292 | 685 | 971 | 1238 | 327 | 712 | 1043 | 1302 | 363 | 705 | 1012 | 1298 | 347 | 761 | 1120 | 1380 | 395 | 827 | 1251 | 1329 | 391 |
| Weg | 521 | 1166 | 1397 | 1883 | 563 | 981 | 1518 | 2049 | 518 | 937 | 1527 | 1919 | 565 | 1169 | 1564 | 1969 | 418 | 1056 | 1192 | 1700 | 500 |

# APPENDIX C: CEF* USING THE PENALTY MODEL

*Table 4. Companies and optimal costs of final inventory (CEF* in millions of Dolars) per period, using the penalty model*

| | Period | | | | | | | | | | | | | | | | | | | | |
|---|---|---|---|---|---|---|---|---|---|---|---|---|---|---|---|---|---|---|---|---|---|
| **Company** | **1** | **2** | **3** | **4** | **5** | **6** | **7** | **8** | **9** | **10** | **11** | **12** | **13** | **14** | **15** | **16** | **17** | **18** | **19** | **20** | **21** |
| Agco | 1581 | 1658 | 1603 | 1560 | 2025 | 2023 | 2014 | 1703 | 2005 | 2043 | 2182 | 1989 | 2415 | 2411 | 2288 | 1723 | 1813 | 1789 | 1672 | 1396 | 1698 |
| Applied Materials | 1647 | 1794 | 1849 | 1701 | 1772 | 1594 | 1380 | 1272 | 1278 | 1318 | 1358 | 1413 | 1533 | 1564 | 1547 | 1472 | 1545 | 1617 | 1643 | 849 | 848 |
| Baker Hughes | 2805 | 2939 | 3053 | 3222 | 3643 | 3811 | 3879 | 3781 | 3880 | 3838 | 3960 | 3884 | 3995 | 4075 | 4150 | 4074 | 3843 | 3535 | 3262 | 2917 | 2789 |
| Bardella | 84 | 89 | 72 | 65 | 67 | 62 | 63 | 57 | 50 | 46 | 46 | 45 | 43 | 41 | 34 | 37 | 34 | 38 | 27 | 30 | 35 |
| Briggs & Stratton | 466 | 427 | 492 | 568 | 480 | 434 | 503 | 537 | 463 | 408 | 469 | 475 | 395 | 376 | 507 | 540 | 447 | 319 | 414 | 447 | 361 |
| Brooks Automation | 127 | 94 | 108 | 103 | 105 | 107 | 103 | 105 | 97 | 93 | 98 | 97 | 94 | 103 | 94 | 95 | 91 | 95 | 72 | 71 | 72 |
| Brunswick | 551 | 527 | 529 | 533 | 558 | 535 | 581 | 576 | 605 | 549 | 580 | 599 | 680 | 667 | 669 | 652 | 680 | 670 | 710 | 354 | 404 |
| Colfax Corp | 77 | 79 | 79 | 56 | 555 | 530 | 519 | 494 | 505 | 467 | 453 | 446 | 473 | 569 | 522 | 443 | 456 | 447 | 450 | 346 | 379 |
| Cummins | 2202 | 2275 | 2295 | 2141 | 2382 | 2581 | 2570 | 2221 | 2387 | 2475 | 2513 | 2381 | 2580 | 2745 | 2833 | 2866 | 2936 | 2986 | 3059 | 1695 | 1738 |
| Deere & Co | 4178 | 4687 | 4688 | 4371 | 5678 | 6112 | 5868 | 5170 | 6243 | 6173 | 5594 | 4935 | 5555 | 5850 | 5439 | 4210 | 4527 | 4624 | 4319 | 3817 | 4250 |
| Donaldson | 231 | 258 | 271 | 274 | 270 | 265 | 256 | 272 | 262 | 240 | 235 | 243 | 241 | 254 | 253 | 279 | 272 | 286 | 260 | 281 | 270 |
| Dover | 845 | 883 | 849 | 803 | 880 | 921 | 923 | 873 | 906 | 917 | 950 | 927 | 840 | 891 | 908 | 864 | 862 | 844 | 811 | 803 | 850 |
| Eaton | 1667 | 1766 | 1769 | 1701 | 1779 | 1756 | 1801 | 2349 | 2394 | 2405 | 2403 | 2382 | 2532 | 2534 | 2498 | 2428 | 2424 | 2439 | 2395 | 1482 | 1551 |
| Flowserve | 1021 | 1091 | 1077 | 902 | 1035 | 1042 | 1049 | 906 | 976 | 1005 | 1005 | 855 | 917 | 956 | 914 | 770 | 888 | 900 | 887 | 506 | 599 |
| FMC Technologies | 617 | 695 | 723 | 712 | 788 | 881 | 963 | 965 | 996 | 1032 | 1051 | 980 | 984 | 1026 | 1046 | 1021 | 1013 | 970 | 875 | 734 | 709 |
| Graco | 103 | 113 | 112 | 105 | 110 | 124 | 121 | 122 | 130 | 133 | 133 | 134 | 147 | 147 | 149 | 124 | 147 | 159 | 161 | 60 | 70 |
| Idex | 216 | 282 | 272 | 254 | 254 | 255 | 250 | 235 | 240 | 232 | 237 | 231 | 244 | 254 | 251 | 238 | 244 | 261 | 253 | 140 | 178 |
| Illinois Tool Works | 1824 | 1920 | 1794 | 1716 | 1824 | 1796 | 1760 | 1585 | 1514 | 1490 | 1308 | 1247 | 1298 | 1305 | 1265 | 1180 | 1187 | 1191 | 1153 | 1086 | 1134 |
| Inds Romi | 148 | 143 | 116 | 101 | 124 | 119 | 107 | 80 | 80 | 64 | 63 | 55 | 68 | 79 | 57 | 31 | 19 | 27 | 14 | 0 | 6 |
| Ingersoll-Rand | 1578 | 1591 | 1513 | 1282 | 1433 | 1475 | 1462 | 1309 | 1466 | 1435 | 1448 | 1166 | 1348 | 1389 | 1419 | 1359 | 1613 | 1635 | 1578 | 1213 | 1414 |
| ITT Industries | 950 | 980 | 1011 | 254 | 268 | 260 | 269 | 304 | 315 | 309 | 297 | 316 | 290 | 309 | 313 | 302 | 288 | 295 | 300 | 293 | 301 |
| Joy Global | 837 | 937 | 1240 | 1086 | 1308 | 1288 | 1296 | 1167 | 1191 | 1118 | 999 | 881 | 870 | 865 | 949 | 840 | 904 | 923 | 913 | 717 | 654 |
| Kennametal | 466 | 461 | 500 | 511 | 571 | 496 | 531 | 526 | 483 | 461 | 513 | 611 | 620 | 548 | 553 | 505 | 475 | 414 | 384 | 313 | 323 |

*continued on following page*

*Table 4. Continued*

| | Period | | | | | | | | | | | | | | | | | | | | |
|---|---|---|---|---|---|---|---|---|---|---|---|---|---|---|---|---|---|---|---|---|---|
| **Company** | **1** | **2** | **3** | **4** | **5** | **6** | **7** | **8** | **9** | **10** | **11** | **12** | **13** | **14** | **15** | **16** | **17** | **18** | **19** | **20** | **21** |
| Lam Research | 356 | 397 | 396 | 373 | 376 | 633 | 568 | 530 | 545 | 559 | 615 | 662 | 717 | 653 | 728 | 825 | 831 | 536 | 513 | 473 | 100 |
| Lennox Intl | 453 | 460 | 394 | 337 | 402 | 427 | 398 | 375 | 464 | 469 | 430 | 379 | 490 | 515 | 502 | 463 | 533 | 541 | 488 | 260 | 358 |
| Manitowo Co | 682 | 735 | 817 | 669 | 777 | 809 | 866 | 708 | 802 | 816 | 855 | 721 | 824 | 831 | 809 | 645 | 695 | 743 | 718 | 594 | 494 |
| Metalfrio | 85 | 77 | 61 | 54 | 67 | 60 | 62 | 59 | 86 | 73 | 75 | 80 | 93 | 95 | 76 | 67 | 74 | 68 | 46 | 42 | 59 |
| Middleby | 116 | 122 | 124 | 124 | 138 | 132 | 147 | 153 | 187 | 204 | 210 | 220 | 260 | 260 | 260 | 226 | 246 | 262 | 321 | 86 | 99 |
| Nordson | 123 | 134 | 138 | 142 | 148 | 155 | 179 | 170 | 183 | 189 | 184 | 198 | 201 | 206 | 210 | 183 | 185 | 212 | 201 | 93 | 90 |
| Oil States Intl | 532 | 593 | 603 | 654 | 728 | 734 | 807 | 702 | 680 | 699 | 257 | 267 | 271 | 248 | 247 | 232 | 238 | 241 | 226 | 213 | 216 |
| Pentair | 412 | 485 | 460 | 450 | 475 | 460 | 1380 | 1380 | 1327 | 1318 | 1296 | 1229 | 1257 | 1250 | 1185 | 1057 | 1103 | 1126 | 1222 | 627 | 651 |
| SPX Corp | 615 | 666 | 695 | 720 | 628 | 613 | 629 | 556 | 586 | 624 | 570 | 502 | 548 | 551 | 564 | 498 | 525 | 547 | 212 | 171 | 189 |
| Timken | 918 | 929 | 949 | 964 | 993 | 943 | 928 | 862 | 834 | 819 | 856 | 810 | 830 | 625 | 618 | 586 | 580 | 574 | 595 | 543 | 551 |
| Toro | 240 | 260 | 232 | 223 | 272 | 251 | 235 | 251 | 336 | 310 | 259 | 240 | 305 | 302 | 294 | 275 | 364 | 341 | 350 | 127 | 219 |
| Weg | 626 | 696 | 685 | 630 | 656 | 627 | 590 | 538 | 554 | 519 | 560 | 519 | 547 | 600 | 583 | 534 | 486 | 520 | 455 | 363 | 373 |

# Chapter 15
# Advances in Analog Integrated Circuit Optimization:
## A Survey

**Prakash Kumar Rout**
*Silicon Institute of Technology, India*

**Debiprasad Priyabrata Acharya**
*NIT Rourkela, India*

**Umakanta Nanda**
*Silicon Institute of Technology, India*

## ABSTRACT

*In a system though the analog circuits occupy very less space but they require far more design time than the digital circuits. This is due to the fact that the number of performance measures of an analog circuit is more than those for digital circuits. Predicting and improving the performance, robustness and overall cost of such systems is a major concern in the process of automation. In the automation process, optimization of performances subjected to a verity of environmental constraints is a central task. In this chapter, efficient analog circuit sizing techniques and their optimization are surveyed.*

## INTRODUCTION

Advances in semiconductor manufacturing technology have resulted in ultra large scale integration (ULSI) of circuits. The complex system-on-chip contains mixed digital and analog circuits. Although analog circuits occupy a small fraction of silicon area but it is highly difficult to design these circuits due to their complexity, noise sensitivity and performance tradeoffs. It is worth noting that the real world is analog and the analog signals need to be processed in integrated circuits (IC). Whatever may be the advancements in digital IC designs the performance of the system is always dictated by the analog part of the integrated circuit. Without automation and optimization the analog IC design suffers from long design time, high complexity, high cost and suboptimal performance. It's no wonder that building ef-

DOI: 10.4018/978-1-5225-2944-6.ch015

ficient analog integrated circuits is said to involve some amount of black magic. Because of this analog design require skilled craftsmen who are in short supply. Since the average analog circuit takes longer to implement than its usually much larger digital counterpart. Problems multiply if the analog design is destined to be a block on a mixed-signal or system chip. Hence there have been great efforts not only for design automation but for performance optimization too. When automation helps in handling the design complexity, optimization helps to attain near-best performance in a very less time which can be accomplished by acceptably moderate skilled designers. Here optimization techniques and their applications to analog integrated circuits are reviewed.

The major building blocks of the analog circuits are operational amplifiers, filters, oscillators, low noise amplifiers, power amplifiers and current and voltage sources. The optimization techniques are generally applied to these circuits to estimate their design parameters for obtaining best performance. Many researchers designate this process as circuit sizing. This has two major purposes: first it replaces cumbersome and adhoc manual tradeoffs by automatic evaluation of design parameters, second, it solves problems which are difficult for hand design. Moreover the optimization algorithms also take into account the constraints in the design space.

Section 2 provides a bird's view on the analog IC design flow and scope for various optimizations therein. The conventional and classical technique based analog IC optimizations are discussed in section 3. The use of evolutionary techniques for the analog IC optimization are presented in section 4. Finally section 5 provides the summary and concluding remarks.

## ANALOG IC DESIGN AND SCOPE FOR OPTIMIZATION

### The Complexity of the Analog IC Design

The performance of analog integrated circuits is very much detrimental in the system performance. The performance specification contains requirements on the various performance metrics of the circuit. Here the performance metric are measures of properties that are used to characterize the behavior of an analog cell. For example an amplifier is characterized by gain, speed, power consumption, linearity and the like. All these performance metrics are very often competing in nature and hence present challenging tradeoffs in the design. This is represented as the analog design octagon which is illustrated in Figure1 (B. Razavi, 2010).

### The Analog IC Design Process

The analog design starts with the specifications and the functionality to be implemented which is mapped onto an architectural description for the design. In this process the decomposition of the required function is carried out until we arrive at easily manageable analog building modules or blocks, usually called as cells.

High level models are used to perform simulations to validate the functionality of the concept. The specifications on the low-level modules or cells are extracted from these simulations. The cells are realized by designing the low-level building blocks which comply with the performance requirements. After the physical design of all the required cells the analog system is assembled. The assembled system layout is released for fabrication. The post fabrication testing and verification confirms the release of the product prototype for field deployment. The analog IC design flow is depicted in Figure 2.

*Figure 1.*

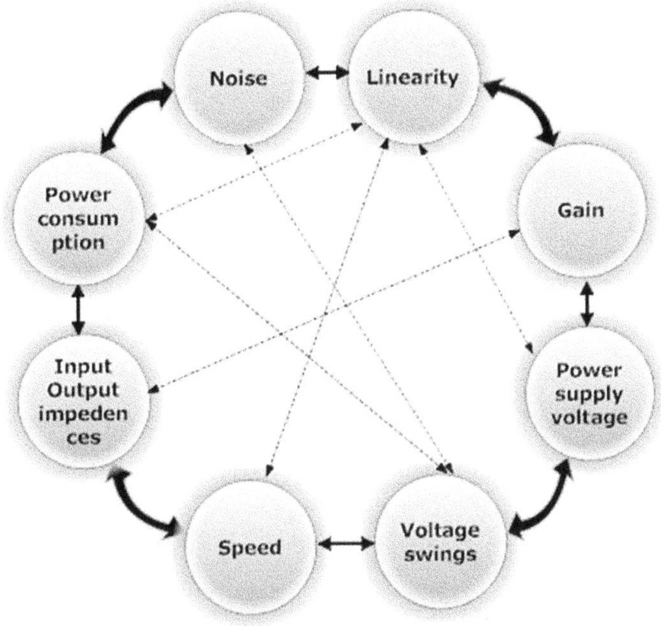


*Figure 2.*

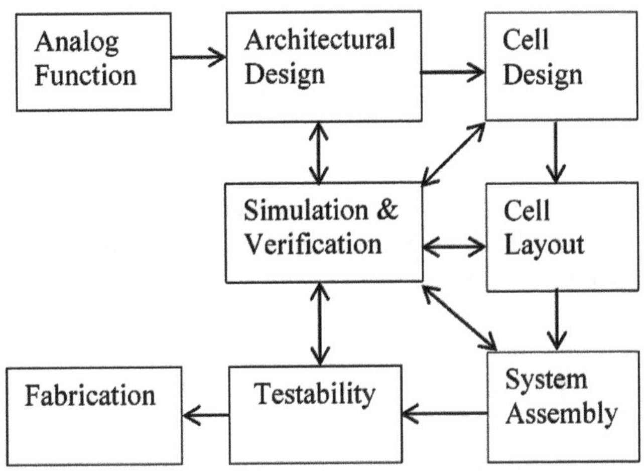


The design of an analog cell first involves the choice of possible topology to implement the functionality in an efficient way. The next step design process (Figure 3) is the circuit sizing. Generally an analog circuit has many real-valued parameters which must be set to meet its specifications. The process of setting these parameters is called circuit sizing. For instance, a two-stage opamp has around twelve parameters including the width and length of all transistors and passive component values which have to be set to achieve the specifications such as gain, bandwidth, power, area, noise, CMMR (common-mode-rejection-ratio), offset, settling time, slew rate and power supply rejection ratio. The simulation experiments are carried out iteratively till the specifications are met. With these circuit parameter val-

*Figure 3.*

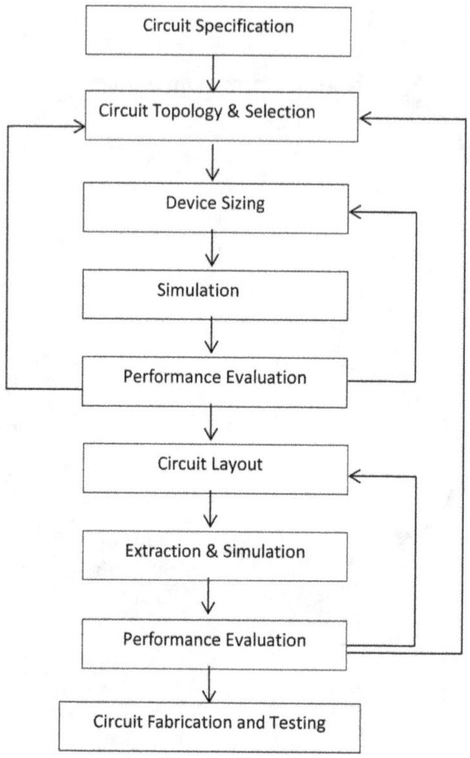


ues the physical layout of the circuit is designed and the circuit parasitics are extracted. The simulation studies are performed by taking into account these extracted parasitics and the performance indices are evaluated. The circuit layout is iteratively redrawn till acceptable performance values are obtained.

Throughout the design process many simulations and validation steps are required. If the circuit fails to meet the specification at some level, the preceeding design steps must be revised. This may include backtracking several steps in the design process.

## SCOPE FOR OPTIMIZATION

A close look at the analog design hexagon reveals that it is highly desirable for the multiple performance objectives to be simultaneously maximized or minimized. In analog and mixed-signal systems very often one uses a single objective function which is a weighted combination of all objectives or the multi objective method.

### Circuit Sizing

In the recent past, circuit sizing has been projected as an optimization problem. This optimization problem has two dimensions: modeling the design problem as an optimization problem and solving the modeled problem. These steps are interdependent and impact each other, for instance, the model of the problem will decide the optimization method that can be used.

The most accurate performance of integrated circuits is measured with the chip fabricated on silicon. Due to the non-availability of the chip, during the design process, the designer uses a simulation engine which models the characteristics of the silicon elements and runs computational algorithms to estimate the performance of the IC. SPICE is the industry standard widely accepted simulator. Hence the correctness of SPICE is the final validation-point in the circuit sizing problem. The highest precision model for optimization uses SPICE as a black-box estimator to which one gives the circuit parameters and gets the IC performance specifications.

Besides this, the designer has access to the governing equations describing the IC behavior. These equations are derived from nodal current and voltage expressions in the circuit with certain assumptions on the transistor behavior. The circuit performances are derived from these equations much faster than SPICE. But their accuracy is less than the SPICE, due to the approximated transistor behavior and approximations made in circuit analysis.

The above two models i.e. SPICE based and equation based models evaluate the IC performance specifications.

## Physical Layout Design

In the design process soon after the circuit sizing the physical layout of the IC is designed. This involves drawing of the geometrical structures of all circuit elements like transistor, diode, resistor, capacitor and inductor and the metal layers for interconnects. For a given requirement there is a possibility of many geometries from which the geometry offering the optimal performance needs to be selected. In addition to this, the placement of the components and the routing of their interconnects has immense requirements for optimization.

In the layout design phase the effect of actual circuit parasitics surfaces. These parasitics have a significant impact on the performance which needs to be included in the process of all optimizations. This is popularly known as parasitic aware optimization. The basic principle behind the parasitic-aware optimization technique is that device and package parasitics are considered as a natural part of the design process from the beginning of the overall design cycle. When all parasitic effects are taken into account, the complete circuit becomes highly complicated for hand analysis, even with help of circuit simulators, so finding the optimum solution is nearly impossible.

## Process Variations

The integrated circuits should be designed in such a way that the manufactured ICs must meet the performance specifications under all operating conditions. The random fluctuations in the fabrication process results in the deviation of the performance. Besides this the variations in the operating conditions like supply voltage and temperature also affect the IC performance. These performance deviations reduce the yield significantly and hence the chip unit cost increases. Therefore one of the major design objectives is to minimize the impact of process variations on the chip performance. This calls for a process variation tolerant IC design methodology.

## OPTIMIZATION METHODOLOGIES

Apart from the performance specification measurement during simulation, the use of optimization engine and setting up the optimization problem also has a great impact on the final IC performance. Optimization is concerned with the finding of minima and maxima of functions, subject to many process and other real constraints. There is no single method available for solving all optimization problems efficiently. Hence a number of optimization methods have been developed for solving different types of optimization problems.

To solve problems, people use algorithms that terminate in a finite number of steps, or iterative methods that converge to a solution (on some specified class of problems), or heuristics that may provide approximate solutions to some problems (although their iterates need not converge).The popular optimization techniques are direct search method, Newton's method, conjugate gradient method, gradient descent method, simplex method (J. Nelder and R. Mead, 1965), neural networks (S. Haykin) and the like. The gradient based techniques most often get trapped in local optima. Apart from this the non-quadratic non-differentiable functions find difficulty in the above mentioned techniques. In such cases heuristic algorithms like Simulated Annealing (S. Kirkpatrick, C. Gelatt and M. Vecchi, 1983) (R.A. Rutenbar, 1989), Genetic Algorithms (D. Goldberg, 1989), Particle Swarm Optimization (J. Kennedy and R. Eberhart, 1995), Differential Evolution (Kenneth Price, Rainer M. Storn and Jouni A. Lampinen, 2005), Artificial Bee Colony Optimization (D. Karaboga, 2005), Bacteria Foraging Optimization (K. M. Passino, 2002)and many hybrids of these are capable of approximate global optimal solutions. These heuristics are also known as the Evolutionary Algorithms.

Optimization problems are many times multi-modal i.e they possess multiple good solutions. They could all be globally good (same cost function value) or there could be a mix of globally good and locally good solutions. Obtaining all (or at least some of) the multiple solutions is the goal of a multi-modal optimizer.

Classical optimization techniques due to their iterative approach do not perform satisfactorily when they are used to obtain multiple solutions, since it is not guaranteed that different solutions will be obtained even with different starting points in multiple runs of the algorithm. Evolutionary Algorithms are however very popular approaches to obtain multiple solutions in a multi-modal optimization task.

There could be many variations to this such as the optimization problem can have multiple objectives and multiple constraints; one objective and multiple constraints; a series of optimization problems with one objective and multiple constraints.

The optimization complexity is increased when more than one objective is added to the problem. The set of trade-off designs that cannot be improved upon according to one criterion without hurting another criterion is known as the Pareto set. The curve generated by plotting the competing objectives of the best designs is known as the Pareto frontier. A design is judged to be "Pareto optimal" if it is not dominated by any other design. If it is worse than another design in some respects and no better in any respect, then it is dominated and is not Pareto optimal. The choice among Pareto optimal solutions to determine the "favorite solution" is left with the decision maker or designer.

### Conventional Techniques for Analog IC Optimization

The conventional methods used in the design optimization of analog integrated circuits include local unconstrained optimization, constrained optimization, stochastic optimization and simulated annealing.

Many algorithms have been developed to estimate the optimal value of the objective functions. Simplex method is used in analog design in (N. S. Nagaraj, 1993) where a new multiple criteria constrained performance optimizer for analog integrated circuits, based on non-linear programming and heuristic techniques is presented. A modified Parkinson Hutchinson Simplex algorithm and Guided Random Search technique are used to perform optimization based only on cost function evaluations. These optimization techniques are combined to combat numerical difficulties faced during circuit simulation and gradient based optimization. The results are well verified with real life circuits at Texas Instruments Inc. to ascertain the consistent improvements in circuit performance. One of the popular tools OPASYN (H.Y. Koh, C.H. Sequin and P.R. Gray, 1990) uses the steepest descent algorithm for optimization of basic two stage operational amplifier, folded cascade operational amplifier and can be extended to any type of analog circuit. The parametric optimization proceeds by developing analytic circuit models, reduction of the dimensionality and size of the search domain, by defining a minimal set of independent design parameters and set reasonable upper and lower bounds on their range. Due to smoothness of the resulting search spaces simple numerical optimization algorithms can be used effectively. The steepest descent optimization algorithm used is simple and efficient but needs a differentiable search space with a continuous first-derivative. Apart from this the tool performs layout generation too. OAC (H. Onodera, H. Kanbara, K. Tamaru, 1990) use gradient based methods for opamp compilation with performance optimization. A parametric optimization consisting of several interactive improvement steps based on circuit simulation and gradient evaluation is given in (J. P. Harvey, M. I. Elmasry and B. Leung, 1992). DELIGHT. SPICE (W. Nye, D.C. Riley, A. Sangiovanni-Vincentelli and A.L. Tits, 1988) is the combination of the DELIGHT interactive optimization based computer aided design system and the SPICE circuit analysis program. Using the DELIGHT.SPICE tool, circuit designers can employ recent powerful optimization algorithms and methodology that emphasizes designer intuition and man-machine interaction in a manner in which designer and computer are complementary to automatically adjust parameters of electronic circuits to improve their performance. They may optimize any performance objective and also study complex tradeoffs between multiple competing objectives, simultaneously satisfying multiple constraint specifications. Jiffy Tune (A.R. Conn, P.K. Coulman, R.A. Haring, G.L. Morrill, C. Visweswariah and C.W. Wu, 1995) is a gradient based approach for circuit optimization. A set theoretic approach for robust design of analog circuits is presented in (O. Altun and M. Bocko, 2006). In the AMGIE system (G.V. Plas, G. Debyser, Francky Leyn, K. Lampaert, J. Vandenbussche, G. G. E. Gielen, W. Sansen, P. Veselinovic, and D. Leenaerts, 2011), there is a provision for the user to select the optimization algorithm as one of the options to be chosen in the specification sheet window. Global-optimization algorithms, like very fast simulated re-annealing (VFSR), and local-optimization algorithms, like Hooke-Jeeves, min-max, or sequential quadratic programming (SQP), can be chosen here. After the sizing optimization in the AMGIE system, the resulting optimal device sizes are automatically back annotated onto the schematic of the circuit under design.

BLADES (F. El-Turky and E. Perry, 1989) is a prototype design environment which uses a divide and conquer method, is capable of designing a wide range of sub-circuit functional blocks as well as a limited class of integrated bipolar operational amplifiers. This is believed to be the first successful design expert system in the analog design domain. It uses different levels of abstraction depending on the complexity of the design task under consideration. The importance of the abstraction level lies in the fact that once design primitives are defined, the problem of extracting the knowledge (design rules) becomes less complex. All circuits designed and tested using BLADES are observed to be stable.

In (K. Matsukawa, T. Morie, Y. Tokunaga, S. Sakiyama, Y. Mitani, M. Takayama, T. Miki, A. Matsumoto, K. Obata and S. Dosho, 2009) convex optimization procedure is used to construct optimization environments for pipelined and delta sigma analog to digital converters. Methods of analog and radio frequency integrated circuit design using optimization with recourse including ellipsoidal uncertainty are provided in (X. Li, Y. Xu, K.L. Hsiung, L. Pileggi and S. Boyd, 2009). Geometric programming is used for device-circuit co-optimization of mixed mode circuit designs in (J. Kim, R. Jhaveri, J. Woo and C. Yang, 2007).The circuit sizing is again modeled as a geometric program (T. Soorapanth, 2007) where op-amp design is implemented in C language using GPGLP library for GP solver (GPGLP Library: ftp//ftp.pitt.edu/dept/ie/GP). The design objectives here are maximization of unity-gain bandwidth, DC gain and minimization of input referred noise and power consumption. In (F. Yang and M. Gan, 2007) improved sigma delta data converter is calibrated through convex optimization. Using the promising methodology of geometric programming and formulation of circuit problems in posynomial form tools like GPCAD (M. Hershenson, S.Boyd and T.Lee, 1998) are developed. Geometric Programming is successfully applied for two stage operational amplifier sizing by Mandal and Viswanathan in (P. Mandal and V. Visvanathan, 2001) where the opamp design is formulated as a sequence of convex programming problems. In this novel work the use of accurate model makes the sizing technique robust. Iterative Sequential Geometric Programming (ISGP) (S. Kundu and P. Mandal, 2014) is used for robust analog circuit sizing.

A sequential quadratic programming technique is used to solve the nonlinear analog circuit optimization problem (E. Hjalmarson, R. Hagglund and L. Wanhammar, 2003). To ensure a good solution the optimization is restarted with different initial values. Here a current mirror operational transconductance amplifier is taken as the design example.

MARS (M. Eick and H. Graeb, 2012) presents a novel approach for automatic computation of matching constraints for analog circuit sizing. It uses a min-max principle for feasibility, nominal and yield optimization. It can be applied with any available optimization method for sizing. This approach firstly detects automatic matching conditions for sizing in analog circuit using a symmetry computation.

Simulated annealing estimates optimal dimensions without the derivatives and hence has been successfully applied to size general analog circuits in VCOs (C.R.C. De Ranter, G. Van der Plas, M.S.J. Steyaert, G.G.E. Gielen and W.M.C. Sansen, 2002), sigma delta modulators (J. Ruiz-Amaya, J. de la Rosa, F.V. Fernandez, F. Medeiro, R. del Rio, B. Perez-Verdu, and A. Rodriguez-Vazquez, 2005), radio frequency receivers (J. Crols, S. Donnay, M. Steyaert and G. Gielen, 1995) and operational amplifiers (A. Torralba, J. Chavez, L.G. Franquelo, 1996). An automatic synthesis tool which uses simulated annealing as its optimizer, for a cascade low noise amplifier (LNA) is proposed in (G. Tulunay and S. Balkir,, 2008).

## Evolutionary Techniques for Analog IC Optimization

Evolutionary techniques are search algorithms that operate by evolving a population of solutions through repeated transformations. These are used to solve big size problems with multiple criteria. Though they do not guarantee to arrive at an optimal solution in an exact way but provide an acceptable approximation in an affordable computing time. Kruiskamp and Leenarts (W. Kruiskamp and D. Leenaerts, 1995) developed DARWIN where GA is used for topology selection and circuit sizing of CMOS operational amplifier. The GA is used in (G. Gielen and R. A. Rutenbar, 2000) for automatic analog synthesis and in (N. Paulino, J. Goes, A. Steiger-Garcao, 2001) for optimization of analog building blocks. The design automation environment GENOM (M. Barros, J. Guilherme and N. Horta, 2010) has been developed by combining optimization algorithm GA along with a supervised learning strategy based on support vector machine (SVM) to create feasibility models in order to reduce the overall number of evaluations. Optimization of a nano-CMOS voltage controlled oscillator using polynomial regression and genetic algorithm is reported in (D. Ghai, S. Mohanty, G. Thakral, 2013). A novel methodology for generation of performance models for sizing of analog high level topology is presented in (S. Pandit, C. Mandal and A. Patra, 2011) where optimal values of the model hyper parameters are determined through a grid search-based technique and a genetic algorithm- (GA) based technique. The high-level models of the individual component blocks are combined analytically to construct the high-level model of a complete system. The accuracy, fastness, genericness and less model construction time are the novelties of the method.

Tawdross and Konig (P. Tawdross and A. Koning, 2005) introduced particle swarm optimization (PSO) in place of GA for field programmable analog scalable device array reconfiguration. An operational amplifier with design constraints was designed using PSO taking into account the external constraints in the above work which was further extended for a three bit flash ADC (P. Tawdross, and A. König, 2006). Current conveyor circuits are optimized in (Y. Cooren, M. Fakhfakh, M. Loulou and P. Siarry, 2007) using particle swarm optimization(PSO). Thakker et. al. (R. A. Thakker, M. S. Baghini and M. B. Patil, 2009)designed low power low voltage analog circuit applying hierarchical PSO. PSO also finds application for analog circuit sizing in (M. Fakhfakh, Y. Cooren, A. Sallem, M. Loulou and P. Siarry, 2010) and (R. Vural and T. Yildirim, 2012). M.Barari et.al. (M. Barari, H. R. Karimi and F. Razaghian, 2014) combined GA with PSO for optimization of operational amplifier circuit.

Differential evolution is a population based evolutionary computation technique which uses a simple differential operator for new candidate solution creation and one-to-one competition scheme for greedy selection of new candidates. B.Liu et.al. (B. Liu, Y. Wang, Z. Yu, L. Liu, M. Li, Z. Wan, J. Lu and F. Fernandez, 2009) proposed competitive co-evolutionary differential evolution for automated sizing of analog integrated circuits with practical user defined specifications. In another work (A. F. Sheta, 2010) analog filter is designed using differential evolution method.

Artificial Bee Colony (ABC) optimization is applied to nano-CMOS phase locked loop(PLL) (O. Garitselov, S. Mohanty, E. Kougianos and Priyadarsan Patra, 2011). Ali Jafari et.al. (A. Jafari, E. Bijami, H. Bana and S. Sadri, 2011) proposed a new hybrid shuffled frog leaping (NHSFL) algorithm to deal with the constraints and obtain the device sizes optimizing the performance of the circuits.

A simulation-based analog circuit synthesis methodology is proposed and validated in (O. Sonmez and G. Dundar, 2011) which optimizes both the simulator and the search algorithm. It uses an accelerated simulator, SPASE, and a modified version of self-adaptive evolutionary strategies for quiker convergence of the algorithm.The performances of genetic algorithm, artificial bee colony optimization and particle swarm optimization in analog active filter design and optimization are evaluated in (R. Vural, T. Yildirim, T. Kadioglu and A. Basargan, 2012) by applying each algorithm to realize two different filter structures.Multi-objective analog circuit design methodology proposed in (K.Ghali, L. Dorie, O. Hammami, 2005) is used in prototyping system on reconfigurable platforms like Field Programmable Analog Arrays (FPAAs) and Field Programmable Gate Arrays (FPGAs) where lot of gains in total design time are achieved compared with simulation based methodologies.The simulation-based analog circuit synthesis tool ANACONDA (R. Phelps, M. Krasnicki, R. Rutenbar, L. R. Carley and J. R. Hellums, 2000) illustrated in Figure 4, combines the population of solutions from evolutionary algorithms with a variant of stochastic pattern search to synthesize a circuit using the same industrial-strength simulation environment created to validate the circuit.

*Figure 4.*

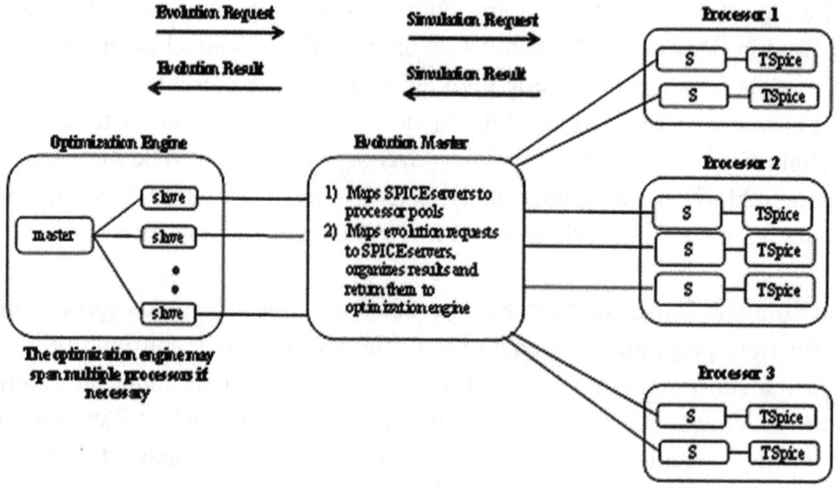


In one of the related work (O. Okobiah, S. Mohanty and E. Kougianos, 2013), a novel optimization methodology incorporating a geostatistical inspired metamodelling technique and a Gravitational Search Algorithm for analog and mixed signal circuit and system design is presented. This proposed methodology is used in the design optimization of a 45 nm CMOS-based thermal sensor. Two nature inspired metaheuristics, differential evolution (DE) and harmony search (HS) algorithms are utilized (R. Vural, U. Bozkurt and T. Yildirim, 2013) for optimal filter design of different topologies and manufacturing series. The feasible solutions provided by a multi-objective evolutionary algorithm (MOEA) in the optimal sizing of analog integrated circuits (ICs) can be very sensitive to process variations. To choose low sensitive optimal MOSFET sizes multi-parameter sensitivity analysis is carried out. The multi-parameter sensitivity analysis verified through the optimization of a recycled folded cascode (RFC) operational transconductance amplifier (OTA) (I. Gomez, E. Cuautle and Luis G. Fraga, 2013) show that the optimal sizes, selected after executing the sensitivity approach, guarantee the lowest sensitivities values while improving the performances of the RFC OTA.

## Parasitic Aware Analog IC Optimization

The performance degradations due to device and package parasitic components are counter acted by use of parasitic aware synthesis for achieving optimum performance. The benefits gained from optimization of RF circuits design by considering chip and package parasitic as an integral part of the design process are demonstrated in (R.Gupta and D.J.Allstot, 1998). In (Brian M. Ballweber, Ravi Gupta, and David J. Allstot, 2000), a 0.6um digital-CMOS technology with three metal layers is considered to design a 0.5-5.5GHz distributed amplifier by considering the package parasitics and using the on-chip inductors as the basis for the delay lines. In this design the parasitic-laden on-chip inductors are considered as an integral part of design from the beginning by using a parasitic-aware optimization methodology based on the simulated annealing technique. The classic details of the parasitic aware optimization is illustrated in (D.J.Allstot, K.Choi and J.Park, 2003) (M. Chu, D. Allstot, J. Huard and K. Wong, 2004). The methodology comprises three major modules linked via a netlist: an optimization core, a parasitic-aware compact model generator, and a standard circuit simulator (Figure 5). The optimization core estimates the design variables in the netlist according to an optimization algorithm. The netlist is simultaneously updated with information from the compact model generator. The parasitic-laden netlist is then simulated by any user-specified circuit simulator like HSPICE or SPECTRE. After the simulation, the outputs are fed back to the optimization core for evaluation and generation of the new netlist variables. The optimization core is the most critical component in parasitic-aware synthesis. Simulated annealing (SA) and particle swarm optimization (PSO) algorithms have been used to implement the core optimizer in (B. Liu, Y. Wang, Z. Yu, L. Liu, M. Li, Z. Wan, J. Lu and F. Fernandez, 2009) (A. F. Sheta, 2010).

## NSGA-II Based Fast Design of CMOS ICs for Performance Optimization

In the nano-scale technology, it is highly desired to design these circuits in high precision in radio frequency operating range offering very low noise and low power consumption. The parasitic components in radio frequency integrated circuits (RFIC) have a significant effect on the device performances. Circuit parasitics affect the speed, power consumption, area and many other performances. In high performance integrated circuits (IC) it is very much needed to consider the parasitic effects during the design phase of the circuit. Inclusion of parasitic components makes, the complete circuit too much complex for hand analysis. Hence finding design parameters manually for optimal circuit performance by the designer is very difficult. The complexity of the problem is further elevated when there is a necessity to optimize multiple competing performance objectives. The complex design landscape not only makes it difficult to arrive at an optimum performance but also consume lot of designer's time to have the first prototype.

In the conventional parasitic cognizant optimization a parasitic aware model library is provided to the netlist on which single objective optimization is carried out. The extracted parasitics for a circuit is provided to the circuit netlist which is subjected to single objective metaheuristic optimization like simulated annealing (R.A. Rutenbar, 1989) or global optimization like swarm intelligence. In most of these cases the parasitics are generated from a circuit initially which may not be yielding optimal performance and hence the parasitics that are considered for further circuit optimization may not be proper in their values. Hence just a parasitic aware optimization of circuits as has been reported in many

*Figure 5.*

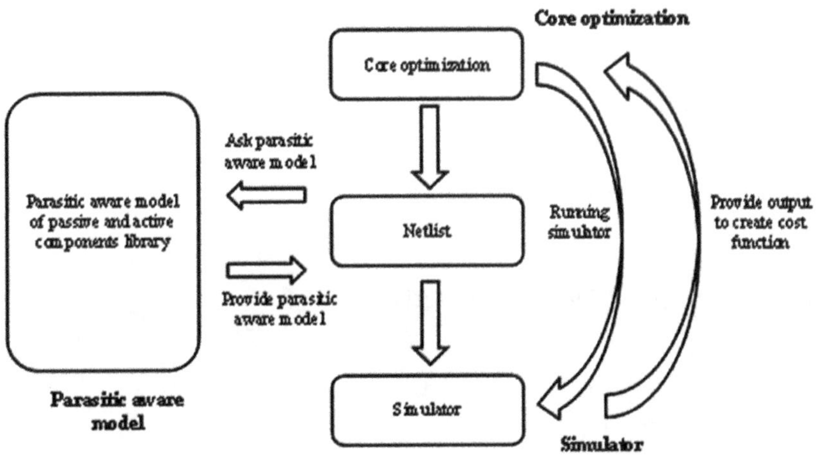

cases may not yield performance close to the optimum value. To circumvent this problem the parasitics should be extracted from an initially optimized circuit so that the parasitic values are more realistic for consideration in further optimization.

Another concern in the design optimization process is the accuracy of the optimization objective models. Ideally one should consider the SPICE device models like BSIM4 (P. K. Rout D. P., 2014)for optimization but they involve hundreds of variables. Directly working with these models is highly impracticable by a designer in industry. A tractable equation based optimization necessitates low-dimensional models with less complexity yet offering sufficient accuracy in the circuit behavior.

Apart from this, it is always desirable that multiple design objectives should be optimized in integrated circuits. For such requirements, the single objective methods are inadequate. So it is motivating to apply efficient multi-objective optimization techniques like Non-dominated Sorting Genetic Algorithm-II (NSGA-II) (P. K. Rout D. P., 2014) to CMOS VCO circuits in a constrained environment.

In the proposed design methodology the above three issues are collectively considered for predictably near optimal performance. Besides this, the technique finds the design parameters in a single run and hence the time to design the first prototype is greatly reduced. The CMOS voltage controlled oscillators are considered here for optimization of phase noise and power consumption with a goal to achieve a targeted frequency of oscillation in a technology constrained environment. Acceptably manageable model equations, which include the parasitics, are considered as optimization objectives. The design parameters obtained from the multi-objective constrained optimization NSGA-II technique are used to design schematic and physical layout level CMOS voltage controlled oscillators in the Cadence Virtuoso Analog Design Environment (ADE). Since the methodology is newly applied to CMOS VCOs, to the best of our knowledge there is no other benchmark result available for direct comparison. Hence, for the demonstration of the methodology, the circuit performance parameters are estimated from the transient and noise analysis in Cadence tool and are compared with their estimated values obtained from optimization. The circuits considered here are CMOS ring oscillators with different number of stages, current starved voltage controlled oscillator (CSVCO) and differential voltage controlled oscillator (DVCO).

## The NSGA-II Algorithm

$NSGA-II\,Algorithm\ \{$

$Random\,Population\,initialization\ P_0$

$Evaluate\,the\,objective\,functions$

$$\left(Phase\,noise\,and\,Power\,consumption\right)$$

$set\,P_0 = \left(F_1, F_2, \ldots\right) = non\_dominated\_sorting\left(P_0\right)$

$for\,all\,F_i \in P_0$

$crowding\_distance\_assignment(F_i)$

$set\, t = 0$

$while\, t = 0 \rightarrow Max\_Gen \quad \{$

$generate\, child\, population\, Q_t \; from \; P_t \,.$

$(by\, crossover\, and\, mutation).$

$set\, R_t = P_t \cup Q_t$

$set\, F = (F_1, F_2, \ldots) = non\_dominated\_sorting(R_t)$

$set\, P_{t+1} = 0$

$set\, i = 1$

$while |P_{t+1}| + |F_i| < N \quad \{$

$crowding\_distance\_assignment(F_i)$

$set\, P_{t+1} = P_{t+1} \cup F_i$

$\}$

$sort\, F_i\, on\, crowding\, distances$

$set\, P_{t+1} = P_{t+1} \cup F_i [1 : (N - |P_{t+1}|)]$

$set\, t = t + 1$

$\}$

$return\,F_1$

}

In the above algorithm $P_0$ is the initial population with size $N$. The $non\_dominated\_sorting$ is a procedure which involves comparing the objective values of every solution to the objective values of all other solutions in the population and $crowding\_distance\_assignment$ procedure involves the sorting of the elements in pareto fronts $F_i$ one time for every objective.

The proposed NSGA-II based analog IC optimization technique in general can be stated as (P. K. Rout a. D., 2015)

$$\{ \; minimize\,F_i\left(x_j\right) \;; i = 1,2,\ldots,M; j = 1,2,\ldots,N$$

$$subject\,to\,G_i\left(x_j\right) > 0$$

$$H_i\left(x_j\right) = 0, \; \}$$

$$x_j\left[x_{min}, x_{max}\right] \tag{1}$$

As a case study, for CMOS voltage controlled oscillator, the objectives $F_i$ are power consumption and phase noise. The inequality constraint $G_i$ is $\left|g_{mn} - g_{mp}\right| - \; < 0$ and the equality constraint $H_i$ is $f_{osc} = 2\text{GHz}$. It may be here noted that the frequency objective is formulated as a constraint. Here $g_{mn}$ and $g_{mp}$ are the transconductance parameters of NMOS and PMOS transistors respectively and $\delta$ is small positive definite real number such that $\left(0{,}10^{-6}\right)$. The design parameters $x_j$ are $\left[W_n, L_n, W_p, L_p\right]$. These parameters are bounded by the maximum and minimum values dictated by the process technology. The required specifications like operating frequency, the design space constraints and the reference circuit model are the inputs to the NSGA-II optimizer block. The main objective of this optimizer is to determine the design parameters of all transistors in the circuit under consideration. The simple equations of power consumption and phase noise are the optimization objectives for the NSGA-II.

This optimization method explores the optimal solutions in a constrained design space with a very marginally tolerable frequency drift around the desired frequency. With these initial optimized design parameters obtained from NSGA-II optimizer, the CMOS VCO schematic and the physical layout are designed in Cadence Virtuoso Analog Design Environment.

In most of the parasitic aware design methods, the parasitics are extracted from a layout, which is not optimized to yield better performance. Hence, there is always a finite probability that the parasitic values considered in the design are far away from the parasitic values of the optimal design. To overcome this inadequacy, in the initial attempt the physical layout generated from the first optimization is subjected to RCLK (Resistance, Capacitance, Inductance and Mutual inductance) parasitic extraction. The circuit model adaptively includes the extracted parasitics. This adapted circuit model is again fed back to the NSGA-II optimizer for second and final optimization. This guides the optimizer with a realistic parasitic model of the circuit, which includes the logic parasitics as well as interconnect parasitics. Hence, the final design parameters for the parasitic aware performance optimized CMOS VCO circuit are obtained. These design parameters are then used to generate the physical layout, which can be taped out for fabrication.

## PERFORMANCE ANALYSIS OF CSVCO AS A CASE STUDY

This design methodology used for fast prototyping, is based on multi-objective evolutionary technique NSGA-II where the parasitic effects are included during the design cycle. The design is fast because with a single run of the algorithm one gets the parameters of the optimized circuit for superior performance. This saves the design cycle time which is normally spent in hit and trial to attain higher performance. CMOS ring oscillators (RO), Current Starved VCO (CSVCO) and Differential VCO (DVCO) are designed with a specification frequency, for minimal phase noise and power consumption by using this methodology. The degrading effects of parasitics on oscillating frequency are taken care of efficiently in the two phases of optimization process to achieve the target frequency while simultaneously minimizing competing objectives, the phase noise and the power consumption with acceptable trade off (Table 1). The application of this methodology on different VCO circuits for design and the subsequent analyses reveal that the proposed design methodology is very efficient and can seamlessly be extended to design any analog integrated circuit. This design approach helps the designer in industry to deliver a product with superior performance in significantly less time.

### IDEA Based Fast Design of CMOS ICs for Performance Optimization

There has been a continuous strive towards development of more efficient evolutionary computing optimization algorithms. Though NSGA-II is a standard multi-objective optimization algorithm, still a better technique available would be an obvious choice among the designers. Infeasibility driven evolutionary algorithm (IDEA) is a recently developed multi-objective optimization algorithm which offers superior performance. Inspired by Moore's law, integrated circuits always need to offer better performance. Under such a situation more efficient optimization technique like IDEA come as a rescue to the designers' burden of achieving a better performance in a given process technology. The optimal solutions of the constrained multi-objective optimization problems very often lie along the constraint boundary. To effectively search along the constraint boundary, the original k objective constrained optimization

*Table 1. Performance summary of the parasitic aware CSVCO*

| Performance Index | NSGA-II Estimated | Post Layout Simulation |
|---|---|---|
| Oscillation Frequency (GHz) | 2.0 | 1.9656 |
| Phase Noise (dBc/Hz at 1 MHz offset) | -91. 92 | -90.29 |
| Power Consumption(μW) | 548 | 564.123 |
| Figure of Merit (FOM) (dBc/Hz) | -160.5527 | -158.6462 |

problem is reformulated as k + 1 objective unconstrained optimization problem as given in (2). The first k objectives are the same as in the original constrained problem where as the additional objective is a measure of constraint violation, referred to as "violation measure".

IDEA differs from NSGA-II mainly in the mechanism for elite preservation. In IDEA, a few infeasible solutions are retained in the population at every generation. Individual solutions in the population are evaluated as per the original problem definition and marked infeasible if any of the constraints are violated. The solutions of the parent and the offspring population are divided into two sets, a feasible set $\left(S_f\right)$ and an infeasible set $\left(S_{inf}\right)$.The solutions in the feasible and the infeasible sets are both ranked using non-dominated sorting and crowding distance sorting of k + 1 objectives. NSGA-II, on the other hand, uses non-dominated sorting and crowding distance for ranking feasible solutions and ranks infeasible solutions in the increasing value of maximum constraint violation. For the feasible solutions, non-dominated sorting using k + 1 objectives is equivalent to the non-dominated sorting the original k objectives as the additional objective value (which is based on the constraint violations) for feasible solutions is always 0.

In the next step the solutions that form the population for the next generation are chosen. In IDEA, a user-defined parameter $\alpha$ is used to identify the proportion of the infeasible solutions to be retained in the population. The numbers $N_f\left(=\left(1-\alpha\right)\times N\right)$ and $N_{inf}\left(=\alpha\times N\right)$ denote the number of feasible and infeasible solutions in the population respectively, where $N$ is the population size. If the infeasible set $S_{inf}$ has more than $N_{inf}$ solutions, then first $N_{inf}$ solutions are selected based on the rank; otherwise all the solutions from $S_{inf}$ are selected. The rest of the solutions are selected from the feasible set $S_f$, provided there are at least $N_f$ number of feasible solutions. If $S_f$ has fewer solutions, all the feasible solutions are selected and the rest are filled with infeasible solutions from $S_{inf}$. The solutions are ranked from 1 to $N$ in order of their selection. That is how; the infeasible solutions that get selected first (at most $S_{inf}$ ), get higher rank than the feasible solutions. In NSGA-II, the elite preservation mechanism weeds out the infeasible solutions from the population. To retain the infeasible solutions in the population, an alternate mechanism is required. In IDEA, the infeasible solutions are ranked higher than the feasible solutions, thus adding selection pressure to generate better infeasible solutions.

## IDEA Algorithm

**Algorithm:** Infeasibility Driven Evolutionary Algorithm (IDEA)
**Require:** $N$ {Population Size}
**Require:** $N_G > 1$ {Number of Generations}
**Require:** $0 < \alpha < 1$ {Proportion of infeasible solutions}

$N_{inf} = \alpha * N$
$N_f = N - N_{inf}$
$pop_1$ = Initialize ()
Evaluate ($pop1$)
**for** $i = 2$ to $N_G$ **do**
*child* $pop_{i-1}$ = Evolve ($pop_{i-1}$)

7. Evaluate ($child\ pop_{i-1}$)
8. ($S_f$, $S_{inf}$) = Split ($pop_{i-1}$ +$child\ pop_{i-1}$)
9. Rank ($S_f$)
10. Rank ($S_{inf}$)
11. $pop_i = S_{inf}(1, N_{inf}) + S_f(1, N_f)$
12. **end for**

$$Minimize\ f_1'(x) = f_1(x), \ldots\ldots, f_k'(x) = f_k(x)$$

$$f_{k+1}'(x) = \textbf{Constraint Violation Measure} \tag{2}$$

## IDEA BASED DESIGN OF CSVCO FOR PERFORMANCE OPTIMIZATION: A CASE STUDY

The required specifications, the design space constraints and the reference circuit model are the inputs to the IDEA processing block (Figure 6). The primary goal of this processor is to determine the design parameters of all transistor elements in the VCO circuits. The implicitly parasitic dependant analytic equations of power consumption and phase noise constitute the optimization objectives of the IDEA processor. This processor is allowed to explore the optimal solutions in a limited design space with a very marginally tolerable frequency drift around the target frequency of the VCO. With these initial optimized design parameters the VCO schematic and subsequent physical layout are designed in Cadence Virtuoso Analog Design Environment (ADE). The physical layout so generated is subjected to RCLK (Resistance, Capacitance, Inductance, and Mutual Inductance) parasitic extraction. The algorithm starts with the reference circuit model (with SPICE parameters), and in every iteration of design, the design parameters are obtained. The layout of the circuit is drawn and the post layout RCLK parasitic extraction is performed. Then the circuit model parameters are modified with the inclusion of the extracted parasitics.

### IDEA Based Design Methodology of VCO for Performance Optimization

**Set:** $N$ {Population Size}
**Set:** $NG > 1$ {Number of Generations}
**Set:** $0 < \alpha < 1$ {Proportion of infeasible solutions}
1. $Ninf = \alpha * N$
2. $Nf = N - Ninf$
3. ***while*** Parameter Constraints C = [$W_{min} < W < W_{max}$, $L_{min} < L < L_{max}$] ***do***
4. pop1 = Initialize () *subject to C*
5. Evaluate $\left[L\{◆f\}\left(pop1\right), P_{avg}\left(pop1\right)\right]$
6. ***for*** $i=2$ to $NG$ ***do***
7. $childpop_{i-1}$ = Evolve ( $pop_{i-1}$ )
8. Evaluate $\left[L\{◆f\}\left(childpop_{i-1}\right), P_{avg}\left(childpop_{i-1}\right)\right]$

*Figure 6.*

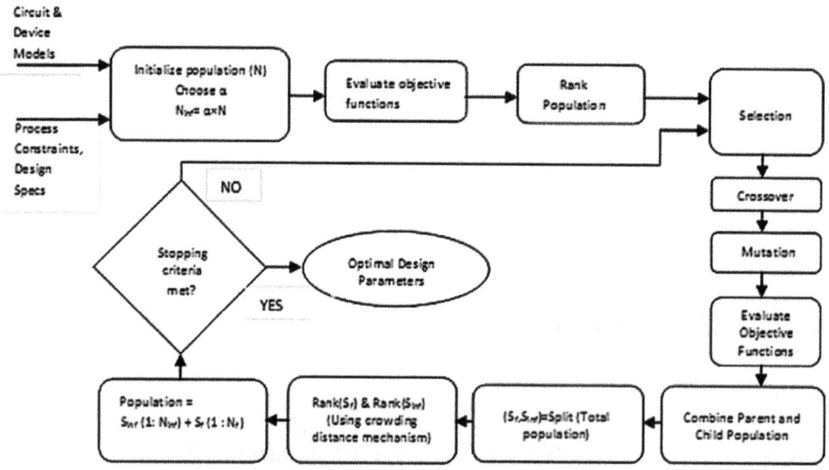


9. $Compute\,D = \left|f_{osc} - f_{target}\right|$
10. ***if*** $D \leq \epsilon$ *then* $Sf$

else $Sinf$

***end if***

11. $\left(Sf, Sinf\right)$ = Split ( $pop_{i-1}$ $+childpop_{i-1}$)
12. Rank$\left(Sf\right)$
13. Rank$\left(Sinf\right)$
14. $pop_i = Sinf\left(1:Ninf\right) + Sf\left(1:Nf\right)$
15. ***end for***
16. *end while*

These modified circuit model parameters are used as the input to the IDEA processor block in place of the reference circuit model in the next iteration of the design. This provides the IDEA processor with a near exact parasitic aware model of the circuit which includes not only the logic parasitics but also the interconnect parasitic estimates. Hence the IDEA algorithm provides the final level parasitic aware performance optimized design parameters for the VCO circuit (Table 2). These design parameters are utilized to generate the physical layout of the circuit with near optimum performance, which can be taped out for fabrication.

Therefore, the final design parameters obtained from this methodology meets the desired specifications along with global best optimal performance parameters.

The IDEA based optimization processing can be stated as:

$$Minimize \underset{P}{\mathcal{L}\{\blacklozenge f\}}$$

$$Subject\,to \quad \begin{aligned} & f_{osc} = f_{Specification} \\ & W_{min} < W < W_{max} \\ & L_{min} < L < L_{max} \\ & \left| g_{mn} - g_{mp} \right| \leq \delta \end{aligned} \tag{3}$$

where $g_{mn}$ and $g_{mp}$ are the transconductance parameters of NMOS and PMOS respectively, δ is a very small positive definite constant. Simulated binary crossover (SBX) and polynomial mutation operators are used in IDEA to generate offspring from a pair of parents selected using binary tournament. Individual solutions in the population are evaluated using the problem definition (3) and the infeasible solutions are identified. The solutions in the parent and offspring population are divided into a feasible set $S_f$ and an infeasible set $S_{inf}$. The solutions in the feasible set and the infeasible set are ranked separately using the non-dominated sorting and crowding distance sorting based on the objectives as per (3). The solutions for the next generation are selected from both the sets to maintain infeasible solutions in the population. The infeasible solutions are ranked higher than the feasible solutions to provide a selection pressure to create better infeasible solutions resulting in an active search through the infeasible search space. The marginally infeasible solutions in IDEA very often prove beneficial trade-offs for the integrated circuit design. Hence the technique is more attractive than NSGA-II for application in IC design optimization problem which is depicted in Table 3.

*Table 2. Performance indices of parasitic aware optimized CMOS CSVCO*

| | Frequency of Oscillation (GHz) | Phase Noise (dBc/Hz at 1 MHz offset) | Power Consumption (μW) | FOM (dBc/Hz) |
|---|---|---|---|---|
| IDEA Estimated | 2 | -87.64 | 498.4632 | -156.6842 |
| Schematic Level | 2.5385 | -85.08 | 494.7281 | -156.2281 |
| Post-layout level | 1.9884 | -87.04 | 496.0658 | -156.0550 |

*Table 3. Comparison of performance parameters of CSVCO*

| Performance Measure | NSGA-II Based | IDEA Based |
|---|---|---|
| Frequency (GHz) | 1.9656 | 1.9884 |
| Phase Noise (dBc/Hz at 1 MHz offset) | -90.29 | -87.04 |
| Power (μW) | 564.1230 | 496.0658 |
| FOM (dBc/Hz) | -158.6462 | -155.8685 |
| Convergence Time (seconds) | 124.910 | 79.873 |

## CONCLUSION

The real world is analog in nature and the analog signals need to be processed by integrated circuits(IC). The performance of the system is always driven and dictated by the analog part of the integrated circuit. It is highly difficult to design analog circuits due to their complexity, noise sensitivity and multiple performance tradeoffs. So there is a desperate need of some algorithms which can help the designer to meet the desired specifications by circuit sizing and optimization of multiple complex and sensitive performance parameters dictated by technology constraints. In general the analog IC design problem is highly complex, multi-objective, multi-modal and multi-constraints based. Along with optimization the design algorithm must be parasitic and process variation aware to make the IC robust enough for targeted application and high yielding for economic viability. Evolutionary algorithm based optimization tools are highly suitable in comparison to conventional gradient based techniques as the problem is multi-modal with multiple competing objectives. Direct search method, Newton's method, conjugate gradient method, gradient descent method, simplex method etc are gradient based algorithms and are very efficient but may lead to sub optimal solutions. IC design automation techniques based on evolutionary algorithms like Genetic Algorithms, Particle Swarm Optimization, Differential Evolution etc are efficient in optimization of performance objectives but they would be more efficient and precise in circuit sizing if better constraint handling method would be incorporated. NSGA-II is a well established efficient multi-objective optimization algorithm which can be used for circuit sizing and performance optimization of the analog ICs. Other recently developed multi-objective algorithms which could handle constraints more efficiently can be tried for optimal analog IC design.

Normally the worst case process is considered while designing to make the analog IC robust against process variations. But practically the IC is used in nominal environment in most of the cases, so to make the IC robust and optimal some different methodology must be thought of.

## REFERENCES

Allstot, D. J., Choi, K., & Park, J. (2003). *Parasitic-Aware Optimization of CMOS RF Circuits*. Kluwer Academic Publishers.

Altun, O., & Bocko, M. (2006). Robust analog circuit design: a set theoretic approach. *IEEE International Symposium on Circuits and Systems*. doi:10.1109/ISCAS.2006.1693245

Ballweber, B. M., Gupta, R., & Allstot, D. J. (2000). A Fully Integrated 0.55.5-GHz CMOS Distributed Amplifier. *IEEE Transactions on Solid-State Circuits*, *35*(2), 231–239. doi:10.1109/4.823448

Barari, M., Karimi, H. R., & Razaghian, F. (2014). *Analog Circuit Design Optimization Based on Evolutionary Algorithms. Mathematical Problems in Engineering*. Hindawi.

M. Barros, J. Guilherme and N. Horta. (2010). Analog circuits optimization based on evolutionary computation techniques. *Integration, the VLSI Journal, 43*, 136-155.

Chu, M., Allstot, D., Huard, J., & Wong, K. (2004). NSGA-Based Parasitic-aware Optimization of a 5 GHz Low-noise VCO. *Design Automation Conference, Asia and South Pacific*.

Conn, A. R., Coulman, P. K., Haring, R. A., Morrill, G. L., Visweswariah, C., & Wu, C. W. (1995). JiffyTune: Circuit optimization using time-domain sensitivities. *IEEE Trans. on Computer-Aided Design of Integrated Circuits and Systems*, *17*(12), 1292–1309. doi:10.1109/43.736569

Cooren, Y., Fakhfakh, M., Loulou, M., & Siarry, P. (2007). Optimizing second generation current conveyors using particle swarm optimization. *The IEEE 19th International Conference on Microelectronics.*

Crols, J., Donnay, S., Steyaert, M., & Gielen, G. (1995). A high-level design and optimization tool for analog RF receiver front-ends. *IEEE/ACM International Conference on Computer-Aided Design (ICCAD)*. doi:10.1109/ICCAD.1995.480170

De Ranter, C. R. C., Van der Plas, G., Steyaert, M. S. J., Gielen, G. G. E., & Sansen, W. M. C. (2002). CYCLONE: Automated design and layout of RF LC-oscillators. *IEEE Trans. on Computer-Aided Design of Integrated Circuits and Systems*, *21*(10), 1161–1170. doi:10.1109/TCAD.2002.802267

Eick, M., & Graeb, H. (2012). MARS: Matching-Driven Analog Sizing. *IEEE Trans. on Computer-Aided Design of Integrated Circuits and Systems*, *31*(8), 1145–1158. doi:10.1109/TCAD.2012.2190069

El-Turky, F., & Perry, E. (1989). BLADES: An artificial intelligence approach to analog circuit design. *IEEE Trans. on Computer-Aided Design of Integrated Circuits and Systems*, *8*(6), 680–692. doi:10.1109/43.31523

Fakhfakh, M., Cooren, Y., Sallem, A., Loulou, M., & Siarry, P. (2010). Analog circuit design optimization through the particle swarm optimization technique. *Analog Integrated Circuits and Signal Processing*, *63*(1), 71–82. doi:10.1007/s10470-009-9361-3

Garitselov, O., Mohanty, S., & Kougianos, E., & Patra, P. (2011). Bee Colony Inspired Metamodeling Based Fast Optimization of a Nano-CMOS PLL. *2011 International Symposium on Electronic System Design (ISED)*. doi:10.1109/ISED.2011.13

Ghai, D., Mohanty, S., & Thakral, G. (2013). Fast optimization of nano-CMOS voltage-controlled oscillator using polynomial regression and genetic algorithm. *Microelectronics Journal*, *44*(8), 631–641. doi:10.1016/j.mejo.2013.04.010

Ghali, K., Dorie, L., & Hammami, O. (2005). Dynamically Reconfigurable Analog Circuit Design Automation Through Multiobjective Optimization and Direct Execution. *The 17th International Conference on Microelectronics.*

Gielen, G., & Rutenbar, R. A. (2000). Computer-aided design of analog and mixed-signal integrated circuits. *Proceedings of the IEEE*. doi:10.1109/5.899053

Goldberg, D. (1989). *Genetic Algorithms in Search, Optimization and Machine Learning*. Reading, MA: Addison-Wesley.

Gomez, I., Cuautle, E., & Luis, G. (2013). Richardson extrapolation-based sensitivity analysis in the multi-objective optimization of analog circuits. *Applied Mathematics and Computation*, *222*, 167–176. doi:10.1016/j.amc.2013.07.059

Gupta, R., & Allstot, D. J. (1998). Parasitic-Aware Design and Optimization of CMOS RF Integrated Circuits. *IEEE Radio Frequency Integrated Circuits Symposium.*

Harvey, Elmasry, & Leung. (1992). STAIC: An Interactive Framework for Synthesizing CMOS and BiCMOS Analog Circuits. *IEEE Trans. on Computer-Aided Design of Integrated Circuits and Systems, 11*(11).

Haykin. (n.d.). *Neural Networks A Comprehensive Foundation.* Prentice Hall International Inc.

Hershenson, M., Boyd, S., & Lee, T. (1998). GPCAD: a tool for CMOS opamp synthesis. *Proc. of IEEE/ACM International Conference on CAD (ICCAD).*

Hjalmarson, E., Hagglund, R., & Wanhammar, L. (2003). An equation-based optimization approach for analog circuit design. *International Symposium on Signals, Circuits and Systems.* doi:10.1109/SCS.2003.1226952

Jafari, A., Bijami, E., Bana, H., & Sadri, S. (2011). A design automation system for CMOS analog integrated circuits using New Hybrid Shuffled Frog Leaping algorithm. *Microelectronics Journal, 43*(11), 908–915. doi:10.1016/j.mejo.2012.05.010

Karaboga, D. (2005). *An idea based on Honey Bee Swarm for numerical optimization.* Dept. of Computer Engineering, Engineering Faculty, Erciyes University.

Kennedy, J., & Eberhart, R. (1995). Particle swarm optimization. *Proc. IEEE international conference on neural networks.* doi:10.1109/ICNN.1995.488968

Kim, J., Jhaveri, R., Woo, J., & Yang, C. (2007). Device-Circuit Co-Optimization for Mixed-mode Circuit Design via Geometric Programming. *IEEE/ACM International Conference on Computer-Aided Design.*

Kirkpatrick, S., Gelatt, C., & Vecchi, M. (1983). Optimization by simulated annealing. *Journal of Science, 220*(4598), 671–680. doi:10.1126/science.220.4598.671 PMID:17813860

Koh, H. Y., Sequin, C. H., & Gray, P. R. (1990). OPASYN: A Compiler for CMOS Operational Amplifiers. *IEEE Trans. on Computer-Aided Design of Integrated Circuits and Systems, 9*(2), 113–125. doi:10.1109/43.46777

Kruiskamp, W., & Leenaerts, D. (1995). DARWIN: CMOS opamp synthesis by means of a genetic algorithm. *Proc. ACM/IEEE Design Automation Conference (DAC).* doi:10.1145/217474.217566

S. Kundu and P. Mandal. (2014). ISGP: Iterative Sequential Geometric Programming for Precise and Robust Analog Circuit Sizing. *Integration, the VLSI Journal, 47*, 510-531.

Li, X., Xu, Y., Hsiung, K. L., Pileggi, L., & Boyd, S. (2009). Regular Analog/RF Integrated Circuits Design Using Optimization With Recourse Including Ellipsoidal Uncertainty. *IEEE Trans. on Computer-Aided Design of Integrated Circuits and Systems, 28*(9), 623–637.

Liu, Wang, Yu, Liu, Li, Wan, Lu, & Fernandez. (2009). Analog circuit optimization system based on hybrid evolutionary algorithms. *Integration, the VLSI Journal, 42*, 137–143.

Mandal, P., & Visvanathan, V. (2001). CMOS Op-Amp Sizing Using a Geometric Programming Formulation. *IEEE Trans. on Computer-Aided Design of Integrated Circuits and Systems, 20*(1), 22–38. doi:10.1109/43.905672

Matsukawa, K., Morie, T., Tokunaga, Y., Sakiyama, S., Mitani, Y., Takayama, M., & Dosho, S. et al. (2009). Design methods for pipeline & delta-sigma A-to-D converters with convex optimization. *Asia and South Pacific Design Automation Conference*. doi:10.1109/ASPDAC.2009.4796560

Nagaraj, N. S. (1993). A new optimizer for performance optimization of analog integrated circuits. *IEEE/ ACM Design Automation Conference (DAC)*. doi:10.1145/157485.164643

Nelder, J., & Mead, R. (1965). A simplex method for function optimization. *The Computer Journal*, *7*(4), 308–313. doi:10.1093/comjnl/7.4.308

Nye, W., Riley, D. C., Sangiovanni-Vincentelli, A., & Tits, A. L. (1988). DELIGHT.SPICE: An optimization-based system for the design of integrated circuits. *IEEE Trans. on Computer-Aided Design of Integrated Circuits and Systems*, *7*(4), 501–519. doi:10.1109/43.3185

Okobiah, O., Mohanty, S., & Kougianos, E. (2013). Geostatistical-inspired fast layout optimisation of a nano-CMOS thermal sensor. *Circuits. Devices & Systems, IET*, *7*(5), 253–262. doi:10.1049/iet-cds.2012.0358

Onodera, H., Kanbara, H., & Tamaru, K. (1990). Operational-amplifier compilation with performance optimization. *IEEE Journal of Solid-State Circuits*, *25*(2), 466–473. doi:10.1109/4.52171

Pandit, S., Mandal, C., & Patra, A. (2011). *A Methodology for Generation of Performance Models for the Sizing of Analog High-Level Topologies. Int. Journal VLSI Design.*

Passino, K. M. (2002). Biomimicry of Bacterial Foraging for distributed optimization and control. *IEEE Control Systems Magazine*, *22*(3), 52–67. doi:10.1109/MCS.2002.1004010

Paulino, N., Goes, J., & Steiger-Garcao, A. (2001). Design methodology for optimization of analog building blocks using genetic algorithms. *The 2001 IEEE International Symposium on Circuits and Systems (ISCAS 2001).*

Phelps, R., Krasnicki, M., Rutenbar, R., Carley, L. R., & Hellums, J. R. (2000). ANACONDA: Simulation-based synthesis of analog circuits via stochastic pattern search. *IEEE Trans.on Computer-Aided Design of Integrated Circuits and Systems*, *19*(6), 703–717. doi:10.1109/43.848091

Plas, Debyser, Francky Leyn, Lampaert, Vandenbussche, Gielen, ... Leenaerts. (2011). AMGIE- A Synthesis Environment for CMOS Analog Integrated Circuits. *IEEE Trans. on Computer-Aided Design of Integrated Circuits and Systems, 20*(9), 1037-1058.

Price, K., Storn, R. M., & Lampinen, J. A. (2005). Differential Evolution: A Practical Approach to Global Optimization. In *Natural Computing Series*. Springer.

Razavi, B. (2010). *Design of Analog CMOS Integrated Circuits*. McGraw Hill Education Private Limited.

Rout, P. K., Acharya, D. P., & Panda, G. (2014). A multiobjective optimization based fast and robost design for low power and low phase noise current starved VCO. *IEEE Transactions on Semiconductor Manufacturing*, *27*(1), 43–50. doi:10.1109/TSM.2013.2295423

Rout, P. K., & Acharya, D. P. (2015). Fast physical design of CMOS ROs for optimal performance using constrained NSGA-II. *AEÜ. International Journal of Electronics and Communications*, *69*(9), 1233–1242. doi:10.1016/j.aeue.2015.05.004

Ruiz-Amaya, J., de la Rosa, J., Fernandez, F. V., Medeiro, F., del Rio, R., Perez-Verdu, B., & Rodriguez-Vazquez, A. (2005). High-level synthesis of switched-capacitor switched-current and continuous-time SD modulators using SIMULINK-based time-domain behavioral models. *IEEE Transactions on Circuits and Systems. I, Regular Papers*, *52*(9), 1795–1810. doi:10.1109/TCSI.2005.852479

Rutenbar. (1989). Simulated annealing algorithms: An overview. *IEEE Circuits and Devices Magazine*, 19-26.

Sheta, A. F. (2010). Analogue filter design using differential evolution. *International Journal of Bio-inspired Computation*, *2*(3-4), 233–241. doi:10.1504/IJBIC.2010.033091

Sonmez & Dundar. (2011). Simulation-based analog and RF circuit synthesis using a modified evolutionary strategies algorithm. *Integration, the VLSI Journal, 44*(2), 144–154.

Soorapanth, T. (2007). Gaining Insights in Analog Design via Geometric Programming. *IEEE International Symposium on Information and Communication Technologies.*

Tawdross, P., & König, A. (2006). Particle swarm optimization for reconfigurable sensor electronics-case study: 3 Bit Flash ADC. *4th IEEE international workshop on intelligent solutions in embedded systems (WISES'06).*

Tawdross, P., & Koning, A. (2005). Investigation of Particle Swarm Optimization for dynamic reconfiguration of field programmable analog circuits. *Proc. of 5th Conference on hybrid intelligent systems.* doi:10.1109/ICHIS.2005.66

Thakker, R. A., Baghini, M. S., & Patil, M. B. (2009). Low-power low-voltage analog circuit design using hierarchical particle swarm optimization. *Proc. IEEE 22nd international conference on VLSI design.* doi:10.1109/VLSI.Design.2009.14

Torralba, A., Chavez, J., & Franquelo, L. G. (1996). FASY: A fuzzy-logic based tool for analog synthesis. *IEEE Trans. on Computer-Aided Design of Integrated Circuits and Systems*, *15*(7), 705–715. doi:10.1109/43.503939

Tulunay, G., & Balkir, S. (2008). A synthesis tool for CMOS RF low noise amplifiers. *IEEE Transactions on Computer-Aided Design of Integrated Circuits and Systems*, *27*(5), 977–982. doi:10.1109/TCAD.2008.917579

Vural, R., Bozkurt, U., & Yildirim, T. (2013). Analog active filter component selection with nature inspired metaheuristics. *Int. J. Electron. Commun.*, *67*(3), 197–205. doi:10.1016/j.aeue.2012.07.009

Vural, R., & Yildirim, T. (2012). Analog Circuit Sizing via Swarm Intelligence. *Int. J. Electron. Commun.*, *66*(9), 732–740. doi:10.1016/j.aeue.2012.01.003

Vural, R., Yildirim, T., Kadioglu, T., & Basargan, A. (2012). Performance evaluation of evolutionary algorithms for optimal filter design. *IEEE Transactions on Evolutionary Computation*, *16*(1), 135–147. doi:10.1109/TEVC.2011.2112664

Yang & Gan. (2007). Robust Calibration of an Improved Delta-Sigma Data Converter Using Convex Optimization. *IEEE Journal of Selected Topics in Signal Processing, 1*(4).

## KEY TERMS AND DEFINITIONS

**ASIC:** Application Specific Integrated Circuit.
**CMOS:** Complementary Metal Oxide Semiconductor.
**CSVCO:** Current Starved Voltage Controlled Oscillator.
**DE:** Differential Evolution.
**DVCO:** Differential Voltage Controlled Oscillator.
**FPAA:** Field Programmable Analog Array.
**FPGA:** Field Programmable Gate Array.
**GA:** Genetic Algorithm.
**IC:** Integrated Circuit.
**IDEA:** Infeasibility Driven Evolusionary Algorithm.
**NSGA-II:** Nondominated Sorting Genetic Algorithm-II.
**PCPVM:** Process Corner Performance Variability Minimization.
**PLL:** Phase Locked Loop.
**PSO:** Particle Swarm Optimization.
**RO:** Ring Oscillator.
**SPICE:** Simulation Program for Integrated Circuit Emphasis.
**VCO:** Voltage Controlled Oscillator.

# Chapter 16
# Lean Manufacturing:
## Principles, Tools, and Practices

**Mousumi Roy**
*University of Connecticut, USA*

## ABSTRACT

*Lean has become a new mantra in today's manufacturing sector. In this millennium, companies are facing a challenge to be economically competitive in manufacturing. Many of them have realized that the old style of mass manufacturing is no longer successful. Hence, lean manufacturing is being embraced by the companies to simultaneously achieve a competitive edge and economic growth. Many studies have shown that lean organizations are capable of meeting customer's expectations consistently, at each step of the production systems. Lean manufacturing also implies efficient use of non-renewable resources in order to maintain a sustainable environment. To reach the full potential of an organization, lean must be embraced as a holistic business strategy. In this chapter, the history of lean innovation will be briefly discussed, followed by the principles of lean manufacturing and various tools in implementing lean practices. Examples of organizations that have experienced significant improvements once transformed to lean manufacturing will also be cited.*

## INTRODUCTION

*Be ready to revise any system, scrap any method, abandon any theory, if the success of the job requires it. (Henry Ford, 1923)*

Henry Ford, the iconic leader of the automobile industry, introduced assembly line and mass production in manufacturing in the late nineteenth century. He revolutionized manufacturing in the West by revising the system, scrapping the existing methods, and abandoning the old theories. Industrial companies in the United States and Europe adopted his innovation of mass manufacturing systems and experienced tremendous success due to abundance of raw materials, energy and low labor costs. Manufacturers from other countries, notably Japan, had high respect for Ford's innovation in auto manufacturing – they had visited and learned the production process from North American companies and had applied the

DOI: 10.4018/978-1-5225-2944-6.ch016

knowledge to upgrade their own manufacturing process. The production strategy was based on a push production system driven by the product demand forecast.

However, after World War II, Japanese companies were faced with many problems such as reduced product demand, shortage of capital, raw materials and space for inventory, high labor and machinery cost. The mass production strategy ceased to be a viable option. Toyota was one of those companies struggling to survive. Out of necessity, the management, supervisors, and workers of the company had to innovate – after a several iterations, they found success under the supervision of Mr. Taichi Ohno. Their decade-long success was noticed by western manufacturers and scholars, which resulted in a five-year, 5 million dollar global study of auto industries by MIT in 1984, called *International Motor Vehicles Program* (IMVP). They recognized *Toyota Production System* (TPS) as a new kind of manufacturing process - it was different from the mass production strategy of the Detroit Auto industry, and it produced superior results. The new production system was coined as "Lean Production" (Krafcik, 1988), since it took considerably less time and cost, and produced less defective parts, compared to the mass production strategies of the Western manufacturing system. TPS was exemplary of the Lean Production.

The auto industries in North America and Europe have been following the success of the Japanese manufacturing strategy. However, they were unsure whether it could be adaptable in the western cultural environment. At the same time, Toyota wanted to expand, and bring its technology to the United States. As an experiment, a joint venture, the *New United Motor Manufacturing, Inc.* (NUMMI) was established in 1984 between General Motors (GM) and Toyota. GM took the opportunity to learn Lean Production, or Lean Manufacturing techniques, while Toyota obtained a chance to implement TPS in a different cultural and labor environment. The success of this project proved that the Lean Manufacturing is not restricted by the cultural differences and can be applied in any country. Since then, many manufacturers have implemented lean strategy and found success in their undertaking. However, they have also realized that implementing lean is neither easy nor fast. A holistic approach and understanding of the principles are necessary to achieve a lean organization (Fullerton, Kennedy, & Widener, 2014).

In this chapter, the principles and the tools of lean strategy will be discussed to provide readers a thorough understanding of the Lean Manufacturing. If applied properly, it has been shown to lead the company in reaching its fullest potential.

## BACKGROUND

Lean Manufacturing (LM) is rooted in the Toyota Production System (TPS). After Japan's defeat in WWII, Toyota was faced with many obstacles as discussed earlier. To be competitive, their only option was to produce many varieties of automobiles in small quantities because of the low market demand for each type. Mr. Taichi Ohno, the chief architect of the TPS, developed and implemented a variety of low-cost techniques to increase the competitive advantage of the company. He described his methods very simply as "a manufacturing strategy that reduces the timeline between the customer order and the shipment by eliminating non-value added waste" (Ohno, 1988) (Box 1).

Mr. Ohno's innovation was however influenced by the American Supermarket System. He was fascinated by the grocery buying process - customers bought exactly what they needed and when they needed, which was very different from the Japanese grocery shopping system at that time. He also noticed that only a few workers were able to run the supermarkets, and it was possible to make a profit even charging low prices for the products.

*Box 1.*

Customer placing the order → Shipping order and collect payment

Reduce timeline by eliminating waste

After the oil crisis of 1973, Toyota continued to be successful with their new manufacturing strategy, and other Japanese manufacturers who were not doing well with mass manufacturing strategy started following the techniques used by TPS and gained positive outcomes. As noted earlier, the continued success of Japanese manufacturers was noticed by auto-industries of the United States in the 1980s. The committee members of the IMVP study (1984-1989) measured the performance gap between Japanese and Western firms and analyzed the factors responsible. They were convinced that the competitive advantage of the Japanese automakers was as a result of their superior organization of the production processes. The final results of the study were summarized, and the evaluation strategies for industries were outlined in the famous book, *The Machine that Changed the World.* (Womack, Jones, & Roos, 1990).

The NUMMI project was the first success of applying lean in the Western manufacturing system. GM had sent many of its managers to NUMMI to gain the knowledge and experience in implementing lean techniques (Liker, 1997). The plant manager of one of the plants of Delphi Saginaw Steering Systems (DSSS), who specialized in steering wheels for GM spent two years in NUMMI project and later adopted Lean Manufacturing in his plant. The plant experienced significant improvement in Quality - the Return/Rejected parts per million (RPPM) reduced from 1917 in 1993 to 93 in 1995 (95% reduction). The productivity (per employee per day) increased from average of 7% to 14% in during that period (Woolson & Husar, 1997).

*Jacobs Vehicle Equipment Company* was one of the early adopters, who achieved unprecedented success by implementing lean principles in the mid-1980s. The company was guided by members of the Toyota Groups' Production Engineering Staff (Messrs. Yoshiki Iwata, Chihiro Nakao, and Akira Takenaka). In three years they achieved the following improvements (Koenigsaecker, 2012):

1. Reducing the lead-time from more than 30 days to one day, with 100% on-time delivery.
2. Reduced quality issues by over 80%.
3. Grew enterprise productivity 86%.
4. They were able to move from monthly production batches to twice in a month batches, to weekly batches, and eventually to daily batches. At the end of the lean transformation, they were able to ship the product the day after the order was placed. The defect rate was reduced to more than 80%.

The journey to lean is neither easy nor fast. However, the benefits of the transformation were found to be remarkable. The following table (Liker, 1997) shows the expected improvements if transforming to the lean production from either the batch production, or Henry Ford–style flow production (Box 2).

Over time, Lean Manufacturing has been adopted in a variety of manufacturing companies (e.g., chemical, computer, electronics, food and beverage, garment, telecommunication, wireless, pharmaceuticals, petroleum, printing, A/C, and heating etc.). Everyone wins when lean works – the management is happier with higher profits and greater employee involvement, the union likes the worker's safety and

*Box 2.*

| | For Batch Production | For Henry Ford style Flow Production |
|---|---|---|
| Productivity increase | +300-400% | +100% |
| Inventory turns | +1000% | +300% |
| Defects | -95% | -80% |
| Lead time | -95% | -75% |

security, and the employees become better trained, happier and more engaged with the manufacturing process (Liker, 1997).

## WHAT IS LEAN MANUFACTURING (LM)?

Defining *Lean* has been a challenge- however, numerous researchers and practitioners have agreed upon the characteristics of the lean concept. Different variations of lean definition are available and have been adopted by the organizations based on their operations and their need. Lean is also a constantly evolving process, which makes it difficult to capture it in a universal definition (Pettersen, 2009). The lean production at Toyota has been studied over decades, so that it can be defined and emulated by other manufacturers. It was suggested that the TPS experienced a natural evolution of scientific methods for manufacturing, where the workers and managers were stimulated in a learning environment and became a part of Toyota's stellar performance (Spear & Brown, 1999).

Lean Manufacturing (also called lean production) has been defined as a production process that emphasizes on making the product *flow* through production without interruption, with the help of a *pull* system that is driven by customer demand, and a culture of striving for *excellence* (Liker, 1997). Thus, the three key elements - *flow*, *pull*, and *excellence* are controlled carefully and efficiently in a lean process. Lean principles help to create a unique environment, where products with the highest quality can be made in the shortest lead time, while maintaining workers' safety, and uplifting workers' morale in the process. The five essential characteristics of a lean production (Krafcik, 1988) are:

1. The maximum number of tasks and responsibilities are transferred to those adding value to the product on the line.
2. There is an efficient system for immediately detecting defects and problems and tracing them to their root cause to make sure they do not occur. Small lot production, JIT, and the zero-defects objective, while avoiding waste, rework and scrap, are essential to uncovering problems within the plant. What Toyota calls the "Five Whys" technique is a key to solving them.
3. There is a comprehensive information system so everyone can respond quickly to any problem and understand the overall situation of the plant.

4. None of this is possible unless the workforce is organized into work teams, that need to be trained to do all the jobs in their area including machine repair, quality checking, material ordering and housekeeping – and carry out their own problem-solving.
5. Such a high level of involvement in proactively solving problems cannot work without a strong reciprocal sense of obligation between the firm and its employees.

Mr. Taichi Ohno, who had been called as the father of Toyota Production System (TPS) visualized lean production system as a structure that is shown in Figure 1. The structure is built on a solid foundation of "elimination of non-value added wastes" and the two robust pillars, *Just-in-time* production and *Autonomation*. It supports a roof that symbolizes the focus on fulfilling the customer need. Another key factor is the *worker involvement* - workers who are motivated, flexible and continuously striving to make improvements are essential in achieving Lean Manufacturing (Groover, 2015).

## Identifying Wastes (Muda)

Identifying wastes is the first step towards building the strong foundation of Lean Manufacturing. According to research conducted by the *Lean Enterprise Research Centre* (*LERC*), fully 60% of production activities in a typical manufacturing operation are the waste (Lean Enterprise Research Center [LERC], n.d.). The activities in a manufacturing process can be categorized in *value-added*, *auxiliary*, and *non-value added* activities. The lean principle demands removal of all *non-value added* activities that do not add any value to customer need. Mr. Ohno identified the following seven wastes (Ohno, 1988) as shown in Figure 2:

1. Defective production (Defects in quality).
2. Production of defective, or poor quality parts which don't meet customer need. Overproduction.
    a. To produce sooner, faster, or in greater quantities than the customer needed.
3. Waiting:
    a. Operators, machines, or parts that wait for a work cycle to be completed.

*Figure 1. The structure of a lean production system*
*Source: Groover, 2015*

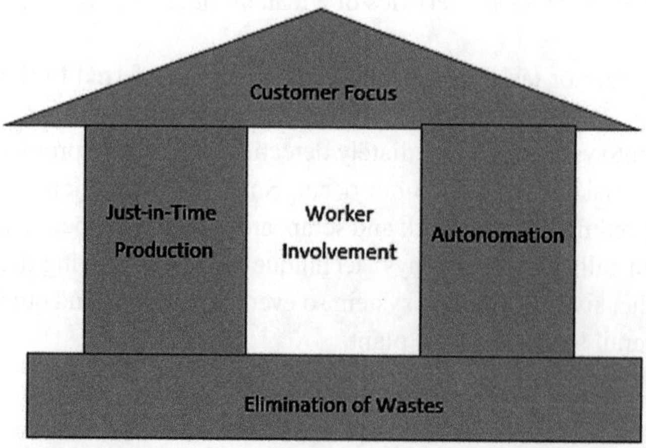

*Figure 2. Seven wastes*

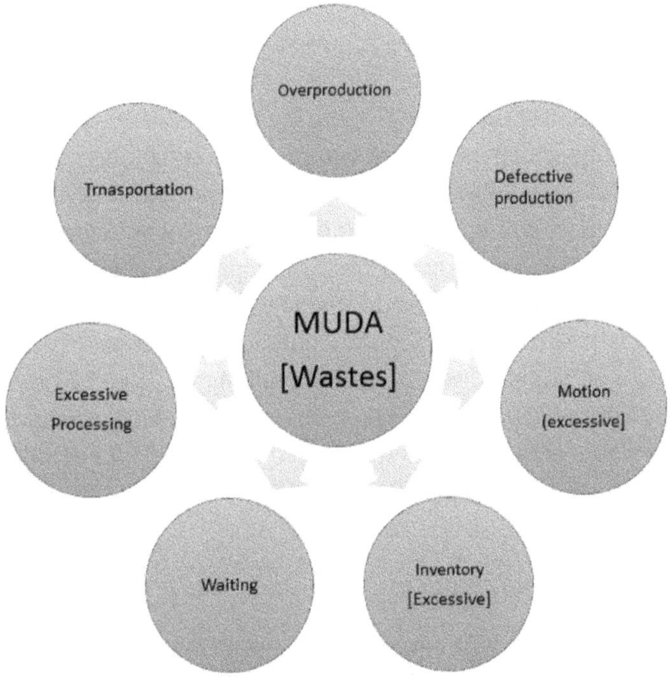


4. Transportation:
    a. Unnecessary movement of people or parts that occur between processes.
5. Inventories (Excessive):
    a. To produce an excess stock of raw material, Work-In-Progress (WIP) or finished goods beyond customer need.
6. Motion (Excessive):
    a. Unnecessary movements of people, parts, or machine within a process.
7. Excessive (Over) processing:
    a. Processing beyond the standard required by the customer.

Another waste has been identified later and added to the list as *non-used employee talent*. The acronym for the eight wastes (Dennis, 2016) can be remembered easily as *DOWNTIME*:

- **D:** Defective Production.
- **O:** Overproduction.
- **W:** Waiting.
- **N:** Non-used Employee Talent.
- **T:** Transportation.
- **I:** Inventory (Excessive).
- **M:** Motion (Excessive).
- **E:** Excessive Processing.

## Value-Stream Mapping (VSM)

VSM is a powerful tool in identifying wastes. It visually indicates the flow of production along the timeline. The map lays out the "current state" of the entire operation starting with the customer and extending back to the suppliers. It shows clearly where the value is added, or waste exists throughout the operation and thus suggests opportunities for eliminating wastes and implementing process improvement to reach the "future state". A true Value-Stream perspective provides the big picture, optimizing not just individual process or part in an individual company, but the whole production from the customer demand to the raw materials, which may include many companies in between (Rother & Shook, 2003)

VSM provides opportunities to identify and eliminate wastes by separating the activities into the value-added, or wastes. It is created by drawing the value-adding steps across the center of the map and the non-value adding steps in vertical lines. A set of symbols are used to map various activities and information of the Value Stream, regarding the process, material, managerial information etc. (Shingo, 1985). Some of those symbols are standardized and used universally, while others are used by certain organizations.

## Elimination of Wastes

The essence of LM is to continuously remove waste from the manufacturing process. A production company of any size can be converted to a lean system by eliminating non-value added wastes. The three core principles that are used to achieve waste-elimination are a) *Just-in-time* production, b) *Autonomation* and c) *Workers' involvement. Just-in-time* production implies making products that fulfill customers' need in exact quantity, and at the exact time when the customers want the products. *Autonomation* follows the principle of "*Jidoka*" which can be translated as "Automation with a human touch." *Worker involvement* throughout the production process is crucial to achieving lean.

There are a variety of soft and hard lean tools (Larteb et al., 2014) that are available to apply those three principles in the process of waste elimination. It is important to understand, however, that a customized and holistic approach is necessary to determine the most appropriate lean tools for a particular manufacturing operation. The three core principles are described next.

### Just-In-Time (JIT) Production

*Just-in-time* (JIT) principle is based on producing the right product at the right time, in the right amount as desired by the customer. In the JIT system, the use of *capital*, *equipment*, and *labor* are optimized to gain competitive advantage. The essence of JIT is to continuously seek the path of simplicity and the ways of waste elimination. It also leads to the reduction of inventory. In fact, another manufacturing strategy called "zero inventory," promoted by the *American Production and Inventory Control Society* (*APICS*), has a similar concept and philosophy as JIT.

Different elements or lean tools are integrated under the umbrella of JIT system, such as: Pull System using Kanban, Inventory Management, Total Quality Control, Set-up Time Reduction, Leveling of the Production, and Partnership with Suppliers, etc. All of these concepts complement each other and are needed to be applied simultaneously to achieve a successful JIT production. It is a company-wide commitment, where the winners are the manufacturers, the suppliers, and ultimately the customers (Lubben, 1988).

## Pull System (Kanban)

The introduction of the pull system is the most important step in LM. The traditional push system of western manufacturing production was based on the demand forecast, which usually led to increased inventories, hence increased expenses for the company. The pull system, on the other hand, follows the actual need for a product - the production does not initiate until a customer places the order. Once the order has been placed, a sequence of Kanbans pulls raw materials through the manufacturing process until the finished product is shipped (Shingo, 1989). It helps to improve the flow of information and materials and reduces inventory. Kanban is the Japanese word for "card you can see," or "Visual signal," or simply, "card."

Many companies in the United States have increased their competitive advantage by using the pull system. For example, one of the Northern California job shops implemented Kanban system for two years, and achieved many improvements; some of which are listed below:

- Eliminated the stockroom.
- Eliminated the Material Planning Department.
- Reduced inventory by 65%.
- Reduced floor shortages by 80%.
- Reduced typical order/receipt/inspect/issue response time by over 85%.

The company became a world-class manufacturer in a short period (Louis, 1992).

Another crucial benefit of the pull system is identifying the bottleneck in the manufacturing system. A bottleneck is a constraint in a manufacturing process since it holds up the production, and slows down the throughput. Hence, it is extremely undesirable. By identifying and removing the bottleneck, the throughput can be increased, the Works-in-Progress (WIPs) reduced, and the flow of the operation restored.

The following rules are strictly followed to maintain a good pull or *Kanban* system (Japan Management Association, 1989):

- Customer (downstream) processes come to withdraw items in the precise amounts specified by the Kanban.
- Supplier (upstream) produces items in the precise amounts and sequences specified by the Kanban.
- No items are made or moved without a Kanban.
- A Kanban must accompany each item, every time.
- Defective products and incorrect amounts are never sent to the next downstream process.
- The number of Kanbans is reduced carefully to lower inventories and to reveal problems.

The number of Kanban cards can be analyzed easily from the *daily demand* of a product, the *lead time* required to produce a container of parts, and the amount of safety stock needed (Heizer, Render, & Munson, 2014).

1. **Inventory Management:** Inventory reduction is a goal as well as the outcome of LM. Inventory is considered as an asset in the push system, whereas excessive inventory is one of the wastes in the pull system, and hence need to be eliminated by preventing the buildup of Inventory. Inventory management is necessary for not only achieving a reduction in the inventory cost but also to find the

causes of variability. Various manufacturing problems such as high set-up time, defective parts, etc. cause variability in the production process, and it prevents the company from reaching an optimal level of performance. Removing variability can restore smooth production flow. Inventory reduction is also dependent upon controlling the lot size. The ideal situation is to have a lot size of one as in one-piece flow, pulled from one process to the next. However, this is often not feasible. The *Economic Order Quantity* (*EOQ*) analysis can be used to calculate desired lot size that minimizes the total holding costs and setup or ordering costs (Heizer, Render, & Munson, 2014). It should be noted that as the lot size gets smaller, the constraint changes and achieving further improvement becomes increasingly challenging.

2. **Total Quality Control:** In today's environment, it is not unusual to see that the cars that are made in the USA are advertised to meet or exceed the quality benchmark provided by cars made by Japanese automakers. To compete in the global market, the function of quality control in JIT has been altered from assurance to prevention. Poor quality, defective parts are exposed quickly because of the shorter lead-time, and preventative measures can be taken quickly to obtain customer satisfaction. Also, fewer buffers (additional quantity of raw materials that are needed for production, or extra inventories of finished goods for shipping) are used in the production, thus reducing costs (Lubben, 1988). In JIT, quality is the considered as the part of the standard operation. The inspection that is done after completion of the part is considered too late, and non-value added work – thus waste. The machine operators who build those parts are held responsible for assuring the quality of the manufactured parts. The *Total Quality Control* in JIT starts with raw materials, since it's hard to produce perfect products or parts without a superior quality of raw materials. Japanese manufacturers held Edward Deming and Joseph Juran's pioneering works on quality in high esteem, and their Plan-Check-Do-Act (PCDA) has been an important tool to improve quality in LM. The six – sigma quality standard has been adopted by many organizations to achieve "*zero defect*" quality in LM (Andersson, Eriksson, & Torstensson, 2006).
3. **Set-up Time Reduction:** Set-up time is one of the most important and commonly monitored elements where significant improvement can be achieved. The competitive advantage of a manufacturing company, particularly with diversified and low volume production, often lies in the speed and accuracy of the die exchange process. *Single Minute Exchange of Dies* (SMED) is a lean tool used in reducing exchange-of-die or set-up time significantly. Mr. Shigeo Shingo, another talented industrial engineer who had contributed to TPS's success by reducing set-up times from hours to minutes, coined the term SMED. The fundamental concept of SMED is to distinguish and separate the whole set-up process in *internal* and *external* elements. The machine has to be stopped for the *internal* elements of the set-up process, whereas *external* elements can be performed while the machine operates. Identifying and performing the *external* elements lead to the reduction of the set up time by 30% to 50% (Shingo, 1985). To achieve a further reduction, the *internal* set-up elements are to be converted to *external* set-up elements by re-examining the actual functions and restructuring the *internal* and *external* elements. SMED process simplifies and streamlines the whole set-up process.
4. **Leveling the Production System (Heijunka):** The production system must be adjusted for peak and valley demand to maintain a smooth production flow. This is called load smoothing or *Heijunka* – it refers to leveling the production by both volume and variety. Keeping a consistent master production schedule helps in achieving a steady demand of resources, reduction in lead-time and a level

work schedule (Liker, 1997). Leveling of the schedule can be accomplished in several ways, some of those as identified by Groover (2015) are:

a. Process frequent small batches (ideal batch size = 1) rather than a few large batches.
b. Freezing of the schedule closest to the due dates.
c. Authorizing overtime during busy periods.
d. Using finished product inventories to absorb extra demand.

5. **Layout:** JIT layout has the objective to reduce waste resulting from lost time due to unnecessary movement of the workers. Workstations that are placed close to each other help the product flow in a linear motion. A "*U*" or "*C*" shaped cellular layout is used to achieve the one-piece flow. This type of cell is most suitable for multipurpose operations – an operator can view the entire process (Louis, 1992). Each cell is run by one or more operators who are capable of operating all of the machines in that cell. The numbers of operators in each cell can be reduced or increased depending on the order quantity. *Kanbans* are used to control the rate of work-flow and the inventory in process at any given time (Lubben, 1988).
6. **Partnership With Suppliers:** Earlier in the mass production system of manufacturing, companies were vertically integrated with suppliers. However, the complexity and sophistication of modern production system resulted in many original equipment manufacturing (OEM) suppliers. Many products are now a combined effort of OEM suppliers and the final assembly company. Thus the relationship between businesses and their suppliers has become more interdependent and required to be mutually beneficial. For example, if the manufacturing company insists on the lowest bid, or provides very little job security, the supplier may present an artificially low bid to get the contract. However, later he may either raise the price, or downgrade the quality. Thus, it will deter long-term, and strong relationship between that particular supplier and the company. In the long run, it results in price increase, slower response, and poorer product quality (Leavy, 1994).

In JIT, a unique relationship is fostered between the company and the suppliers to develop trust and commitment towards a long time partnership. The objective is to remove waste and drive down costs by working together. To create a lean ecosystem, all partners in the supply chain have to develop along with the manufacturer (Sanders, Elangeswaran, & Wulfsberg, 2016). A true partnership with a supplier can be built by sharing the knowledge and skills in improving quality, eliminating long set-ups, reducing inventory, and monitoring the progress. Supplier's ability to perform depends on the trust, communication, linearity of production, and the time and visibility to make the changes (Lubben, 1988). Trust and collaboration are critical in the global partnership to successfully achieving Lean Manufacturing. The interdependency between the manufacturer and the suppliers in China has been studied extensively and its vital importance in JIT practice has been noted (Chen & Chen, 1997). It was found that a more open and reliable communication between the company and the supplier led to improved quality, cost and delivery performances (Richeson, Lackey, & Starner Jr., 1995). Sharing company's manufacturing schedules with suppliers are one of the ways to strengthen communication with them.

Another important criterion is the location of the vendors. The proximity of the suppliers leads to many advantages since they could be involved with the project at an early stage, communications become easier, and cooperation grows in solving manufacturing problems if and when they arise. The overall cost of business is also lessened by reduced transportation cost. To facilitate JIT, many automakers such as Toyota and GM have built a local suppliers' base near their plants. (Heizer, Render, & Munson, 2016)

## Autonomation (Jidoka)

Autonomation or *Jidoka* stands for "automation with a human touch." – Mr. Sakichi Toyoda created the term. It implies designing and building operations and equipment to free people from the machines in order to perform value-added work that is appropriate for humans (Liker, 1997). The goal of autonomation is to build partially automated equipment and manufacturing process, which can be stopped when the defects are detected. This concept provides the equal right to every employee to stop the machine immediately if anything goes wrong.

Two of the lean tools that are used to achieve autonomation are *Error Proofing* and *Total Productive Maintenance* (TPM).

1. **Error Proofing (Poka-Yoke):** Error Proofing or *Poka-Yoke* is a tool for quality control in preventing defects. It authorizes every worker in the production process not to accept, create, or pass along a defective part. The *Poke- Yoke* technique was developed by Mr. Shigeo Shingo with a goal to reach zero breakdowns (Shingo, 1986). The *Poka-Yoke* responses to error are as follows:
    a. Stop the process.
    b. Provide an audible or visible warning to alert operator and other workers.

It is one of the most effective ways to prevent a defective part from reaching the customer. Each machine is equipped with simple automatic stop devices that prevent faulty products from being sent on to the next process. The location of the malfunctioning machine or workstation is notified on the *Andon* board (a centrally located bulletin board) to inform other employees. The Poke-Yoke response also includes audible and/or visible warning. Supervisors and workers gather immediately to find the cause of a breakdown, and find a solution to the problem - the repair work is done right away. Poke-Yoke devices are designed to ensure the two fundamental aspects of manufacturing – *Safety,* and *Quality.* A scoring system for evaluating the effectiveness of those devices has been attempted in a recent study (Saurin, Ribeiro, & Vidor, 2012).

2. **Total Productive Maintenance (TPM):** Total Productive Maintenance (TPM) is a system, which integrates *preventive* maintenance with *predictive* maintenance to avoid *emergency* maintenance. *Preventive* maintenance is a routine repair procedure to avoid breakdown of the machines, whereas in *predictive* maintenance, precautions are taken to avoid malfunctions that are anticipated ahead of time. Finally, *emergency* maintenance is necessary to repair the breakdown of the machine during the production process. Mr. Seiichi Nakajima is regarded as the father of TPM since he had developed this system thoroughly. His proactive and preventative maintenance techniques for improving equipment stability were based on eight principles as follows (Nakajima, 1988):
    a. Autonomous maintenance.
    b. Planned Maintenance.
    c. Quality maintenance.
    d. Focused Improvement.
    e. Early Equipment Management.
    f. Training and Education.
    g. Safety Health Environment.
    h. Administration/Office TPM.

The use of a TPM program builds a shared responsibility that inspires greater participation by plant floor workers. In the appropriate environment, this can be very useful in improving productivity. A positive and significant relationship between TPM and low cost, high quality and strong delivery performance in manufacturing has been found in an earlier study (McKone, Schroeder, & Cua, 2001).

Often TPM and Total Quality Management (TQM) are used interchangeably since both of them are geared to improve the quality. However, the goal of TQM is to enhance the product quality, services, and the customer satisfaction, whereas TPM additionally focuses on excelling the process of making the product, the work environment, and the leadership in the organization.

## Worker Involvement

The talent and co-operation of the employees make lean production possible. It starts with showing respect for people at the workplace. In LM, employees are treated as knowledge workers (Heizer, Render, & Munson, 2014). They are empowered to apply JIT and Autonomation principles and make a continuous improvement at all times during the manufacturing process. Workers are encouraged to enhance their technical and problem-solving skills. The peers and their supervisors recognize their talents. Building trust in employees is an important part of the lean philosophy. It increases not only employee morale but also the productivity of a company (Melohn, 1983). The elements or lean tools used by the employees include *Visual Management and 5S*, *Standard Work Practice*, and *Continuous Improvement*, which are described next.

1. **Visual Management and 5S:** Visual Management implies that the status of work at the factory floor should be evident just by looking at it. Workers practice various regulations to meet visual management criteria – some of which are listed below:
    a. Objects that block the view are not permitted inside the plant.
    b. The build-up of *Work-in-Process* (*WIP*) is limited only up to a certain height.
    c. Andon boards that are located above the assembly line highlights the status of the workstations.
    d. Worker training includes the use of pictures and diagrams to document work instructions.

Another tool of visual management is the 5S system. It is a straightforward and inexpensive housekeeping tool, which assists workers in maintaining a clean and organized work environment. The five elements of this 5S system are:

a. **Sort (*Seiri*):** Reducing waste by eliminating anything not required in the work area.
b. **Set in order (*Seiton*):** Organizing the work items into clearly designated storages.
c. **Shine (*Seiso*):** Cleaning the work area thoroughly – it is part of an inspection.
d. **Standardize (*Seiketsu*):** Ensuring that the first three steps are done in a standardized manner by using visual aids, and 5S manuals.
e. **Sustain (*Shitsuke*):** Practicing the standards as per step 4 so that it becomes a natural part of your work.

This system not only helps to find the right tools and parts quickly, but it also helps to locate problems in the machines and the workplace such as fluid leaks, cracks in the devices, etc. The importance of both the technical (visible) and philosophical (invisible) aspects of the 5S system has been explored by researchers in a managerial framework (Gapp, Fisher, & Kobayashi, 2008).

2. **Standard Work Practice:** Work Practice is the key to producing high-quality parts consistently. Standardization eliminates wastes by removing the possibility of defects as well as over-processing. Tools, processes and workplace arrangements should be as simple and standardized as possible. Some of the standard work practices that were followed at Toyota Production System (TPS) (Heizer, Render, & Munson, 2014) were:
   a. Work shall be completely specified as to content, sequence, timing, and outcome,
   b. Product and service flows must be simple and direct,
   c. Any improvement must be made by the scientific method at the lowest possible level of the organization.

The standardization of work ensures best practices, and the current best practice becomes the baseline for further improvement. As a part of continuous improvement, work standards are often changed to incorporate a new idea or a better method.

3. **Continuous Improvement (Kaizen):** Continuous Improvement is an integral part of the lean culture, also called as *Kaizen*. It is usually implemented with the help of worker teams, sometimes called *Quality Circle* (Groover, 2015). These groups are organized to address specific problems related to quality, productivity, cost, safety, maintenance and other areas of interests that may occur in the workplace. Its goal is to improve the production process using the knowledge and talent of the employees. Workers from all levels participate in a *Kaizen* event in small or large groups to share ideas and develop innovative ways to implement improvements or resolve problems. Participants also learn to use scientific methods to perform experiments, to identify wastes and to improve the manufacturing process. Two of the techniques that are used for continuous improvement are:
   a. ***Plan-Do-Check-Act* (PDCA) Cycle:** It is also known as *Shewhart cycle,* or *Deming cycle.* It represents an ongoing effort to improve the production process of an organization. Dr. Edwards Deming, the father of modern quality control, popularized this scientific method of quality control and was regarded highly in Japan for manufacturing innovations after World War II. According to Deming (1982), the PDCA cycle emphasizes the following steps:
      i. *Plan* = Establish process to reach Target output.
      ii. *Do* = Implement *Plan* and measure/ collect data.
      iii. *Check* =Study/Analyze the data collected.
      iv. *Act* = If the results are better than existing process, change the new plan as the new standard, otherwise keep the old standard.

The team members work together in improving the production process in incremental steps, rather than initiating radical changes. The third step of PCDA always provides some insight into how to improve the process, and the cycle can be repeated over time with the new ideas.

b. **Five Whys:** This is a form of *root cause analysis*, where answers to the 5 "why" questions lead to finding the root causes of a problem. Mr. Sakichi Toyoda developed this technique as a part of problem-solving training at TPS. The application of this method in the manufacturing industry provides a fact-based and structured approach to identify and eliminate defects (Murugaiah, Jebaraj, & Srikamaladevi, 2010).

*Figure 3. Deming cycle of continuous improvement*

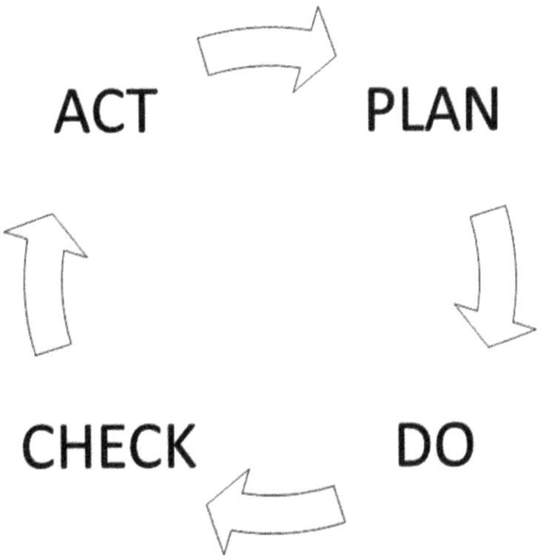


## FUTURE RESEARCH DIRECTIONS

Lean Manufacturing (LM) has been adopted by a variety of manufacturing systems, such as product, process, and fixed layout; batch and mass production; discrete and continuous production. It has become a system composed of highly integrated elements and a wide variety of management practices (Bhamu & Singh, 2014). There are numerous publications on lean practice, which have however created fragmented views on the various issues regarding implementation of lean. There is also a deficiency of unified theories and practices to manage the systemic, human and organizational dimensions of Lean Manufacturing (Marodin & Saurine, 2013). Future research should focus on integrating the earlier studies to provide a better understanding of critical issues such as, why companies fail or succeed in implementing lean practices, or what the various factors of Lean Manufacturing are that are responsible for improving performance metrics. The effect of specific variables such as company size, product or process type, etc. on the success of implementing lean should also be investigated.

Several additional areas of LM can be explored in future research. For example, how to improve the communication between manufacturers and suppliers? Also, timely delivery of products are not guaranteed in the present logistics systems due to various reasons such as the inadequate status of goods being shipped, the discrepancy between the required and transported goods and unforeseen time delays during transfer of goods. Innovative ways are to be explored to eliminate such wastes in these areas. The key concept of LM is to maintain a smooth flow of processes to create products at the required pace of customers with a minimum waste (Shah & Ward, 2003). However, production flow can be disrupted any time due to machine breakdown, capacity shortage, errors in inventory counting, delays in decision making, etc. Alternative ways are to be investigated to maintain a smooth flow of production and reduce waste build-up (Sanders, Elangeswaran, & Wulfsberg, 2016). Identifying the key variables which affect the performance metrics of LM and defining reliable scales for measurement are of utmost importance. Some of the current research on this topic are being explored using fuzzy logic based methods (Sushilawati, Tan, Bell, & Sarwar, 2015).

At the juncture of the fourth industry revolution (also called Industry 4.0), automation and cloud computing are in the forefront of the future in manufacturing. The goal of Industry 4.0 is to enhance the human-machine interaction by using principles of Cyber-Physical Systems (CPS), Internet of Things (IoT), and smart systems with future-oriented technologies (Sanders, Elangeswaran, & Wulfsberg, 2016). As noted earlier, implementation of lean has been challenging and time consuming due to lack of communication between various stakeholders in the manufacturing sector. Industry 4.0 initiatives can be used to explore possible solutions. For example, IoT can be used to network the entire factory to form a smart environment. Production facilities, smart machines, suppliers, and customers could be connected digitally to result in a seamless and integrated communication system. This production network will also reduce wastes to a minimum.

Automation has been an integral part of LM from the beginning. Future advancement will include the application of robotics and other automation technologies that are being integrated for upgrading production facilities suitable for Computer-Integrated-Manufacturing (CIM). 3D printing and Virtual Manufacturing are also being explored in reducing wastes, and thus reducing production costs. For example, 3D printing helps to reduce the time, and raw materials that are needed to make prototypes and Virtual Manufacturing contributes to visualizing and modifying the various attributes of the manufacturing process on the computer-based environment. Integration of other technologies such as Radio-Frequency Identification (RFID), IoT, and Advance Analytics should be explored to transform any industry to a LM without the need to struggle over time to achieve lean manufacturing goals.

## CONCLUSION

When Mr. Ohno started at Toyota in the 1970s, the productivity of the company was one-ninth that of their Detroit counterparts. In recent years, it has consistently ranked amongst the top two of the world's automakers. Lean Manufacturing was the key to their success. It took decades for them to perfect their manufacturing process. The manufacturing firms in the United States who had adapted lean philosophy had to struggle and make a long term commitment before achieving success. However, they had experienced the tremendous improvements in the key metrics such as cost, throughput time, inventories, defects, capital spending, space utilization, job-related injuries, responsiveness to changing customer needs, etc. (Liker, 1997).

The manufacturing sector in the US had been experiencing no or very little growth in the last few decades. In 2008, the US economy faced with a major economic crisis – the largest auto company in the United States, General Motors had to declare bankruptcy and was helped by the US Government to survive. Since 2008, the US economy somewhat recovered - however, the manufacturing sector is still faced with challenges ahead. With globalization, low population growth, aging demography, and moderate economic growth in the developed world (Roy & Roy, 2015), Western manufacturers are looking carefully to find a way to gain competitive advantage. Many manufacturing organizations have adopted lean practices successfully in the United States, Europe, South Korea, and Mexico, as well as in India (Ghosh, 2012), and in China (Taj, 2008). In a survey in the USA, when asked whether their companies are implementing lean operating principles, 56 percent of manufacturing executives answered yes in 2008, and 61 percent answered yes in 2009. (Black & Phillips, 2010). Automation, digitization and integration of resources will continue to transform the manufacturing industries to lean until the fourth industrial revolution is fully realized.

In modern times, sustainability is a major concern for the customers. Many organizations are searching for ways to create environmentally renewable products, which would produce minimal waste to the landfill. Lean practices are considered green, since the principles of LM are based on sustainable practices, such as driving out waste, maximizing the use of resources and economic efficiency, and focusing on issues outside the immediate firm (Heizer, Render, & Munson, 2014). Lean Manufacturing, if applied correctly will produce environment-friendly products that will enhance customer satisfaction, and increase competitive advantage for the manufacturers.

## REFERENCES

Andersson, R., Eriksson, H., & Torstensson, H. (2006). Similarities and differences between TQM, six sigma and lean. *The TQM Magazine*, *18*(3), 282–296. doi:10.1108/09544780610660004

Bhamu, J., & Singh, S. K. (2014). Lean Manufacturing: Literature review and research issues. *International Journal of Operations & Production Management*, *34*(7), 940–876. doi:10.1108/IJOPM-08-2012-0315

Black, I. T., & Phillips, D. T. (2010). The Lean to Green Evolution. *Industrial Engineer*, *42*(6), 46–51.

Chen, S., & Chen, R. (1997). Manufacturer-supplier relationship in a JIT environment. *Production and Inventory Management Journal*, *38*(1), 58–70.

Deming, W. E. (1982). *Out of the Crisis*. Cambridge, MA: MIT Press.

Dennis, P. (2016). *Lean Production implified: A plain-language guide to the world's most powerful production system*. Boca Raton, FL: CRC Press.

Ford, H. (1923). *Popular Research Topics Henry Ford Quotations*. Retrieved October 13, 2001, from https://www.thehenryford.org/collections-and-research/digital-resources/popular-topics/henry-ford-quotes/

Fullerton, R. R., Kennedy, F. A., & Widener, S. K. (2014). Lean manufacturing and firm performance: The incremental contribution of lean management accounting practices. *Journal of Operations Management*, *32*(7), 414–428. doi:10.1016/j.jom.2014.09.002

Gapp, R., Fisher, R., & Kobayashi, K. (2008). Implementing 5S within a Japanese context: An integrated management system. *Management Decision*, *46*(4), 565–579. doi:10.1108/00251740810865067

Ghosh, M. (2012). Lean manufacturing performance in Indian manufacturing plants. *Journal of Manufacturing Technology Management*, *24*(1), 113–122. doi:10.1108/17410381311287517

Groover, M. P. (2015). Automation, Production Systems, and Computer-Integrated Manufacturing (4th ed.). Upper Saddle River, NJ: Pearson Higher Ed.

Heizer, J., Render, B., & Munson, C. (2014). Principles of operations management: sustainability and supply chain management (9th ed.). Upper Saddle River, NJ: Pearson Higher Ed.

Japan Management Association (Ed.). (1989). *Kanban Just-In-Time at Toyota Management: Begins at the Workplace*Lu, D. J., Trans.). Cambridge, MA: Productivity Press.

Koenigsaecker, G. (2012). *Leading the lean enterprise transformation*. Boca Raton, FL: CRC Press.

Krafcik, J. F. (1988). Triumph of the lean production system. *MIT Sloan Management Review, 30*(1), 41–50.

Larteb, Y., Haddout, A., Benhadou, M., Manufacturing, L., Yang, C., Yeh, T., & Valero, M. (2015). Successful lean implementation: The systematic and simultaneous consideration of soft and hard lean practices. *International Journal of Engineering and General Science, 3*(2), 1258–1270.

Lean Enterprise Research Center (LERC). (n.d.). *The Key Lean Thinking Principles*. Retrieved from http://www.leanenterprise.org.uk/what-is-lean-thinking/what-is-lean-thinking-and-key-lean-thinking-principles.html

Leavy, B. (1994). Two strategic perspectives on the buyer-supplier relationship. *Production and Inventory Management Journal, 35*(2), 47–60.

Liker, J. K. (Ed.). (1997). *Becoming lean: Inside stories of US manufacturers*. Portland, OR: CRC Press.

Louis, R. (1992). *How to Implement Kanban for American Industry*. Cambridge, MA: Productivity Press.

Lubben, R. T. (1988). *Just-in-Time manufacturing: an aggressive manufacturing strategy*. New York, NY: McGraw-Hill Companies.

Marodin, G. A., & Saurin, T. A. (2013). Implementing lean production systems: Research areas and opportunities for future studies. *International Journal of Production Research, 51*(22), 6663–6680. doi:10.1080/00207543.2013.826831

McKone, K. E., Schroeder, R. G., & Cua, K. O. (2001). The impact of total productive maintenance practices on manufacturing performance. *Journal of Operations Management, 19*(1), 39–58. doi:10.1016/S0272-6963(00)00030-9

Melohn, T. H. (1983). How to build employee trust and productivity. *Harvard Business Review, 61*(1), 56–68.

Murugaiah, U., Jebaraj, B. S., Srikamaladevi, M., & Muthaiyah, S. (2010). Scrap loss reduction using the 5-whys analysis. *International Journal of Quality & Reliability Management, 27*(5), 527–540. doi:10.1108/02656711011043517

Nakajima, S. (1988). *Introduction to TPM: Total Productive Maintenance (Translation)*. Cambridge, MA: Productivity Press.

Ohno, T. (1988). *Toyota production system: beyond large-scale production*. Boca Raton, FL: CRC Press.

Pettersen, J. (2009). Defining lean production: Some conceptual and practical issues. *The TQM Journal, 21*(2), 127–142. doi:10.1108/17542730910938137

Richeson, L., Lackey, C. W., & Starner, J. W. Jr. (1995). The effect of communication on the linkage between manufacturers and suppliers in Just-in Time Environment. *Journal of Supply Chain Management, 31*(1), 21–34.

Rother, M., & Shook, J. (2003). *Learning to see: value stream mapping to add value and eliminate muda*. Cambridge, MA: Lean Enterprise Institute.

Roy, A., & Roy, M. (2015). Antecedents and consequences of impending population implosion in the developed world: Implications for business systems. *International Journal of Sustainable Society*, *7*(2), 151–172. doi:10.1504/IJSSOC.2015.069913

Sanders, A., Elangeswaran, C., & Wulfsberg, J. (2016). Industry 4.0 Implies Lean Manufacturing: Research Activities in Industry 4.0 Function as Enablers for Lean Manufacturing. *Journal of Industrial Engineering and Management*, *9*(3), 811–833. doi:10.3926/jiem.1940

Saurin, T. A., Ribeiro, J. L. D., & Vidor, G. (2012). A framework for assessing poka-yoke devices. *Journal of Manufacturing Systems*, *31*(3), 358–366. doi:10.1016/j.jmsy.2012.04.001

Shah, S. R., & Ward, P. T. (2003). Lean manufacturing: Context, practice bundles, and performance. *Journal of Operations Management*, *21*(2), 129–149. doi:10.1016/S0272-6963(02)00108-0

Shingo, S. (1985). *A Revolution in Manufacturing: The SMED System*. Cambridge, MA: Productivity Press.

Shingo, S. (1986). *Zero quality control: Source inspection and the poka-yoke system*. Boca Raton, FL: CRC Press.

Shingo, S. (1989). *A Study of the Toyota Production System from an Industrial Engineering Viewpoint*. Portland, OR: Productivity Press.

Spear, S., & Brown, H. K. (1999, September). *Decoding the DNA of the Toyota Production System. Harvard Business Review*.

Susilawati, A., Tan, J., Bell, D., & Sarwar, M. (2015). Fuzzy logic based method to measure degree of lean activity in manufacturing industry. *Journal of Manufacturing Systems*, *34*, 1–11. doi:10.1016/j.jmsy.2014.09.007

Taj, S. (2008). LM performance in China: Assessment of 65 manufacturing plants. *Journal of Manufacturing Technology Management*, *19*(2), 217–234. doi:10.1108/17410380810847927

Womack, J. P., Jones, D. T., & Roos, D. (1990). *The Machine that Changed the World*. New York, NY: Simon and Schuster.

Woolson, D., & Husar, M. (1997). Transforming a plant to lean in a large, traditional company: Delphi Saginaw Steering Systems, GM. In J. K. Liker (Ed.), *Becoming Lean: Inside Stories of U.S. Manufacturers* (pp. 120–158). Portland, OR: Productivity Press.

## KEY TERMS AND DEFINITIONS

**5S:** It consists of five housekeeping tools that help to manage the factory floor visually.

**Autonomation:** Use of automation with human intelligence in the manufacturing process.

**Continuous Improvement:** Everyone in the company takes part in improving the manufacturing process continuously over time. It is also called Kaizen in Japanese.

**Just-In-Time:** It means producing the right product in the right amount and in the right time when it is needed by the customer.

**Lean:** It is a philosophy of eliminating or reducing wastes.

**Quality:** It defines how well the manufactured product meets its design and ultimately the customer's need.

**SMED:** It represents the time needed to change machine set-up.

**Waste Elimination:** Anything which does not add value to the manufacturing process is a waste, and it needs to be removed to make the process lean.

# Section 4

# Smart Factories and Industry 4.0

*Smart Factory concepts describe fully networked, autonomous factories and form an essential part of flexible, however still highly efficient manufacturing systems. From a business perspective, the term industry 4.0 stands for a new organizational step of controlling the entire value chain along the product life cycle. The requirements for the further development of existing manufacturing systems towards a Smart Factory are analyzed and studied in this section.*

# Chapter 17
# A Maturity Model to Organize the Multidimensionality of Digitalization in Smart Factories

**Peter Schott**
*FAU Erlangen-Nuernberg, Germany*

**Matthias Lederer**
*Friedrich-Alexander-Universität Erlangen-Nürnberg, Germany*

**Sina Niedermaier**
*FAU Erlangen-Nuernberg, Germany*

**Freimut Bodendorf**
*Friedrich-Alexander-Universität Erlangen-Nürnberg, Germany*

**Matthias Hafner**
*FAU Erlangen-Nuernberg, Germany*

## ABSTRACT

*Smart Factory concepts describe fully networked, autonomous factories and form an essential part of flexible, but still highly efficient production systems. The requirements for the further development of existing production environments towards a Smart Factory are multidimensional and vastly complex. Many companies therefore fail in the structured realization of a holistic Smart Factory concept. They either focus one dimension of the challenge or merely address the maximum penetration of powerful technologies. This chapter addresses this issue and describes a systematic development path towards a Smart Factory by means of a domain specific maturity model. Based on the analysis of existing maturity models, requirements are derived which must be considered when realizing a Smart Factory. In total, 20 design fields (e.g., degree of intelligence, communication protocols, human-machine-interface and IT security) and respective detail descriptions result from this research. They holistically structure the relevant fields of action to pursue a Smart Factory.*

DOI: 10.4018/978-1-5225-2944-6.ch017

## INTRODUCTION

The industry is in a fundamental shift. With the opening of new markets and increasing international competition, the environment for manufacturing companies is becoming increasingly dynamic as well as unpredictable (Bauer et al., 2014, p. 6). The trend towards product individualization up to lot size one and the overall volatile market conditions require production systems which are able to flexibly adapt to new environmental circumstances (Roth, 2016, p. 5). In order to meet this demand for customer-specific products in developed industrialized countries, solutions are needed which allow a cost-effective, complex and varied production (Baumer, 2014, p. 264). In this context, the fourth industrial revolution, known as "Industry 4.0", is currently heavily discussed. Industry 4.0 represents an approach in which information and communication technology (ICT) enables networked production in an entirely new way (Roth, 2016, p. 5; Bischoff et al., 2015, p. 1). An elementary component of this visionary form of manufacturing is the fully self-controlled, modular and intelligent factory, the so-called "Smart Factory". Powered by the opportunities of IT paradigms and technologies, physical production processes merge with the digital data recorded within smart factories (Russwurm, 2013, p. 21). By applying technologies for information generation, networking and processing, many elements of a production system can be digitalized and thus automated within processes. For example, the traditional planning functions of production systems that base on statistical models and assumptions may be replaced by real-time data driven production control (Bischoff et al., 2015, pp. 3, 8).

Thus, the interconnection of heterogeneous system elements such as sensors, actuators, workpieces, machines, and planning and control systems requires a network of previously decoupled and proprietary information and production systems (Wolff & Schulze, 2013, p. 11; Siepmann, 2016, pp. 726).

The combination of already established, independent technologies and methods from different application areas (such as internet technologies, bio-informatics, etc.) to reach a uniformly effective solution is often described as the essential innovation and most challenging task of Smart Factory initiatives (Siepmann, 2016, p. 37).

For companies, it is a major challenge to see which technologies and methods are needed and how they can be orchestrated to accomplish a Smart Factory. Due to the complexity and the high investment of such projects, many companies are implementing delimited components to gradually approach a Smart Factory according to the principles of Industry 4.0 (Spath et al., 2013, p. 120). Keeping this incremental implementation plan in mind, first a clear identification of the current system state and then a profound development plan become necessary for Smart Factory initiatives.

The relevant literature currently offers only limited support to manage such challenging projects. Individual development areas (e.g. employee competences and processes) as well as single technologies (e.g. sensors) are either viewed in isolation or emanate significantly from the overall goal of a complete penetration of Industry 4.0 technologies. However, a comprehensive and standardized development path is still not available (Bischoff et al., 2015, p. 69). Due to the lack of a holistic consideration of these heterogeneous aspects, strategic conclusions and concrete recommendations for actions cannot be easily identified (BDI 2015, p. 27). In a nutshell, a structured and defined approach including all fields relevant for a Smart Factory is missing. This hinders many implementation projects focusing on the implementation of a Smart Factory (Bischoff et al., 2015).

For companies, initiatives and projects in the context of intelligent factories come with numerous problems. They mainly arise from the complexity of the overall task. So it is necessary to provide a procedure model for the gradual introduction. Such a model should offer paths for technologies and methods how to develop the current production environment towards a Smart Factory. In this contribution, requirements for such a model are developed (section 3) and tested against existing models (section 4). Therefore, basics on Industry 4.0 and Smart Factories are discussed in section 2. By combining the relevant enablers of a Smart Factory from the angle of different dimensions, the requirements can serve as a basis for assessing the maturity of production resources and the company for practitioners. Consequently, a multidimensional maturity model for smart factories is introduced (section 5). The model divides the requirements that come along with the transformation of a traditional production environment into a Smart Factory into different design fields. These fields and their subdivision into different capability levels serve as basis for a systematic implementation path towards a Smart Factory.

## BASICS OF A SMART FACTORY

This section presents theoretical and practical foundations for projects in the field of Smart Factories. Experience shows that Smart Factory projects are organized in interdisciplinary teams. In order to get to know the specificities which are also important for the requirements and different design levers of the maturity model, the basic principles are laid down in the following.

### Industry 4.0

Projects in the field of Industry 4.0 are supported and forced by the federal governments of many developed countries (e.g., Germany). Oftentimes, these programs are launched to take the advantages of the fourth industrial revolution.

From a business perspective the term industry 4.0 stands for a new organizational step of controlling the entire value chain along the product life cycle. The link of production with information and communication technology enables the production of individual and at the same time cost-effective products. Typical requirements for this future development are technological ones, such as the availability of relevant information in real-time and the continuous networking of the entities involved in the manufacturing process. Furthermore, the ability to derive the relevant information from the obtained data within the manufacturing process is crucial. Industry 4.0 has the potential to create new business models by making use of intensive networking along the value chain or within entire value networks. Business models may also arise from the coupling of products and intelligent services to "hybrid products" (BMBF, 2014, p. 16; Bauernhansl et al., 2015, p. 1; Kaufmann, 2015, p. 2; BMWi, 2015, p. 19).

Despite the current discussion in research and practice, the different perceptions and characterizations of Industry 4.0 show that there is currently no clear definition (Bauer et al., 2014, p. 18).

To come to a conceptual understanding for Industry 4.0 initiatives, this contribution is based on the definition by Bischoff et al. (2015). They describe that the term stands for the further development of production systems by linking the real and the digital world. This connection is created by self-controlling cyber-physical systems (CPS) which are equipped with embedded systems. Industry 4.0 describes explicitly the vertical (within a company) and the horizontal (across different business sectors as well as across several companies along the supply chain) linking of CPSs. As a result, all partners of a network profit from an efficient, decentralized and flexible production of products or services (Bischoff et al., 2015, p. 12).

A CPS is characterized by the connection of physical with virtual objects. Their networking is realized via open and global information networks such as the internet. Equipped with embedded systems, sensors and actuators, a CPS is able to capture and evaluate physical data and thus actively react to environmental influences (Geisberger et al., 2011, p. 13; Bauernhansl, 2014, p. 15 f .; Hertel, 2015, p. 726). Embedded systems are computer systems which are integrated into objects such as products, machines or materials and enable the necessary data processing (Simon, 2013, p. 38).

By linking the individual CPSs to a complete system, an intelligent factory – the Smart Factory – is finally created. It acts as a core element of industry 4.0 and will be discussed in the next section.

## Smart Factory

The customers' demand for individualized products and for a high availability leads to an increasingly complex production environment for companies. This complexity can only be controlled by an increase in decentralization and autonomy of the production modules involved in manufacturing (Bauernhansl, 2014, pp. 14, 17 f.). An intelligent factory is prepared for the future if it provides a production environment in which products, entire production lines and logistics systems communicate with each other and in which they control themselves autonomously without any human intervention or monolithic, rigid software systems (Bauernhansl, 2014, p. 16).

This factory of the future can be characterized by the following elements (Bauernhansl, 2014, p. 17):

- **Intelligent Objects:** Intelligent or even so-called smart objects include items which are equipped with advanced functions, such as the acquisition, processing and storage of data, as well as the ability to interact with their environment. An intelligent object can be a single product, which stores and delivers information on its processing steps as well as an intelligent system in a whole. In the last case, this is understood in the sense of a CPS.
- **Communication Technology Network:** The ability to communicate and interconnect with and between intelligent objects is a fundamental prerequisite for a Smart Factory.
- **User Interaction and Information Provision:** Due to the increasing complexity of processes and exponentially increasing data volumes it is necessary to support people with new technologies. This is why the interaction between man and machine – called Human Machine Interaction (HMI) – plays an important role of todays' discussions. A context-adaptive support proactively delivers information to humans. Managers as well as front line workers have access to production data and facilities at all times, but the system should, however, be able to act independently.
- **Vertical Integration and Data Continuity:** The hierarchical control architectures of the classical automation pyramids (see next section) are resolved by stronger vertical integration.
- **Changeable Production Systems:** A flexible production system can be realized by the flexible combination of intelligent objects with self-controlled production systems (Zühlke et al., 2012, p. 31).

In summary, the combination of the elements mentioned enables a fast and flexible response to customer requirements and an automated production with simultaneously a high number of variants for small lot sizes. The basis for the last two points (vertical integration and data continuity and changeable production systems) has its basis in the automation of service-orientation.

## Automation

A major foundation for a continuous information technology network and the implementation of a Smart Factory is service orientation. Instead of a detailed and rigid hierarchical planning and control, the increasing complexity of the production environment is to be controlled in decentralized and flexible optimizations of single business areas (Bischoff et al., 2015, p. 116). The architectural paradigm of service orientation, the so-called service-oriented architecture (SOA), has its origin in the technological field of distributed systems. Its goal is to divide different applications and components into single modules which are called services (Melzer, 2010, p. 13). A distributed IT system is a combination of independent functional modules which are connected via a transport system and perform valuable task within a given business process. These modules are located on different computers and typically do not have a common memory. However, they are used together in workflows to work on a common goal (Bengel, 2014, p. 4; Schill & Springer, 2007, p. 3 f.).

Each service represents a complete functional area, which is identifiable throughout the network and can be used by other services in this network (Bischoff et al., 2015, p. 117). By aggregating services – called orchestration – functionalities can be bundled and new services are provided (Schill & Springer, 2007, p. 3 f.). As a result, the concept of service orientation makes it possible to map entire machines or systems by bundling individual functionalities into services (Bischoff et al., 2015, p. 116).

The traditional automation pyramid (c.f. Figure 1) represents a hitherto rigid hierarchical model of the functional system structure. In addition, the data and information are concentrated on individual levels of the system (VDI, 2013, p. 4) and operate with isolated data silos. The pyramid is divided into a total of six levels, which are presented below.

The company's top management uses an Enterprise Resource Planning (ERP) system to carry out production planning and order processing. The operational management is one level lower and uses a Manufacturing Execution System (MES) to handle production planning as well as the collection of product data. The measured values of the lower levels are collected for monitoring purposes in supervisory control and data acquisition (SCADA) systems. On the control level, Programmable Logic Controllers (PLC) control the individual systems. The field level represents the interface to the production process

*Figure 1. Traditional automation pyramid (own illustration, derived from Schöning & Dorchain, 2014, p. 550)*

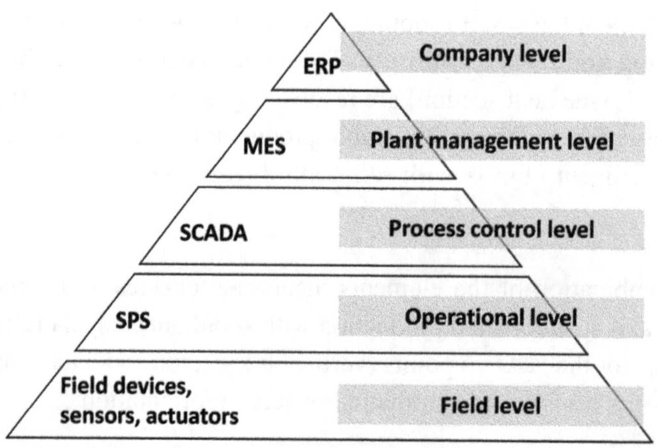

through input and output signals. At the lowest level, the production process is represented on a sensor and actuator level which serves the data collection for all of the superordinate layers mentioned (Fallenbeck & Eckert, 2014, p. 405).

By implementing CPSs the information constraint between the layers is lifted. It results in a flattening and gradual dissolution of the traditional and rigid automation pyramid (c.f. Figure 2). Assets like PLCs, ERP and MES systems, or field devices (c.f. Figure 2 on the right, depicted as white, grey and black network nodes) from different hierarchical layers can connect directly to data streams on the field level (c.f. Figure 2 on the right, depicted as grey edges between network nodes and physical machinery on the shop floor (grey boxes)). The network-like structure consequently allows access to data, information, and services across the layers. This increases the amount of information within the system enormously and enables direct feedback loops without the limitations of multi-level data integration and transportation (Schöning & Dorchain, 2014, p. 550).

For example, the mapping of data from different levels of detail may result in intelligent monitoring views which offer an easy drill-down into production details. Interdisciplinary data exchange between departments and companies can also be easily realized.

Summarizing, the service orientation enables the realization of an adaptive configuration. Company divisions or departments can be modularized as autonomous performance units, even if they work together in processes targeting the same objectives. Service units can be made available to the company or to partners in a value network. In consequence, this leads to the idea to define "Everything as a Service" (XaaS) (Bischoff et al., 2015, p. 116). The implementation of a network structure according to the principle of service orientation and the introduction of CPS in the production takes place step by step. Smart Factory initiatives aim to implement the concept into historically grown and already existing production environments.

## Interim Summary

In summary, the large number of dimensions, stakeholders, and technologies regarding a Smart Factory project call for development steps to structure the way to reach a Smart Factory.

In order to measure a company's capabilities concerning their efforts in different application areas (e.g., process management, lean management), maturity models have found wide practical acceptance

*Figure 2. Resolution of the rigid automation pyramid towards an information network (own illustration, derived from Schöning & Dorchain, 2014, p. 550)*

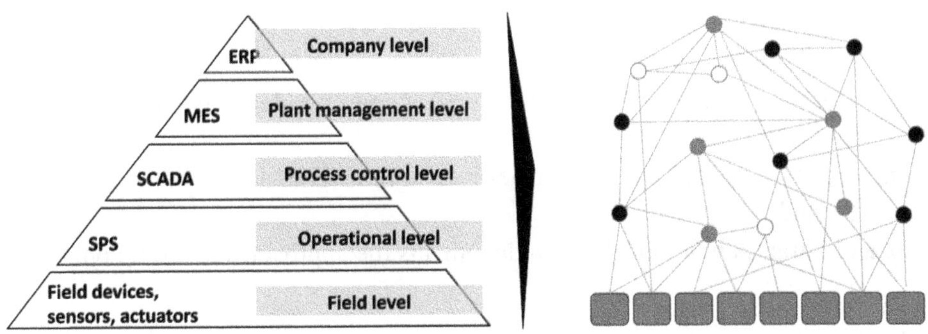

and enjoy strong academic interest. Such a model may represent a valuable means to track a company's Smart Factory capabilities, unveil eventual shortcomings and indicate where to initiate targeted improvement measures, hence contribute to prepare production systems for the future.

## Requirements for Smart Factory Maturity Models

It can be noted that the majority of existing maturity models focuses on the one-dimensional optimization of given problem domains. This fact strongly emphasizes on the process perspective and neglects further design levels as presented in the last section. A maturity model for Smart Factory projects has to describe characteristics of objects, functionalities, persons, abilities, IT infrastructures as well as technologies.

This section defines various requirements for a maturity model for Smart Factories. The requirements are then used to evaluate already existing maturity models for their suitability for the present problem situation (section 4).

The requirements (R) for the model are based on a systematic literature analysis. It combines the methods described by Ahlemann et al. (2005, p. 22), Hecht (2014, pp. 129-131), and Daniel (2008, p. 92-95) for assessing a maturity model.

### Degree of Adaptability (R1)

This requirement ensures that an existing model can be adapted to other areas. An adaptation can be made by removing or adapting specific questions or evaluations of the model. In order to ensure adaptability, a systematic design of the model and flexible model instruments are decisive (Ahlemann et al., 2005, p. 22). To assess if this requirement is met the following levels of satisfaction can be differentiated:

- Adaptation to other areas of interest is possible and is actively promoted by a pre-defined approach (+).
- Adaptation is generally possible, but no methodology is shown to implement an adaptation (0).
- Adaptation is not possible (-).

### Transparency and Low Complexity (R2)

Transparency and low complexity play a key role in the successful implementation of the model in practice (Daniel, 2008, p. 94). The creation of a complete and at the same time a complexity-reduced maturity model is, however, challenging. Different design fields need to be considered in detail and at the same time transparently, without losing clarity and comprehensibility. The following levels of satisfaction can be differentiated for R2:

- Complexity-reduced model which clearly describes relevant elements and documents the procedure (+),
- Complexity-reduced model which describes relevant model contents, but does not document the procedure (0), and
- Complex and intransparent model which documents the approach and model only very rough (-).

### Consideration of Dependencies (R3)

Since the challenges of a Smart Factory are often about combining different technologies, it is important to avoid the isolated development of single areas. Rather, a holistic view needs to be taken by analyzing the dependencies within as well as between the design fields (Siepmann, 2016, p. 37). R3 can be assessed on the following levels:

- Development of design-field-specific development paths with integrated approach for cross-departmental views on dependencies (+),
- Development of design-field-specific development paths without consideration of dependencies between design fields (0), and
- No development of different development stages for design fields (-).

### Flexible Number of Development Stages (A4)

In the context of a Smart Factory, technologies, methods and functions of production resources are considered. These do not necessarily follow the same development stages. Therefore, a flexible and individual number of development stages for different design fields should be supported. However, an over-simplification of the areas must be avoided (van Steenbergen et al., 2010, p. 318). For this purpose, the following features are differentiated:

- Full support of a flexible number of development stages (+),
- Partial support for a flexible number of development stages (0), and
- Fixed number of development stages (-).

### Consideration of the Specific Aspects of a Smart Factory (A5)

It is required that the maturity model focusses on the specific consideration presented in section 2. The following levels can be distinguished:

- Comprehensive consideration of relevant aspects of the Smart Factory and Industry 4.0 (+),
- Partial consideration of relevant aspects of the Smart Factory and Industry 4.0 (0), and
- No consideration of relevant aspects of Smart Factory and Industry 4.0 (-).

## Suitability of existing maturity models

For the selection of a suitable maturity model that may have the potential to support Smart Factory initiatives, a literature research was carried out. An analysis of existing maturity models for the industrial sector was performed. In total five maturity levels result. The aim of the detailed assessment is to test if existing maturity models already offer suitable support for organizing a Smart Factory project.

### Existing Models

In a short-list, four maturity models can be identified for a detailed analysis:

- **Capability Maturity Model Integration (CMMI):** The CMMI is a maturity model of the Software Engineering Institute at Carnegie Mellon University. This model follows the optimization of existing business processes as well as the entire organization. The CMMI contains a total of 22 process areas which cover four categories: Process Management, Project Management, Engineering and Support. In the case of the continuous representation, a development path for single processes along the capability grades of zero to three is possible. In the end, a process-spanning development path of maturity one to five is possible. For the step-like variant, all process areas of this maturity degree as well as the process areas of lower maturity degrees must be mastered to achieve a specific maturity level (SEI, 2002, p. 18 f.).
- **Software Process Improvement and Capability Determination (SPICE):** Similar to CMMI, SPICE is widely used. It serves as an international standard for assessing the functionality of software as well as for initiating process improvements (Becker, 2008, p. 178). The standard is also known under the designation ISO/IEC 15504. The structure of SPICE is based on the architecture of the CMMI. In addition to the identification of a process maturity, it also allows to identify improvement potentials for individual processes. For each process a development path is specified along the levels zero (incomplete) to five (optimizing). In addition to the maturity model the standard ISO/IEC 15504 includes partial publications for carrying out an assessment which addresses the possibility of self-assessment (DIN, 2011, p.12).
- **Software Process Management Maturity Matrix:** SPM is a model for structuring and improving the software product management process for companies. The SPM matrix is widely accepted both in science and business practice (Bekkers and van de Weerd, 2010, pp. 3, 20). This model includes all relevant dimension of optimization in so-called focus areas. In addition, the model provides an overview of the required skill levels and capabilities depicted in a best practice order for implementation. The SPM Maturity Matrix is one of the Focus Area Maturity models which are characterized by an individual number and degree of developmental capacities for the different Focus Areas (Van Steenbergen et al., 2010, p. 318).
- **Lean Capability Maturity Model (LCM):** LCM, developed by the Center for Industrial Production Aalborg, supports the implementation and sustainable implementation of a Lean Concept. It offers the possibility to describe the development path of slender and smooth structures. The model is a combination of theoretical and practical approaches to solve an overall optimization problem by incorporating both literary research and practical experience in the implementation of lean structures. The LCM includes five levels of maturity which enable companies to assess the current state of implementation of lean structures. In addition, there are fields of action that are necessary to work on to achieve a better maturity stage (Jørgensen et al., 2007, p. 372-376).

## Assessment

This section evaluates whether the identified maturity models meet the requirements for Smart Factory implementation projects completely (+), partly (0) or not at all (-). The results are outlined in the assessment matrix (c.f. Table 1).

Regarding the degree of adaptability (R1) it can be stated that in principle each of the presented maturity models can be adapted to other subject areas. Particularly the SPICE model and the CMMI meet this important requirement. The transferability of these established models has already been extensively discussed in scientific literature. In addition, both models have been successfully applied to other areas

*Table 1. Assessment of existing maturity models*

| | Maturity Models | | | |
|---|---|---|---|---|
| **Requirements** | **CMMI** | **SPICE** | **SPM** | **LCM** |
| R1 | + | + | 0 | 0 |
| R2 | + | + | + | + |
| R3 | - | - | + | 0 |
| R4 | - | - | + | - |
| R5 | - | - | - | 0 |

of interest (Ahlemann et al., 2005). Due to the fact that the SPM Maturity Matrix as well as the LCM represent more recent models, a comprehensive transferability cannot be fully demonstrated. High transparency and low complexity (R2) are fully met by each of the models. All contents are made clear and can intelligibly be described by the models. However, not all models sufficiently fulfill the requirement to consider dependencies between different design fields (R3). Only the SPM Maturity Matrix allows to display dependencies within a focus area as well as across different focus areas by dynamically positioning the respective capabilities. Thus, a holistic approach to the analysis of dependencies between different design fields and development stages can be identified across the model. The LCM also takes the dependencies within and between different process areas into account but does not provide sufficient guidance in performing this task.

The examined models predominantly have fixed level structures. Only the SPM Maturity Matrix allows for a flexible arrangement of levels (R4), because the focus areas may have a different number as well as different developmental stages compared to the original framework. In this case, too much simplification of the model can be avoided.

With regard to the content requirements for the consideration of specific aspects of a Smart Factory (R5) it is remarkable that none of the models meet the required functional scope. Domain specific adoptions of these reference models listed above exist (e.g., National BIM Standard Capability Maturity Model, based on CMMI). Due to their strict limitations in terms of the functional scope of their model content they also lack applicability for the Smart Factory initiatives (Knackstedt et al., 2009).

In summary, it can be concluded that none of the presented models can be fully adopted as a maturity model for the organization of Smart Factory initiatives. However, structures of existing models can be used to transfer them to a new maturity model specially designed for Smart Factories. For example, the SPM Maturity Matrix offers the possibility to meet the heterogeneous dimensions of Smart Factory initiatives. For this reason, the following section will outline fundamentals of a new Smart Factory maturity model.

## TOWARDS A SMART FACTORY MATURITY MODEL

In this section, the dimensions, design fields and their respective capability levels which are fundamental for the networking of production resources towards a Smart Factory are derived. Therefore, the relevant technologies, methods and aspects that allow for a holistic view on a Smart Factory are identified and grouped into categories. Facing a lack of scientific foundation and documentation of existing maturity model approaches (Becker et al., 2009, p. 12) the categorization is carried out following the methodology of a qualitative content analysis according to Mayring (2010, p. 605). This enables a systematic approach and transparent documentation of results, whereby the traceability of the category is to be ensured. The approach includes the following basic steps:

Firstly, a comprehensive literature search is carried out to identify the relevant topics. A search for keywords in academic databases as well as databases from different institutions like the Fraunhofer Institute, VDI, VDMA, ZVI or Acatech and other search engines is done. The keyword used as well as databases used are listed in table 7.

Secondly, the selection criteria and the abstraction levels for category formation are defined. The smallest selection criterion is a sentence and the largest unit is a paragraph. This is to ensure a profound understanding of the individual context. The categories' abstraction level should be industry and company independent according to the breadth and depth of the model (as defined in section 3). Furthermore, a generic transfer to different application domains such as production and logistics should be possible.

For example, the so-called CP classification of objects is used in order to assign the identified technologies and aspects to a respective dimension, design field or capability level (DIN, 2016, p. 17 f.). The classification scheme is provided by the Platform Industry 4.0 and is established in practice.

Thirdly, a step-by-step inductive classification of the dimensions, design fields and capability levels is documented. The content extracted from the work material is either subsumed under an existing category, or a high level category (dimension, D), a next smaller category (design field, F), or the smallest category (capability level) is created.

Table 3 shows a matrix (according to Webster & Watson (2002, p. xvii)) that indicates which of the analyzed publication deals with which dimensions and design fields.

With regards to the digital networking of production resources the result of the qualitative content analysis is a cross-disciplinary and structured illustration of the possible dimensions, design fields and capability levels.

To meet the requirements of transparency and complexity reduction as stated in section 3, not all of the design fields illustrated in Table 3 are taken into account for the Smart Factory maturity model. So, the following design fields are not considered either for thematic reasons or because they can be represented by (combinations of) other design fields:

*Table 2. Used databases and keywords*

| Data Bases | Keywords |
|---|---|
| • Business Source Premier (EBSCO)<br>• IEEE XploreScienceDirect<br>• Scopus<br>• SpringerLink<br>• TEMA Technology and Management<br>• The ACM Digital Library<br>• Google Scholar | • Industry 4.0<br>• Smart Factory<br>• Smart Manufacturing<br>• Networked Production<br>• Networked Manufacturing<br>• Cyber Physical Systems<br>• Internet of Things, IoT<br>• Smart Product<br>• Smart object<br>• Intelligent object<br>• Autonomy<br>• Decentralization |

*Table 3. Design field matrix*

| | Dimensions | | | | | | | | | | | | | | | | | | | | | | |
|---|---|---|---|---|---|---|---|---|---|---|---|---|---|---|---|---|---|---|---|---|---|---|---|
| | Objects | | | | | | Communication | | Information | | | | System | | Inter action | | Architecture | | | | Other | | |
| | Systemic Awareness | Communication Capap. | Object Intelligence | Localizability | Mobility | Degree of Automation | Interfaces | Protocols | Data Storage | Data Processing | Data Analytics | Data Generation | Control System | Information Model | Information Provision | HMI | Loose Coupling | Nestability | Encapsulation | IT Security | Energy Efficiency | Individualization | Condition Monitoring |
| 1 | X | X | X | X | | | | | | | | | | | | | | | | X | | | |
| 2 | X | X | | | | | | | | | | | | | | | | | | | | | |
| 3 | | | X | | X | | | | | | | | | | | | | | | | | X | |
| 4 | | | X | | X | | | | X | X | | | X | | | | | | | | | | |
| 5 | | X | | | | | X | | X | | | X | | | | X | | | | | | | |
| 6 | X | X | X | | | | | | | | | | | | | | | | | | | | |
| 7 | | | X | | | | X | | x | X | | | | | | | | | | | | | |
| 8 | X | | X | | | | | | X | X | | | X | | | | | | | | | | |
| 9 | | | X | | | | | | | X | | | | | | | | | | | | | |
| 10 | | | | | | | X | | | | | | | | | X | | | | | | | |
| 11 | X | | X | X | X | | | | X | X | | | X | | | | | | | | | | |
| 12 | | | X | | X | X | X | X | | | | X | X | | | X | | | | | | | |
| 13 | | | | | | | | | | | X | X | X | X | | | | | | | | X | |
| 14 | | | | | | | X | X | X | | | | | X | X | | | | | | | | |
| 15 | | | | | | | | | | | | | | | X | | | | | | | | |
| 16 | | | | | | | X | X | | | | | | | | | | | | | | | |
| 17 | | | | | | | X | X | | | | | | | | | | | | X | | | X |
| 18 | | | X | X | | X | | | | | X | | | | | | X | | | X | | X | |
| 19 | | X | | | | | | X | | | X | X | | | | | X | | | | | | |
| 20 | | X | | | X | | X | X | X | X | X | | | | | X | | | | X | | X | |
| 21 | | X | X | | | | | | | | | | X | | | | | | | | | | |
| 22 | X | X | X | X | X | | | | | | | | X | | | | | | | | | | |
| 23 | | | X | | | | | | | | | | X | | | | | | | | | | |
| 24 | | | | | | X | | | | | | | | | | | | | | | | | |
| 25 | | | X | | | | | X | | | | | X | | X | X | | | | | | | |
| 26 | X | X | X | X | X | | X | X | | | | X | | | X | X | | | | X | X | X | X |
| 27 | | | | | | | X | X | | | X | X | | | X | X | | | | X | | X | |
| 28 | | | | | | | | | | | | | | | | | | | | X | | | |
| 29 | X | X | | | | | | | X | X | | | | X | | X | | | | X | | | |
| 30 | X | | X | X | | | X | X | | | X | X | X | | X | X | X | X | X | X | | | |
| 31 | | | X | | X | | X | X | | | X | | | X | X | X | | | | X | | | |
| 32 | | | | | X | | X | X | X | | | | | | X | X | | | | X | | | |
| 33 | | | | | X | | | | | | X | X | | | X | X | | | | X | | | |
| 34 | | | X | | X | X | | | X | | X | | X | | | | | | | | | | |
| 35 | | | X | | | | | X | | | X | X | X | X | | | | | | | | | |
| 36 | X | X | X | X | X | X | | | | | | | X | | | | | | | | | | |

*continued on following page*

*Table 3. Continued*

| | Dimensions | | | | | | | | | | | | | | | | | | | | | | |
|---|---|---|---|---|---|---|---|---|---|---|---|---|---|---|---|---|---|---|---|---|---|---|---|
| | Objects | | | | | | Communication | | Information | | | | System | | Inter action | | Architecture | | | | Other | | |
| | Systemic Awareness | Communication Capap. | Object Intelligence | Localizability | Mobility | Degree of Automation | Interfaces | Protocols | Data Storage | Data Processing | Data Analytics | Data Generation | Control System | Information Model | Information Provision | HMI | Loose Coupling | Nestability | Encapsulation | IT Security | Energy Efficiency | Individualization | Condition Monitoring |
| 37 | X | X | | | | | X | X | | | | | | | | | | | | X | | | |
| 38 | | X | | | X | | X | | X | X | | | | | | | | | | | | | |
| 39 | | X | X | | | | | X | | X | X | | | | | | | | | X | | | |
| 40 | | | X | | | | | | | | X | | | | | | | | | | | | |
| 41 | X | X | | | | | | X | | | | | | | | | | | | | | | |
| 42 | | X | X | | | | | | | | | | | X | | X | | | | | X | | X |
| 43 | | X | X | | | | | X | X | X | | | | | X | X | | | | | | | |

Sources in numerical order:
1: Hern&ez & Reiff-Marganiec (2014), Classifying Smart Objects using Capabilities; 2: Epple et al. (2015), Statusreport Industrie 4.0 – Gegenstände, Entitäten, Komponenten; 3: Rammert (2009), Hybride H&lungsträgerschaft - Ein soziotechnisches Modell verteilten H&elns; 4: Trentesaux (2009), Distributed control of production systems; 5: Diekmann & Hagenoff (2007), Ubiquitous Computing-Technologien zur Integration der realen Welt in betriebliche Informationssysteme; 6: Brintrup et al. (2011), Will Intelligent Assets Take Off? Towards Self-Serving Aircraft; 7: Windelb& et al. (2010), Internet der Dinge in der Logistik - Qualitätsanforderungen durch das Internet der Dinge in der Logistik; 8: Meyer et al. (2009), Intelligent Products: A Survey; 9: Jordan (2015), Studienergebnisse "Cyber-Physical Systems in der Produktionspraxis" Anwendungsfälle und Leuchtturmprojekte in NRW; 10: Seitz et al. (2007), Digitale Sprach- und Datenkommunikation, Netze – Protokolle – Vermittlung; 11: Böse (2012), Selbststeuerung in der Fahrzeuglogistik, Modellierung und Analyse selbststeuernder logistischer Prozesse in der Auftragsabwicklung von Automobilterminals; 12: Bischoff et al. (2015), Erschließen der Potenziale von Industrie 4.0 im Mittelst&; 13: Friedli et al. (2015), Industrie 4.0 - ein Beitrag zur Entwicklung von "Smart Networks"; 14: Zühlke et al. (2012), Produktion 2020 – Auf dem Weg zur 4. industriellen Revolution; 15: Wang & Chen (2009), Usability Issues of an Augmented Virtuality Environment Design; 16: Graf et al. (2016), Netzkommunikation für Industrie 4.0. Diskussionspapier; 17: Heer et al. (2015), Communication Technologies for the Smart Factory of the Future. Building a communication infrastructure for Industry 4.0 & the Internet of Things; 18: Bauer et al. (2014), Industrie 4.0 – Volkswirtschaftliches Potenzial für Deutschl&; 19: Minguez (2013), The Manufacturing Service Bus; 20: Wolff & Schulze (2013), Industrie 4.0 – Cyber Physical Systems in der Produktion; 21: Bongaerts (2000), Hierarchy in distributed shop floor control; 22: McFarlane et al. (2003), Auto ID systems & intelligent manufacturing control; 23: Kuprat et al. (2015), Aufgaben der Produktionsplanung im Kontext von Industrie 4.0; 24: Weck & Brecher (2009), Werkzeugmaschinen 4 – Automatisierung von Maschinen und Anlagen; 25: Kleinjohann et al. (2013), Cyber Physical Devices - Die Schnittstelle zwischen Cyber Space und realer Welt; 26: Adolphs et al. (2015), Statusreport. Referenzarchitekturmodell Industrie 4.0 (RAMI4.0); 27: Bauernhansl et al. (2015), Geschäftsmodell-Innovation durch Industrie 4.0 - Chancen und Risiken für den Maschinen- und Anlagenbau; 28: Jänicke et al. (2016), IT-Security in Industrie 4.0 - Erste Schritte zu einer sicheren Produktion; 29: Bangemann et al. (2015), Statusreport. Industrie 4.0 – Technical Assets. Grundlegende Begriffe, Konzepte, Lebenszyklen und Verwaltung; 30: Bauer et al. (2014), Industrie 4.0 – Volkswirtschaftliches Potenzial für Deutschl&; 31: Vogel-Heuser (2014), Herausforderungen und Anforderungen aus Sicht der IT und der Automatisierungstechnik; 32: Schmitt & Zühlke (2013), Smartphones und Tablets in der industriellen Produktion, Nutzerfreundliche Bedienung von Feldgeräten; 33: Peissner & Hipp (2013), Potenziale der Mensch-Technik Interaktion für die effiziente und vernetzte Produktion von morgen; 34: Felix (2015), Dezentrale Produktionssteuerung für die Automobilindustrie; 35: Stich & Hering (2015), Daten und Software als entscheidender Wettbewerbsfaktor; 36: Böse & Windt (2007), Catalogue of Criteria for Autonomous Control in Logistics; 37: Br& et al. (2009), Internet der Dinge – Übersichtsstudie; 38: Nochta (2008), Smart Items in Real Time Enterprises; 39: Wortmann & Flüchter (2015), Internet of Things - Technology & Value Added; 40: Lee (2015), Smart Factory Systems; 41: Krückhans & Meier (2013), Industrie 4.0 – H&lungsfelder der Digitalen Fabrik zu Optimierung der Ressourceneffizienz in der Produktion; 42: VDI/VDE-GMA (2013), Cyber-Physical Systems: Chancen und Nutzen aus Sicht der Automation; 43: Schmitt & Orfgen (2015), Dynamische Funktionsskalierung mittels Apps auf intelligenten Feldgeräten.

- Product individualization, product related services
- Condition monitoring
- Energy efficiency

Despite these design fields play an essential role for Industry 4.0, they are no longer explicitly considered in the Smart Factory maturity model. The model is focused on the information and communication technology aspects for the networking of production resources and thus the realization of autonomous systems. Topics such as product individualization, the direct integration of users into the production process and the provision of additional services are, however, the subject of business model innovations. They are not explicitly considered in the context of this maturity model since such business model innovations are strategic innovations (Bauernhansl et al., 2015, p. 8 f.). One aspect which is not directly addressed but covered by the joint implementation of other design fields is the subject of monitoring. For example, during condition monitoring the condition of an asset is permanently recorded by means of measuring and assessing sensor data. Like this, machine wear can be detected in time, production losses can be avoided and the lifespan of the machine increases (Wang, 2003, p. 1; Davis, 1998, p. 4). Since this topic area can be expressed by other design fields and thus does not meet the quality criteria of the categories defined in section 3, monitoring is excluded for the Smart Factory maturity model. However, monitoring can be considered as an Industry 4.0 application case in more detail in subsequent work. Similarly, energy-efficient production is also considered to be an application that can be expressed by the combination of other design fields. Energy efficiency applications focus on the synchronization of energy supply and energy consumption and control energy flows by means of monitoring and prognosticating energy needs (Richter, 2015, p. 20).

Table 4 depicts the resulting capabilities (C) in 20 design fields (F) which are arranged in six dimensions (D).

*Table 4. Dimensions and design fields*

| Dimensions | Design Fields | Description |
|---|---|---|
| D1: Object dimension | F1.1: Degree of system familiarity | The existence as well as the identity of a physical object are not originally known to an information system. In order to store the information of an asset in an information system, digital management shells must be created and provided with information (Epple et al., 2014, p. 4 f.). This virtual representation of real objects is a prerequisite for networking within a Smart Factory (Reboredo, 2014, p. 81).<br>C1 = unknown, C2 = anonymously known, C3 = individually known, C4 = managed as an entity |
| | F1.2: Communication capabilities | Communication capability is the ability of an asset to share information with other assets across a digital communications network via, for example, a fieldbus, industrial Ethernet or a wireless communication network (Epple et al., 2014, p. 6; Hernandez & Reiff-Marganiec, 2014, p. 312)<br>C1 = none, C2 = passive, C3 = active, C4 = compliant to the service system |
| | F1.3: Degree of intelligence | Degree of intelligence describes the abilities of an asset to deal with data. The intelligence of an asset ranges from simple data handling to decision-making (Meyer et al., 2009, p. 140).<br>C1 = none, C2 = data handling, C3 = information recognition, C4 = decision ability |
| | F1.4: Localization | Within a value chain, products go through several transport, manufacturing, and assembly steps. In order to control and monitor these processes (especially in the fields of logistics) the localization of equipment, materials, assemblies and finished products is crucial (Bauer et al., 2013, p. 109; Nochta, 2008, p. 2013).<br>C1 = none, C2 = localizable |
| | F1.5: Mobility | Mobility describes the spatial mobility of assets. This offers an increase in flexibility by the local decoupling of a stationary asset towards a mobile asset (Rammert, 2009, p. 24; Böse & Windt, 2007, p. 63).<br>C1 = stationary, C2 = partly mobile, C3 = mobile |
| | F1.6: Degree of automation | Automation defines measures for independent process execution. The degree of automation provides information on the proportion of the automated functions on the overall function of a system (Weck & Brecher, 2006, p. 1). Increasingly, the focus is on the realization of workflows with varying tasks.<br>C1 = manual, C2 = partly automated, C3 = fully automated, C4 = modularly automated |
| D2: Communication dimension | F2.1: Communication interfaces | A central aspect in the realization of a Smart Factory is the continuous communication between all objects involved as well as the connection of the object dimension with the information dimension. In this design field, the different interfaces are differentiated as capability levels of the underlying communication network (Fuchß, 2009, p. 122).<br>C1 = none, C2 = cable based, C3 = wireless |
| | F2.2: Communication protocols | In order to ensure the seamless interaction of manufacturer-specific assets, the availability and the use of standardized communication protocols is necessary (Adolphs et al., 2016, p. 12; Hertel, 2015, p. 726).<br>C1 = none, C2 = proprietary, C3 = standardized |
| D3: Information dimension | F3.1: Point of information storage | Data storage can be organized centrally (outsourced in a higher-level information system) as well as decentrally and thus embedded (Nochta, 2008, pp. 212 f.; Böse, 2012, p. 20). This design field is about the storage of data that go beyond a pure information identification. This includes additional object-related data or environmental data collected by sensors (Diekmann & Hagenhoff, 2007, p. 23).<br>C1 = none, C2 = outsourced, C3 = combined, C4 = embedded |
| | F3.2: Point of information processing | Assets can increasingly process data and information themselves to manage the growing data volumes. For example, captured sensor data can be evaluated directly embedded in the asset for analysis and monitoring purposes or for decision-making without forwarding the data to a higher-level system (Nochta, 2008, p. 215 f.). Thus, for this design field the function of data processing can be carried out either in a superordinate information system or embedded in an asset itself (Böse, 2012, p. 20; Jordan, 2015, p. 559; Schmitt & Orfgen, 2015, p. 54).<br>C1 = none, C2 = outsourced, C3 = combined, C4 = embedded |
| | F3.3: Degree of data collection | Data collection includes the systematic recording of machine, plant and product data. Their goal is the generation of a delay-free, virtual mapping of the processes in production. This real-time picture of the production process is the basis for a real-time production control (Bischoff et al., 2015, p. 76; Spath et al., 2013, p. 133).<br>C1 = none, C2 = historically, C3 = real-time |

*continued on following page*

*Table 4. Continued*

| Dimensions | Design Fields | Description |
|---|---|---|
| | F3.4: Data analytics | The intelligent processing and analysis of data will lead to significant increases in manufacturing productivity (Stich & Hering, 2015, p. 9 f.). The design field data analytics therefore assesses the ability to extract relevant information and knowledge from the collected data (Agarwal & Dhar, 2014, p. 444).<br>C1 = none, C2 = descriptive, C3 = diagnostic, C4 = predictive, C5 = prescriptive |
| D4: System dimension | F4.1: Organizational structure of the control system | The organizational structure of a control system defines the decision-making competences as well as the type of coordination of different elements within a production system. This can range from a hierarchical organizational structure with central control to a heterarchical organizational structure with decentralized and self-controlling systems (Böse, 2012, p. 18).<br>C1 = no control system, C2 = centralized, C3 = hybrid, C4 = decentralized |
| | F4.2: Continuity of the information model | The continuous collection and storage of data and information across the entire value chain and the entire product and production lifecycle are important prerequisites for a Smart Factory (Adolphs et al., 2015, p. 9). In reality, several different IT systems with closed information pools exist side by side without allowing reliable networking or data integration (Vogel-Heuser, 2014, p. 39).<br>C1 = no continuity, C2 = partly continuous, C3 = continuous |
| D5: Interaction dimension | F5.1: Style of information provision | The exponentially growing data volume and increasing complexity within a Smart Factory must be visualized for humans. This includes, for example, the display of work instructions on monitors, and the context-based information provision through smart glasses using augmented reality (Bischoff et al., 2015, p. 90). For this reason, this design field includes the different forms of information provision for humans within the manufacturing environment.<br>C1 = none, C2 = analogous, C3 = digital (stationary), C4 = digital (mobile), C5 = Augmented / Virtual reality |
| | F5.2: Human-machine interface | The user interface describes the interaction between man and machine. The interface defines the way in which human workforce can transmit instructions to a machine. The objective is that humans perceive the interaction with a machine less as a traditional machine control, but as natural and intuitive as possible (Seitz et al., 2007, p. 15; Bauernhansl et al., 2015, pp. 16, 24).<br>C1 = none, C2 = technical, C3 = natural |
| D6: Architec-tural dimension | F6.1: Loose coupling | Loose coupling origins from software architecture and describes how closely software components are interconnected. Thus, it represents a measure of the dependency between these components. Loose coupling facilitates the integration, maintainability and exchangeability of components.<br>C1 = no loose coupling, C2 = loose coupling |
| | F6.2: Nestability | A physical asset, such as machinery or sensors, can consist of several subordinate assets. In the sense of a system modularization, this dependency should also be represented virtually. Consequently, the management shell of a physical asset can refer to one or more other assets. The nestability thus offers, for example, the possibility to map a complete system consisting of individual modules as a staggered arrangement of virtual assets (Adolphs et al., 2016, p. 11).<br>C1 = no nestability, C 2 = nestability |
| | F6.3: Encapsulation | The increasing interconnection of assets and their mutual provision of information and functionalities via a Service-oriented architecture (SOA) has to ensure that the respective core functionalities can be maintained in the event of an external disturbance. In particular, the management shell and the information contained in it should be capable of being used as a safeguard (DIN, 2016, p. 28).<br>C1 = no encapsulation, C2 = encapsulation |
| | F6.4: IT security | With increasing digital networking of the production environment, the importance of data, information and communication security increases. IT security describes the protection of a system from an impermissible external influence. Securing the confidentiality, integrity, and availability of stored and transmitted data must be fully guaranteed (Jänicke et al., 2016, p. 4; Adolphs et al., 2016, p. 13; Heer et al., 2015, p. 5).<br>C1 = no IT security, C2 = IT security |

## CONCLUSION

The intelligent production of the future is not based on a single revolutionary technology, but rather on a steady transformation and further development of various technologies and the networking of many heterogeneous entities involved in the production process. Numerous different organizational functions within a company, opposing objectives or thrusts between different departments or technological concepts as well as a multitude of processual, market-driven and business boundary conditions need to be considered. As a consequence, companies need an instrument to organize their development projects across these multidimensional challenges and to determine the implementation status for the complete networking of production resources towards a Smart Factory. It is a typical problem in which the application of a suitable maturity model may help.

For this purpose, four existing maturity models from the literature were analyzed with regards to their suitability for the domain of the Smart Factory. The comparison took place with regard to defined requirements for the model architecture, as well as the model content of a maturity model for a Smart

Factory. Already established maturity models, such as the CMMI or SPICE, as well as models from related domains such as the LCM were analyzed in depth. The comparison shows that none of the considered maturity models meets the requirements that come along with the specifics of smart factories. Nevertheless, structures of existing models can be transferred to the model to be developed in this book chapter. Thus, the maturity model development strategy transfers structures of existing maturity models to the new Smart Factory application field. These fundamental insights about the architecture, the content, and technological as well as conceptual relationships within a Smart Factory maturity model form the outcome of this chapter. Scientists can use this as a starting point for further research in terms of determining industry-specific capabilities within the individual design fields that result from this research and work towards a systematic Smart Factory implementation support for companies.

The practical relevance of the findings lies in the clear representation of the 20 individual design fields for Smart Factories and their individual subdivision into specific capability levels. Practitioners can use these as a basis for organizing their Smart Factory initiatives, for example by identifying existing technological residuals within one or several design fields or forming an appropriate project team with reasonable know-how and expertise to enhance a system's performance within one or several design fields onto a higher capability level. They can better evaluate the advantages and disadvantages of existing maturity models and make a selection for their own projects. Like this, necessary technological and conceptual developments on different maturity levels can be assessed and organized under consideration of specific dependencies within a Smart Factory.

## REFERENCES

Adolphs, P., Auer, S., Bedenbender, H., Billmann, M., Hankel, M., Heidel, R., & Waser, B. (2016). *Struktur der Verwaltungsschale – Fortentwicklung des Referenzmodells für Industrie 4.0-Komponente.* Retrieved June 23, 2016, from https://www.plattform-i40.de/I40/Redaktion/DE/Downloads/Publikation/struktur-der-verwaltungsschale.pdf?__blob=publicationFile&v=8

Adolphs, P., Bedenbender, H., Dirzus, D., Ehlich, M., Epple, U., Hanke, M., & Wollschlaeger, M. (2015). *Statusreport. Referenzarchitekturmodell Industrie 4.0 (RAMI4.0).* Retrieved June 14, 2016 from http://www.vdi.de/fileadmin/user_upload/VDI-GMA_Statusreport_Referenzarchitekturmodell-Industrie40.pdf

Agarwal, R., & Dhar, V. (2014). Big Data, Data Science, and Analytics: The Opportunity and Challenge for IS Research. *Information Systems Research*, *25*(3), 443–448. doi:10.1287/isre.2014.0546

Ahlemann, F., Teuteberg, F., & Schöder, C. (2005). *Kompetenz- und Reifegradmodelle für das Projektmanagement – Grundlagen, Vergleich und Einsatz.* Osnabrück: Universität Osnabrück.

Bauer, K., Diegner, B., Diemer, J., Dorst, W., Ferber, S., Glatz, R., & Stumpf, V. (2013). *Deutschlands Zukunft als Produktionsstandort sichern. Umsetzungsempfehlungen für das Zukunftsprojekt Industrie 4.0. Abschlussbericht des Arbeitskreises Industrie 4.0.* Retrieved June 20, 2016 from https://www.bmbf.de/files/Umsetzungsempfehlungen_Industrie4_0.pdf

Bauer, W., Schlund, S., Marrenbach, D., & Ganschar, O. (2014). *Industrie 4.0 – Volkswirtschaftliches Potenzial für Deutschland.* Stuttgart: Fraunhofer Institut IAO.

Bauernhansl, T. (2014). Die Vierte industrielle Revolution – Der Weg in ein wertschaffendes Produktionsparadigma. In T. Bauernhansl, M. ten Hompel, & B. Vogel-Heuser (Eds.), *Industrie 4.0 in der Produktion, Automatisierung und Logistik* (pp. 5–35). Wiesbaden: Springer. doi:10.1007/978-3-658-04682-8_1

Bauernhansl, T., Paulus-Rohmer, D., Schatz, A., Weskamp, M., Emmrich, V., & Döbele, M. (2015). *Geschäftsmodell-Innovation durch Industrie 4.0 – Chancen und Risiken für den Maschinen- und Anlagenbau*. München: Wieselhuber & Partner, Fraunhofer Institut IPA.

Baumer, S (2014). Verlässliche eingebettete Systeme für Industrie 4.0. *Zeitschrift für wirtschaftlichen Fabrikbetrieb, 109*(4), 264.

BDI. (2015). *Die Digitale Transformation der Industrie – Eine europäische Studie von Roland Berger Strategy Consultants im Auftrag des BDI*. München: BDI.

Becker, J., Knackstedt, R., & Pöppelbuß, J. (2009). Dokumentationsqualität von Reifegradmodellen. In J. Becker (Ed.), *Arbeitsberichte des Instituts für Wirtschaftsinformatik. Arbeitsbericht Nr. 123*. Münster: Universität Münster.

Becker, T. (2008). *Prozesse in Produktion und Supply Chain optimieren*. Heidelberg, Germany: Springer.

Bekkers, W., & van de Weerd, I. (2010). *SPM Maturity Matrix*. Technical Report UU-CS-2010-013. Department of Information and Computing Sciences. Utrecht: Utrecht University.

Bengel, G. (2014). *Grundkurs Verteilte Systeme. Grundlagen und Praxis des Client-Server und Distributed Computing*. Wiesbaden: Springer.

Bischoff, J., Taphorn, C., Wolter, D., Braun, N., Fellbaum, M., Goloverov, S. L., & Scheffler, D. (2015). *Erschließen der Potenziale von Industrie 4.0 im Mittelstand*. Berlin: BMWi (Bundesministerium für Wirtschaft und Energie).

BMBF. (2014). *Die neue Hightech-Strategie – Innovationen für Deutschland*. Retrieved November 9, 2016 from https://www.bmbf.de/pub_hts/HTS_Broschure_Web.pdf

BMWi. (2015). *Industrie 4.0 und Digitale Wirtschaft, Impulse für Wachstum, Beschäftigung und Innovation*. Retrieved November 9, 2016 from https://www.bmwi.de/BMWi/Redaktion/PDF/I/industrie-4-0-und-digitale-wirtschaft,property=pdf,bereich=bmwi2012,sprache=de,rwb=true.pdf

Böse, F. (2012). Selbststeuerung in der Fahrzeuglogistik, Modellierung und Analyse selbststeuernder logistischer Prozesse in der Auftragsabwicklung von Automobilterminals. In B. Scholz-Reiter (Ed.), *Informationstechnische Systeme und Organisation von Produktion und Logistik*. Berlin: GITO-Verlag.

Böse, F., & Windt, K. (2007). Catalogue of Criteria for Autonomous Control in Logistics. In M. Hülsmann & K. Windt (Eds.), *Understanding Autonomous Cooperation and Control Logistics. The Impact of Autonomy on Management, Information, Communication and Material Flow* (pp. 57–72). Berlin: Springer. doi:10.1007/978-3-540-47450-0_5

Daniel, K. (2008). *Managementprozesse und Performance – Ein Konzept zur reifegradbezogenen Verbesserung des Managementhandelns*. Wiesbaden: Gabler.

Davis, A. (1998). *Handbook of Condition Monitoring – Techniques and Technology*. Amsterdam: Springer Netherlands. doi:10.1007/978-94-011-4924-2

Diekmann, T., & Hagenhoff, S. (2007). *Ubiquitous Computing – Technologien zur Integration der realen Welt in betriebliche Informationssysteme. Arbeitsbericht Nr. 15/2006*. Göttingen: Universität Göttingen.

DIN e. V. (2011). DIN ISO/IEC 15504-3. Informationstechnik – Prozess-Assessment. Berlin: Beuth Verlag.

DIN e.V. (2016). *DIN SPEC 91345:2016-04. Referenzarchitekturmodell Industrie 4.0 (RAMI4.0)*. Berlin: Beuth Verlag.

Epple, U., Bangemann, T., Barbian, M., Bauer, C., Braune, A., Diesner, M., & Wollschläger, M. (2014). *Statusreport Industrie 4.0 – Gegenstände, Entitäten, Komponenten*, Retrieved June 21, 2016 from https://www.vdi.de/fileadmin/vdi_de/redakteur_dateien/sk_dateien/VDI_Industrie_4.0_Komponenten_2014.pdf

Fallenbeck, N., & Eckert, C. (2014). IT-Sicherheit und Cloud Computing. In T. Bauernhansl, M. ten Hompel, & B. Vogel-Heuser (Eds.), *Industrie 4.0 in der Produktion, Automatisierung und Logistik* (pp. 397–431). Wiesbaden: Springer. doi:10.1007/978-3-658-04682-8_20

Fuchß, T. (2009). *Mobile Computing: Grundlagen und Konzepte für mobile Anwendungen*. München: Carl Hanser Verlag. doi:10.3139/9783446405134

Geisberger, E., Cengarle, M. V., Keil, P., Niehaus, J., Thiel, C., & Thönnißen-Fries, H. J. (2011). Cyber Physical Systems. *Innovationsmotor für Mobilität, Gesundheit, Energie und Produktion*. Retrieved November 9, 2016 from http://www.acatech.de/fileadmin/user_upload/Baumstruktur_nach_Website/Acatech/root/de/Material_fuer_Sonderseiten/Cyber-Physical-Systems/acatech_POSITION_CPS_web.pdf

Hecht, S. (2014). *Ein Reifegradmodell für die Bewertung und Verbesserung von Fähigkeiten im ERP-Anwendungsmanagement*. Wiesbaden: Gabler. doi:10.1007/978-3-658-05523-3

Heer, T., Kleineberg, O., & Bagchi, S. (2015). *Communication Technologies for the Smart Factory of the Future. Building a communication infrastructure for Industry 4.0 and the Internet of Things*. St. Louis, MO: Belden Inc.

Hernández, M. E., & Reiff-Marganiec, S. (2014). Classifying Smart Objects using Capabilities. *Smart Computing, 2014*, 309–316.

Hertel, M. (2015). Risiken der Industrie 4.0 – Eine Strukturierung von Bedrohungsszenarien der Smart Factory. *HMD Praxis der Wirtschaftsinformatik, 52*(5), 724–738. doi:10.1365/s40702-015-0161-1

Jänicke, L., Jochem, M., Klasen, W., Kosch, B., Krammel, M., Linke, L., & Zimmermann, S. (2016). IT-Security in Industrie 4.0 - Erste Schritte zu einer sicheren Produktion. Berlin: Bundesministerium für Wirtschaft und Energie (BMWi).

Jordan, F. (2015). Studienergebnisse "Cyber-Physical Systems in der Produktionspraxis" Anwendungsfälle und Leuchtturmprojekte in NRW. *Zeitschrift für wirtschaftlichen Fabrikbetrieb, 109*(9), 558-561.

Jørgensen, F., Matthiesen, R., Nielsen, J., & Johansen, J. (2007). Lean Maturity, Lean Sustainability. In J. Olhager & F. Persson (Eds.), IFIP - The International Federation for Information Processing (pp. 371–378). Academic Press.

Kaufmann, T. (2015). *Geschäftsmodelle in Industrie 4.0 und dem Internet der Dinge. Der Weg vom Anspruch in die Wirklichkeit.* Wiesbaden: Springer. doi:10.1007/978-3-658-10272-2

Knackstedt, R., Pöppelbuß, J., & Becker, J. (2009). Vorgehensmodell zur Entwicklung von Reifegradmodellen. In H. R. Hansen (Ed.), *Business Services. Konzepte, Technologien, Anwendungen* (pp. 535–544). Vienna: Österreichische Computer Gesellschaft.

Mayring, P. (2010). Qualitative Inhaltsanalyse. In G. Mey & K. Mruck (Eds.), *Handbuch Qualitative Forschung in der Psychologie* (pp. 601–613). Wiesbaden: VS Verlag für Sozialwissenschaften. doi:10.1007/978-3-531-92052-8_42

Melzer, I. (2010). *Service-orientierte Architektur mit Web Services, Konzepte –Standards – Praxis.* Heidelberg, Germany: Spektrum Akademischer Verlag. doi:10.1007/978-3-8274-2550-8

Meyer, G. G., Främling, K., & Holmström, J. (2009). Intelligent Products: A Survey. *Computers in Industry, 60*(3), 137–148. doi:10.1016/j.compind.2008.12.005

Nochta, Z. (2008). Smart Items in Real Time Enterprises. In M. Mühlhäuser, & I. Gurevych (Eds.), Handbook of Research on Ubiquitous Computing Technology for Real Time Enterprises (pp. 211-218). Academic Press. doi:10.4018/978-1-59904-832-1.ch010

Rammert, W. (2009). Hybride Handlungsträgerschaft - Ein soziotechnisches Modell verteilten Handelns. In O. Herzog & T. Schildhauer (Eds.), *Intelligente Objekte. Technische Gestaltung, Wirtschaftliche Verwertung, Gesellschaftliche Wirkung* (pp. 23–32). Berlin: Springer. doi:10.1007/978-3-642-02220-3_3

Reboredo, P. (2014). Laufzeitvalidierung einer Plattform zur semantischen Integration von Feldgeräten. In W. A. Halang & H. Unger (Eds.), *Industrie 4.0 und Echtzeit* (pp. 81–90). Berlin: Springer.

Richter, M. (2015). *Industrie 4.0 und Energieeffizienz in der Produktion. Aktuelle Entwicklungen und Anwendungsbeispiele*. Retrieved August 20, 2016 from http://industrie40.vdma.org/documents/4214230/5356229/IWU%20Vortrag%20Mark%20Richter/d1060b17-9c8b-450b-8dcf-cb66ddfa95be

Roth, A. (2016). *Einführung und Umsetzung von Industrie 4.0: Grundlagen, Vorgehensmodell und Use Cases aus der Praxis*. Berlin: Springer. doi:10.1007/978-3-662-48505-7

Russwurm, S. (2013). Software: Die Zukunft der Industrie. In U. Sendler (Ed.), *Industrie 4.0 – Beherrschung der industriellen Komplexität mit SysLM (Systems Lifecycle Management)* (pp. 21–36). Berlin: Springer.

Schill, A., & Springer, T. (2007). *Verteilte Systeme, Grundlagen und Basistechnologien*. Berlin: Springer.

Schmitt, M., & Orfgen, M. (2015). Dynamische Funktionsskalierung mittels Apps auf intelligenten Feldgeräten. *Zeitschrift für wirtschaftlichen Fachbetrieb (ZWF), 110*, 54-58.

Schöning, H., & Dorchain, M. (2014). Data Mining und Analyse. In T. Bauernhansl, M. ten Hompel, & B. Vogel-Heuser (Eds.), *Industrie 4.0 in der Produktion, Automatisierung und Logistik* (pp. 543–554). Wiesbaden: Springer. doi:10.1007/978-3-658-04682-8_27

SEI. (2002). *Capability Maturity Model Integration (CMMI$^{SM}$), Version 1.1. CMMI$^{SM}$ for Software Engineering (CMMI-SW, V1.1), Continuous Representation, CMU/SEI-2002-TR-028*. Pittsburgh, PA: SEI.

Seitz, J., Debes, M., Heubach, M., & Tosse, R. (2007). *Digitale Sprach- und Datenkommunikation, Netze – Protokolle – Vermittlung*. Leipzig: Carl Hanser Verlag.

Siepmann, D. (2016). Industrie 4.0 – Fünf zentrale Paradigmen. In A. Roth (Ed.), *Einführung und Umsetzung von Industrie 4.0: Grundlagen, Vorgehensmodell und Use Cases aus der Praxis* (pp. 35–45). Berlin: Springer.

Simon, W. (2013). Blick in die Zukunft: Industrie 4.0. *Industrial Engineering (American Institute of Industrial Engineers)*, (2): 38–40.

Spath, D., Ganschar, O., Gerlach, S., Hämmerle, M., Krause, T., & Schlund, S. (2013). *Produktionsarbeit der Zukunft-Industrie 4.0*. Stuttgart: Fraunhofer Verlag.

Stich, V., & Hering, N. (2015). Daten und Software als entscheidender Wettbewerbsfaktor. *Industrie 4.0 Magazin - Zeitschrift für integrierte Produktionsprozesse,* 8-13.

van Steenbergen, M., Bos, R., Brinkkemper, S., van de Weerd, I., & Bekkers, W. (2010). The Design of Focus Area Maturity Models. In R. Winter (Ed.), *Global Perspectives on Design Science Research* (pp. 317–332). Berlin: Springer. doi:10.1007/978-3-642-13335-0_22

VDI/VDE. (2013). *Cyber-Physical Systems: Chancen und Nutzen aus Sicht der Automation*. Retrieved June 27, 2016 from https://www.vdi.de/uploads/media/Stellungnahme_Cyber-Physical_Systems.pdf

Vogel-Heuser, B. (2014). Herausforderungen und Anforderungen aus Sicht der IT und der Automatisierungstechnik. In T. Bauernhansl, M. ten Hompel, & B. Vogel-Heuser (Eds.), *Industrie 4.0 in der Produktion, Automatisierung und Logistik* (pp. 37–48). Wiesbaden: Springer. doi:10.1007/978-3-658-04682-8_2

Wang, K. (2003). *Intelligent Condition Monitoring and Diagnosis Systems – A computational Intelligence Approach*. Amsterdam: IOS Press.

Webster, J., & Watson, R. T. (2002). Analyzing the past to prepare for the future: Writing a literature review. *Management Information Systems Quarterly*, *26*(2), xiii–xxiii.

Weck, M., & Brecher, C. (2009). *Werkzeugmaschinen 4 – Automatisierung von Maschinen und Anlagen*. Berlin: Springer.

Wolff, I., & Schulze, S. (2013). *Industrie 4.0 – Cyber Physical Systems in der Produktion. Clustermanagement IKT.NWR*. Wuppertal: Bergische Universität.

Zühlke, D., Schlick, J., & Stephan, P. (2012). Produktion 2020 – Auf dem Weg zur 4. industriellen Revolution. *IM Information Management und Consulting*, 26-33.

## KEY TERMS AND DEFINITIONS

**Automation Pyramid:** The hierarchical model of the automation pyramid represents the standard in manufacturing automation. Hierarchical structures play a special role in handling complex systems. A hierarchical system exists when individual subsystems (levels) with different priorities can be differentiated against subordinate or superordinate levels. The superordinate subsystems depend directly on the function filling of subordinate levels. The higher the hierarchy level, the greater the understanding and responsibility for the performance of the overall system. Deeper levels are characterized by increasing detail knowledge about individual processes and technologies.

**Complexity:** Complexity describes the totality of all interdependent features and elements that stand in a diverse but holistic relationship (structure) within a system. A system's complexity is composed of the elements' specific behavior and their variability of their course of action. Complexity thus refers to the diversity of individual system elements and their (dynamic) interactions over time.

**Cyber-Physical Systems:** A cyber-physical system (CPS) is characterized by the connection of physical objects with virtual objects. Their networking is realized via open and global information networks such as the Internet. Equipped with embedded systems, sensors and actuators, a CPS is able to capture and evaluate physical data and thus actively react to environmental influences.

**Industry 4.0:** Industry 4.0 describes the vision of a future production environment consisting of intelligent, self-organizing system elements. The basis for Industry 4.0 is the availability of all relevant system information in real-time through the networking of all entities involved in the value creation process as well as the ability to derive the best value flow from these data at any time.

**Maturity Model:** Maturity models describe the development stages of processes, objects, organizations, and technologies within a specific application domain. The concept of maturity implicates a development path that systematically describes the development of individual viewing objects and areas in different discrete maturity levels.

**Smart Factory:** A Smart Factory provides a production environment in which products, complete production lines and logistics systems communicate with each other over powerful networks and are largely autonomously controlled without any human intervention. Smart factories form a core area of the Industry 4.0 and promise the realization of mass customization (fast and flexible manufacture of customer-specific products with maximum efficiency).

**Smart Objects:** Intelligent or even so-called smart objects include items that are equipped with advanced functions, such as the acquisition, processing and storage of data, as well as the ability to interact with their environment. An intelligent object can be a single product, which stores and delivers information on its processing steps as well as an intelligent system in a whole.

# Chapter 18
# Data Analytics in Industry 4.0:
## In the Perspective of Big Data

**Mahir Oner**
*Istanbul Technical University, Turkey*

**Sultan Ceren Oner**
*Istanbul Technical University, Turkey*

## ABSTRACT

*The new form of future generation machines and automated systems could be synchronized by IoT adaptation. By this way, a very large size data can be carefully stored in data repositories and have to be analyzed for extracting knowledge. Thus, optimization techniques are becoming invaluable tools for finding patterns from parallel distributed machines. On the other hand, statistical methods and optimization models could not be utilized efficiently due to excessive dimension of data. Additionally, data analytics should be applied and results should be gathered by using practical approaches especially for security, access control and fault detection issues. In this study, optimization techniques are evaluated in the perspective of big data analytics and both mathematical and statistical methods will be extensively analyzed for different versions of problem solving and decision making in Industry 4.0 era.*

## INTRODUCTION

Manufacturing management paradigms have been increasingly applied all over the world because of the global competition of trade organizations and rapid changes in technology. In recent years, thanks to the communication improvements, customers have become more conscious about purchasing goods or services. Naturally, customer requirements are changing day by day, increasingly high expectations of customers are appeared - and these changes force companies to be speedier to satisfy customer orders with more qualified products in acceptable prices. Furthermore, organizations have to be customer oriented and more flexible against the dynamism of manufacturing environment which increases uncertainties in critical parameters. Besides flexibility, this customer oriented attitude helps organizations to have a better chance of making more profit but also brings pressure to take risks such as inventory shortages,

DOI: 10.4018/978-1-5225-2944-6.ch018

decreasing demand, late shipments, quality loses etc. Within this context, manufacturing management applications are sought after following dramatic losses by some world re-known companies. Many organizations try to cope with this continuously changing environment only with decision making tools such as ERP, CRM etc. Generally, these tools are not sufficient enough for detecting faults and making adjustments according to these new realities. In this context, Industry 4.0 was first declared by German government in terms of interaction between intelligent and communicative machines that enable self-organizing and centralized structure for sustainable manufacturing.

The transformation process from steam engine to Industry 4.0 is realized by the improvements in both information technology and industry. The first conversion was the mechanical production as seen from textile looms. Second transformation occurred in assembly lines in terms of division of labor and mass production. At the end of 1970s, automation and industrial robots were introduced and finally, cyber-physical systems, which are the combination of digital and physical systems, and digitalization of industry has emerged using smart machines in production processes.

The improvements in information technology also provide the infrastructure of Industry 4.0. The initial development is in the management of essential business processes such as financial analysis, order placement and production planning via digital platforms. The progress is continuing as the increasing usage of Internet and penetration of personal computers in daily life. The final complement in the digitalization is the integration of products with digital systems such as the adaptation of digital sensors, software and wireless devices. With these revolutionary changes, new business models are emerged in order to maintain the fierce competition and value creation.

In this study, we focused on explaining advantageous properties and critical issues of Industry 4.0 in the context of manufacturing environment. Furthermore, opportunities and threats are defined from the operational perspective. The evaluation of intelligent cross linked machines (Cyber- Physical Systems), end-to-end engineering (product development) and digitalized control systems are analyzed. Finally, some of the business models, value creation networks and changes that should be adapted to organizations and sectors are briefly explained.

The rest of the paper is organized as follows: a literature review of Industry 4.0 is presented in background section. Optimization techniques for data analytics are explained with respect to business need and methodology. Additionally, big data and fuzzy techniques are interpreted for the extraction of valuable solutions. Finally, conclusions and recommendations are provided in the last section.

## BACKGROUND

The main idea of Industry 4.0 is first declared by Kagermann in 2011 and supported by German National Academy of Science and Engineering in 2013. The context of Industry 4.0 was introduced by Industrial Internet Consortium (ICC) and wide range of applications could be found in Bosch, Siemens, Apple etc. (Stock and Seliger, 2016). The main idea of Industry 4.0 is the installation of smart products and smart services with smart factories using Internet of Things, Cyber-Physical Systems in order to provide communication of each objects and decentralized systems. (Weyer et al., 2015) The four design principles are listed as follows:

Interoperability: The connection and communication of machines, devices, sensors, and people using Internet of Things (IoT) or Internet of People (IoP).

Information transparency: The accomplishment of information systems to create a virtual sample of the physical world by providing digital models with raw data. This requires the transformation of gathered data to knowledge using cloud computing, big data analytics etc.

Technical assistance: The integration of assistance systems using machine-machine interaction and human-machine interaction to support decision making and eliminating unpleasant, too exhausting, or unsafe actions using cyber physical systems in manufacturing.

Decentralized decision making: The autonomous decision making and interpretation ability with respect to conflicting goals of other systems (Hermann et al, 2016).

Thinking in this concept, it is obvious that an increasing number of publications have appeared in the literature about Industry 4.0 and its applications as seen from Figure 1. For instance, Zezulka et al. (2016) proposed a strategic roadmap for vertical and horizontal integration of Industry 4.0 components and offered a technical solution of Industry 4.0 components that can enable the creation of Industry 4.0 case studies and first I4.0 applications. Wittenberg (2016) also indicated human machine interaction of I4.0. Long et al. (2016) used stochastic Petri nets (ECSPN) for modeling the production systems in Industry 4.0. Gaub (2016) mentioned the customization of mass-produced parts by combining injection molding and additive manufacturing with Industry 4.0 technologies. In addition, Thames and Schaefer (2016) proposed a Software-defined ClouD manufacturing for Industry 4.0. Recognizing the comprehensive perspective of Industry 4.0, the book by Gilchrist (2016) explores the current state of the production, processing, and manufacturing industries and discover what it will take to achieve re-industrialization of the former industrial powerhouses that can counterbalance the benefits of cheap labor providers dominating the industry. Finally, Quinn et al. (2016) gave a categorical framework of manufacturing for the perspective of Industry 4.0.

According to Global Industry 4.0 Survey that is conducted by PWC, the main components of Industry 4.0 are digitization and integration of vertical and horizontal value chains, Digitization of product and service offerings and digital business models and customer access. They pointed out that data analytics is the core capability of Industry 4.0 as seen in Figure 2.

*Figure 1. Number of publications appeared in Science Direct database from 2012-2017*

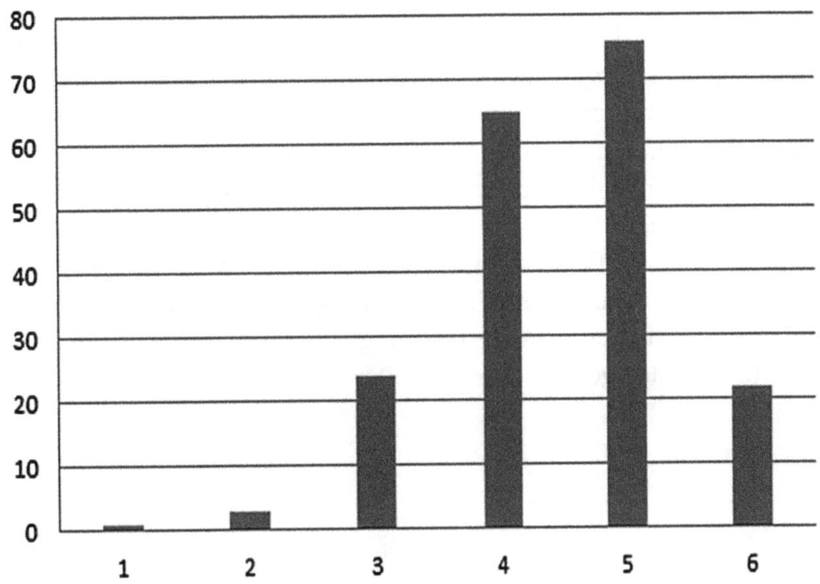

*Figure 2. Issues and solutions of the issues appeared in Industry 4.0 (Inspired by Berger, 2014)*

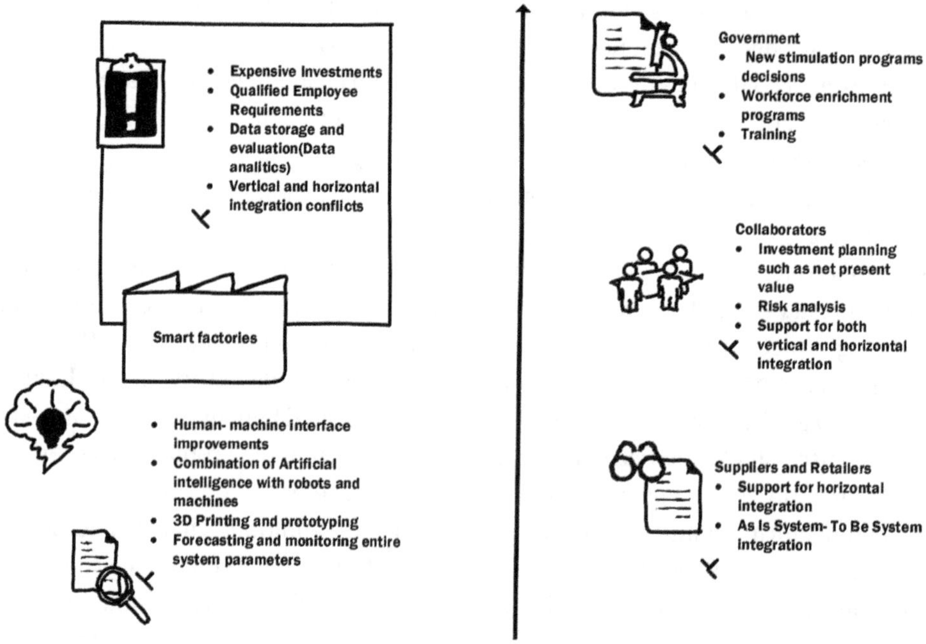


The triggers that directly affect Industry 4.0 are big data analytics, smart robots, Internet of Things, augmented reality, additive production, cyber security, simulation, vertical and horizontal integration and cloud computing. The brief explanations are given in the following:

Big data analytics: For the evaluation of raw data gathered from diversified sources, both internal and external sources, and obtaining knowledge from the raw data enable real time decision making, tracing and tracking decentralized services.

Smart Robots: Adaptive and flexible robots that are combined with the usage of artificial intelligence thereby providing easier manufacturing of different products by recognizing the lower segments of each parts. This, in turn, helps in decreasing production costs, reducing production time and waiting time.

*Internet of Things*: Objects will be "smart" by the integration of microprocessors, embedded software, wireless connections and data storage units. In production lines, this smart connection enables intelligence in manufacturing in terms of monitoring, controlling, optimization and autonomy. Monitoring implies the reflection of the machine gathered from other machines and its operations and giving warnings according to the records from these reflections. Controlling denotes the management of capability of the machines and defining new functions to the machines according to the specialized tasks. Finally, optimization provides utilizing artificial intelligence and predictive methods to improve overall system ability. Autonomy supports the combination of monitoring, optimization and controlling of the machines in order to coordinate joint activities and self-organizing of the functions of the machines.

The concept of artificial intelligence is listed as follows:

- **Genetic Algorithms:** For the improvement of rule extraction and selection problem, genetic algorithm the best approach for defining patterns, i.e., those patterns which minimize/maximize the value of the function to be optimized. Generally, genetic algorithms obtains iterative steps as follows:
  - Selection of the parent,
  - Reintegration,
  - Mutation.

In manufacturing, selection of proper rules for machines considering related job and also, assigning jobs to machines have become important tasks for Industry 4.0 applications. For instance, there is an example of the rule extraction method for manufacturing.

Let f be a function maximized. The function f is explained over all binary strings of length $l$ and is named as the fitness of the strings. Two parent strings from the current population are extracted for the recreation of a current string. The probability of a parent string $H_j$ will be extracted from N strings $H_1$, $H_2$,...,$H_N$ is

$$p(H_j) = \frac{f(H_j)}{\sum_{i=1}^{N} f(H_i)}$$

This implies that strings with greater fitness are more likely to be selected. Let $f_\mu$ be the average fitness appeared in the population,

$$f_\mu = \frac{1}{N}\sum_{i=1}^{N} f(H_i)$$

The probability $p(H_j)$ can be rewritten as follows in Wang et al.(2009)'s study.

$$p(H_j) = \frac{f(H_j)}{Nf_\mu}$$

- **Tabu Search:** Tabu search is based on the idea of realizing flexible structures in coincidence with strategic restrictions and diversified aspiration levels in order to exploit search. Meta-heuristic that integrates a local heuristic search procedure to find out the solution space considering local optima by use of a Tabu list. This provides production method selection among diversified rules which is a crucial concept in Industry 4.0 applications (Euchi and Chabchoub, 2010).
- **Heuristic Greedy Search:** A greedy algorithm generally assists the selection of proper choice that seems to be the best at a specific time. This implies that it searches for locally-optimal solutions considering this solution will lead to a globally-optimal solution. This property will be an inevitable tool for real time decision making in terms of digital manufacturing and robotics adaptation in manufacturing. The main concept of Greedy-algorithm problem is explained as follows:

- Let currentTime and numberOfRules be the basic component for greedy search that defines time period and number of considered solutions.
- Sort the array *G* in a non-decreasing order.
- Select each to-do item one-by-one.
- Add the time that it will take to complete that to-do item into currentTime.
- Add one to numberOfRules.
- In each iteration, select the things which will take the minimum amount of time to complete while maintaining two variables currentTime and numberOfRules (Grant, and Venayagamoorthy, 2009).

*Augmented reality*: Virtual and physical elements could be combined with each other such as smart wears such as Google Glass to ensure decreasing the number of faults especially in order picking, quality control, maintenance and repairing of machines and product placement. In this concept, stochastic optimization is utilized.

*Cloud computing*: Data storage and data processing via cloud will be useful in the accessibility of processed data on request in anytime and anywhere. Additionally, sophisticated algorithms could be activated on cloud structure which provides easier data management and knowledge extraction.For instance,

*Additive production*: Manufacturing of complex parts of product will be carried out by 3D printers which support companies in prototyping, designing and production stages especially for lower volumes of production that require risky investments. Additionally, supply chain management and logistics costs will evolve according to these changes.

*Cyber security:* Data transmission security is essential while connected machines and products are working together. The security infrastructure that provides reliable data interchange between virtual servers and connected devices via wireless network. Additionally, new user identification, access privilege and secure data storage can be available.

*Simulation*: Before the implementation of Industry 4.0 tools, analyzers should model the physical system and the model should be tested according to the changing conditions and different scenarios in order to evaluate the practicability of Industry 4.0 project or tool and determine the optimum parameters of the system.

*Vertical and horizontal integration*: Vertical integration requires the intelligent cross-linking and digitalization of business units in different hierarchal levels within the organization. On the other hand, horizontal integration enables entire value creation between organizations for enriching product life cycle. (Acatech, 2015). The horizontal and vertical integration enable real-time data sharing, productivity in resource allocation, coherent working business units and accurate planning which are crucial for connected devices in Industry 4.0.

Additionally, German government introduced a standardization roadmap on Industry 4.0. They mentioned that "The future-oriented project Industry 4.0 presented by the German Federal Government is intended to reflect the importance of manufacturing technology and the ICT sector [URL-1]. The Federal Ministries of Education and Research (BMBF) and Economic Affairs and Energy (BMWi) are coordinating their funding activities in this regard." [URL-2]. The concept refers to the use of information and communication technologies for machines and systems, based on a symbiosis of information technology and engineering and adapted to the environmental conditions and requirements of the user. Additionally, Siemens's position in Industry 4.0 has a significant effect that integrates cloud technology with automated machines and the company declared that 75 percent of the PLC production process at

Amberg is automated as presented in Hannover Messe 2016. Additionally, With its valve systems from the Advanced Valve, (AV) series and matching valve electronics in the Advanced Electronic System (AES), Aventics makes a dedicated contribution to the development of the Internet of Things. Axoom also integrates the components made by different manufacturers within the production value. They are networked together and are thereby able to work together intelligently. The roadmap for this purpose is presented in five steps: order entry, order management, procurement, production planning and KPI reporting. [URL-3] The final example is from Beckhoff which integrates automated maintenance and cloud system as a project of "Machine diagnostics and predictive maintenance" enabling seamless online and offline analysis of machines and production data.

## Optimization Techniques for Data Analytics

Examples for the intersection between operations research and data mining are presented below (Olafsson et al, 2008).

- **Support Vector Machines:** Determination of linear and nonlinear discrimination of the models between data points is carried out with a "hyperplane" which is found by the maximizing the distance to the closest positive example and negative example.
- **Metaheuristic for Clustering Problems:** Metaheuristic approaches are useful for obtaining adequate number of clusters, determination of the optimal framework of Bayesian network in classification and determination of the best combination of the set of the attributes by implementing initial solution and gathering search result based on fundamentals such as genetic algorithm and particle swarm optimization.
- **Attribute Selection Problems:** The elimination of irrelevant features from database simplifies training and inference procedure. For this problem, a simple combinatorial optimization model is suggested by Olafsson and Yang (2004) that investigates whether the feature should be included for the decision problem or not.
- **Data Visualization Problems:** High dimensional data such as at least 2 or 3 attributes requires the reservation of the relationships between instances. This problem could be presented using quadratic assignment problem formulation by a dimensional matrix including the distance between *n* instances in the set of $R^n x R^n$. The minimization of the deviation between different instances is the objective function of the related problem.
- **Classification:** One of the most common applications in data mining are in classification tasks. The main goal in this task is proposing an accurate model for proper classification by analyzing the training data. The comparison between the model gathered from training data and the validation of this model using test data is the critical point for a better description of the classes. For classification problems, decision tree induction, Bayesian networks and neural networks are generally discussed both in practice and literature.

Optimization could be adopted for the specification of proactive customer-product interaction strategy such as estimation of customer's future actions, product error probability, building product profiles, personalization for tailored product and services and assigning appropriate actions to the targeted product segment. Additionally, personalization on recommendation systems-especially in collaborative and content based filtering- could be modeled with the maximization of perceived utility of each item by each user. In this respect, optimization methods in production is a promising area and OR methods such as mixed integer programming, heuristic approaches etc. could be integrated with data mining techniques.

Due to the major aim of explaining relationship in data sets, data mining could be enhanced with classical statistical methods. According to Hosking et al. (1997), many statistical approaches such as clustering, discriminant analysis and nonparametric regression etc. could be utilized in data mining problems. On the other hand, computation time and construction of the problems require much more attention. The main goals for making inferences could be listed as follows:

- Understanding the patterns of correlation between data values.
- Making predictions according to training data.

Most frequently appeared issues in the formulation of statistical models are:

- **Complex Structure of the Models:** For instance, character recognition problems need much more arrays of pixels than classical statistical models. Thus neural networks based models and classifiers are useful in these cases.
- **Large Scale Problems:** Large datasets cause computational complexity and scalability problems.
- **Validation:** Minimization of prediction error or misclassification rate is required for the validation of the preferred method and selected attributes. Additionally, direct comparison of data mining methods and statistical approaches are needed especially for classification and clustering techniques.

Some of the frequently used methods are given in the following:

*Decision trees:* Top down induction of decision trees obtains the selection of appropriate attributes for classification problem and division of the data with respect to these attributes. Using nonlinear programming, multi category based decision tree problems could be solved efficiently. Additionally, decision tree size is another issue for large datasets. At this point, decision tree growth could be expressed as a NP complete problem to reduce tree size such as genetic algorithm etc. Note that there is generally no method for optimal decision tree size and some of the researchers combine heuristic methods with association rule algorithm and utilized fitness function.

*Bayesian Networks:* This approach uses learning from the conditional probability of each attribute from training data and by the application of Bayes rules. Classification is done by defining the probability of a class and predicting the class value with the highest probability. The critical point is the estimation of marginal probabilities of attribute combinations and dependencies between the attributes. If some of the nodes in the network could not be presented, these conditional probabilities could be calculated via nonlinear programming. If the structure is unknown, model could be formulated by combinatorial optimization tools. The objective function for these problems could be the minimization of the distance between attributes.

*Neural Networks:* A neural network consists of three or more layers including input and output layers. Input layer has one node for each attribute and transforms nodes to output nodes using intermediate layer. Each node is connected with arcs including associated weights for the determination of output value. The optimization processes could be adapted in the determination of arc weights with respect to minimizing squared error of between actual and targeted output. Dependencies between input and output variables are generally nonlinear. Thus, NLP based algorithms and steepest descent approaches could be preferred in order to train neural network. The performance of these algorithms are highly depend on the size of the training data.

*Clustering:* If the data has not a label, learning from the training data will be unsupervised. Grouping data under appropriate classes necessitates using clustering algorithms. Two types could be listed: Hierarchical clustering and partitional clustering. In hierarchical clustering instances are grouped according to similarities between instances. Here, dendrogram presents relationships between instances in a hierarchical way. On the other hand, partitional clustering could be defined as the division of data into one partition and each instance falls into one class. For instance, in k means clustering, instances are separated according to a certain number of clusters and assigned to closest center. According to the assignments, cluster centers are recalculated until reaching the acceptable limits of error. For representing cluster compactness, objective function could be defined as the maximization of distance between different clusters and minimization of the distance within each cluster. Recently, objective function could be taken as the minimization of the distance to the cluster center. However, this approach only focuses on assigning instances to a cluster center as soon as possible, so the clusters could not be separated effectively. Additionally, determination of clusters from farthest point to nearest point is another approach for clustering. From this point of view, assigning instances to best clusters indicates "set covering" and "facility location" problems.

In a clustering problem, sometimes, data cannot be assigned in a specific class. Thus, fuzzy c-means clustering is a useful approach that assists clustering with imprecise data or information. In fuzzy clustering, the input is a set of observations or objects each of which consists of different attributes. The result of cluster analysis produces the clusters and membership of each data point to these clusters. These outputs are represented by the partition matrix. Ruspini (1970) defines the conditions for a fuzzy partition matrix as follows:

$$\mu_{ik} \in \left[0,1\right], 1 \leq i \leq c, 1 \leq k \leq N, \tag{1a}$$

$$\sum\nolimits_{i}^{c} = 1\mu_{ik} = 1, 1 \leq k \leq N, \tag{1b}$$

$$0 < \sum\nolimits_{k}^{N} = 1\mu ik < N, 1 \leq i \leq c \tag{1c}$$

Equation (1b) constrains the sum of each column to 1, and thus the total membership of each equals one.

One of the most popular fuzzy clustering methods is fuzzy c-means (FCM) which is based on minimization of the following objective function:

$$J(Z,U,V) = \sum\nolimits_{i=1}^{c} \sum\nolimits_{j=1}^{N} (\mu_{ij})^{m} \parallel z_j - v_i \parallel \tag{2}$$

where Z is the data set to be partitioned, U is the fuzzy partition matrix, V is the vector of cluster centers. In the formula, N represents the number of observations, c is the number of clusters and μ shows the membership value, m is the parameter called fuzzifier which determines the fuzziness of the resulting clusters. The fuzzifier parameter can get values 1 and more. When the fuzzifier parameter equals to one, then the clusters are formed in crisp format. In the formula, $z_k - v_i$ shows the distance between observation k and the center of cluster i.

The minimization of the mention objective function represents a nonlinear optimization problem that can be solved by using a variety of methods such as iterative minimization, simulated annealing or genetic algorithms. Babuska (2009) gives the steps of fuzzy c-means (FCM) algorithm as follows:

1. Initialize $U=[u_{ij}]$ matrix, $U^{(0)}$
2. At k-step: calculate the centers vectors $V^{(k)}=[v_i]$ with $U^{(k)}$

$$v_i = \frac{\sum_{i=1}^{N} \mu_{ij}^{m} \cdot z_j}{\sum_{i=1}^{N} \mu_{ij}^{m}}$$

3. Update $U^{(k)}$, $U^{(k+1)}$

$$u_{ij} = \frac{1}{\sum_{k=1}^{c} \left( \frac{\| z_j - v_i \|}{\| z_j - v_k \|} \right)^{\frac{2}{m-1}}}$$

*If* $\| U^{(k+1)} - U^{(k)} \| < \delta$ *then STOP; otherwise return to step 2.*

Recommendation Systems: Recommender systems have been attracted after the first papers published by Resnick et al. (1994) and Shardanand and Maes (1995). The main idea in recommendation system is classifying users related to similarities and trying to find out the most appropriate choice among different alternatives in accordance with the users' preferences (Bobadilla et al., 2013). Thus, recommendation systems have been used in many areas such as movie, news, book, learning style, product recommendations as well as task, content and document recommendation (Park et al., 2012).

Over the last twenty years, researchers have investigated recommender systems in terms of two basic methodology: content based filtering which groups related to similar items for offering new items in accordance with past preferences and collaborative filtering which infers finding the most convenient option depending on users' earlier preferences in order to predict new users' preferences (Zhang and Jiao, 2007). As seen from the literature, most of the studies are focused on clustering issues for collaborative filtering and content based applications. Different tools of operations research such as genetic algorithm, MCDM and utility models are utilized in order to make correct user- item matching.

The most emphasized point in literature is the cold start problem that the recommendation system could not make any inferences due to the insufficient data and this problem especially arises when a new item is added to the system. Furthermore, data sparsity reduces recommendation accuracy (Lika et al., 2014). In order to solve these problems, hybrid models are preferred with the assistance of cloud computing and machine learning algorithms.

Association Rule Mining: Association rule discovery aims to find out interesting correlations and relationship between attributes and especially applied in market basket and shopping cart analysis for the investigation of the factors in purchasing behavior of an "item". Most frequently applied approach for association rule mining is Apriori algorithm which emphasis on that if itemset is not frequent, any superset of this dataset will not be frequent and if minimum support level is satisfied, the itemset will be frequent. Additionally, if determined rules are supported by minimum confidence level, the rules could be identified as candidate rules. Consequently, objective function could be specified as the maximization of the difference between minimum support level and minimum confidence. On the other hand, maximization of support level causes extraction of a great many of rules and maximizing confidence leads to the extraction of specific rules. Thus, more research should be performed for finding appropriate rules.

As a special form of association rule mining, frequent itemset mining, first introduced by Agrawal, Imielinski, and Swami (1993), has become a popular data mining technique and plays an fundamental role in substantial data mining problems such as mining associations, correlations, episodes, etc. Frequent itemset mining was initially proposed for market basket analysis in dealing with the problem of mining association rule. Formally, it can be described as follows. Assume I = $\{i_1, i_2, ..., i_m\}$ is the universal item set and DB = $\{T_1, T_2, ..., T_n\}$ is a transaction database, where each $T_k$ ($1 <= k <= n$) is a transaction which is a set of items such that $T_k$ is involved in *I*. *P* is called an itemset if *P* is a set of items. Let *P* be an itemset. In these conditions, transaction *T* contains *P* if and only if *P* # *T*. The support of itemset *P* is the number of transactions in DB that contain *P*. Let n be the predefined minimum support threshold and |DB| be the number of transactions in DB. An itemset *P* is frequent if *P*'s support is not less than *n* x |DB|.For the gathered transaction database DB and threshold n, the task of mining frequent itemsets is to discover the set of all itemsets with support not less than n |DB|.

For the prediction of the locations of the users, frequent itemset mining algorithm proposed by Hong-Deng and Lv (2015) is used for generating tree structure. The assumptions of tree structure are as follows: (1) It consists of one root labeled as ''null'', and a set of item pre- fix subtrees as the children of the root. (2) Each node in the item prefix subtree consists of five fields: item-name, count, children-list, pre-order, and post-order. item-name registers which item this node represents. Count registers the number of transactions presented by the portion of the path reaching this node. Children-list registers all children of the node. Pre-order is the pre-order rank of the node. Post-order is the post-order rank of the node. For a node, its pre-order is the sequence number of the node when scanning the tree by pre-order traversal and its post-order is the sequence number of the node when scanning the tree by post-order traversal.

## THE TERMINOLOGY "BIG DATA"

As a result of the innovations in information science and communication technologies, data generation began to appear in various types including sound, graphic, word, picture, GPS data, entries in search engines etc., i$^{th}$ petabytes and eksabytes of storage needs complicating data storage and data management for companies. These massive amount of data is captured by RFIDs, wireless sensors, mobile devices and digital cameras. For instance, 120,000 tweets, 2 billion searches in Google, 277,000 entries in Facebook arise per minute in Web. The term "big data" is first declared by Michael Cox and David Ellsworth in the article named as "Application-controlled demand paging for out-of-core visualization" appeared in Proceedings of the IEEE 8th conference on Visualization in 1997 and consequently, both academic and practitioner based publications and reports increased dramatically. The usage of this term in marketing and retailing appeared not long after Cox and Ellsworth: Roger E. Bohn and James E. Short published "How Much Information? 2009 Report on American Consumers". Their work emphasizes on the consumption of the size of the information is approximately 1.3 trillion hours. The sum of the consumption reached to 3.6 Zettabytes and 10,845 trillion words, translating to 100,500 words and 34 gigabytes for an average person on an average day. In the World Economic Forum held in Davos, 2015 is selected as a "data driven" year that "big data" term will become widespread. This situation implies that both invaluable and "trash" data is accumulated and sorted out in order to provide the advantage of extracting knowledge that will point out the misunderstandings. For instance, World Bank declared that $1.5M valued projects are ongoing and they recently finished 100 projects in various sectors and regions in order to find out valuable patterns of information.

According to Gartner Group, big data means "high velocity, high volume and high variety of information assets that require new forms of processing to enable enhanced decision making, insight discovery and process optimization". Therefore, big data consists of the terms listed below:

1. **Variety:** Big data obtains various types of data collected from various resources. The data type could be text, graphic, picture, audio etc. and generally unstructured. The execution, auditability and summarization of data are necessary for realistic analysis.
2. **Volume:** Amount of the data is another property of big data term. By the increasing of the data size, data becomes lower density such as tweets, clicks on a web page, geo data etc. Thus, data should be transformed to meaningful data.
3. **Velocity:** The interchange of the data required to be faster due to the real time data transfer as realized in IoT, mobile applications and mobile advertising. For example, location based mobile advertising needs the combination of geo data and user preferences with time adjusted recommendations.
4. **Value:** Effective data analysis is required for extracting knowledge from data. On the other hand, doing computations on huge amount of data for recognition patterns or predicting behavior sometimes results in misunderstanding of information caused by statistical sampling.
5. **Verification:** Inconsistencies in data analysis are caused by irrelevant data usage. Thus, verification is implemented by using different data sources or analysis tools for making comparisons especially with real life cases. Verification provides the determination of unnecessary data and unproductive data analysis methods for future big data based analytics.

Boyd and Crawford (2012) propose implementation of "big data" in following three stages: (1) Technology: Minimizing computation time and maximizing algorithmic support to gather, analyze, link, and compare large data sets. (2) Integration: collecting necessary data and separating "trashes" from data warehouses. (3) Analysis: determination of patterns for economic, social, technical, and legal interpretations. From this point of view, this procedure is similar to the "known" data analysis processes except separation of unnecessary data from storages and the required computational time.

Methods that will be used in big data analytics implementation of manufacturing actions should also be determined according to the properties listed below:

- **Objective of the Analysis:** In order to decide the proper strategy that will be adopted in each group, organizational resources and environmental factors should be included in the big data analytics model.

- **Sampling:** Inadequate number of observations will lead to gathering meaningless information from the manufacturing transactions that will cause unproductive usage of company's budget.
- **Data Type and Source:** The selection of big data analytics method highly identifies the intended use. A priori segmentation (classification) is based on the selected variables and classifying production needs according to these variables with a certain number of classes. Post hoc segmentation mainly applies when number of classes is not known before clustering. In dynamic classification, production needs are generally segmented as alternative segments. Conjoint analysis is another method that manufacturing needs are grouped with respect to the preferred types of production or services.
- **Determination of Variables:** Geographic and economic variables are commonly used for grouping production activities in job shop manufacturing. On the other hand, product or service usage frequency, manufacturing velocity and preferences should also be considered for the determination of appropriate big data analytics method. Additionally, they should be combined with geographic and economic variables for identification of different production characteristics.
- **Optimization Techniques:** For the data analysis, clustering, CHAID, multiple regression, discriminant analysis, latent class structure, ANN, SOM, metaheuristic methods such as genetic algorithm are preferred in literature. Hiziroglu (2013) mentioned that previous studies demonstrated that hierarchical approaches could not deal with large data sets because of the partitioning ineffectiveness of dendrogram. To overcome this problem, hybrid methods should be applied to large datasets such as hierarchical approaches, non-hierarchal approaches (K means) and association rule mining.
- **Standardization and Normalization:** Different variables could have diversified number of data types. To make the calculations easily, standardization and normalization is necessary prior to implementation.
- **Validation:** Multiple executions, cross validation, adopting completely different data to test purposed segmentation method, evaluating compactness (closeness of the elements in the same group) and separation (distance between different clusters) are critical measures for proving the validation of segmentation technique.

Big data implementation is generally assisted with an infrastructure such as Hadoop, MapReduce, Hive, HBase, MongoDB or Cassandra. Naturally, programming languages like Java, Linux, PHP is necessary before and after the big data solution adaptation. Additionally, while working with multiple data warehouses, NoSQL platform knowledge provides the advantage of the extension of clearing "trash" data in big data solution platform. Among big data infrastructure solutions, MapReduce is introduced by Google for separating data to smaller parts thereby providing ability for quick pre-analysis by reducing the combination of extractions according to the business logic. In order to cope with the enormous amount of unstructured data storage, ingestion, processing and filtering, Hodoop infrastructure is a promising solution introduced by Apache Software Developers and developed in Java and especially useful for parallel processing. It is not just simply a "data warehouse". The usage of the interactive and iterative processing of business intelligence based operations in web or mobile platforms is realized such as web log analysis, visitor behavior, text analyzing, search analysis by integrating machine learning approaches and data mining methods. In other words, Hadoop structure is useful for fast data loading, ETL, image processing and online data integration including distributed file systems YARN and Map Reduce platform. In Hadoop, YARN tool clusters resources according to the relation of obtained data.

As a part of Hadoop framework, Map Reduce platform is similar to Google's solution. In addition to that Hadoop framework could be described as an "ecosystem" serving with diversified software packages as Apache Pig, Apache Hive, Apache HBase, Apache Phoenix, Apache Spark, Apache ZooKeeper, Cloudera Impala, Apache Flume, Apache Sqoop, Apache Oozie, and Apache Storm.

The dissemination of Bluetooth 4.0 supported devices such as Beacons will ensure the location determination in indoor locations. This situation provides the development of recommendation systems framework in terms of presenting special actions via mobile devices to the "smart" products. On the other hand, irrelevant offerings and wrong directions given automatically will affect customer attitude towards the firm in a negative way. Thus, intentional algorithms should be more personalized and speedy enough in computing the results.

## Fuzzy Logic Based Techniques

Diversified data mining techniques including clustering, classification, pattern mining, association rule mining have been widely performed with the assistance of artificial neural networks, machine learning and genetic algorithms for more correct and precise forecasting in order to submit options to decision makers. These methods generally aim to simplify finding out valuable information and knowledge hiding in relational, object oriented, transactional and spatial databases.

As Soltani and Navimipour (2016) mentioned in their review paper, especially customer loyalty, behaviors, purchasing tendencies and potential customers' identification could not be specified in deterministic way. Thus, data mining methods could be adopted in imprecise environment under the name of "fuzzy data mining" algorithms. Also mentioned in Hou et al. (2007)'s study, fuzzy data mining techniques are being preferred in terms of simplifying of the explanation of social terms and human beings with imprecise data. Therefore, one could conclude that fuzzy data mining techniques can be utilized in real life applications such as in manufacturing, especially for production planning. For the purpose of representing the realistic situation, production studies that dealt with social practices are being adopted to fuzzy environment in accordance with their problem domains. For instance, production classification, segmentation and determination of product groups are the critical subjects in production planning especially when there is difficulty of assigning an item to a specific group. A similar example could be given in IoT based systems for avoiding "sharp" discrimination of people needs. For all of these purposes, "fuzzy clustering", especially fuzzy c-means clustering is included as clustering method for production planning.

Since increasing number of innovations in information technology and communication have being appeared, the intersection of statistics, data management and knowledge management has become more substantial especially in the applications of data mining and machine learning algorithms. In extracting information and patterns from large data sets, fuzzy set theory based algorithms are adopted widely for establishing intelligent, self-organizing and self-learning systems (Hüllermeier, 2011). The steps that fuzzy set based approaches to data mining techniques could be presented as marked with starts as seen in Figure 3.

*Figure 3. Steps of data mining process (Inspired from Han and Kamber, 2001)*

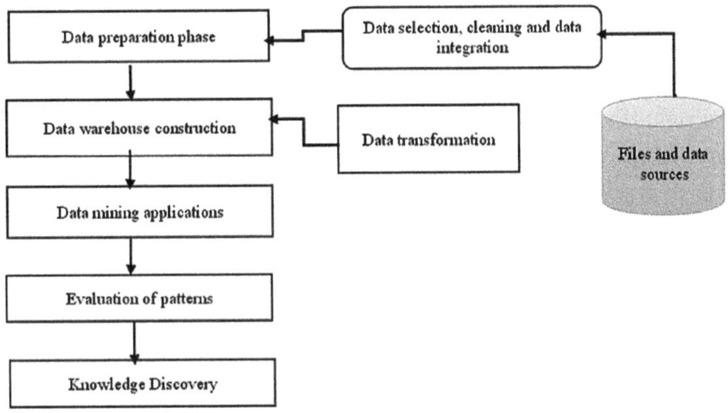


Hüllermeier (2011) also declared that fuzzy set theory based approaches are particularly used in data preparation, data transformation and data mining application phases. In data preparation phase, crisp data could be "fuzzified" for mapping data for the data warehouse form. Complex mathematical spaces and vectors are converted to fuzzy metric spaces in data transformation phase for performing data analysis. Finally, data mining process can be progressed with fuzzy set theory based analysis is used rather than dealing with forming fuzzy numbers. (Here, "fuzzy" implies the analysis is performed with "fuzzy" approaches and extensions). In this respect, fuzzy data analysis refers to the modeling processes of data in imprecise and vague environment with respect to converting "crisp" summaries of data into "fuzzy" representation. This approach includes both fuzzy extensions of standard data analysis tools and transforming data into a sophisticated environment as fuzzy metric spaces.

## SOLUTIONS AND RECOMMENDATIONS

Clearly Industry 4.0 has significant advantages, however problems could appear in implementation stage. The most crucial issue is the expensiveness of the initial investments. Thus, governments should introduce new stimulus packages for diversified sectors (Ustundag, 2016). Additionally, initial implementation could not be conducted if collaborators do not support the company in terms of both technically and financially. In this respect, risk analysis and net present values should be considered before making the investments. Furthermore, qualified employees are required especially for monitoring and controlling the entire system operations. Governments should support training programs in order to simplify the implementation stage of Industry 4.0.

In this study, we focused on explaining advantages, value creation aimed properties and critical issues of Industry 4.0 for manufacturing environment both in academic and practical aspects. Additionally, opportunities and threats are defined from the operational perspective. The evaluation of intelligent cross linked machines (Cyber- Physical Systems), end-to-end engineering (product development) and digitalized control systems are discussed. Finally, some of the business tools, value creation tips and changes that should be adopted to different organizations and sectors are briefly explained.

## CONCLUSION

Current global competition forces companies to deal with difficulties such as increasing costs and decreasing number of resources. Thus, companies are prone to focus on industrial value creation based activities especially in manufacturing and logistics. Recently, value creation based industrialization is reached in the form of tracking, controlling, optimization and autonomy of manufacturing that is named as Industry 4.0, the fourth stage of industrialization. In this study, opportunities and threats of Industry 4.0 are discussed within the concept of the relationship between big data analytics, smart robots, Internet of Things, 3D printing and cloud computing. Additionally, further improvements are outlined with respect to the developments in information systems and manufacturing.

## REFERENCES

Acatech. (2015). *Umsetzungsstrategie Industrie 4.0 Ergebnisbericht der Plattform Industrie 4.0*. Acatech2015.

Agrawal, R., Imielinski, T., & Swami, A. (1993). Mining association rules between sets of items in large databases. *Proceeding SIGMOD '93 Proceedings of the 1993 ACM SIGMOD international conference on Management of data*, 207-216. doi:10.1145/170035.170072

Babuska, R. (2009). *Fuzzy And Neural Control Disc Course Lecture Notes*. Retrieved October 5, 2012. from http://www.dcsc.tudelft.nl/~disc_fnc/transp/fncontrol.pdf

Berger, R. (2014). *Industry 4.0-The new industrial revolution-How Europe will succeed?* Roland Berger Strategy Consultants.

Bobadilla, J., Ortega, F., Hernando, A., & Gutierrez, A. (2013). Recommender systems survey. *Knowledge-Based Systems*, *46*, 109–132. doi:10.1016/j.knosys.2013.03.012

Boyd, D., & Crawford, K. (2012). Critical Questions For Big Data, Journal Information. *Communicatio Socialis*, 662–679.

Deng, Ma, & Liu. (2015). *An Efficient Data Structure for Fast Mining High Utility Itemsets*. arXiv:1510.02188

Euchi, J., & Chabchoub, H. (2010). A hybrid tabu search to solve the heterogeneous fixed fleet vehicle routing problem. *Logistics Research*, *2*(1), 3–11. doi:10.1007/s12159-010-0028-3

Gaub, H. (2016). Customization of mass-produced parts by combining injection molding and additive manufacturing with Industry 4.0 technologies. *Reinforced Plastics*, *60*(6), 401–404. doi:10.1016/j.repl.2015.09.004

Gilchrist, A. (2016). *Industry 4.0: The Industrial Internet of Things, 2016*. Apress. doi:10.1007/978-1-4842-2047-4

Grant, L. L., & Venayagamoorthy, G. K. (2009). Swarm Intelligence for collective Robotic Search. In *Design and Control of Intelligent Robotic Systems* (pp. 29–47). Academic Press.

Han, J., & Kamber, M. (2001). *Data Mining Concepts and Techniques*. Morgan Kauffman Publishers.

Hermann, Pentek, & Otto. (2016). *Design Principles for Industrie 4.0 Scenarios*. Academic Press.

Hiziroglu, A. (2013). A neuro-fuzzy two-stage clustering approach to customer segmentation. *Journal of Marketing Analytics*, *1*(4), 202–221. doi:10.1057/jma.2013.17

Hosking, J. R. M., Pednault, E. P. D., & Sudan, M. (1997). A Statistical Perspective on Data Mining. *Future Generation Computer Systems*, *13*(2-3), 117–134. doi:10.1016/S0167-739X(97)00016-2

Hüllermeier, E. (2011). Fuzzy Sets in Machine Learning and Data Mining. *Applied Soft Computing*, *11*(2), 1493–1505. doi:10.1016/j.asoc.2008.01.004

Lika, B., Kolomvatsos, K., & Hadjiefthymiades, S. (2014). Facing the cold start problem in recommender systems. *Expert Systems with Applications, 41*(4), 2065–2073.

Long, F., Zeiler, P., & Bertsche, B. (2016). Modelling the production systems in industry 4.0 and their availability with high-level Petri nets. *IFAC-PapersOnLine*, *49*(12), 145–150. doi:10.1016/j.ifacol.2016.07.565

Olafsson, S., Li, X., & Wu, S. (2008). Operations Research and Data Mining. *European Journal of Operational Research*, *187*(3), 1429–1448. doi:10.1016/j.ejor.2006.09.023

Park, D. H., Kim, H. K., Choi, Y., & Kim, J. K. (2012). A literature review and classification of recommender systems research. *Expert Systems with Applications*, *39*(11), 10059–10072. doi:10.1016/j.eswa.2012.02.038

Quinn, J., Liu, Y., & Grosvenor, R. (2016). Author links open the author workspace.Roger Grosvenor., A Categorical Framework of Manufacturing for Industry 4.0 and Beyond. *Procedia CIRP*, *52*, 173–178. doi:10.1016/j.procir.2016.08.005

Ruspini, E. H. (1970). Numerical methods for fuzzy clustering. *Information Sciences*, *2*(3), 319–350. doi:10.1016/S0020-0255(70)80056-1

Soltani, Z., & Navimipour, N. J. (2016). Customer Relationship Management mechanisms: A systematic review of the state of art literature and recommendations for future research. *Computers in Human Behavior*, *61*, 667–688. doi:10.1016/j.chb.2016.03.008

Stock, T., & Seliger, G. (2016). Opportunities of Sustainable Manufacturing in Industry 4.0, 13th Global Conference on Sustainable Manufacturing – Decoupling Growth from Resource Use. *Procedia CIRP, 40*, 536-541.

Thames, L., & Schaefer, D. (2016). Software-defined Cloud Manufacturing for Industry 4.0. *Procedia CIRP*, *52*, 12–17. doi:10.1016/j.procir.2016.07.041

Ustundağ, A. (2016). *Endüstri 4.0 ve Akıllı Sistemler*. Retrieved from http://alpustundag.blogspot.com.tr/

Wang, W., Cha, K., Cheng, C., & And Qiu, L. (2009). A comparison of performance of several artificial intelligence methods for forecasting monthly discharge time series. *Journal of Hydrology (Amsterdam)*, *374*(3-4), 294–306. doi:10.1016/j.jhydrol.2009.06.019

Weyer, S., Schmitt, M., Ohmer, M., & Gorecky, D. (2015). Towards Industry 4.0 - Standardization as the crucial challenge for highly modular, multi-vendor production systems. *IFAC-Papers OnLine, 48*(3), 579-584

Wittenberg, C. (2016). Human-CPS Interaction - requirements and human-machine interaction methods for the Industry 4.0. *IFAC-Papers OnLine*, *49*(19), 420–425. doi:10.1016/j.ifacol.2016.10.602

Zezulka, F., Marcon, P., Vesely, I., & Sajdl, O. (2016). Industry 4.0 – An Introduction in the phenomenon. *IFAC-Papers OnLine*, *49*(25), 8–12. doi:10.1016/j.ifacol.2016.12.002

Zhang, Y., & Jiao, J. (2007). An associative classification-based recommendation system for personalization in B2C e-commerce applications. *Expert Systems with Applications*, *33*(2), 357–367. doi:10.1016/j.eswa.2006.05.005

## KEY TERMS AND DEFINITIONS

**Beacon:** A technology that assists indoor location based system using Bluetooth Low Energy.

**Bluetooth:** A technology that connects mobile devices via wireless connection infrastructure.

**Data Mining:** Methodologies that contribute the extraction of information among large scale data.

**Fuzzy Clustering:** Decomposition of data elements with respect to each element could be belong to one or more classes via a set of membership functions.

**Fuzzy Logic:** A form of explanations of crisp values to 0-1 values in order to represent the "truth" via determining membership function.

**Hierarchical Clustering:** A data mining method for gathering hierarchical structure when grouping related data via top down or bottom up approach.

**Partitional Clustering:** Dissociating a data set into a set of disjoint clusters.

**Segmentation:** Grouping individuals according to the similar properties.

# Compilation of References


Acatech. (2015). *Umsetzungsstrategie Industrie 4.0 Ergebnisbericht der Plattform Industrie 4.0*. Acatech2015.

Açıl, A. (2008). Solving the profitable tour problem using ant colony system [Doctoral Thesis].

Adolphs, P., Auer, S., Bedenbender, H., Billmann, M., Hankel, M., Heidel, R., & Waser, B. (2016). *Struktur der Verwaltungsschale – Fortentwicklung des Referenzmodells für Industrie 4.0-Komponente*. Retrieved June 23, 2016, from https://www.plattform-i40.de/I40/Redaktion/DE/Downloads/Publikation/struktur-der-verwaltungsschale.pdf?__blob=publicationFile&v=8

Adolphs, P., Bedenbender, H., Dirzus, D., Ehlich, M., Epple, U., Hanke, M., & Wollschlaeger, M. (2015). *Statusreport. Referenzarchitekturmodell Industrie 4.0 (RAMI4.0)*. Retrieved June 14, 2016 from http://www.vdi.de/fileadmin/user_upload/VDI-GMA_Statusreport_Referenzarchitekturmodell-Industrie40.pdf

Afshar-Nadjafi, B., Rahimi, A., & Karimi, H. (2013). A genetic algorithm for mode identity and the resource constrained project scheduling problem. *Scientia Iranica*, *20*(3), 824–831.

Agarwal, A., Shankar, R., & Tiwari, M. K. (2006). Modeling the metrics of lean, agile and leagile supply chain: An ANP-based approach. *European Journal of Operational Research*, *173*(1), 211–225. doi:10.1016/j.ejor.2004.12.005

Agarwal, R., & Dhar, V. (2014). Big Data, Data Science, and Analytics: The Opportunity and Challenge for IS Research. *Information Systems Research*, *25*(3), 443–448. doi:10.1287/isre.2014.0546

Aghezzaf, E. H., Jamali, M. A., & Ait-Kadi, D. (2007). An integrated production and preventive maintenance planning model. *European Journal of Operational Research*, *181*(2), 679–685. doi:10.1016/j.ejor.2006.06.032

Aghezzaf, E. H., Khatab, A., & Le Tam, P. (2016). Optimizing production and imperfect preventive maintenance planning' s integration in failure-prone manufacturing systems. *Reliability Engineering & System Safety*, *145*, 190–198. doi:10.1016/j.ress.2015.09.017

Aghezzaf, E. H., & Najid, N. M. (2008). Integrated production planning and preventive maintenance in deteriorating production systems. *Information Sciences*, *178*(17), 3382–3392. doi:10.1016/j.ins.2008.05.007

Aglan, C., & Durmusoglu, M. B. (2015). Lot-splitting approach of a hybrid manufacturing system under CONWIP production control: A mathematical model. *International Journal of Production Research*, *53*(5), 1561–1583. doi:10.1080/00207543.2014.957873

Agrawal, R., Imielinski, T., & Swami, A. (1993). Mining association rules between sets of items in large databases. *Proceeding SIGMOD '93 Proceedings of the 1993 ACM SIGMOD international conference on Management of data*, 207-216. doi:10.1145/170035.170072

Ahlemann, F., Teuteberg, F., & Schöder, C. (2005). *Kompetenz- und Reifegradmodelle für das Projektmanagement – Grundlagen, Vergleich und Einsatz*. Osnabrück: Universität Osnabrück.

Ahmad, R., & Kamaruddin, S. (2012). An overview of time-based and condition-based maintenance in industrial application. *Computers & Industrial Engineering*, *63*(1), 135–149. doi:10.1016/j.cie.2012.02.002

Aitken, J., Childerhouse, P., & Towill, D. (2003). The impact of product life cycle on supply chain strategy. *International Journal of Production Economics*, *85*(2), 127–140. doi:10.1016/S0925-5273(03)00105-1

Akay, B., & Karaboga, D. (2012). Artificial bee colony algorithm for large-scale problems and engineering design optimization. *Journal of Intelligent Manufacturing*, *23*(4), 1001–1014. doi:10.1007/s10845-010-0393-4

Akbari, M., & Rashidi, H. (2016). A multi-objectives scheduling algorithm based on cuckoo optimization for task allocation problem at compile time in heterogeneous systems. *Expert Systems with Applications*, *60*, 234–248. doi:10.1016/j.eswa.2016.05.014

Akbari, R., Zeighami, V., & Ziarati, K. (2011). Artificial Bee Colony for Resource Constrained Project Scheduling Problem. *International Journal of Industrial Engineering Computations*, *2*(1), 45–60. doi:10.5267/j.ijiec.2010.04.004

Akgündüz, O. S., & Tunalı, S. (2010). An adaptive genetic algorithm approach for the mixed-model assembly line sequencing problem. *International Journal of Production Research*, *48*(17), 5157–5179. doi:10.1080/00207540903117857

Akturk, M. S. (2011). Joint cell loading and scheduling approach to cellular manufacturing systems. *International Journal of Production Research*, *49*(21), 6321–6341. doi:10.1080/00207543.2010.532165

Alaswad, S., & Xiang, Y. (2017). A review on condition-based maintenance optimization models for stochastically deteriorating system. *Reliability Engineering & System Safety*, *157*, 54–63. doi:10.1016/j.ress.2016.08.009

Allahverdi, A. (2015). The third comprehensive survey on scheduling problems with setup times/costs. *European Journal of Operational Research*, *246*(2), 345–378. doi:10.1016/j.ejor.2015.04.004

Allahverdi, A., Ng, C. T., Cheng, T. C. E., & Kovalyov, M. (2009). A survey of scheduling problems with setup times or costs. *European Journal of Operational Research*, *187*(3), 985–1032. doi:10.1016/j.ejor.2006.06.060

Allstot, D. J., Choi, K., & Park, J. (2003). *Parasitic-Aware Optimization of CMOS RF Circuits*. Kluwer Academic Publishers.

Almeida, A. T., & Bohoris, G. A. (1995). Decision theory in maintenance decision making. *Journal of Quality in Maintenance Engineering*, *1*(1), 39–45. doi:10.1108/13552519510083138

Aloise, D., Deshpande, A., Hansen, P., & Popat, P. (2009). NP-hardness of Euclidean sum-of-squares clustering. *Machine Learning*, *75*(2), 245–248. doi:10.1007/s10994-009-5103-0

Alomar, M., & Pasek, Z. J. (2014). Linking supply chain strategy and processes to performance improvement. *Procedia CIRP*, *17*, 628–634. doi:10.1016/j.procir.2014.01.144

Alrabghi, A., & Tiwari, A. (2015). State of the art in simulation-based optimisation for maintenance systems. *Computers & Industrial Engineering*, *82*, 167–182. doi:10.1016/j.cie.2014.12.022

Alrabghi, A., & Tiwari, A. (2016). A novel approach for modelling complex maintenance systems using discrete event simulation. *Reliability Engineering & System Safety*, *154*, 160–170. doi:10.1016/j.ress.2016.06.003

Al-Sultan, K. (1995). A Tabu search approach to the clustering problem. *Pattern Recognition*, *9*(28), 1443–1451. doi:10.1016/0031-3203(95)00022-R

Altun, O., & Bocko, M. (2006). Robust analog circuit design: a set theoretic approach. *IEEE International Symposium on Circuits and Systems*. doi:10.1109/ISCAS.2006.1693245

Andersson, R., Eriksson, H., & Torstensson, H. (2006). Similarities and differences between TQM, six sigma and lean. *The TQM Magazine*, *18*(3), 282–296. doi:10.1108/09544780610660004

Andijani, A., & Duffuaa, S. (2002). Critical evaluation of simulation studies in maintenance. *Production Planning and Control*, *13*(4), 336–341. doi:10.1080/09537280110077578

Araújo, A.M.P., & Neto, A.A. (2010). *Aprendendo a contabilidade: Como entender o processo contábil, como interpretar as deonstrações contábeis, como aplicar a contabilidade em negócios*. RibeirãoPreto: Inside Books.

Areal, J. J., Martín, R. M., & Campos, J. G. (2011). Simulated annealing vs. genetic algorithms applied using a new cost function for the car sequencing problem. *International Journal of Manufacturing Technology and Management*, *23*(1-2), 113–136. doi:10.1504/IJMTM.2011.042111

Arunraj, N. S., & Maiti, J. (2010). Risk-based maintenance policy selection using AHP and goal programming. *Safety Science*, *48*(2), 238–247. doi:10.1016/j.ssci.2009.09.005

Ashby, J. R., & Uzsoy, R. (1995). Scheduling and order release in a single-stage production system. *Journal of Manufacturing Systems*, *14*(4), 290–306. doi:10.1016/0278-6125(95)98881-6

Askin, R. G., & Huang, Y. (2001). Forming effective worker teams for cellular manufacturing. *International Journal of Production Research*, *39*(11), 2431–2451. doi:10.1080/00207540110040466

Atashpaz Gargari, E. (n. d.). Retrieved September 1st, 2014, from http://atashpaz.com/

Atashpaz-Gargari, E., & Lucas, C. (2007). Imperialist competitive algorithm: An algorithm for optimization inspired by imperialistic competition. In *Proceedings of the IEEE Congress on Evolutionary Computation*.

Aversa, R., Botter, R. C., Haralambides, H. E., & Yoshizaki, H. T. Y. (2005). A Mixed Integer Programming Model on the Location of a Hub Port in the East Coast of South America. *Maritime Economics & Logistics*, *7*(1), 1–18. doi:10.1057/palgrave.mel.9100121

Azadeh, A., Sheikhalishahi, M., & Koushan, M. (2015). An integrated fuzzy DEA–Fuzzy simulation approach for optimization of operator allocation with learning effects in multi products CMS. *Applied Mathematical Modelling*, *37*(24), 9922–9933. doi:10.1016/j.apm.2013.05.039

Azaron, A., Brown, K. N., Tarim, S. A., & Modarres, M. (2007). A multi-objective stochastic programming approach for supply chain design considering risk. *International Journal of Production Economics*, *116*(1), 129–138. doi:10.1016/j.ijpe.2008.08.002

Babuska, R. (2009). *Fuzzy And Neural Control Disc Course Lecture Notes*. Retrieved October 5, 2012. from http://www.dcsc.tudelft.nl/~disc_fnc/transp/fncontrol.pdf

Bagchi, S., Uckun, S., Miyabe, Y., & Kawamura, K. (1991).Exploring problem-specific recombination operators for job-shop scheduling. In *Proceedings of the Fourth International Conference on Genetic Algorithms and their Applications* (pp. 10-17).

Bag, S., Chakraborty, D., & Roy, A. R. (2009). A production inventory model with fuzzy random demand and with flexibility and reliability considerations. *Comput. Ind. Eng. J.*, *56*(1), 411–416. doi:10.1016/j.cie.2008.07.001

Ba, H. T., Cholette, M. E., Borghesani, P., Zhou, Y., & Ma, L. (2016). Opportunistic Maintenance Considering Non-Homogenous Opportunity Arrivals and Stochastic Opportunity Durations. *Reliability Engineering & System Safety*. doi: 10.1016/j.ress.2016.12.011

Bailey, A., Ornbuki-Berrnan, B., & Asobiela, S. (2013). Discrete PSO for the uncapacitated single allocation hub location problem. In *Proceedings of the 2013 IEEE Symposium on Computational Intelligence in Production and Logistics Systems, CIPLS 2013 - 2013 IEEE Symposium Series on Computational Intelligence SSCI '13* (pp. 92–98). doi:10.1109/CIPLS.2013.6595205

Bakhshi-Jooybari, M. (2002). A theoretical and experimental study of friction in metal forming by the use of the forward extrusion process. *Journal of Materials Processing Technology*, *125*, 369–374. doi:10.1016/S0924-0136(02)00343-6

Bal, A., & Satoglu, S. I. (2013) Maintenance Management of Production Systems with Sensors and RFID: A Case Study. In *Proceedings of the Global Conference on Engineering Technology and Management.*

Balaji, A. N., & Porselvi, S. (2014). Artificial Immune System Algorithm and Simulated Annealing Algorithm for Scheduling Batches of Parts based on Job Availability Model in a Multi-cell Flexible Manufacturing System. *Procedia Engineering*, *97*, 1524–1533. doi:10.1016/j.proeng.2014.12.436

Ball, G. H. (1971). Classification analysis (Technical Report SRI Project 5533). Stanford Research Institute.

Ballweber, B. M., Gupta, R., & Allstot, D. J. (2000). A Fully Integrated 0.55.5-GHz CMOS Distributed Amplifier. *IEEE Transactions on Solid-State Circuits*, *35*(2), 231–239. doi:10.1109/4.823448

Banaszak, Z. A., & Zaremba, M. B. (2006). Project-Driven Planning and Scheduling Support for Virtual Manufacturing. *Journal of Intelligent Manufacturing*, *17*(6), 641–651. doi:10.1007/s10845-006-0034-0

Banerjee, A. (1986). A joint economic-lot-size model for purchaser and vendor. *Decision Sciences*, *17*(3), 292–311. doi:10.1111/j.1540-5915.1986.tb00228.x

Barari, M., Karimi, H. R., & Razaghian, F. (2014). *Analog Circuit Design Optimization Based on Evolutionary Algorithms. Mathematical Problems in Engineering*. Hindawi.

Barata, J., Soares, C. G., Marseguerra, M., & Zio, E. (2002). Simulation modelling of repairable multi-component deteriorating systems for on condition maintenance optimisation. *Reliability Engineering & System Safety*, *76*(3), 255–264. doi:10.1016/S0951-8320(02)00017-0

Barbati, M., Bruno, G., & Genovese, A. (2011). Applications of agent-based models for optimization problems: A literature review. *Expert Systems with Applications*, *39*(5), 6020–6028. doi:10.1016/j.eswa.2011.12.015

Barlow, R., & Hunter, L. (1960). Optimum preventive maintenance policies. *Operations Research*, *8*(1), 90–100. doi:10.1287/opre.8.1.90

Barrios, A., Ballestin, F., & Valls, V. (2011). A double genetic algorithm for the MRCPSP/max. *Computers & Operations Research*, *38*(1), 33–43. doi:10.1016/j.cor.2009.09.019

Bassett, M. (2000). Assigning projects to optimize the utilization of employees time and expertise. *Computers & Chemical Engineering*, *24*(2-7), 1013–1021. doi:10.1016/S0098-1354(00)00534-2

Bauer, K., Diegner, B., Diemer, J., Dorst, W., Ferber, S., Glatz, R., & Stumpf, V. (2013). *Deutschlands Zukunft als Produktionsstandort sichern. Umsetzungsempfehlungen für das Zukunftsprojekt Industrie 4.0. Abschlussbericht des Arbeitskreises Industrie 4.0.* Retrieved June 20, 2016 from https://www.bmbf.de/files/Umsetzungsempfehlungen_Industrie4_0.pdf

Bauer, W., Schlund, S., Marrenbach, D., & Ganschar, O. (2014). *Industrie 4.0 – Volkswirtschaftliches Potenzial für Deutschland.* Stuttgart: Fraunhofer Institut IAO.

Bauernhansl, T. (2014). Die Vierte industrielle Revolution – Der Weg in ein wertschaffendes Produktionsparadigma. In T. Bauernhansl, M. ten Hompel, & B. Vogel-Heuser (Eds.), *Industrie 4.0 in der Produktion, Automatisierung und Logistik* (pp. 5–35). Wiesbaden: Springer. doi:10.1007/978-3-658-04682-8_1

Bauernhansl, T., Paulus-Rohmer, D., Schatz, A., Weskamp, M., Emmrich, V., & Döbele, M. (2015). *Geschäftsmodell-Innovation durch Industrie 4.0 – Chancen und Risiken für den Maschinen- und Anlagenbau.* München: Wieselhuber & Partner, Fraunhofer Institut IPA.

Baumer, S (2014). Verlässliche eingebettete Systeme für Industrie 4.0. *Zeitschrift für wirtschaftlichen Fabrikbetrieb, 109*(4), 264.

Bazaraa, M. S., Sherali, H. D., & Shetty, C. M. (1993). *Nonlinear Programming: Theory and Algorithms.* New York: John Wiley.

BDI. (2015). *Die Digitale Transformation der Industrie – Eine europäische Studie von Roland Berger Strategy Consultants im Auftrag des BDI.* München: BDI.

Becker, J., Knackstedt, R., & Pöppelbuß, J. (2009). Dokumentationsqualität von Reifegradmodellen. In J. Becker (Ed.), *Arbeitsberichte des Instituts für Wirtschaftsinformatik. Arbeitsbericht Nr. 123.* Münster: Universität Münster.

Becker, T. (2008). *Prozesse in Produktion und Supply Chain optimieren.* Heidelberg, Germany: Springer.

Bekkers, W., & van de Weerd, I. (2010). *SPM Maturity Matrix.* Technical Report UU-CS-2010-013. Department of Information and Computing Sciences. Utrecht: Utrecht University.

Ben-Daya, M., Darwish, M., & Ertogral, K. (2008). The joint economic lot sizing problem: Review and extensions. *European Journal of Operational Research, 185*(2), 726–742. doi:10.1016/j.ejor.2006.12.026Chen, X., & Simchi-Levi, D. (2012). *Pricing and Inventory Management. Handbook of Pricing* (R. Philips & O. Ozalp, Eds.). Oxford University Press.

Ben-Daya, M., & Duffuaa, S. O. (1995). Maintenance and quality: The missing link. *Journal of Quality in Maintenance Engineering, 1*(1), 20–26. doi:10.1108/13552519510083110

Ben-Daya, M., & Duffuaa, S. O. (2000). *Overview of maintenance modelling areas* (pp. 3–35). Springer, US: Maintenance, Modeling and Optimization. doi:10.1007/978-1-4615-4329-9_1

Bengel, G. (2014). *Grundkurs Verteilte Systeme. Grundlagen und Praxis des Client-Server und Distributed Computing.* Wiesbaden: Springer.

Berger, R. (2014). *Industry 4.0-The new industrial revolution-How Europe will succeed?* Roland Berger Strategy Consultants.

Berman, O., Drezner, Z., & Wesolowsky, G. O. (2007). The transfer point location problem. *European Journal of Operational Research, 179*(3), 978–989. doi:10.1016/j.ejor.2005.08.030

Berrichi, A., & Yalaoui, F. (2013). Efficient bi-objective ant colony approach to minimize total tardiness and system unavailability for a parallel machine scheduling problem. *International Journal of Advanced Manufacturing Technology, 68*(9-12), 2295–2310. doi:10.1007/s00170-013-4841-0

Berrichi, A., Yalaoui, F., Amodeo, L., & Mezghiche, M. (2010). Bi-objective ant colony optimization approach to optimize production and maintenance scheduling. *Computers & Operations Research, 37*(9), 1584–1596. doi:10.1016/j.cor.2009.11.017

Bertesekas, D. P. (2005). *Dynamic programming and optimal control.* Belmont: Athena Scientific.

Betrand, J. W. M., & Van Ooijen, H. P. G. (2008). Optimal work order release for make-to-order job shops with customer order lead-time costs, tardiness costs and work-in-process costs. *International Journal of Production Economics, 116*(2), 233–241. doi:10.1016/j.ijpe.2008.08.055

Beutel, A. L., & Minner, S. (2011). Safety stock planning under casual demand forecasting. *International Journal of Production Economics, 140*(2), 637–645. doi:10.1016/j.ijpe.2011.04.017

Beyer, H. G., & Deb, K. (2001). On self-adaptive features in real-parameter evolutionary algorithms. *Evolutionary Computation. IEEE Transactions on, 5*(3), 250–270.

Beyer, H. G., & Deb, K. (2001). On self-adaptive features in real-parameter evolutionary algorithms. *IEEE Transactions on Evolutionary Computation, 5*(3), 250–270. doi:10.1109/4235.930314

Bhamu, J., & Singh, S. K. (2014). Lean Manufacturing: Literature review and research issues. *International Journal of Operations & Production Management, 34*(7), 940–876. doi:10.1108/IJOPM-08-2012-0315

Birhanu, D., Lanka, K., & Rao, A. N. (2014). A survey of classifications in supply chain strategies. *Procedia Engineering, 97*, 2289–2297. doi:10.1016/j.proeng.2014.12.473

Bischoff, J., Taphorn, C., Wolter, D., Braun, N., Fellbaum, M., Goloverov, S. L., & Scheffler, D. (2015). *Erschließen der Potenziale von Industrie 4.0 im Mittelstand.* Berlin: BMWi (Bundesministerium für Wirtschaft und Energie).

Black, I. T., & Phillips, D. T. (2010). The Lean to Green Evolution. *Industrial Engineer, 42*(6), 46–51.

Bloechliger, I., & Zufferey, N. (2013). Multi-Coloring and Project-Scheduling with Incompatibility and Assignment Costs. *Annals of Operations Research, 211*(1), 83–101. doi:10.1007/s10479-013-1397-1

Blum, C., & Roli, A. (2003). Metaheuristics in combinatorial optimization: Overview and conceptual comparison. *ACM Computing Surveys, 35*(3), 268–308. doi:10.1145/937503.937505

BMBF. (2014). *Die neue Hightech-Strategie – Innovationen für Deutschland.* Retrieved November 9, 2016 from https://www.bmbf.de/pub_hts/HTS_Broschure_Web.pdf

BMWi. (2015). *Industrie 4.0 und Digitale Wirtschaft, Impulse für Wachstum, Beschäftigung und Innovation.* Retrieved November 9, 2016 from https://www.bmwi.de/BMWi/Redaktion/PDF/I/industrie-4-0-und-digitale-wirtschaft,property=pdf,bereich=bmwi2012,sprache=de,rwb=true.pdf

Bobadilla, J., Ortega, F., Hernando, A., & Gutierrez, A. (2013). Recommender systems survey. *Knowledge-Based Systems, 46*, 109–132. doi:10.1016/j.knosys.2013.03.012

Bollapragada, R., Camm, J., Rao, U. S., & Junying, W. (2005). A two-phase greedy algorithm to locate and allocate hubs for fixed-wireless broadband access. *Operations Research Letters, 33*(2), 134–142. doi:10.1016/j.orl.2004.05.007

Bonabeau, E., Dorigo, M., & Theraulaz, G. (1999). *Swarm intelligence: from natural to artificial systems.* New York, NY, USA: Oxford University Press, Inc.

Borzabadi, A. H., & Mehne, H. H. (2009). Ant colony optimization for optimal control problems. *Journal of Information and Computing Science, 4*(4), 259–264.

Böse, F. (2012). Selbststeuerung in der Fahrzeuglogistik, Modellierung und Analyse selbststeuernder logistischer Prozesse in der Auftragsabwicklung von Automobilterminals. In B. Scholz-Reiter (Ed.), *Informationstechnische Systeme und Organisation von Produktion und Logistik.* Berlin: GITO-Verlag.

Böse, F., & Windt, K. (2007). Catalogue of Criteria for Autonomous Control in Logistics. In M. Hülsmann & K. Windt (Eds.), *Understanding Autonomous Cooperation and Control Logistics. The Impact of Autonomy on Management, Information, Communication and Material Flow* (pp. 57–72). Berlin: Springer. doi:10.1007/978-3-540-47450-0_5

Boukas, E. K. (1998). Hedging point policy improvement. *Journal of Optimization Theory and Applications*, *97*(1), 47–70. doi:10.1023/A:1022670932536

Bousdekis, A., Magoutas, B., Apostolou, D., & Mentzas, G. (2015a). A proactive decision making framework for condition-based maintenance. *Industrial Management & Data Systems*, *115*(7), 1225–1250. doi:10.1108/IMDS-03-2015-0071

Bousdekis, A., Magoutas, B., Apostolou, D., & Mentzas, G. (2015b). Review, analysis and synthesis of prognostic-based decision support methods for condition based maintenance. *Journal of Intelligent Manufacturing*. doi: 10.1007/s10845-015-1179-5

Boyd, D., & Crawford, K. (2012). Critical Questions For Big Data, Journal Information. *Communicatio Socialis*, 662–679.

Brandimarte, P., & Zotteri, G. (2007). *Introduction to distribution logistics* (Vol. 21). John Wiley & Sons. doi:10.1002/9780470170052

Briant, O., Naddef, D., & Mounie, G. (2008), Greedy approach and multi-criteria simulated annealing for the car sequencing problem. *European Journal of Operational Research*, *191*, 993–1003

Budai, G., Dekker, R., & Nicolai, R. P. (2008). Maintenance and production: a review of planning models. In *Complex system maintenance handbook* (pp. 321–344). Springer London. doi:10.1007/978-1-84800-011-7_13

Buddhakulsomsiri, J., & Kim, D. S. (2007). Priority rule-based heuristic for multi-mode resource-constrained project scheduling problems with resource vacations and activity splitting. *European Journal of Operational Research*, *178*(2), 374–390. doi:10.1016/j.ejor.2006.02.010

Burgess, A. G., Morgan, I., & Vollmann, T. E. (1993). Cellular manufacturing: Its impact on the total factory. *International Journal of Production Research*, *31*(9), 2059–2077. doi:10.1080/00207549308956844

Byon, S., & Hwang, S. (2003). Die shape optimal design in cold and hot extrusion. *Journal of Materials Processing Technology*, *138*(1), 316–324. doi:10.1016/S0924-0136(03)00092-X

Caballé, N. C., Castro, I. T., Pérez, C. J., & Lanza-Gutiérrez, J. M. (2015). A condition-based maintenance of a dependent degradation-threshold-shock model in a system with multiple degradation processes. *Reliability Engineering & System Safety*, *134*, 98–109. doi:10.1016/j.ress.2014.09.024

Cámara, M., de Toro, F., & Ortega, J. (2013). An analysis of multiobjective evolutionary algorithms for optimization problems with time constraints. *Applied Artificial Intelligence*, *27*(9), 851–879. doi:10.1080/08839514.2013.835237

Campbell, J. F. (2009). Hub location for time definite transportation. *Computers & Operations Research*, *36*(12), 3107–3116. doi:10.1016/j.cor.2009.01.009

Campuzano, F., & Mula, J. (2011). *Supply Chain Simulation*. London: Springer-Verlag. doi:10.1007/978-0-85729-719-8

Canetta, L., Cheikhrouhou, N., & Glardon, R. (2010). Modelling hybrid demand (e- commerce ''+'' traditional) evolution: A scenario planning approach. *International Journal of Production Economics*. doi:10.1016/j.ijpe.2010.06.003

Caniato, F., Caridi, M., Castelli, C., & Golini, R. (2011). Supply chain management in the luxury industry: A first classification of companies and their strategies. *International Journal of Production Economics*, *133*(2), 622–633. doi:10.1016/j.ijpe.2011.04.030

Cánovas, L., García, S., & Marín, A. (2007). Solving the uncapacitated multiple allocation hub location problem by means of a dual-ascent technique. *European Journal of Operational Research*, *179*(3), 990–1007. doi:10.1016/j.ejor.2005.08.028

Castro, I. T. (2009). A model of imperfect preventive maintenance with dependent failure modes. *European Journal of Operational Research*, *196*(1), 217–224. doi:10.1016/j.ejor.2008.02.042

Castro, I. T., Caballé, N. C., & Pérez, C. J. (2015). A condition-based maintenance for a system subject to multiple degradation processes and external shocks. *International Journal of Systems Science*, *46*(9), 1692–1704. doi:10.1080/00207721.2013.828796

Celano, G., Costa, A., & Fichera, S. (2008). Scheduling of unrelated parallel manufacturing cells with limited human resources. *International Journal of Production Research*, *46*(2), 405–427. doi:10.1080/00207540601138452

Cerny, V. (1985). Thermodynamical approach to the traveling salesman problem an efficient simulation algorithm. *Journal of Optimization Theory and Applications*, *45*(1), 41–51. doi:10.1007/BF00940812

Ceryan, O., Duenyas, I., & Koren, Y. (2012). Optimal control of an assembly system with demand for the end-product and intermediate components. *IIE Transactions*, *44*(5), 386–403. doi:10.1080/0740817X.2011.609525

Cesaní, V. I., & Steudel, H. J. (2005). A study of labor assignment flexibility in cellular manufacturing systems. *Computers & Industrial Engineering*, *48*(3), 571–591. doi:10.1016/j.cie.2003.04.001

Cevikcan, E., & Durmusoglu, M. B. (2011). Minimising utility work and utility worker transfers for a mixed-model assembly line. *International Journal of Production Research*, *49*(24), 7293–7314. doi:10.1080/00207543.2010.537385

Cevikcan, E., & Durmusoglu, M. B. (2014). An integrated job release and scheduling approach on parallel machines: An application in electric wire-harness industry. *Computers & Industrial Engineering*, *76*, 318–332. doi:10.1016/j.cie.2014.08.012

Chakri, A., Khelif, R., Benouaret, M., & Yang, X. S. (2017). New directional bat algorithm for continuous optimization problems. *Expert Systems with Applications*, *69*, 159–175. doi:10.1016/j.eswa.2016.10.050

Chandrasekaran, K., & Simon, S. P. (2012). Multi-objective scheduling problem: Hybrid approach using fuzzy assisted cuckoo search algorithm. *Swarm and Evolutionary Computation*, *5*, 1–16. doi:10.1016/j.swevo.2012.01.001

Chatzidimitriou, C., Symeonidis, A. L., Kontogounis, I., & Pericles, A. M. (2008). Agent Mertacor: A robust design for dealing with uncertainty and variation in SCM environments. *Expert Systems with Applications*, *35*(3), 591–603. doi:10.1016/j.eswa.2007.07.050

Cheng, Q., Sun, B., & Zhao, Y. (2016). A Method to Analyse The Machining Accuracy Reliability Sensitivity Of Machine Tools Based On Fast Markov Chain Simulation *Eksploatacja I Niezawodnosc, 18*(4), 552.

Chen, H., Ihlow, J., & Lehmann, C. (1999). A genetic algorithm for flexible job shop scheduling, In *Proceedings of IEEE International Conference on Robotics and Automation* (Vol. 2, pp. 1120-1125). doi:10.1109/ROBOT.1999.772512

Chen, S., & Chen, R. (1997). Manufacturer-supplier relationship in a JIT environment. *Production and Inventory Management Journal*, *38*(1), 58–70.

Che, Z. H. (2012). Clustering and selecting suppliers based on simulated annealing algorithms. *Computers & Mathematics with Applications (Oxford, England)*, *1*(63), 228–238. doi:10.1016/j.camwa.2011.11.014

Chibba, A. (2007). *Measuring supply chain performance measures: prioritizing performance measures* [Doctoral dissertation]. Luleå Tekniska Universitet.

Chilan, C. M., & Conway, B. A. (2015). A Reachable Set Analysis Method for Generating Near-Optimal Trajectories of Constrained Multiphase Systems. *Journal of Optimization Theory and Applications*, *167*(1), 161–194. doi:10.1007/s10957-014-0651-2

Ching, H. Y. (1999). *Gestão de estoques na cadeia de logística integrada: Supply chain*. São Paulo: Atlas.

Choi, K., Kim, S., Lee, H., & Kwon, I. (1997). An operation scheme for make-to-order job-shop production systems. *Computers & Industrial Engineering*, *33*(3), 765–768. doi:10.1016/S0360-8352(97)00248-9

Chong, C. S., Low, M. Y. H., Sivakumar, A. I., & Gay, K. L. (2006). A Bee Colony Optimization Algorithm to Job Shop Scheduling. In *Proceedings of the 2006 Winter Simulation Conference* (pp. 1954–1961). doi:10.1109/WSC.2006.322980

Chu, M., Allstot, D., Huard, J., & Wong, K. (2004). NSGA-Based Parasitic-aware Optimization of a 5 GHz Low-noise VCO. *Design Automation Conference, Asia and South Pacific*.

Chutima, P., & Naruemitwong, W. (2014). A Pareto biogeography-based optimisation for multi-objective two-sided assembly line sequencing problems with a learning effect. *Computers & Industrial Engineering*, *69*, 89–104. doi:10.1016/j.cie.2014.01.001

Clerc, M. (2012). Standard Particle Swarm Optimisation.

Coello, C. A. C., Pulido, G. T., & Lechuga, M. S. (2004). Handling multiple objectives with particle swarm optimization. *IEEE Transactions on Evolutionary Computation*, *8*(3), 256–279. doi:10.1109/TEVC.2004.826067

Conn, A. R., Coulman, P. K., Haring, R. A., Morrill, G. L., Visweswariah, C., & Wu, C. W. (1995). JiffyTune: Circuit optimization using time-domain sensitivities. *IEEE Trans. on Computer-Aided Design of Integrated Circuits and Systems*, *17*(12), 1292–1309. doi:10.1109/43.736569

Contreras, I., Fernández, E., & Marín, A. (2010). The Tree of Hubs Location Problem. *European Journal of Operational Research*, *202*(2), 390–400. doi:10.1016/j.ejor.2009.05.044

Cooren, Y., Fakhfakh, M., Loulou, M., & Siarry, P. (2007). Optimizing second generation current conveyors using particle swarm optimization. *The IEEE 19th International Conference on Microelectronics*.

Cordeau, J. F., Laporte, G., & Pasin, F. (2008). Iterated tabu search for the car sequencing problem. *European Journal of Operational Research*, *191*, 945–956.

Corne, D. W., Jerram, N. R., Knowles, J. D., & Oates, M. J. (2001, July). PESA-II: Region-based selection in evolutionary multiobjective optimization. In *Proceedings of the 3rd Annual Conference on Genetic and Evolutionary Computation* (pp. 283-290). Morgan Kaufmann Publishers Inc.

Correa, H. L., Gianesi, I. G. N., & Caon, M. (2009). *Planejamento, programação e controle da produção. 5*. São Paulo: Atlas.

Crols, J., Donnay, S., Steyaert, M., & Gielen, G. (1995). A high-level design and optimization tool for analog RF receiver front-ends. *IEEE/ACM International Conference on Computer-Aided Design (ICCAD)*. doi:10.1109/ICCAD.1995.480170

Cui, Z., & Cai, X. (2013). Artificial Plant Optimization Algorithm. In *Swarm Intelligence and Bio-Inspired Computation: Theory and Applications* (p. 351).

Cui, W. W., Lu, Z., & Pan, E. (2014). Integrated production scheduling and maintenance policy for robustness in a single machine. *Computers & Operations Research*, *47*, 81–91. doi:10.1016/j.cor.2014.02.006

Cura, T. (2008). *Modern sezgisel teknikler ve uygulamaları*. İstanbul: Papatya Yayıncılık.

Cura, T. (2012). A particle swarm optimization approach to clustering. *Expert Systems with Applications*, *1*(39), 1582–1588. doi:10.1016/j.eswa.2011.07.123

Dahal, K. P., & Chakpitak, N. (2007). Generator maintenance scheduling in power systems using metaheuristic-based hybrid approaches. *Electric Power Systems Research*, *77*(7), 771–779. doi:10.1016/j.epsr.2006.06.012

Dalfard, V. M., & Mohammadi, G. (2012). Two meta-heuristic algorithms for solving multi-objective flexible job-shop scheduling with parallel machine and maintenance constraints. *Computers & Mathematics with Applications (Oxford, England)*, *64*(6), 2111–2117. doi:10.1016/j.camwa.2012.04.007

Damodaran, P., & Vélez-Gallego, M. (2012). A simulated annealing algorithm to minimize makespan of parallel batch processing machines with unequal job ready times. *Expert Systems with Applications*, *39*(1), 1451–1458. doi:10.1016/j.eswa.2011.08.029

Daniel, K. (2008). *Managementprozesse und Performance – Ein Konzept zur reifegradbezogenen Verbesserung des Managementhandelns*. Wiesbaden: Gabler.

Das, S., & Panigrahi, B. K. (2009). Multi-objective evolutionary algorithms. In Encyclopedia of artificial intelligence (pp. 1145-1151). Hershey, PA: IGI Global. doi:10.4018/978-1-59904-849-9.ch167

Dasgupta, S. (2007). *The hardness of k-means clustering* (Technical Report CS2007-0890). University of California, San Diego.

Dasgupta, P., & Das, S. (2015). A discrete inter-species cuckoo search for flowshop scheduling problems. *Computers & Operations Research*, *60*, 111–120. doi:10.1016/j.cor.2015.01.005

Das, S. R., & Canel, C. (2005). An algorithm for scheduling batches of parts in a multi-cell flexible manufacturing system. *International Journal of Production Economics*, *97*(3), 247–262. doi:10.1016/j.ijpe.2004.07.006

Das, S., & Suganthan, P. N. (2011). Differential evolution: A survey of the state-of-the-art. *IEEE Transactions on Evolutionary Computation*, *15*(1), 4–31. doi:10.1109/TEVC.2010.2059031

Davis, A. (1998). *Handbook of Condition Monitoring – Techniques and Technology*. Amsterdam: Springer Netherlands. doi:10.1007/978-94-011-4924-2

de Jonge, B., Dijkstra, A. S., & Romeijnders, W. (2015). Cost benefits of postponing time-based maintenance under lifetime distribution uncertainty. *Reliability Engineering & System Safety*, *140*, 15–21. doi:10.1016/j.ress.2015.03.027

de la O, G., & Matis, T. (2014). Supply Chain Strategies for the International Interoceanic Mazatlan Houston Logistic Corridor. *Journal of Applied Research and Technology*, *12*(4), 666–673. doi:10.1016/S1665-6423(14)70083-5

De Ranter, C. R. C., Van der Plas, G., Steyaert, M. S. J., Gielen, G. G. E., & Sansen, W. M. C. (2002). CYCLONE: Automated design and layout of RF LC-oscillators. *IEEE Trans. on Computer-Aided Design of Integrated Circuits and Systems*, *21*(10), 1161–1170. doi:10.1109/TCAD.2002.802267

Deb, K., Agrawal, S., Pratap, A., & Meyarivan, T. (2002). A fast elitist non-dominated sorting genetic algorithm for multi-objective optimization: NSGA-II. In *Proceedings of the International Conference on Parallel Problem Solving From Nature* (pp. 849-858). Springer Berlin Heidelberg.

Deb, K. (2001). *Multi-Objective Optimization Using Evolutionary Algorithms*. New York, NY: John Wiley& Sons. Inc.

Dekker, R. (1996). Applications of maintenance optimization models: A review and analysis. *Reliability Engineering & System Safety*, *51*(3), 229–240. doi:10.1016/0951-8320(95)00076-3

Deming, W. E. (1982). *Out of the Crisis*. Cambridge, MA: MIT Press.

Deng, Ma, & Liu. (2015). *An Efficient Data Structure for Fast Mining High Utility Itemsets.* arXiv:1510.02188

Dennis, P. (2016). *Lean Production implified: A plain-language guide to the world's most powerful production system.* Boca Raton, FL: CRC Press.

Diekmann, T., & Hagenhoff, S. (2007). *Ubiquitous Computing – Technologien zur Integration der realen Welt in betriebliche Informationssysteme. Arbeitsbericht Nr. 15/2006.* Göttingen: Universität Göttingen.

DIN e. V. (2011). DIN ISO/IEC 15504-3. Informationstechnik – Prozess-Assessment. Berlin: Beuth Verlag.

DIN e.V. (2016). *DIN SPEC 91345:2016-04. Referenzarchitekturmodell Industrie 4.0 (RAMI4.0).* Berlin: Beuth Verlag.

Ding, S. H., & Kamaruddin, S. (2015). Maintenance policy optimization—literature review and directions. *International Journal of Advanced Manufacturing Technology, 76*(5-8), 1263–1283. doi:10.1007/s00170-014-6341-2

Dong, M. (2013). Parallel machine scheduling with limited controllable machine availability. *International Journal of Production Research, 51*(8), 2240–2252. doi:10.1080/00207543.2012.714002

Doostparast, M., Kolahan, F., & Doostparast, M. (2015). Optimisation of PM scheduling for multi-component systems–a simulated annealing approach. *International Journal of Systems Science, 46*(7), 1199–1207. doi:10.1080/00207721.2013.815822

Do, P., Voisin, A., Levrat, E., & Iung, B. (2015). A proactive condition-based maintenance strategy with both perfect and imperfect maintenance actions. *Reliability Engineering & System Safety, 133*, 22–32. doi:10.1016/j.ress.2014.08.011

Doran, D., & Giannakis, M. (2011). An examination of a modular supply chain: A construction sector perspective. *Supply Chain Management: An International Journal, 16*(4), 260–270. doi:10.1108/13598541111139071

Dorigo, M. (1992). *Optimization, learning and natural algorithms* [PhD thesis]. Politecnico di Milano, Italy.

Dorigo, M., & Caro, G. D. (1999). The ant colony optimization meta-heuristic. In New ideas in optimization (pp. 11-32). Maidenhead, UK: McGraw-Hill Ltd.

Dorigo, M., Caro, G. D., & Gambardella, L. M. (1999). Ant algorithms for discrete optimization. *Artificial Life, 5*(2), 137–172. doi:10.1162/106454699568728 PMID:10633574

Dowsland, K. A. (1991). *Simulated annealing. C. R. Revees içinde, Modern heuristic techniques (s. 20).* New York: John Wiley&Sons, Inc.

Drexl, A., & Kimms, A. (2001). Sequencing jit mixed-model assembly lines under station load and part-usage constraints. *Management Science, 47*(3), 480–491.

Drineas, P., Frieze, A., Kannan, R., Vempala, S., & Vinay, V. (2004). Clustering large graphs via the singular value decomposition. *Machine Learning, 2*(9-33), 56.

Duarte, C. S. R., Carvalho, J. V., & Barbosa-Póvoa, A. P. (2012). A new heuristic for car sequencing based on an integer programming approach. *IFAC Proceedings, 45*(6), 164-169. doi:10.3182/20120523-3-RO-2023.00073

Dumitrescu, I., & Stuetzle, T. (2003). Combinations of local search and exact algorithms. *Lecture Notes in Computer Science, 2611*, 211–223. doi:10.1007/3-540-36605-9_20

Durillo, J. J., Nebro, A. J., Luna, F., Coello Coello, C. A., & Alba, E. (2010). Convergence speed in multi-objective metaheuristics: Efficiency criteria and empirical study. *International Journal for Numerical Methods in Engineering, 84*(11), 1344–1375. doi:10.1002/nme.2944

Durmusoglu, M. B., & Kaya, G. (2012). Lean Thinking Based Investment Planning at Design Stage of Cellular/Hybrid Manufacturing Systems. In *Industrial Engineering: Concepts, Methodologies, Tools, and Applications: Concepts, Methodologies, Tools, and Applications* (p. 409).

Durmusoglu, M. B., & Kaya, G. (2012). Lean Thinking Based Investment Planning at Design Stage of Cellular/Hybrid Manufacturing Systems. In *Industrial Engineering: Concepts, Methodologies, Tools, and Applications: Concepts, Methodologies, Tools, and Applications* (Vol. 1, p. 409).

Durmusoglu, M. B., & Satoglu, S. I. (2011). Axiomatic design of hybrid manufacturing systems in erratic demand conditions. *International Journal of Production Research*, *49*(17), 5231–5261. doi:10.1080/00207543.2010.510487

EghbaliZarch, M., Abedzadeh, M., & Setak, M. (2013). Differential evolution algorithm for multi-commodity and multi-level of service hub covering location problem. *International Journal of Industrial Engineering Computations*, *4*(1).

Egilmez, G., Erenay, B., & Süer, G. A. (2014). Stochastic skill-based manpower allocation in a cellular manufacturing system. *Journal of Manufacturing Systems*, *33*(4), 578–588. doi:10.1016/j.jmsy.2014.05.005

Ehrgott, M., & Gandibleux, X. (2000). A survey and annotated bibliography of multiobjective combinatorial optimization. *OR-Spektrum*, *22*(4), 425–460. doi:10.1007/s002910000046

Eick, M., & Graeb, H. (2012). MARS: Matching-Driven Analog Sizing. *IEEE Trans. on Computer-Aided Design of Integrated Circuits and Systems*, *31*(8), 1145–1158. doi:10.1109/TCAD.2012.2190069

Ellegard, C. (2008). Supply risk management in a small company perspective, *Supply Chain Management. International Journal (Toronto, Ont.)*, *13*(6), 425–434.

Elmi, A., Solimanpur, M., Topaloglu, S., & Elmi, A. (2011). A simulated annealing algorithm for the job shop cell scheduling problem with intercellular moves and reentrant parts. *Computers & Industrial Engineering*, *61*(1), 171–178. doi:10.1016/j.cie.2011.03.007

El-Turky, F., & Perry, E. (1989). BLADES: An artificial intelligence approach to analog circuit design. *IEEE Trans. on Computer-Aided Design of Integrated Circuits and Systems*, *8*(6), 680–692. doi:10.1109/43.31523

Epple, U., Bangemann, T., Barbian, M., Bauer, C., Braune, A., Diesner, M., & Wollschläger, M. (2014). *Statusreport Industrie 4.0 – Gegenstände, Entitäten, Komponenten*, Retrieved June 21, 2016 from https://www.vdi.de/fileadmin/vdi_de/redakteur_dateien/sk_dateien/VDI_Industrie_4.0_Komponenten_2014.pdf

Erbas, C., Cerav-Erbas, S., & Pimentel, A. D. (2006). Multiobjective optimization and evolutionary algorithms for the application mapping problem in multiprocessor system-on-chip design. *IEEE Transactions on Evolutionary Computation*, *10*(3), 358–374. doi:10.1109/TEVC.2005.860766

Eskandari, H., Geiger, C., & Lamont, G. (2007). FastPGA: A dynamic population sizing approach for solving expensive multiobjective optimization problems. In Evolutionary Multi-Criterion Optimization (pp. 141-155). Springer Berlin/Heidelberg. doi:10.1007/978-3-540-70928-2_14

Eskandari, H., & Geiger, C. D. (2008). A fast Pareto genetic algorithm approach for solving expensive multiobjective optimization problems. *Journal of Heuristics*, *14*(3), 203–241. doi:10.1007/s10732-007-9037-z

Eskandari, H., & Geiger, C. D. (2009). Evolutionary multiobjective optimization in noisy problem environments. *Journal of Heuristics*, *15*(6), 559–595. doi:10.1007/s10732-008-9077-z

Estellon, B., Gardi, F., & Nouioua, K. (2008). Two local search approaches for solving real-life car sequencing problems. *European Journal of Operational Research*, *191*, 928–944.

Estellon, B., & Gardi, F. (2013). Car sequencing is NP-hard: A short proof. *The Journal of the Operational Research Society*, *64*(10), 1503–1504. doi:10.1057/jors.2011.165

Euchi, J., & Chabchoub, H. (2010). A hybrid tabu search to solve the heterogeneous fixed fleet vehicle routing problem. *Logistics Research*, *2*(1), 3–11. doi:10.1007/s12159-010-0028-3

Fakhfakh, M., Cooren, Y., Sallem, A., Loulou, M., & Siarry, P. (2010). Analog circuit design optimization through the particle swarm optimization technique. *Analog Integrated Circuits and Signal Processing*, *63*(1), 71–82. doi:10.1007/s10470-009-9361-3

Fallenbeck, N., & Eckert, C. (2014). IT-Sicherheit und Cloud Computing. In T. Bauernhansl, M. ten Hompel, & B. Vogel-Heuser (Eds.), *Industrie 4.0 in der Produktion, Automatisierung und Logistik* (pp. 397–431). Wiesbaden: Springer. doi:10.1007/978-3-658-04682-8_20

Fan, J., Cao, M., & Feng, D. (2010, June). Multi-objective dual resource-constrained model for cell formation problem. In *Proceedings of the 2010 IEEE International Conference on Management of Innovation and Technology (ICMIT)* (pp. 1031-1036). IEEE. doi:10.1109/ICMIT.2010.5492881

Farahani, R., Abedian, M., & Sharahi, S. (2009). *Dynamic Facility Location Problem.*

Farahani, R. Z., Hekmatfar, M., Arabani, A. B., & Nikbakhsh, E. (2013). Hub location problems: A review of models, classification, solution techniques, and applications. *Computers & Industrial Engineering*, *64*(4), 1096–1109. doi:10.1016/j.cie.2013.01.012

Federgruen, A., & Zheng, Y. S. (1992). Efficient algorithm for computing an optimal (r, Q) policy in continuous review stochastic inventory systems. *INFORMS Operations Research*, *40*(4), 808–813. doi:10.1287/opre.40.4.808

Ferreira, R. J., de Almeida, A. T., & Cavalcante, C. A. (2009). A multi-criteria decision model to determine inspection intervals of condition monitoring based on delay time analysis. *Reliability Engineering & System Safety*, *94*(5), 905–912. doi:10.1016/j.ress.2008.10.001

Feyzioglu, O., & Pierreval, H. (2009). Hybrid organization of functional departments and manufacturing cells in the presence of imprecise data. *International Journal of Production Research*, *47*(2), 343–368. doi:10.1080/00207540802425898

Fioriolli, J. C. (2002). *Modelagem estocástica de sistemas hierárquicos de estoques.* Porto Alegre: UFRGS.

Fisher, M. (1997). What is the right supply chain for your product? *Harvard Business Review*, *75*(2), 62–68.

Fister, I., Rauter, S., Yang, X. S., Ljubič, K., & Fister, I. Jr. (2015). Planning the sports training sessions with the bat algorithm. *Neurocomputing*, *149*, 993–1002. doi:10.1016/j.neucom.2014.07.034

Fitouhi, M. C., & Nourelfath, M. (2012). Integrating noncyclical preventive maintenance scheduling and production planning for a single machine. *International Journal of Production Economics*, *136*(2), 344–351. doi:10.1016/j.ijpe.2011.12.021

Fonseca, C. M., & Fleming, P. J. (1993, June). Genetic Algorithms for Multiobjective Optimization: FormulationDiscussion and Generalization. Icga, 93(July), 416-423.

Ford, H. (1923). *Popular Research Topics Henry Ford Quotations.* Retrieved October 13, 2001, from https://www.thehenryford.org/collections-and-research/digital-resources/popular-topics/henry-ford-quotes/

Forrester, J. W. (1961). *Industrial Dynamics.* Massachusetts: The M.I.T. Press.

Fowler, J. W., Hogg, G. L., & Mason, S. J. (2002). Workload control in the semiconductor industry. *Production Planning and Control*, *13*(7), 568–578. doi:10.1080/0953728021000026294

Fowler, J. W., Wirojanagud, P., & Gel, E. S. (2008). Heuristics for workforce planning with worker differences. *European Journal of Operational Research*, *190*(3), 724–740. doi:10.1016/j.ejor.2007.06.038

Frazelle, E. H. (2015). *Inventory Strategy: Maximizing Financial, Service and Operations Performance with Inventory Strategy*. Mc Graw-Hill.

Fuchß, T. (2009). *Mobile Computing: Grundlagen und Konzepte für mobile Anwendungen*. München: Carl Hanser Verlag. doi:10.3139/9783446405134

Fullerton, R. R., Kennedy, F. A., & Widener, S. K. (2014). Lean manufacturing and firm performance: The incremental contribution of lean management accounting practices. *Journal of Operations Management*, *32*(7), 414–428. doi:10.1016/j.jom.2014.09.002

Gagne, C., Gravel, M., & Price, W. L. (2006). Solving real car sequencing problems with ant colony optimization. *European Journal of Operational Research*, *174*, 1427-1448.

Gandomi, A. H., Yang, X. S., & Alavi, A. H. (2011). Mixed variable structural optimization using firefly algorithm. *Computers & Structures*, *89*(23), 2325–2336. doi:10.1016/j.compstruc.2011.08.002

Gao, K. Z., Suganthan, P. N., Chua, T. J., Chong, C. S., Cai, T. X., & Pan, Q. K. (2015). A two-stage artificial bee colony algorithm scheduling flexible job-shop scheduling problem with new job insertion. *Expert Systems with Applications*, *42*(21), 7652–7663. doi:10.1016/j.eswa.2015.06.004

Gao, K. Z., Suganthan, P. N., Pan, Q. K., Chua, T. J., Chong, C. S., & Cai, T. X. (2016b). An improved artificial bee colony algorithm for flexible job-shop scheduling problem with fuzzy processing time. *Expert Systems with Applications*, *65*, 52–67. doi:10.1016/j.eswa.2016.07.046

Gao, M. L., Shen, J., Yin, L. J., Liu, W., Zou, G. F., Li, H. T., & Fu, G. X. (2016a). A novel visual tracking method using bat algorithm. *Neurocomputing*, *177*, 612–619. doi:10.1016/j.neucom.2015.11.072

Gao, W. (2016). Improved ant colony clustering algorithm and its performance study. *Computational Intelligence and Neuroscience*, *2016*, 1–14. doi:10.1155/2016/4835932 PMID:26839533

Gapp, R., Fisher, R., & Kobayashi, K. (2008). Implementing 5S within a Japanese context: An integrated management system. *Management Decision*, *46*(4), 565–579. doi:10.1108/00251740810865067

Garey, M. R., Johnson, D. S., & Sethi, R. (1976). The Complexity of Flow Shop and Job-Shop Scheduling. *Mathematics of Operations Research*, *1*(2), 117–129. doi:10.1287/moor.1.2.117

Garey, M., & Johnson, D. S. (1979). *Computer and Intractability: A Guide to the Theory of NP-Completeness*. San Francisco: Freeman.

Garg, A., & Deshmukh, S. G. (2006). Maintenance management: Literature review and directions. *Journal of Quality in Maintenance Engineering*, *12*(3), 205–238. doi:10.1108/13552510610685075

Garitselov, O., Mohanty, S., & Kougianos, E., & Patra, P. (2011). Bee Colony Inspired Metamodeling Based Fast Optimization of a Nano-CMOS PLL. *2011 International Symposium on Electronic System Design (ISED)*. doi:10.1109/ISED.2011.13

Gaub, H. (2016). Customization of mass-produced parts by combining injection molding and additive manufacturing with Industry 4.0 technologies. *Reinforced Plastics*, *60*(6), 401–404. doi:10.1016/j.repl.2015.09.004

Gavronovic, H. (2008). Local search and suffix tree for car-sequencing problem with colors. *European Journal of Operational Research*, *191*(3), 972–980. doi:10.1016/j.ejor.2007.04.051

Geisberger, E., Cengarle, M. V., Keil, P., Niehaus, J., Thiel, C., & Thönnißen-Fries, H. J. (2011). Cyber Physical Systems. *Innovationsmotor für Mobilität, Gesundheit, Energie und Produktion.* Retrieved November 9, 2016 from http://www.acatech.de/fileadmin/user_upload/Baumstruktur_nach_Website/Acatech/root/de/Material_fuer_Sonderseiten/Cyber-Physical-Systems/acatech_POSITION_CPS_web.pdf

Gelareh, S., & Nickel, S. (2011). Hub location problems in transportation networks. *Transportation Research Part E, Logistics and Transportation Review, 47*(6), 1092–1111. doi:10.1016/j.tre.2011.04.009

Gendreau, M., & Potvin, J.-Y. (2005). Metaheuristics in Combinatorial Optimization. *Annals of Operations Research, 140*(1), 189–213. doi:10.1007/s10479-005-3971-7

Gendreau, M., & Potvin, J.-Y. (2010). *Handbook of Metaheuristics. International Series in Operations Research & Management Science, 146.* Springer.

Georgiadis, P., Vlachos, D., & Iakovou, E. (2005). A system dynamics modeling framework for the strategic supply chain management of food chains. *Journal of Food Engineering, 70*(3), 351–364. doi:10.1016/j.jfoodeng.2004.06.030

Gertsbakh, I. (2013). *Reliability theory: with applications to preventive maintenance.* Springer.

Ghai, D., Mohanty, S., & Thakral, G. (2013). Fast optimization of nano-CMOS voltage-controlled oscillator using polynomial regression and genetic algorithm. *Microelectronics Journal, 44*(8), 631–641. doi:10.1016/j.mejo.2013.04.010

Ghali, K., Dorie, L., & Hammami, O. (2005). Dynamically Reconfigurable Analog Circuit Design Automation Through Multiobjective Optimization and Direct Execution. *The 17th International Conference on Microelectronics.*

Ghasemy Yaghin, R., Fatemi Ghomi, S. M. T., & Torabi, S. A. (2013). A possibilistic multiple objective pricing and lot-sizing model with multiple demand classes. *Fuzzy Sets and Systems, 231*(16), 26–44. doi:10.1016/j.fss.2012.11.012

Ghasemy Yaghin, R., Fatemi Ghomi, S. M. T., & Torabi, S. A. (2014). Enhanced joint pricing and lotsizing problem in a two-echelon supply chain with logit demand function. *International Journal of Production Research, 52*(17), 4967–4983. doi:10.1080/00207543.2014.885665

Ghedjati, F. (1994). Résolution par des heuristiques dynamiques et des algorithmes génétiques du problème d'ordonnancement de type Job-Shop généralisé (à machines non identiques en parallèle et contraintes de précédence). Unpublished doctoral dissertation, University of Paris.

Ghosh, A., Das, S., Chowdhury, A., & Giri, R. (2011). An ecologically inspired direct search method for solving optimal control problems with Bézier parameterization. *Engineering Applications of Artificial Intelligence, 24*(7), 1195–1203. doi:10.1016/j.engappai.2011.04.005

Ghosh, M. (2012). Lean manufacturing performance in Indian manufacturing plants. *Journal of Manufacturing Technology Management, 24*(1), 113–122. doi:10.1108/17410381311287517

Giachetti, R. E., Martinez, L. D., Sáenz, O. A., & Chen, C. S. (2003). Analysis of the structural measures of flexibility and agility using a measurement theoretical framework. *International Journal of Production Economics, 86*(1), 47–62. doi:10.1016/S0925-5273(03)00004-5

Giannakis, M., & Louis, M. (2010). A multi-agent based framework for supply chain risk management. *Journal of Purchasing and Supply Management, 17*(1), 23–31. doi:10.1016/j.pursup.2010.05.001

Giard, V., & Jeunet, J. (2010). Optimal sequencing of mixed models with sequence-dependent setups and utility workers on an assembly line. *International Journal of Production Economics, 123*(2), 290–300. doi:10.1016/j.ijpe.2009.09.001

Gielen, G., & Rutenbar, R. A. (2000). Computer-aided design of analog and mixed-signal integrated circuits. *Proceedings of the IEEE*. doi:10.1109/5.899053

Gilchrist, A. (2016). *Industry 4.0: The Industrial Internet of Things, 2016*. Apress. doi:10.1007/978-1-4842-2047-4

Glock, C. H. (2012). The joint economic lot size problem: A review. *International Journal of Production Economics, 135*(2), 671–686. doi:10.1016/j.ijpe.2011.10.026

Glover, F. (1986). Future paths for integer programming and links to artificial intelligence. *Computers & Operations Research, 13*(5), 533–549. doi:10.1016/0305-0548(86)90048-1

Glover, F., & Kochenberger, G. A. (2003). *Genetic algorithms. Hand book of metaheuristics (s. 59). içinde New York.* Boston, Dordrecht, London, Moscow: Kluwer Academic Publishers.

Godinho Filho, M., & Fernandes, F. C. F. (2010). *Planejamento e controle da produção: dos fundamentos ao essencial.* São Paulo: Atlas.

Goldberg, D. (1989). *Genetic Algorithms in Search, Optimization and Machine Learning*. Reading, MA: Addison-Wesley.

Goldberg, D. E. (1989). *Genetic algorithms in search, optimization, and machine learning*. Boston, MA, USA: Addison-Wesley Longman Publishing Co., Inc.

Goldberg, D. E. (1989). *Genetie Algorithms in Seareh*. Optirnization and Machine Leaming.

Golle, U., Rothlauf, F., & Boysen, N. (2014). Car sequencing versus mixed-model sequencing: A computational study. *European Journal of Operational Research, 237*(1), 50–61. doi:10.1016/j.ejor.2014.01.012

Gomez, I., Cuautle, E., & Luis, G. (2013). Richardson extrapolation-based sensitivity analysis in the multi-objective optimization of analog circuits. *Applied Mathematics and Computation, 222*, 167–176. doi:10.1016/j.amc.2013.07.059

Gorener, A. (2013). Supply Chain Strategy Selection: An Application in Manufacturing Industry by Fuzzy VIKOR Method. *International Journal of Alanya Faculty of Business, 5*(3), 47–62.

Gottlieb, J., Puchta, M., & Solnon, C. (2003). A study of greedy, local search and ant colony optimization approaches for car sequencing problems. In Applications of Evolutionary Computing, LNCS (Vol. 2611, pp. 246–257).

Goyal, S. K. (1976). An integrated inventory model for a single supplier–single customer problem. *International Journal of Production Research, 15*(1), 107–111. doi:10.1080/00207547708943107

Grant, L. L., & Venayagamoorthy, G. K. (2009). Swarm Intelligence for collective Robotic Search. In *Design and Control of Intelligent Robotic Systems* (pp. 29–47). Academic Press.

Gravel, M., Gagne, C., & Price, W. L. (2005). Review and comparison of three methods for the solution of the car sequencing problem. *The Journal of the Operational Research Society, 56*(11), 1287–1295. doi:10.1057/palgrave.jors.2601955

Gravel, M., Price, W., & Gagne, C. (2000). An interactive tool for designing manufacturing cells for an assembly job-shop. *International Journal of Production Research, 38*(2), 309–322. doi:10.1080/002075400189428

Graves, G. H., & Lee, C. Y. (1999). Scheduling maintenance and semiresumable jobs on a single machine. *Naval Research Logistics, 46*(7), 845–863. doi:10.1002/(SICI)1520-6750(199910)46:7<845::AID-NAV6>3.0.CO;2-#

Grizelj, B., & Jurković, Z. (2009). Optimization of extrusion process by genetic algorithms and conventional techniques. *Tehnički vjesnik: znanstveno-stručni časopis tehničkih fakulteta Sveučilišta u Osijeku, 16*(4), 27-33.

Groover, M. P. (2015). Automation, Production Systems, and Computer-Integrated Manufacturing (4th ed.). Upper Saddle River, NJ: Pearson Higher Ed.

Gungor, Z., & Unler, A. (2008). K-Harmonic means data clustering with tabu-search method. *Applied Mathematical Modelling*, *6*(32), 1115–1125. doi:10.1016/j.apm.2007.03.011

Gupta, R., & Allstot, D. J. (1998). Parasitic-Aware Design and Optimization of CMOS RF Integrated Circuits. *IEEE Radio Frequency Integrated Circuits Symposium.*

Hamdan, M. (2010). On the disruption-level of polynomial mutation for evolutionary multi-objective optimisation algorithms. *Computing and Informatics*, *29*(5), 783–800.

Hamdan, M. (2012). On the disruption-level of polynomial mutation for evolutionary multi-objective optimisation algorithms. *Computing and Informatics*, *29*(5), 783–800.

Hammami, R., & Frein, Y. (2014). Integration of the profit-split transfer pricing method in the design of global supply chains with a focus on offshoring context. *Computers & Industrial Engineering*, *76*, 243–252. doi:10.1016/j.cie.2014.07.021

Hanafizadeh, P., & Sherkat, M. H. (2009). Designing fuzzy-genetic learner model based on multi-agent systems in supply chain management. *Expert Systems with Applications*, *36*(6), 10120–10134. doi:10.1016/j.eswa.2009.01.008

Han, J., & Kamber, M. (2001). *Data Mining Concepts and Techniques*. Morgan Kauffman Publishers.

Hansen, P. (1986). The steepest ascent mildest descent heuristic for combinatorial programming. In Proceedings of the Congress on Numerical Methods in Combinatorial Optimization, Capri, Italy.

Han, Y. Y., Gong, D. W., Sun, X. Y., & Pan, Q. K. (2014). An improved NSGA-II algorithm for multi-objective lot-streaming flow shop scheduling problem. *International Journal of Production Research*, *52*(8), 2211–2231. doi:10.1080/00207543.2013.848492

Harhalakis, G., Lu, T., Minis, I., & Nagi, R. (1996). A practical method for design of hybrid-type production facilities. *International Journal of Production Research*, *34*(4), 897–918. doi:10.1080/00207549608904942

Harvey, Elmasry, & Leung. (1992). STAIC: An Interactive Framework for Synthesizing CMOS and BiCMOS Analog Circuits. *IEEE Trans. on Computer-Aided Design of Integrated Circuits and Systems, 11*(11).

Haskose, A., Kingsman, B. G., & Worthington, D. (2004). Performance analysis of make-to-order manufacturing systems under different workload control regimes. *International Journal of Production Economics*, *90*(2), 169–186. doi:10.1016/S0925-5273(03)00052-5

Haykin. (n.d.). *Neural Networks A Comprehensive Foundation.* Prentice Hall International Inc.

Hecht, S. (2014). *Ein Reifegradmodell für die Bewertung und Verbesserung von Fähigkeiten im ERP-Anwendungsmanagement*. Wiesbaden: Gabler. doi:10.1007/978-3-658-05523-3

Heer, T., Kleineberg, O., & Bagchi, S. (2015). *Communication Technologies for the Smart Factory of the Future. Building a communication infrastructure for Industry 4.0 and the Internet of Things*. St. Louis, MO: Belden Inc.

Heizer, J., Render, B., & Munson, C. (2014). Principles of operations management: sustainability and supply chain management (9th ed.). Upper Saddle River, NJ: Pearson Higher Ed.

Hermann, Pentek, & Otto. (2016). *Design Principles for Industrie 4.0 Scenarios.* Academic Press.

Hernández, M. E., & Reiff-Marganiec, S. (2014). Classifying Smart Objects using Capabilities. *Smart Computing*, *2014*, 309–316.

Hershenson, M., Boyd, S., & Lee, T. (1998). GPCAD: a tool for CMOS opamp synthesis. *Proc. of IEEE/ACM International Conference on CAD (ICCAD).*

Hertel, M. (2015). Risiken der Industrie 4.0 – Eine Strukturierung von Bedrohungsszenarien der Smart Factory. *HMD Praxis der Wirtschaftsinformatik*, *52*(5), 724–738. doi:10.1365/s40702-015-0161-1

Hertz, A., Schindl, D., & Zufferey, N. (2005). Lower Bounding and Tabu Search Procedures for the Frequency Assignment Problem with Polarization Constraints. *4OR*, *3*(2), 139-161.

Hertz, A., & Widmer, M. (2003). Guidelines for the use of meta-heuristics in combinatorial optimization. *European Journal of Operational Research*, *151*(2), 247–252. doi:10.1016/S0377-2217(02)00823-8

He, Y., Wu, T., Zhang, C., & Liang, Z. (2015). An improved MIP heuristic for the intermodal hub location problem. *Omega (United Kingdom)*, *57*, 203–211.

Hiziroglu, A. (2013). A neuro-fuzzy two-stage clustering approach to customer segmentation. *Journal of Marketing Analytics*, *1*(4), 202–221. doi:10.1057/jma.2013.17

Hjalmarson, E., Hagglund, R., & Wanhammar, L. (2003). An equation-based optimization approach for analog circuit design. *International Symposium on Signals, Circuits and Systems*. doi:10.1109/SCS.2003.1226952

Hnaien, F., Yalaoui, F., Mhadhbi, A., & Nourelfath, M. (2016). A mixed-integer programming model for integrated production and maintenance. *IFAC-PapersOnLine*, *49*(12), 556–561. doi:10.1016/j.ifacol.2016.07.694

Ho, N. B., & Tay, J. C. (2004, June 19-23). GENACE: An Efficient Cultural Algorithm for Solving the Flexible Job-Shop Problem. In *Proceedings of the IEEE Congress on Evolutionary Computation* (*Vol. 2*, pp. 1759-1766). doi:10.1109/CEC.2004.1331108

Hobbs, J. A. (1976). *Control Over Inventory and Production*. São Paulo: McGraw-Hill.

Ho, C. (2011). The optimal integrated inventory policy with price-and-credit-linked demand under two-level trade credit. *Computers & Industrial Engineering*, *60*(1), 117–126. doi:10.1016/j.cie.2010.10.009

Ho, L. L., & Quinino, R. C. (2012). Integrating on-line process control and imperfect corrective maintenance: An economical design. *European Journal of Operational Research*, *222*(2), 253–262. doi:10.1016/j.ejor.2012.04.019

Holland, J. H. (1962). Outline for a logical theory of adaptive systems. *Journal of the ACM*, *9*(3), 297–314. doi:10.1145/321127.321128

Holland, J. H. (1975). *Adaptation in natural and artificial systems*. Ann Arbor, MI: University of Michigan Press.

Holt, C. C., Modigliani, F., Muth, J. F., & Simon, H. A. (1960). *Planning Production, Inventories, and Work Force*. Prentice-Hall.

Hormozi, A. M. (2001). Agile manufacturing: The next logical step. *Benchmarking: An International Journal*, *8*(2), 132–143. doi:10.1108/14635770110389843

Horn, J., Nafpliotis, N., & Goldberg, D. E. (1994, June). A niched Pareto genetic algorithm for multiobjective optimization. In *Proceedings of the First IEEE Conference on Evolutionary Computation* (pp. 82-87). Ieee. doi:10.1109/ICEC.1994.350037

Hosking, J. R. M., Pednault, E. P. D., & Sudan, M. (1997). A Statistical Perspective on Data Mining. *Future Generation Computer Systems*, *13*(2-3), 117–134. doi:10.1016/S0167-739X(97)00016-2

Hosseini, S., & Al Khaled, A. (2014). A survey on the Imperialist Competitive Algorithm metaheuristic: Implementation in engineering domain and directions for future research. *Applied Soft Computing*, *24*, 1078–1094. doi:10.1016/j.asoc.2014.08.024

Hsieh, T. P., & Dye, C. Y. (2013). A production-inventory model incorporating the effect of preservation technology investment when demand is fluctuating with time. *Journal of Computational and Applied Mathematics*, *239*, 25–36. doi:10.1016/j.cam.2012.09.016

Hsu, C.-C., & Chen, Y.-C. (2007). Mining of mixed data with application to catalog marketing. *Expert Systems with Applications*, *32*(1), 12–23. doi:10.1016/j.eswa.2005.11.017

Huang, Y., & Lin, J. (2011). A new bee colony optimization algorithm with idle-time-based filtering scheme for open shop-scheduling problems. *Expert Systems with Applications*, *38*(5), 5438–5447. doi:10.1016/j.eswa.2010.10.010

Hüllermeier, E. (2011). Fuzzy Sets in Machine Learning and Data Mining. *Applied Soft Computing*, *11*(2), 1493–1505. doi:10.1016/j.asoc.2008.01.004

Hwang, H., & Koh, S. G. (1992). Optimal control of work-in-process with dual limit switches in two-stage transfer line. *European Journal of Operational Research*, *63*(1), 66–75. doi:10.1016/0377-2217(92)90055-E

Hyun, C. J., Kim, Y., & Kim, Y. K. (1998). A genetic algorithm for multiple objective sequencing problems in mixed model assembly lines. *Computers & Operations Research*, *25*(7), 675–690. doi:10.1016/S0305-0548(98)00026-4

Ignaciuk, P., & Bartoszewicz, A. (2012). Linear-Quadratic optimal control of periodic-review perishable inventory systems. *IEEE Transactions on Automatic Control*, *20*, 1400–1407.

Ioannou, G. (2006). Time-phased creation of hybrid manufacturing systems. *International Journal of Production Economics*, *102*(2), 183–198. doi:10.1016/j.ijpe.2005.03.005

Iranmanesh, H., Skandari, M. R., & Allahverdiloo, M. (2008). Finding Pareto optimal front for the multi mode time, cost quality trade-off in project scheduling. *Issues, 59.*

Ivanov, D., Dolgui, A., & Sokolov, B. (2011). Applicability of optimal control theory to adaptive supply chain planning and scheduling. *Annual Reviews in Control*, *36*(1), 73–84. doi:10.1016/j.arcontrol.2012.03.006

Iyoob, I. M. (2003). Analysis of equipment availability under varying corrective maintenance models. In *Proceedings of the IIE annual conference* (p. 1). Institute of Industrial and Systems Engineers (IISE).

Jaddi, N. S., Abdullah, S., & Hamdan, A. R. (2015). Optimization of neural network model using modified bat-inspired algorithm. *Applied Soft Computing*, *37*, 71–86. doi:10.1016/j.asoc.2015.08.002

Jafari, A., Bijami, E., Bana, H., & Sadri, S. (2011). A design automation system for CMOS analog integrated circuits using New Hybrid Shuffled Frog Leaping algorithm. *Microelectronics Journal*, *43*(11), 908–915. doi:10.1016/j.mejo.2012.05.010

Jänicke, L., Jochem, M., Klasen, W., Kosch, B., Krammel, M., Linke, L., & Zimmermann, S. (2016). IT-Security in Industrie 4.0 - Erste Schritte zu einer sicheren Produktion. Berlin: Bundesministerium für Wirtschaft und Energie (BMWi).

Japan Management Association (Ed.). (1989). *Kanban Just-In-Time at Toyota Management: Begins at the Workplace*Lu, D. J., Trans.). Cambridge, MA: Productivity Press.

Jardine, A. K., Lin, D., & Banjevic, D. (2006). A review on machinery diagnostics and prognostics implementing condition-based maintenance. *Mechanical Systems and Signal Processing*, *20*(7), 1483–1510. doi:10.1016/j.ymssp.2005.09.012

Jin, X., Siegel, D., Weiss, B. A., Gamel, E., Wang, W., Lee, J., & Ni, J. (2016). The present status and future growth of maintenance in US manufacturing: results from a pilot survey. *Manufacturing review, 3.*

Joly, A., & Frein, Y. (2008). Heuristics for an industrial car sequencing problem considering paint and assembly shop objectives. *Computers & Industrial Engineering*, *55*(2), 295–310. doi:10.1016/j.cie.2007.12.014

Jones, D. F., Mirrazavi, S. K., & Tamiz, M. (2002). Multi-objective meta-heuristics: An overview of the current state-of-the-art. *European Journal of Operational Research, 137*(1), 1–9. doi:10.1016/S0377-2217(01)00123-0

Jong, K. A. (1985). Genetic algorithms: A 10 year perspective. In *Proceedings of the 1st International Conference on Genetic Algorithms* (pp. 169-177). Hillsdale, NJ, USA: L. Erlbaum Associates Inc.

Jordan, F. (2015). Studienergebnisse "Cyber-Physical Systems in der Produktionspraxis" Anwendungsfälle und Leuchtturmprojekte in NRW. *Zeitschrift für wirtschaftlichen Fabrikbetrieb, 109*(9), 558-561.

Jørgensen, F., Matthiesen, R., Nielsen, J., & Johansen, J. (2007). Lean Maturity, Lean Sustainability. In J. Olhager & F. Persson (Eds.), IFIP - The International Federation for Information Processing (pp. 371–378). Academic Press.

Jurković, M., Barišić, B., & Jurković, Z. (2000). *Mathematical modelling and optimization of machining processes.*

Jurković, M., Jurković, Z., & Cukor, G. (2005). Genetic algorithm application in optimization of extrusion forming process. *Paper presented at the 10th International Scientific Conference on Production Engineering CIM '05.*

Jurković, Z., Jurković, M., & Buljan, S. (2006). Optimization of extrusion force prediction model using different techniques. *Journal of Achievements in materials and manufacturing engineering, 17*(1-2), 353-356.

Jurković, Z., Brezočnik, M., Grizelj, B., & Mandić, V. (2009). Optimization of extrusion process by genetic algorithms and conventional techniques. *Technical Gazette, 16*(4), 27–33.

Kacem, I. (2003). Ordonnancement mutlticritère des Job-Shop Flexible: formulation, bornes inférieures et approche évolutionniste coopérative. Unpublished doctoral dissertation, University of Lille.

Kacem, I., Hammadi, F., & Borne, P. (2001a, October 7-10). Approach by Localisation And Genetic Manipulation algorithm for Flexible Job-shop Problems. In *Proceedings of the International IEEE Conference on systems & cybernetics*, Tucson, Arizona (pp. 2599-2604).

Kacem, I., Hammadi, F., & Borne, P. (2001c). Multi-objective Optimization for Flexible Job-Shop Scheduling Problem: Hybridization of Genetic Algorithms with Fuzzy Logic. In *Proceedings of IFDICON'2001, European Workshop on Intelligent Forecasting, Diagnosis and Control*, Santorini, Greece.

Kacem, I., Hammadi, F., & Borne, P. (2001b). Direct Chromosome Representation and Advanced Genetic Operators for Job-shop Problems. In *Proceedings of the International Conference on Computational Intelligence for Modelling, Control and Automation (CIMCA'01)*, Las Vegas, NV.

Kanthababu, M. (2013). Multi-Objective Optimization of Manufacturing Processes Using Evolutionary Algorithms. In *Industrial Engineering: Concepts, Methodologies, Tools, and Applications.* doi:10.4018/978-1-4666-1945-6.ch021

Kar, A. K. (2016). Bio inspired computing–A review of algorithms and scope of applications. *Expert Systems with Applications, 59*, 20–32. doi:10.1016/j.eswa.2016.04.018

Karaboga, D. (2005). *An idea based on Honey Bee Swarm for numerical optimization.* Dept. of Computer Engineering, Engineering Faculty, Erciyes University.

Karaboga, D., & Akay, B. (2010). Proportional-Integral-Derivative Controller Design by Using Artificial Bee Colony, Harmony Search, and the Bees Algorithms. *Journal of Systems and Control Engineering, 224*(1), 869–883.

Karaboga, D., & Baştürk, B. (2008). On the Performance of Artificial Bee Colony (ABC) Algorithm. *Applied Soft Computing, 8*(1), 687–697. doi:10.1016/j.asoc.2007.05.007

Karaboga, D., Gorkemli, B., Ozturk, C., & Karaboga, N. (2014). A comprehensive survey: Artificial bee colony (ABC) algorithm and applications. *Artificial Intelligence Review, 42*(1), 21–57. doi:10.1007/s10462-012-9328-0

Karimi, H., & Bashiri, M. (2011). Hub covering location problems with different coverage types. *Scientia Iranica*, *18*(6), 1571–1578. doi:10.1016/j.scient.2011.09.018

Kashan, H. (2013). A particle swarm optimizer for grouping problems. *Information Sciences*, *252*, 81–95. doi:10.1016/j.ins.2012.10.036

Kaufmann, T. (2015). *Geschäftsmodelle in Industrie 4.0 und dem Internet der Dinge. Der Weg vom Anspruch in die Wirklichkeit*. Wiesbaden: Springer. doi:10.1007/978-3-658-10272-2

Keblis, M. F., & Feng, Y. (2012). Optimal pricing and production control in na assembly system with a general stockout cost. *IEEE Transactions on Automatic Control*, *57*(7), 1821–1826. doi:10.1109/TAC.2011.2178335

Keizer, M. C. O., Teunter, R. H., & Veldman, J. (2017). Joint condition-based maintenance and inventory optimization for systems with multiple components. *European Journal of Operational Research*, *257*(1), 209–222. doi:10.1016/j.ejor.2016.07.047

Kennedy, J., & Eberhart, R. (1995). Particle swarm optimization. *Proc. IEEE international conference on neural networks*. doi:10.1109/ICNN.1995.488968

Kennedy, J., & Eberhart, R. C. (1995). A new optimizer using particle swarm theory. In *Proceedings of the Sixth International Symposium on Micromachine and Human Science* (pp. 39-43).

Kennedy, J., Eberhart, R. C., & Yuhui, S. (2001). *Swarm Intelligence*. San Francisco, CA: Morgan Kaufmann.

Kenné, J. P., Boukas, E. K., & Gharbi, A. (2003). Control of production and corrective maintenance rates in a multiple-machine, multiple-product manufacturing system. *Mathematical and Computer Modelling*, *38*(3-4), 351–365. doi:10.1016/S0895-7177(03)90093-8

Kenné, J. P., & Nkeungoue, L. J. (2008). Simultaneous control of production, preventive and corrective maintenance rates of a failure-prone manufacturing system. *Applied Numerical Mathematics*, *58*(2), 180–194. doi:10.1016/j.apnum.2006.11.010

Kher, H. V., & Jensen, J. B. (2002). Shop Performance Implications of Using Cells, Partial Cells, and Remainder Cells*. *Decision Sciences*, *33*(2), 161–190. doi:10.1111/j.1540-5915.2002.tb01641.x

Kim, J., Jhaveri, R., Woo, J., & Yang, C. (2007). Device-Circuit Co-Optimization for Mixed-mode Circuit Design via Geometric Programming. *IEEE/ACM International Conference on Computer-Aided Design*.

Kim, S., & Jeong, B. (2007). Product sequencing problem in Mixed-Model Assembly Line to minimize unfinished works. *Computers & Industrial Engineering*, *53*(2), 206–214. doi:10.1016/j.cie.2007.06.011

Kirkpatrick, S., Gelatt, C. D., & Vecchi, M. P. (1983). Optimization by simulated annealing. *Science*, *220*(4598), 671–680. doi:10.1126/science.220.4598.671 PMID:17813860

Kis, T. (2004). On the complexity of the car sequencing problem. *Operations Research Letters*, *32*(4), 331–335. doi:10.1016/j.orl.2003.09.003

Kleijn, M. J., & Dekker, R. (1998). *An overview of inventory systems with several demand classes. Econometric Institute Report 9838/A*. Rotterdam, The Netherlands: Erasmus University.

Klutke, G. A., Kiessler, P. C., & Wortman, M. A. (2003). A critical look at the bathtub curve. *IEEE Transactions on Reliability*, *52*(1), 125–129. doi:10.1109/TR.2002.804492

Knackstedt, R., Pöppelbuß, J., & Becker, J. (2009). Vorgehensmodell zur Entwicklung von Reifegradmodellen. In H. R. Hansen (Ed.), *Business Services. Konzepte, Technologien, Anwendungen* (pp. 535–544). Vienna: Österreichische Computer Gesellschaft.

Knowles, J. D., & Corne, D. W. (2000). Approximating the nondominated front using the Pareto archived evolution strategy. *Evolutionary Computation*, *8*(2), 149–172. doi:10.1162/106365600568167 PMID:10843519

Kobayashi, S., Ono, I., & Yamamoura, M. (1995). An efficient genetic algorithm for Job-Shop scheduling Problem. In *Proceedings of the 6th International Conference on Genetic Algorithms* (pp. 506-511).

Koenigsaecker, G. (2012). *Leading the lean enterprise transformation*. Boca Raton, FL: CRC Press.

Koh, H. Y., Sequin, C. H., & Gray, P. R. (1990). OPASYN: A Compiler for CMOS Operational Amplifiers. *IEEE Trans. on Computer-Aided Design of Integrated Circuits and Systems*, *9*(2), 113–125. doi:10.1109/43.46777

Kolbe, K. Pure Instinct, 1994.

Konak, A., Coit, D. W., & Smith, A. E. (2006). Multi-objective optimization using genetic algorithms: A tutorial. *Reliability Engineering & System Safety*, *91*(9), 992–1007. doi:10.1016/j.ress.2005.11.018

Kotani, S., Ito, T., & Ohno, K. (2004). Sequencing problem for a mixed-model assembly line in the Toyota production system. *International Journal of Production Research*, *42*(23), 4955-4974.

Krafcik, J. F. (1988). Triumph of the lean production system. *MIT Sloan Management Review*, *30*(1), 41–50.

Kruiskamp, W., & Leenaerts, D. (1995). DARWIN: CMOS opamp synthesis by means of a genetic algorithm. *Proc. ACM/IEEE Design Automation Conference (DAC)*. doi:10.1145/217474.217566

Kubzin, M. A., & Strusevich, V. A. (2006). Planning machine maintenance in two-machine shop scheduling. *Operations Research*, *54*(4), 789–800. doi:10.1287/opre.1060.0301

Kull, T. J., & Talluri, S. (2008). A Supply Risk Reduction Model Using Integrated Multi Criteria Decision Making. *IEEE Transactions on Engineering Management*, *55*(3), 409–419. doi:10.1109/TEM.2008.922627

Kumar, P., & Deshmukh, S. G. (2006). A model for flexible supply chain through flexible manufacturing. *Global Journal of Flexible Systems Management*, *7*(3/4), 17.

Kurt, L. (1985). *Handbook of metal forming*. Society of Manufacturing Engineers.

Kuzman, K. (2001). Problems of accuracy control in cold forming. *Journal of Materials Processing Technology*, *113*(1), 10–15. doi:10.1016/S0924-0136(01)00688-4

Lahmar, M., & Kulkarni, N. R. (2006). Optimal control policies for production/inventorysystems with exchange transactions. *Proceedings of the IIE Annual Conference and Exhibition.*

Lambert, D. M., Cooper, M. C., & Pagh, J. D. (1998). Supply chain management: implementation issues and research opportunities. *The international journal of logistics management*, *9*(2), 1-20.

Langberg, N. A. (1988). Comparisons of replacement policies. *Journal of Applied Probability*, *25*(04), 780–788. doi:10.1017/S0021900200041577

Lange, K. (1985). Handbook of metal forming. McGraw-Hill Book Company.

Larteb, Y., Haddout, A., Benhadou, M., Manufacturing, L., Yang, C., Yeh, T., & Valero, M. (2015). Successful lean implementation: The systematic and simultaneous consideration of soft and hard lean practices. *International Journal of Engineering and General Science*, *3*(2), 1258–1270.

Lean Enterprise Research Center (LERC). (n.d.). *The Key Lean Thinking Principles*. Retrieved from http://www.lean-enterprise.org.uk/what-is-lean-thinking/what-is-lean-thinking-and-key-lean-thinking-principles.html

Leavy, B. (1994). Two strategic perspectives on the buyer-supplier relationship. *Production and Inventory Management Journal*, *35*(2), 47–60.

Lee, K.-M., & Yamakawa, T., & Lee, Keon-Myung., (1998).A genetic algorithm for general machine scheduling problems. In *Knowledge-Based Intelligent Electronic Systems, Proceedings KES '98. Second International Conference* (Vol. 2, pp. 60-66), 21-23 Apr 1998.

Lee, H. L. (2002). Aligning supply chain strategies with product uncertainties. *California Management Review*, *44*(3), 105–119. doi:10.2307/41166135

Lenskold, J. (2003). *Marketing ROI: The Path to Campaign, Customer, and Corporate Profitability*. San Diego, CA, U.S.A.: McGraw-Hill.

Leong, W. F., & Yen, G. G. (2008). PSO-based multiobjective optimization with dynamic population size and adaptive local archives. *Systems, Man, and Cybernetics, Part B: Cybernetics. IEEE Transactions on*, *38*(5), 1270–1293.

Leu, Y.-Y., Matheson, L., & Rees, L. (1996). Sequencing mixed-model assembly lines with genetic algorithms. *Computers & Industrial Engineering*, *30*(4), 1027–1036. doi:10.1016/0360-8352(96)00050-2

LeVeque, R. J. (2007). *Finite Difference Methods for Ordinary and Partial Differential Equations: Steady-State and Time-Dependent Problems*. Philadelphia: Society for Industrial and Applied Mathematics. doi:10.1137/1.9780898717839

Levitin, G., & Lisnianski, A. (2000). Optimization of imperfect preventive maintenance for multi-state systems. *Reliability Engineering & System Safety*, *67*(2), 193–203. doi:10.1016/S0951-8320(99)00067-8

Liang, W., Huang, C., Tseng, T., Lin, Y., & Tseng, J. (2012). The evaluation of intelligent agent performance: An example of B2C e-commerce negotiation. *Computer Standards & Interfaces*, *34*(5), 439–446. doi:10.1016/j.csi.2012.02.003

Liao, C. J. (1992). Optimal control of jobs for production systems. *Computers & Industrial Engineering*, *22*(2), 163–169. doi:10.1016/0360-8352(92)90043-J

Li, D., & OBrien, C. (2001). A quantitative analysis of relationships between product types and supply chain strategies. *International Journal of Production Economics*, *73*(1), 29–39. doi:10.1016/S0925-5273(01)00094-9

Li, H. (2005). Stochastic comparison of age-dependent block replacement policies. *Methodology and Computing in Applied Probability*, *7*(4), 473–488. doi:10.1007/s11009-005-5004-z

Li, H., Deloux, E., & Dieulle, L. (2016). A condition-based maintenance policy for multi-component systems with Lévy copulas dependence. *Reliability Engineering & System Safety*, *149*, 44–55. doi:10.1016/j.ress.2015.12.011

Li, H., & Zhang, H. (2013). Ant Colony Optimization-Based Multi-Mode Scheduling Under Renewable and Nonrenewable Resource Constraints. *Automation in Construction*, *35*, 431–438. doi:10.1016/j.autcon.2013.05.030

Li, J. Q., Pan, Q. K., & Tasgetiren, M. F. (2014). A discrete artificial bee colony algorithm for the multi-objective flexible job-shop scheduling problem with maintenance activities. *Applied Mathematical Modelling*, *38*(3), 1111–1132. doi:10.1016/j.apm.2013.07.038

Li, J., Min, K. J., Otake, T., & Voorhis, T. M. (2008). Inventory and investment in setup and quality operations under return on investment maximization. *European Journal of Operational Research*, *185*(2), 593–605. doi:10.1016/j.ejor.2006.11.045

Lika, B., Kolomvatsos, K., & Hadjiefthymiades, S. (2014). Facing the cold start problem in recommender systems. *Expert Systems with Applications*, *41*(4), 2065–2073.

Liker, J. K. (Ed.). (1997). *Becoming lean: Inside stories of US manufacturers*. Portland, OR: CRC Press.

Lin, S. W., Ying, K. C., Lu, C. C., & Gupta, J. N. (2011). Applying multi-start simulated annealing to schedule a flowline manufacturing cell with sequence dependent family setup times. *International Journal of Production Economics, 130*(2), 246–254. doi:10.1016/j.ijpe.2011.01.004

Lin, T. W., & Wang, C. H. (2012). A hybrid genetic algorithm to minimize the periodic preventive maintenance cost in a series-parallel system. *Journal of Intelligent Manufacturing, 23*(4), 1225–1236. doi:10.1007/s10845-010-0406-3

Li, R., & Ryan, J. K. (2011). A Bayesian Inventory Model Using Real-Time Condition Monitoring Information. *Production and Operations Management, 20*(5), 754–771. doi:10.1111/j.1937-5956.2010.01200.x

Liu, Wang, Yu, Liu, Li, Wan, Lu, & Fernandez. (2009). Analog circuit optimization system based on hybrid evolutionary algorithms. *Integration, the VLSI Journal, 42*, 137–143.

Liu, Y. F., & Liu, S. Y. (2013). A Hybrid Discrete Artificial Bee Colony Algorithm for Permutation Flowshop Scheduling Problem. *Applied Soft Computing, 13*(3), 1459–1463. doi:10.1016/j.asoc.2011.10.024

Liu, Z., & Cheng, T. C. E. (2002). Scheduling with job release dates, delivery times and preemption penalties. *Information Processing Letters, 82*(2), 107–111. doi:10.1016/S0020-0190(01)00251-4

Li, X., Xu, Y., Hsiung, K. L., Pileggi, L., & Boyd, S. (2009). Regular Analog/RF Integrated Circuits Design Using Optimization With Recourse Including Ellipsoidal Uncertainty. *IEEE Trans. on Computer-Aided Design of Integrated Circuits and Systems, 28*(9), 623–637.

Li, Y., Li, X., & Gupta, J. N. (2015). Solving the multi-objective flowline manufacturing cell scheduling problem by hybrid harmony search. *Expert Systems with Applications, 42*(3), 1409–1417. doi:10.1016/j.eswa.2014.09.007

Loganathan, M. K., & Gandhi, O. P. (2016). Maintenance cost minimization of manufacturing systems using PSO under reliability constraint. *International Journal of System Assurance Engineering and Management, 7*(1), 47–61. doi:10.1007/s13198-015-0374-2

Long, F., Zeiler, P., & Bertsche, B. (2016). Modelling the production systems in industry 4.0 and their availability with high-level Petri nets. *IFAC-PapersOnLine, 49*(12), 145–150. doi:10.1016/j.ifacol.2016.07.565

Lototskii, V. A., & Mandel, A. S. (1979). Models and methods of multiproduct inventory control. *Automation and Remote Control, 40*(6), 888–896.

Louis, R. (1992). *How to Implement Kanban for American Industry*. Cambridge, MA: Productivity Press.

Loukil, T., Teghem, J., & Tuyttens, D. (2005). Solving multiobjective production scheduling problems using metaheuristics. *European Journal of Operational Research, 161*(1), 42–61. doi:10.1016/j.ejor.2003.08.029

Love, R. F., & Moris, J. G. G. W. (Ed.). (1988). Facility location: Models and methods Publication in Operations Research (7th ed.). New York: Elsevier.

Lubben, R. T. (1988). *Just-in-Time manufacturing: an aggressive manufacturing strategy*. New York, NY: McGraw-Hill Companies.

Luke, S. (2013). *Essentials of Metaheuristics* (Vol. II). Fairfax: Lulu.com.

Luke, S. (2009). *Simulated annealing. S. Luke içinde, Essentials of metaheuristics (s. 23)*. Raleigh, North Carolina: Lulu.

Lu, Z., Zhang, Y., & Han, X. (2013). Integrating run-based preventive maintenance into the capacitated lot sizing problem with reliability constraint. *International Journal of Production Research, 51*(5), 1379–1391. doi:10.1080/00207543.2012.693637

M. Barros, J. Guilherme and N. Horta. (2010). Analog circuits optimization based on evolutionary computation techniques. *Integration, the VLSI Journal, 43*, 136-155.

Maddah, B., Yassine, A. A., Salameh, M. K., & Chatila, L. (2014). Reserve stock models: Deterioration and preventive replenishment. *European Journal of Operational Research, 232*(1), 64–71. doi:10.1016/j.ejor.2013.06.043

Madu, C. N. (2000). Competing through maintenance strategies. *International Journal of Quality & Reliability Management, 17*(9), 937–949. doi:10.1108/02656710010378752

Magutu, P. O., Aduda, J., & Nyaoga, R. B. (2015). Does Supply Chain Technology Moderate the Relationship between Supply Chain Strategies and Firm Performance? Evidence from Large-Scale Manufacturing Firms in Kenya. *International Strategic Management Review, 3*(1), 43–65. doi:10.1016/j.ism.2015.07.002

Majumder, A., & Laha, D. (2016). A new cuckoo search algorithm for 2-machine robotic cell scheduling problem with sequence-dependent setup times. *Swarm and Evolutionary Computation, 28*, 131–143. doi:10.1016/j.swevo.2016.02.001

Mandal, P., & Visvanathan, V. (2001). CMOS Op-Amp Sizing Using a Geometric Programming Formulation. *IEEE Trans. on Computer-Aided Design of Integrated Circuits and Systems, 20*(1), 22–38. doi:10.1109/43.905672

Marianov, V., & Serra, D. (2003). Location models for airline hubs behaving as M/D/c queues. *Computers & Operations Research, 30*(7), 983–1003. doi:10.1016/S0305-0548(02)00052-7

Marichelvam, M. K., Prabaharan, T., & Yang, X. S. (2014). Improved cuckoo search algorithm for hybrid flow shop scheduling problems to minimize makespan. *Applied Soft Computing, 19*, 93–101. doi:10.1016/j.asoc.2014.02.005

Marín, A. (2005). Formulating and solving splittable capacitated multiple allocation hub location problems. *Computers & Operations Research, 32*(12), 3093–3109. doi:10.1016/j.cor.2004.04.008

Marini, F., & Walczak, B. (2015). Particle swarm optimization (PSO). A tutorial. *Chemometrics and Intelligent Laboratory Systems, 149*, 153–165. doi:10.1016/j.chemolab.2015.08.020

Marodin, G. A., & Saurin, T. A. (2013). Implementing lean production systems: Research areas and opportunities for future studies. *International Journal of Production Research, 51*(22), 6663–6680. doi:10.1080/00207543.2013.826831

Marseguerra, M., & Zio, E. (2000). Optimizing maintenance and repair policies via a combination of genetic algorithms and Monte Carlo simulation. *Reliability Engineering & System Safety, 68*(1), 69–83. doi:10.1016/S0951-8320(00)00007-7

Marseguerra, M., Zio, E., & Podofillini, L. (2002). Condition-based maintenance optimization by means of genetic algorithms and Monte Carlo simulation. *Reliability Engineering & System Safety, 77*(2), 151–165. doi:10.1016/S0951-8320(02)00043-1

Martín, J. C., & Román, C. (2003). Hub location in the South-Atlantic airline market A spatial competition game. *Transportation Research Part A, Policy and Practice, 37*(10), 865–888. doi:10.1016/S0965-8564(03)00060-0

Martin, K. F. (1994). A review by discussion of condition monitoring and fault diagnosis in machine tools. *International Journal of Machine Tools & Manufacture, 34*(4), 527–551. doi:10.1016/0890-6955(94)90083-3

Martorell, S., Villanueva, J. F., Carlos, S., Nebot, Y., Sánchez, A., Pitarch, J. L., & Serradell, V. (2005). RAMS+ C informed decision-making with application to multi-objective optimization of technical specifications and maintenance using genetic algorithms. *Reliability Engineering & System Safety, 87*(1), 65–75. doi:10.1016/j.ress.2004.04.009

Mason-Jones, R., Naylor, B., & Towill, D. R. (2000). Engineering the leagile supply chain. *International Journal of Agile Management Systems, 2*(1), 54–61. doi:10.1108/14654650010312606

Matsukawa, K., Morie, T., Tokunaga, Y., Sakiyama, S., Mitani, Y., Takayama, M., & Dosho, S. et al. (2009). Design methods for pipeline & delta-sigma A-to-D converters with convex optimization. *Asia and South Pacific Design Automation Conference*. doi:10.1109/ASPDAC.2009.4796560

Ma, Y., Chu, C., & Zuo, C. (2010). A survey of scheduling with deterministic machine availability constraints. *Computers & Industrial Engineering*, *58*(2), 199–211. doi:10.1016/j.cie.2009.04.014

Mayring, P. (2010). Qualitative Inhaltsanalyse. In G. Mey & K. Mruck (Eds.), *Handbuch Qualitative Forschung in der Psychologie* (pp. 601–613). Wiesbaden: VS Verlag für Sozialwissenschaften. doi:10.1007/978-3-531-92052-8_42

McKone, K. E., Schroeder, R. G., & Cua, K. O. (2001). The impact of total productive maintenance practices on manufacturing performance. *Journal of Operations Management*, *19*(1), 39–58. doi:10.1016/S0272-6963(00)00030-9

Mechefske, C. K., & Wang, Z. (2001). Using fuzzy linguistics to select optimum maintenance and condition monitoring strategies. *Mechanical Systems and Signal Processing*, *15*(6), 1129–1140. doi:10.1006/mssp.2000.1395

Meerschaert, M. M. (2013). *Mathematical Modeling*. East Lansing: Academic Press.

Melohn, T. H. (1983). How to build employee trust and productivity. *Harvard Business Review*, *61*(1), 56–68.

Melzer, I. (2010). *Service-orientierte Architektur mit Web Services, Konzepte –Standards – Praxis*. Heidelberg, Germany: Spektrum Akademischer Verlag. doi:10.1007/978-3-8274-2550-8

Merendino, S., & Celebi, M. E. (2013). A simulated annealing clustering algorithm based on center perturbation using gaussian mutation. In *Proceedings of the Twenty-Sixth International Florida Artificial Intelligence Research Society Conference*, Florida.

Mesghouni, K., Hammadi, S., & Borne, P. (2004). Evolutionary Algorithms for Job-Shop scheduling. *International Journal of Applied Mathematics and Computer Science*, *14*(1), 91–103.

Meyer, G. G., Främling, K., & Holmström, J. (2009). Intelligent Products: A Survey. *Computers in Industry*, *60*(3), 137–148. doi:10.1016/j.compind.2008.12.005

Michalewicz, Z., Janikow, C. Z., & Krawczyk, J. B. (1992). A modified genetic algorithm for optimal control problems. *Computers & Mathematics with Applications (Oxford, England)*, *23*(12), 83–94. doi:10.1016/0898-1221(92)90094-X

Miguel, P. A. C. (2010). *Metodologia de pesquisa em engenharia de produção e gestão de operações*. Rio de Janeiro: Elsevier.

Mirjalili, S. (2015). Moth-flame optimization algorithm: A novel nature-inspired heuristic paradigm. *Knowledge-Based Systems*, *89*, 228–249. doi:10.1016/j.knosys.2015.07.006

Mobley, R. K. (2002). *An introduction to predictive maintenance*. Butterworth-Heinemann.

Mohamed, A. A. A., Mohamed, Y. S., El-Gaafary, A. A., & Hemeida, A. M. (2017). Optimal power flow using moth swarm algorithm. *Electric Power Systems Research*, *142*, 190–206. doi:10.1016/j.epsr.2016.09.025

Mohammadi, M., Tavakkoli-Moghaddam, R., Siadat, A., & Rahimi, Y. (2016). A game-based meta-heuristic for a fuzzy bi-objective reliable hub location problem. *Engineering Applications of Artificial Intelligence*, *50*, 1–19. doi:10.1016/j.engappai.2015.12.009

Moori, R. G., Marcondes, R. C., & Ávila, R. T. (2002). A análise de agrupamentos como instrumento de apoio à melhoria de qualidade dos serviços aos cliente. *J. Admin. Contemp*, *6*(1), 63–84. doi:10.1590/S1415-65552002000100005

Moreira, M. R. A., & Alves, R. A. F. (2009). A methodology for planning and controlling workload in a job-shop: A four-way decision-making problem. *International Journal of Production Research, 47*(10), 2805–2821. doi:10.1080/00207540701725083

Morin, S., Gagné, C., & Gravel, M. (2009). Ant colony optimization with a specialized pheromone trail for the car-sequencing problem. *European Journal of Operational Research, 197*(3), 1185–1191. doi:10.1016/j.ejor.2008.03.033

Morita, M., Machuca, J. A., Flynn, E. J., & de los Ríos, J. L. P. (2015). Aligning product characteristics and the supply chain process–A normative perspective. *International Journal of Production Economics, 161*, 228–241. doi:10.1016/j.ijpe.2014.09.024

Mosadegh, H., Zandieh, M., & Fatemi Ghomi, S. M. T. (2012). Simultaneous solving of balancing and sequencing problems with station-dependent assembly times for mixed-model assembly lines. *Applied Soft Computing, 12*(4), 1359–1370. doi:10.1016/j.asoc.2011.11.027

Murthy, C. V., & Srinivasan, G. (1995). Fractional cell formation in group technology. *The International Journal of Production Research, 33*(5), 1323-1337.

Murthy, C. V., & Srinivasan, G. (1995). Fractional cell formation in group technology. *International Journal of Production Research, 33*(5), 1323–1337. doi:10.1080/00207549508930212

Murugaiah, U., Jebaraj, B. S., Srikamaladevi, M., & Muthaiyah, S. (2010). Scrap loss reduction using the 5-whys analysis. *International Journal of Quality & Reliability Management, 27*(5), 527–540. doi:10.1108/02656711011043517

Murugan, P., Kannan, S., & Baskar, S. (2009). NSGA-II algorithm for multi-objective generation expansion planning problem. *Electric Power Systems Research, 79*(4), 622–628. doi:10.1016/j.epsr.2008.09.011

Naeij, J., & Shavandi, H. (2010). An optimal lot sizing and pricing planning in two-echelon supply chain. *International Journal of Industrial Engineering Computations, 1*(1), 11–32. doi:10.5267/j.ijiec.2010.01.002

Nagaraj, N. S. (1993). A new optimizer for performance optimization of analog integrated circuits. *IEEE/ACM Design Automation Conference (DAC)*. doi:10.1145/157485.164643

Nagurney, A., Cruz, J., Dong, J., & Zhang, D. (2005). Supply chain networks, electronic commerce, and supply side and demand side risk. *European Journal of Operational Research, 164*(1), 120–142. doi:10.1016/j.ejor.2003.11.007

Naifar, F., Gzara, M., Moukrim, A., & Loukil, T. (2006). Hybrid Evolutionary Algorithm With Insertion Heuristics For The Flexible Job Shop Problem. In *Proceedings of the International Conference on Service Systems and Service Management*. doi:10.1109/ICSSSM.2006.320681

Nakajima, S. (1988). *Introduction to TPM: Total Productive Maintenance (Translation)*. Cambridge, MA: Productivity Press.

Nakawaga, T. (2005). *Maintenance Theory of Reliability*. Springer-Verlag.

Nakrani, S., & Tovey, C. (2004). On Honey Bees and Dynamic Server Allocation in Internet Hosting Centers. *Adaptive Behavior, 12*(3-4), 223–240. doi:10.1177/105971230401200308

Naylor, J. B., Naim, M. M., & Berry, D. (1999). Leagility: Integrating the lean and agile manufacturing paradigms in the total supply chain. *International Journal of Production Economics, 62*(1), 107–118. doi:10.1016/S0925-5273(98)00223-0

Nelder, J., & Mead, R. (1965). A simplex method for function optimization. *The Computer Journal, 7*(4), 308–313. doi:10.1093/comjnl/7.4.308

Ngai, E. W. T., & Wat, F. K. T. (2004). Fuzzy decision support system for risk analysis in e-commerce development. *Decision Support Systems, 40*(2), 235–255. doi:10.1016/j.dss.2003.12.002

Nguyen, T. T., & Vo, D. N. (2015). Modified cuckoo search algorithm for short-term hydrothermal scheduling. *International Journal of Electrical Power & Energy Systems, 65*, 271–281. doi:10.1016/j.ijepes.2014.10.004

Nguyen, T. T., & Vo, D. N. (2016). *An efficient cuckoo bird inspired meta-heuristic algorithm for short-term combined economic emission hydrothermal scheduling*. Ain Shams Engineering Journal.

Nguyen, T. T., Vo, D. N., & Truong, A. V. (2014). Cuckoo search algorithm for short-term hydrothermal scheduling. *Applied Energy, 132*, 276–287. doi:10.1016/j.apenergy.2014.07.017

Niakan, F., Baboli, A., Moyaux, T., & Botta-Genoulaz, V. (2016). A bi-objective model in sustainable dynamic cell formation problem with skill-based worker assignment. *Journal of Manufacturing Systems, 38*, 46–62. doi:10.1016/j.jmsy.2015.11.001

Nicholas, J., & Soni, A. (2005). *The portal to lean production: Principles and practices for doing more with less*. CRC Press.

Nochta, Z. (2008). Smart Items in Real Time Enterprises. In M. Mühlhäuser, & I. Gurevych (Eds.), Handbook of Research on Ubiquitous Computing Technology for Real Time Enterprises (pp. 211-218). Academic Press. doi:10.4018/978-1-59904-832-1.ch010

Norman, B. A., Tharmmaphornphilas, W., Needy, K. L., Bidanda, B., & Warner, R. C. (2002). Worker assignment in cellular manufacturing considering technical and human skills. *International Journal of Production Research, 40*(6), 1479–1492. doi:10.1080/00207540110118082

Nuruddin, M., & Bayuaji, R. (2009). Application of Taguchi's approach in the optimization of mix proportion for Microwave Incinerated Rice Husk Ash foamed concrete. *International Journal of Cyber Ethics in Education, 9*, 121–129.

Nye, W., Riley, D. C., Sangiovanni-Vincentelli, A., & Tits, A. L. (1988). DELIGHT.SPICE: An optimization-based system for the design of integrated circuits. *IEEE Trans. on Computer-Aided Design of Integrated Circuits and Systems, 7*(4), 501–519. doi:10.1109/43.3185

Ohno, T. (1988). *Toyota production system: beyond large-scale production*. Boca Raton, FL: CRC Press.

Okada, I., Zhang, F., Yang, Y., & Fujimura, S. (2010). A Random Key-based Genetic Algorithm for Resource-constrained Project Scheduling Problem with Multiple Modes. In *Proceedings of the International Multi Conference of Engineers and Computer Scientists 2010* (Vol. 1, pp. 106-111).

Okobiah, O., Mohanty, S., & Kougianos, E. (2013). Geostatistical-inspired fast layout optimisation of a nano-CMOS thermal sensor. *Circuits. Devices & Systems, IET, 7*(5), 253–262. doi:10.1049/iet-cds.2012.0358

Olafsson, S., Li, X., & Wu, S. (2008). Operations Research and Data Mining. *European Journal of Operational Research, 187*(3), 1429–1448. doi:10.1016/j.ejor.2006.09.023

Oliva, G. M. (2005). *Planejamento conjunto e colaborativo da cadeia de suprimentos: Modelo de controle ótimo multiobjetivo com custos de transporte*. Porto Alegre: UFRGS.

Öner, C., & Oztayşi, B. (2017). Sustainable Supply Chains and Risk Management for E-Commerce Companies Using Fuzzy Inference System. In Kahraman, Cengiz, Ucal Sari, İrem (Eds.), Intelligence Systems in Environmental Management: Theory and Applications. Springer.

Ong, P. (2014). Adaptive Cuckoo Search Algorithm for Unconstrained Optimization. *TheScientificWorldJournal, 8*. doi:10.1155/2014/943403 PMID:25298971

Onodera, H., Kanbara, H., & Tamaru, K. (1990). Operational-amplifier compilation with performance optimization. *IEEE Journal of Solid-State Circuits, 25*(2), 466–473. doi:10.1109/4.52171

Onwubolu, G., & Mutingi, M. (2001). A genetic algorithm approach to cellular manufacturing systems. *Computers & Industrial Engineering, 39*(1-2), 125–144. doi:10.1016/S0360-8352(00)00074-7

Osaki, S. (2002). *Stochastic Models in Reliability and Maintenance*. Berlin: Springer. doi:10.1007/978-3-540-24808-8

Osman, I. H., & Laporte, G. (1996). Metaheuristics: A bibliography. *Annals of Operations Research, 63*(5), 513–623. doi:10.1007/BF02125421

Otake, T., & Min, K. J. (2001). Inventory and investment in quality improvement under return on investment maximization. *Computers & Operations Research, 28*(10), 113–124. doi:10.1016/S0305-0548(00)00022-8

Otake, T., Min, K. J., & Chen, C. (1991). Inventory and investment in setup operations under return on investment maximization. *Computers & Operations Research, 26*(9), 883–899. doi:10.1016/S0305-0548(98)00095-1

Ovalle, O. R., & Marquez, A. C. (2002). The effectiveness of using e-collaboration tools in the supply chain: An assessment study with system dynamics. *Journal of Purchasing and Supply Management, 9*(4), 151–163. doi:10.1016/S1478-4092(03)00005-0

Özbayrak, M., Papadopoulou, T. C., & Akgün, M. (2007). Systems dynamics modeling of a manufacturing supply chain system. *Simulation Modelling Practice and Theory, 15*(10), 1338–1355. doi:10.1016/j.simpat.2007.09.007

Öztemel, E., & Selam, A. A. (2010). Project Scheduling Using Bee Colony Optimization. In *Proceedings of the 7th International Conference on Intelligent and Manufacturing Systems* (pp. 709-716).

Özturk, C., & Karaboga, D. (2008, October). Classification by neural networks and clustering with artificial bee colony (abc) algorithm. In *Proceedings of the 6th International Symposium on Intelligent and Manufacturing Systems, Features, Strategies and Innovation* (pp. 110-124).

Pacini, E., Mateos, C., & Garino, C. G. (2014). Distributed job scheduling based on Swarm Intelligence: A survey. *Computers & Electrical Engineering, 40*(1), 252–269. doi:10.1016/j.compeleceng.2013.11.023

Pamuk, F. S., & Sepil, C. (2001). A solution to the hub center problem via a single-relocation algorithm with tabu search. *IIE Transactions, 33*(5), 399–411. doi:10.1080/07408170108936838

Pandit, S., Mandal, C., & Patra, A. (2011). *A Methodology for Generation of Performance Models for the Sizing of Analog High-Level Topologies. Int. Journal VLSI Design.*

Pan, Q. K., Tasgetiren, M. F., Suganthan, P. N., & Chua, T. J. (2011). A Discrete Artificial Bee Colony Algorithm for the Lot-Streaming Flowshop Scheduling Problem. *Information Sciences, 181*(12), 2455–2468. doi:10.1016/j.ins.2009.12.025

Pan, Q. K., Wang, L., Li, J. Q., & Duan, J. H. (2014). A novel discrete artificial bee colony algorithm for the hybrid flowshop scheduling problem with makespan minimisation. *Omega, 45*, 42–56. doi:10.1016/j.omega.2013.12.004

Paredis, J. (1992). *Exploiting constraints as background knowledge for genetic algorithms: A case-study for scheduling*. The Netherlands: Ever Science Publishers.

Parello, B. D. & Kabat, W. C. (1986). Job-Shop Scheduling Using Automated Reasoning: A Case Study of the Car-Sequencing Problem. *Journal of Automated Reasoning, 2*(1), 1-42.

Park, D. H., Kim, H. K., Choi, Y., & Kim, J. K. (2012). A literature review and classification of recommender systems research. *Expert Systems with Applications*, *39*(11), 10059–10072. doi:10.1016/j.eswa.2012.02.038

Passino, K. M. (2002). Biomimicry of bacterial foraging for distributed optimization and control. *IEEE control systems, 22*(3), 52-67.

Passino, K. M. (2002). Biomimicry of Bacterial Foraging for distributed optimization and control. *IEEE Control Systems Magazine*, *22*(3), 52–67. doi:10.1109/MCS.2002.1004010

Pathak, K., & Ramakrishnan, N. (2007). Optimization of die angle and ram velocity for rod extrusion using dynamic material modeling and genetic algorithm. *Indian Journal of Engineering and Materials Sciences*, *14*(6), 399.

Paulino, N., Goes, J., & Steiger-Garcao, A. (2001). Design methodology for optimization of analog building blocks using genetic algorithms. *The 2001 IEEE International Symposium on Circuits and Systems (ISCAS 2001).*

Peng, Y., Dong, M., & Zuo, M. J. (2010). Current status of machine prognostics in condition-based maintenance: A review. *International Journal of Advanced Manufacturing Technology*, *50*(1-4), 297–313. doi:10.1007/s00170-009-2482-0

Pereira, C. M., Lapa, C. M., Mol, A. C., & Da Luz, A. F. (2010). A particle swarm optimization (PSO) approach for non-periodic preventive maintenance scheduling programming. *Progress in Nuclear Energy*, *52*(8), 710–714. doi:10.1016/j.pnucene.2010.04.009

Perez, H. D. (2013). Supply chain strategies: Which one hits the mark. *CSCMP Supply Chain Quarterly*, 1.

Perez-Franco, R., Phadnis, S., Caplice, C., & Sheffi, Y. (2016). Rethinking supply chain strategy as a conceptual system. *International Journal of Production Economics*, *182*, 384–396. doi:10.1016/j.ijpe.2016.09.012

Pettersen, J. (2009). Defining lean production: Some conceptual and practical issues. *The TQM Journal*, *21*(2), 127–142. doi:10.1108/17542730910938137

Pham, D. T., & Sholedolu, M. (2008). Using a hybrid PSO-bees algorithm to train neural networks for wood defect classification. In *Proceedings of the 4th International Virtual Conference on Intelligent Production Machines and Systems, IPROMS.*

Pham, D. T., Al-Jabbouli, H., Mahmuddin, M., Otri, S., & Darwish, A. H. (2008c). Application of the Bees algorithm to fuzzy clustering. In *Proceedings of the 4th International Virtual Conference on Intelligent Production Machines and Systems*, Dunbeath, Scotland.

Pham, D. T., Castellani, M., & Fahmy, A. A. (2008, July). Learning the inverse kinematics of a robot manipulator using the bees algorithm. In *Proceedings of the 6th IEEE International Conference on Industrial Informatics INDIN '08* (pp. 493-498). IEEE. doi:10.1109/INDIN.2008.4618151

Pham, D. T., Ghanbarzadeh, A., Koc, E., Otri, S., Rahim, S., & Zaidi, M. (2005). Bee Algorithm A Novel Approach to Function Optimization (Technical Note: MEC 0501). The Manufacturing Engineering Center, Cardiff University.

Pham, D. T., Koç, E., Kalyoncu, M., & Tınkır, M. (2008b). Hierarchical PID controller design for a flexible link robot manipulator using the bees algorithm. *methods, 25*, 32.

Pham, D. T., Otri, S., Ghanbarzadeh, A., & Koc, E. (2006c, April). Application of the bees algorithm to the training of learning vector quantisation networks for control chart pattern recognition. In Proceedings of the Information and Communication Technologies ICTTA'06 (Vol. 1, pp. 1624-1629). IEEE. doi:10.1109/ICTTA.2006.1684627

Pham, D. T., Soroka, A. J., Ghanbarzadeh, A., Koc, E., Otri, S., & Packianather, M. (2006d, August). Optimising neural networks for identification of wood defects using the bees algorithm. In *Proceedings of the 2006 IEEE International Conference on Industrial Informatics* (pp. 1346-1351). IEEE. doi:10.1109/INDIN.2006.275855

Pham, D.T., Ghanbarzadeh, A., Koc, E., Otri, S., Rahim, S., Zaidi, M. (2006a). The Bees Algorithm – A Novel Tool for Complex Optimization Problems. In *Intelligent Production Machines and Systems* (pp. 454–459).

Pham, D. T., Darwish, A. H., Eldukhri, E. E., & Otri, S. (2007a, July). Using the bees algorithm to tune a fuzzy logic controller for a robot gymnast. In *Proceedings of IPROMS 2007 conference*, Cardiff, UK.

Pham, D. T., & Darwish, H. A. (2010). Using the bees algorithm with Kalman filtering to train an artificial neural network for pattern classification. *Proceedings of the Institution of Mechanical Engineers. Part I, Journal of Systems and Control Engineering, 224*(7), 885–892. doi:10.1243/09596518JSCE1004

Pham, D. T., Ghanbarzadeh, A., Koc, E., & Otri, S. (2006b, January). Application of the bees algorithm to the training of radial basis function networks for control chart pattern recognition. In *Proceedings of 5th CIRP international seminar on intelligent computation in manufacturing engineering (CIRP ICME'06), Ischia, Italy* (pp. 711-716).

Pham, D. T., & Koc, E. (2008). Flowshop Sequencing Using the Bees Algorithm. *In Proceedings of the IPROMS Innovative Production Machines and Systems Virtual Conference*, Cardiff, Wales, UK.

Pham, D. T., Koc, E., Lee, J. Y., & Phrueksanant, J. (2007a). Using the Bees Algorithm to Schedule Jobs for a Machine. In *Proceedings of the 8th International Conference on Laser Methodology, CMM and Machine Tool Performance* (pp. 430–439).

Pham, D. T., Otri, S., Afify, A., Mahmuddin, M., & Al-Jabbouli, H. (2007c, May). Data clustering using the bees algorithm. In *Proceedings of 40th CIRP international manufacturing systems seminar.*

Pham, H. (2003). *Handbook of Reliability Engineering*. London: Springer. doi:10.1007/b97414

Pham, H. V., Cooper, E. W., Cao, T., & Kamei, K. (2012). Hybrid Kansei-SOM model using risk management and company assessment for stock trading. *Inform. Sci.* doi:10.1016/j.ins.2011.11.036

Pham, H., & Wang, H. (1996). Imperfect maintenance. *European Journal of Operational Research, 94*(3), 425–438. doi:10.1016/S0377-2217(96)00099-9

Phelps, R., Krasnicki, M., Rutenbar, R., Carley, L. R., & Hellums, J. R. (2000). ANACONDA: Simulation-based synthesis of analog circuits via stochastic pattern search. *IEEE Trans.on Computer-Aided Design of Integrated Circuits and Systems, 19*(6), 703–717. doi:10.1109/43.848091

Phillips, R. L. (2005). *Pricing and Revenue Optimization.* Stanford, CA: Stanford University Press.

Pinedo, M. (2016). *Scheduling: Theory, Algorithms and Systems.* New York: Springer. doi:10.1007/978-3-319-26580-3

Pires, D. F., Antunes, C. H., & Martins, A. G. (2012). NSGA-II with local search for a multi-objective reactive power compensation problem. *International Journal of Electrical Power & Energy Systems, 43*(1), 313–324. doi:10.1016/j.ijepes.2012.05.024

Plas, Debyser, Francky Leyn, Lampaert, Vandenbussche, Gielen, … Leenaerts. (2011). AMGIE- A Synthesis Environment for CMOS Analog Integrated Circuits. *IEEE Trans. on Computer-Aided Design of Integrated Circuits and Systems, 20*(9), 1037-1058.

Ponnambalam, S. G., Aravindan, P., & Rao, M. S. (2003). Genetic algorithms for sequencing problems in mixed model assembly lines. *Computers & Industrial Engineering, 45*(4), 669–690. doi:10.1016/j.cie.2003.09.001

Porn, R., Harjunkoski, I., & Westerlund, T. (1999). Convexification of different classes of non-convex MINLP problems. *Computers & Chemical Engineering*, *23*(3), 439–448. doi:10.1016/S0098-1354(98)00305-6

Pozo, H. (2004). *Administração de recursos materiais e patrimoniais: Uma abordagem logística*. São Paulo: Atlas.

Prajapati, A., Bechtel, J., & Ganesan, S. (2012). Condition based maintenance: A survey. *Journal of Quality in Maintenance Engineering*, *18*(4), 384–400. doi:10.1108/13552511211281552

Prandstetter, M. & Raidl, G. R. (2008). An integer linear programming approach and a hybrid variable neighborhood search for the car sequencing problem. *European Journal of Operational Research*, *191*, 1004-1022.

Prandstetter, M. (2005). *Exact and heuristic methods for solving the car sequencing Problem*. Diplomarbeit, Viyana Technical Univercity.

Prats, X., Puig, V., Quevedo, J., & Nejjari, F. (2010). Lexicographic optimisation for optimal departure aircraft trajectories. *Aerospace Science and Technology*, *14*(1), 26–37. doi:10.1016/j.ast.2009.11.003

Price, K., Storn, R. M., & Lampinen, J. A. (2005). Differential Evolution: A Practical Approach to Global Optimization. In *Natural Computing Series*. Springer.

Puchinger, J., & Raidl, G. R. (2005). Combining metaheuristics and exact algorithms in combinatorial optimization: A survey and classification. *Lecture Notes in Computer Science*, *3562*, 41–53. doi:10.1007/11499305_5

Puerto, J., Ramos, A. B., Rodriguez-Chia, A. M., & Sanchez-Gil, M. C. (2016). Ordered median hub location problems with capacity constraints. *Transportation Research Part C, Emerging Technologies*, *70*, 142–156. doi:10.1016/j.trc.2015.05.012

Qin, Y., Wang, J., & Wei, C. (2014). Joint pricing and inventory control for fresh produce and foods with quality and physical quantity deteriorating simultaneously. *International Journal of Production Economics*, *152*, 42–48. doi:10.1016/j.ijpe.2014.01.005

Quinn, J., Liu, Y., & Grosvenor, R. (2016). Author links open the author workspace.Roger Grosvenor., A Categorical Framework of Manufacturing for Industry 4.0 and Beyond. *Procedia CIRP*, *52*, 173–178. doi:10.1016/j.procir.2016.08.005

Rabiee, M., Zandieh, M., & Ramezani, P. (2012). Bi-objective partial flexible job shop scheduling problem: NSGA-II, NRGA, MOGA and PAES approaches. *International Journal of Production Research*, *50*(24), 7327–7342. doi:10.1080/00207543.2011.648280

Rahman, M. A., & Islam, M. Z. (2014). A hybrid clustering technique combining a novel genetic algorithm with K-Means. *Knowledge-Based Systems*, *71*, 345–365. doi:10.1016/j.knosys.2014.08.011

Ramesh, S., Kannan, S., & Baskar, S. (2012). Application of modified NSGA-II algorithm to multi-objective reactive power planning. *Applied Soft Computing*, *12*(2), 741–753. doi:10.1016/j.asoc.2011.09.015

Rammert, W. (2009). Hybride Handlungsträgerschaft - Ein soziotechnisches Modell verteilten Handelns. In O. Herzog & T. Schildhauer (Eds.), *Intelligente Objekte. Technische Gestaltung, Wirtschaftliche Verwertung, Gesellschaftliche Wirkung* (pp. 23–32). Berlin: Springer. doi:10.1007/978-3-642-02220-3_3

Razavi, B. (2010). *Design of Analog CMOS Integrated Circuits*. McGraw Hill Education Private Limited.

Reboredo, P. (2014). Laufzeitvalidierung einer Plattform zur semantischen Integration von Feldgeräten. In W. A. Halang & H. Unger (Eds.), *Industrie 4.0 und Echtzeit* (pp. 81–90). Berlin: Springer.

Reeve, J. M., & Srinivasan, M. M. (2005). Which supply chain design is right for you?.

Reeves, C. R. (1995). *Modern heuristic techniques for combinatorial problems*. London: McGraw-Hill.

Respen, J., Zufferey, N., & Amaldi, E. (2016). Metaheuristics for a Job Scheduling Problem with Smoothing Costs Relevant for the Car Industry. *Networks*, *67*(3), 246–261. doi:10.1002/net.21656

Revees, C. R., & Rowe, J. E. (2003). *Genetic algorithmsp principles and perspectives*. Dordrecht: Kluwer Academic Publishers.

Richeson, L., Lackey, C. W., & Starner, J. W. Jr. (1995). The effect of communication on the linkage between manufacturers and suppliers in Just-in Time Environment. *Journal of Supply Chain Management*, *31*(1), 21–34.

Richter, M. (2015). *Industrie 4.0 und Energieeffizienz in der Produktion. Aktuelle Entwicklungen und Anwendungsbeispiele*. Retrieved August 20, 2016 from http://industrie40.vdma.org/documents/4214230/5356229/IWU%20Vortrag%20Mark%20Richter/d1060b17-9c8b-450b-8dcf-cb66ddfa95be

Ritchie, B., & Brindley, C. (2007). An emergent framework for supply chain risk management and performance measurement. *The Journal of the Operational Research Society*, *58*(11), 1398–1411. doi:10.1057/palgrave.jors.2602412

Ritthipakdee, A., Thammano, A., Premasathian, N., & Uyyanonvara, B. (2014). An Improved Firefly Algorithm for Optimization Problems. In *Proceedings of the 5th International Symposium on Advanced Control of Industrial Processes*.

Rodríguez-Martín, I., & Salazar-González, J. J. (2008). Solving a capacitated hub location problem. *European Journal of Operational Research*, *184*(2), 468–479. doi:10.1016/j.ejor.2006.11.026

Rogers, P., & Nandi, A. (2007). Judicious order acceptance and order release in make-to-order manufacturing systems. *Production Planning and Control*, *18*(7), 610–625. doi:10.1080/09537280701582422

Roh, J., Hong, P., & Min, H. (2014). Implementation of a responsive supply chain strategy in global complexity: The case of manufacturing firms. *International Journal of Production Economics*, *147*, 198–210. doi:10.1016/j.ijpe.2013.04.013

Rokach, L., & Maimon, O. (2005). Clustering methods. In Data mining and knowledge discovery handbook (pp. 321-352). Springer US. doi:10.1007/0-387-25465-X_15

Rosa, H. (2010). *Um modelo conjunto de localização e operação de estoque em redes dinâmicas*. Florianópolis: UFSC.

Rosenberg, D. (1991). Optimal price-inventory decisions: Profit vs. ROII. *IIE Transactions*, *23*(1), 17–22. doi:10.1080/07408179108963837

Roth, A. (2016). *Einführung und Umsetzung von Industrie 4.0: Grundlagen, Vorgehensmodell und Use Cases aus der Praxis*. Berlin: Springer. doi:10.1007/978-3-662-48505-7

Rothenbächer, A.-K., Drexl, M., & Irnich, S. (2016). Branch-and-Price-and-Cut for a Service Network Design and Hub Location Problem. *European Journal of Operational Research*, *255*(3), 935–947. doi:10.1016/j.ejor.2016.05.058

Rother, M., & Shook, J. (2003). *Learning to see: value stream mapping to add value and eliminate muda*. Cambridge, MA: Lean Enterprise Institute.

Rotting, T. A., & Gjelsvik, A. (1992). Stochastic dual dynamic programming for seasonal scheduling in the norwegian power system. *IEEE Transactions on Power Systems*, *7*(1), 273–279. doi:10.1109/59.141714

Rout, P. K., & Acharya, D. P. (2015). Fast physical design of CMOS ROs for optimal performance using constrained NSGA-II. *AEÜ. International Journal of Electronics and Communications*, *69*(9), 1233–1242. doi:10.1016/j.aeue.2015.05.004

Rout, P. K., Acharya, D. P., & Panda, G. (2014). A multiobjective optimization based fast and robost design for low power and low phase noise current starved VCO. *IEEE Transactions on Semiconductor Manufacturing*, *27*(1), 43–50. doi:10.1109/TSM.2013.2295423

Roy, A., & Roy, M. (2015). Antecedents and consequences of impending population implosion in the developed world: Implications for business systems. *International Journal of Sustainable Society*, *7*(2), 151–172. doi:10.1504/IJSSOC.2015.069913

Roy, B., & Sussmann, B. (1964). *Les problèmes d'ordonnancement avec contraintes disjonctives (Note DS N°.9 bis)*. SEMA Montrouge.

Ruiz-Amaya, J., de la Rosa, J., Fernandez, F. V., Medeiro, F., del Rio, R., Perez-Verdu, B., & Rodriguez-Vazquez, A. (2005). High-level synthesis of switched-capacitor switched-current and continuous-time SD modulators using SIMULINK-based time-domain behavioral models. *IEEE Transactions on Circuits and Systems. I, Regular Papers*, *52*(9), 1795–1810. doi:10.1109/TCSI.2005.852479

Ruiz-Castro, J. E. (2015). A preventive maintenance policy for a standby system subject to internal failures and external shocks with loss of units. *International Journal of Systems Science*, *46*(9), 1600–1613. doi:10.1080/00207721.2013.827258

Ruiz, R., García-Díaz, J. C., & Maroto, C. (2007). Considering scheduling and preventive maintenance in the flowshop sequencing problem. *Computers & Operations Research*, *34*(11), 3314–3330. doi:10.1016/j.cor.2005.12.007

Ruspini, E. H. (1970). Numerical methods for fuzzy clustering. *Information Sciences*, *2*(3), 319–350. doi:10.1016/S0020-0255(70)80056-1

Russwurm, S. (2013). Software: Die Zukunft der Industrie. In U. Sendler (Ed.), *Industrie 4.0 – Beherrschung der industriellen Komplexität mit SysLM (Systems Lifecycle Management)* (pp. 21–36). Berlin: Springer.

Rutenbar. (1989). Simulated annealing algorithms: An overview. *IEEE Circuits and Devices Magazine*, 19-26.

S. Kundu and P. Mandal. (2014). ISGP: Iterative Sequential Geometric Programming for Precise and Robust Analog Circuit Sizing. *Integration, the VLSI Journal*, *47*, 510-531.

Saboury, A., Ghaffari-Nasab, N., Barzinpour, F., & Saeed Jabalameli, M. (2013). Applying two efficient hybrid heuristics for hub location problem with fully interconnected backbone and access networks. *Computers & Operations Research*, *40*(10), 2493–2507. doi:10.1016/j.cor.2013.01.008

Sadollah, A., & Bahreininejad, A. (2012). Optimization of die design using metaheuristic methods in cold forward extrusion process. *Neural Computing & Applications*, *21*(8), 2071–2076. doi:10.1007/s00521-011-0630-6

Sajadieh, M. S., & Jokar, M. R. A. (2009). Optimizing shipment, ordering and pricing policies in a two stage supply chain with price-sensitive demand. *Transportation Research Part E, Logistics and Transportation Review*, *45*(4), 564–571. doi:10.1016/j.tre.2008.12.002

Salem, H., & Hassine, A. B. (2015). Meeting Scheduling Based on Swarm Intelligence, *Procedia. Computer Science*, *60*, 1081–1091.

Samrout, M., Yalaoui, F., Châtelet, E., & Chebbo, N. (2005). New methods to minimize the preventive maintenance cost of series–parallel systems using ant colony optimization. *Reliability Engineering & System Safety*, *89*(3), 346–354. doi:10.1016/j.ress.2004.09.005

Sana, S. S. (2010). A production-inventory model in an imperfect production process. *European Journal of Operational Research*, *200*(2), 451–464. doi:10.1016/j.ejor.2009.01.041

Sanchez, L. M., & Nagi, R. (2001). A review of agile manufacturing systems. *International Journal of Production Research*, *39*(16), 3561–3600. doi:10.1080/00207540110068790

Sanders, A., Elangeswaran, C., & Wulfsberg, J. (2016). Industry 4.0 Implies Lean Manufacturing: Research Activities in Industry 4.0 Function as Enablers for Lean Manufacturing. *Journal of Industrial Engineering and Management*, *9*(3), 811–833. doi:10.3926/jiem.1940

Sarker, R., Omar, M., Hasan, S. K., & Essam, D. (2013). Hybrid Evolutionary Algorithm for job scheduling under machine maintenance. *Applied Soft Computing*, *13*(3), 1440–1447. doi:10.1016/j.asoc.2012.04.032

Sasaki, M., Campbell, J. F., Krishnamoorthy, M., & Ernst, A. T. (2014). A Stackelberg hub arc location model for a competitive environment. *Computers & Operations Research*, *47*, 27–41. doi:10.1016/j.cor.2014.01.009

Satoglu, S. I., Durmusoglu, M. B., & Ertay, T. (2010). A mathematical model and a heuristic approach for design of the hybrid manufacturing systems to facilitate one-piece flow. *International Journal of Production Research*, *48*(17), 5195–5220. doi:10.1080/00207540903089544

Satoglu, S. I., & Suresh, N. C. (2009). A goal-programming approach for design of hybrid cellular manufacturing systems in dual resource constrained environments. *Computers & Industrial Engineering*, *56*(2), 560–575. doi:10.1016/j.cie.2008.06.009

Satoglu, S. I., & Ustundag, A. (2013). Value of RFID Enhanced Maintenance in Aerospace Industry. In *The value of RFID* (pp. 141–153). Springer London. doi:10.1007/978-1-4471-4345-1_11

Saurin, T. A., Ribeiro, J. L. D., & Vidor, G. (2012). A framework for assessing poka-yoke devices. *Journal of Manufacturing Systems*, *31*(3), 358–366. doi:10.1016/j.jmsy.2012.04.001

Sawhney, R. (2006). Interplay between uncertainty and flexibility across the value-chain: Towards a transformation model of manufacturing flexibility. *Journal of Operations Management*, *24*(5), 476–493. doi:10.1016/j.jom.2005.11.008

Schaller, J. E., Gupta, J. N., & Vakharia, A. J. (2000). Scheduling a flowline manufacturing cell with sequence dependent family setup times. *European Journal of Operational Research*, *125*(2), 324–339. doi:10.1016/S0377-2217(99)00387-2

Schill, A., & Springer, T. (2007). *Verteilte Systeme, Grundlagen und Basistechnologien*. Berlin: Springer.

Schmitt, M., & Orfgen, M. (2015). Dynamische Funktionsskalierung mittels Apps auf intelligenten Feldgeräten. *Zeitschrift für wirtschaftlichen Fachbetrieb (ZWF)*, *110*, 54-58.

Schöning, H., & Dorchain, M. (2014). Data Mining und Analyse. In T. Bauernhansl, M. ten Hompel, & B. Vogel-Heuser (Eds.), *Industrie 4.0 in der Produktion, Automatisierung und Logistik* (pp. 543–554). Wiesbaden: Springer. doi:10.1007/978-3-658-04682-8_27

Schroeder, R. G., & Krishnan, R. (1976). Return on investment as a criterion for inventory model. *Decision Sciences*, *7*(4), 697–704. doi:10.1111/j.1540-5915.1976.tb00713.x

Segic, M. R., Marinovic, M., & Potokar, M. (2007). Modification of production-inventory control model with quadratic and linear costs. *ISOR*, *9*, 369–376.

SEI. (2002). *Capability Maturity Model Integration (CMMI[SM]), Version 1.1. CMMI[SM] for Software Engineering (CMMI-SW, V1.1), Continuous Representation, CMU/SEI-2002-TR-028*. Pittsburgh, PA: SEI.

Seitz, J., Debes, M., Heubach, M., & Tosse, R. (2007). *Digitale Sprach- und Datenkommunikation, Netze – Protokolle – Vermittlung*. Leipzig: Carl Hanser Verlag.

Selim, S. Z., & Alsultan, K. (1991). A simulated annealing algorithm for the clustering problem. *Pattern Recognition*, *24*(10), 1003–1008. doi:10.1016/0031-3203(91)90097-O

Selim, S. Z., & Ismail, M. A. (1984). K-means-type algorithms: A generalized convergence theorem and characterization of local optimality. *IEEE Transactions on Pattern Analysis and Machine Intelligence*, *1*(6), 81–87. doi:10.1109/TPAMI.1984.4767478 PMID:21869168

Sen, A., & Zhang, A. (1999). The newsboy problem with multiple demand classes. *IIE Transactions*, *31*(5), 431–444. doi:10.1080/07408179908969846

Sethi, S. P., & Thompson, G. L. (1980). Turnpike Horizons for Production Planning. *Management Science*, *26*(3), 229–241. doi:10.1287/mnsc.26.3.229

Sethi, S. P., & Thompson, G. L. (2006). *Optimal control theory: applications to management science and economics*. New York: Springer.

Seyed-Alagheband, S. A., Fatemi Ghomi, S. M. T., & Zandieh, M. (2011). A simulated annealing algorithm for balancing the assembly line type II problem with sequence-dependent setup times between tasks. *International Journal of Production Research*, *49*(3), 805–825. doi:10.1080/00207540903471486

Shachnai, H. T., & Woeginger, G. (2002). Minimizing makespan and preemption costs on a system of uniform machines. Springer Berlin / Heidelberg. doi:10.1007/3-540-45749-6_74

Shah, S. R., & Ward, P. T. (2003). Lean manufacturing: Context, practice bundles, and performance. *Journal of Operations Management*, *21*(2), 129–149. doi:10.1016/S0272-6963(02)00108-0

Shambu, G., & Suresh, N. C. (2000). Performance of hybrid cellular manufacturing systems: A computer simulation investigation. *European Journal of Operational Research*, *120*(2), 436–458. doi:10.1016/S0377-2217(98)00371-3

Sharifi, H., Ismail, H. S., Qiu, J., & Tavani, S. N. (2013). Supply chain strategy and its impacts on product and market growth strategies: A case study of SMEs. *International Journal of Production Economics*, *145*(1), 397–408. doi:10.1016/j.ijpe.2013.05.005

Sharma, A., Yadava, G. S., & Deshmukh, S. G. (2011). A literature review and future perspectives on maintenance optimization. *Journal of Quality in Maintenance Engineering*, *17*(1), 5–25. doi:10.1108/13552511111116222

Sharma, R. K., Kumar, D., & Kumar, P. (2005). FLM to select suitable maintenance strategy in process industries using MISO model. *Journal of Quality in Maintenance Engineering*, *1*(11), 359–374. doi:10.1108/13552510510626981

Shelokar, P., Jayaraman, V., & Kulkarni, B. (2004). An ant colony approach for clustering. *Analytica Chimica Acta*, *509*(2), 187–195. doi:10.1016/j.aca.2003.12.032

Sherwin, D. (2000). A review of overall models for maintenance management. *Journal of Quality in Maintenance Engineering*, *6*(3), 138–164. doi:10.1108/13552510010341171

Sheta, A. F. (2010). Analogue filter design using differential evolution. *International Journal of Bio-inspired Computation*, *2*(3-4), 233–241. doi:10.1504/IJBIC.2010.033091

Shingo, S. (1985). *A Revolution in Manufacturing: The SMED System*. Cambridge, MA: Productivity Press.

Shingo, S. (1986). *Zero quality control: Source inspection and the poka-yoke system*. Boca Raton, FL: CRC Press.

Shingo, S. (1989). *A Study of the Toyota Production System from an Industrial Engineering Viewpoint*. Portland, OR: Productivity Press.

Sidibe, I. B., Khatab, A., Diallo, C., & Kassambara, A. (2017). Preventive maintenance optimization for a stochastically degrading system with a random initial age. *Reliability Engineering & System Safety*, *159*, 255–263. doi:10.1016/j.ress.2016.11.018

Siepmann, D. (2016). Industrie 4.0 – Fünf zentrale Paradigmen. In A. Roth (Ed.), *Einführung und Umsetzung von Industrie 4.0: Grundlagen, Vorgehensmodell und Use Cases aus der Praxis* (pp. 35–45). Berlin: Springer.

Silver, E. A., Pyke, D. F., & Peterson, R. (1998). *Inventory management and production planning and scheduling*. New York: Wiley.

Silver, E. A., & Zufferey, N. (2005). Inventory Control of Raw Materials under Stochastic and Seasonal Lead Times. *International Journal of Production Research*, *43*(24), 5161–5179. doi:10.1080/00207540500219866

Silver, E. A., & Zufferey, N. (2011). Inventory Control of an Item with a Probabilistic Replenishment Lead Time and a Known Supplier Shutdown Period. *International Journal of Production Research*, *49*(4), 923–947. doi:10.1080/00207540903449888

Simon, W. (2013). Blick in die Zukunft: Industrie 4.0. *Industrial Engineering (American Institute of Industrial Engineers)*, (2): 38–40.

Singh, S. K., & Yadav, S. P. (2015). Modeling and optimization of multi objective non-linear programming problem in intuitionistic fuzzy environment. *Applied Mathematical Modelling*, *39*(16), 4617–4629. doi:10.1016/j.apm.2015.03.064

Slack, N., Chambers, S., & Johnston, R. (2010). *Operations management*. Pearson education.

Sodhi, M. S., & Tang, C. S. (2009). Modeling supply- chain planning under demand uncertainty using stochastic programming: A survey motivated by asset–liability management. *International Journal of Production Economics*, *121*(2), 728–738. doi:10.1016/j.ijpe.2009.02.009

Solimanpur, M., & Elmi, A. (2013). A tabu search approach for cell scheduling problem with makespan criterion. *International Journal of Production Economics*, *141*(2), 639–645. doi:10.1016/j.ijpe.2012.10.001

Solnon, C. (2008). Combining two pheromone structures for solving the car sequencing problem with Ant Colony Optimization. *European Journal of Operational Research*, *191*, 1043–1055

Solnon, C., Cung, V. D., Nguyen, A., & Artigues, C. (2008). The car sequencing problem: Overview of state-of-the-art methods and industrial case-study of the ROADEF2005 challenge problem. *European Journal of Operational Research*, *191*(3), 912–927. doi:10.1016/j.ejor.2007.04.033

Soltani, Z., & Navimipour, N. J. (2016). Customer Relationship Management mechanisms: A systematic review of the state of art literature and recommendations for future research. *Computers in Human Behavior*, *61*, 667–688. doi:10.1016/j.chb.2016.03.008

Song, D. P. (2009). Stability and optimization of a production inventory system under prioritized base-stock control. *IMA J. Manag. Math*, *20*(1), 59–79. doi:10.1093/imaman/dpn010

Sonmez & Dundar. (2011). Simulation-based analog and RF circuit synthesis using a modified evolutionary strategies algorithm. *Integration, the VLSI Journal*, *44*(2), 144–154.

Soorapanth, T. (2007). Gaining Insights in Analog Design via Geometric Programming. *IEEE International Symposium on Information and Communication Technologies*.

Sörensen, K., & Glover, F. (2013). Metaheuristics. In S. I. Gass & M. C. Fu (Eds.), *Encyclopedia of Operations Research and Management Science* (3rd ed., pp. 960–970). London: Springer. doi:10.1007/978-1-4419-1153-7_1167

Spath, D., Ganschar, O., Gerlach, S., Hämmerle, M., Krause, T., & Schlund, S. (2013). *Produktionsarbeit der Zukunft-Industrie 4.0*. Stuttgart: Fraunhofer Verlag.

Spear, S., & Brown, H. K. (1999, September). *Decoding the DNA of the Toyota Production System. Harvard Business Review*.

Stanimirović, Z. (2010). A genetic algorithm approach for the capacitated single allocation p-hub median problem. *Computing and Informatics, 29*(1), 117–132.

Sterman, J. D. (2000). *Business Dynamics: Systems Thinking and Modeling for a Complex World.* Mc Graw Hill.

Stich, V., & Hering, N. (2015). Daten und Software als entscheidender Wettbewerbsfaktor. *Industrie 4.0 Magazin - Zeitschrift für integrierte Produktionsprozesse,* 8-13.

Stock, T., & Seliger, G. (2016). Opportunities of Sustainable Manufacturing in Industry 4.0, 13th Global Conference on Sustainable Manufacturing – Decoupling Growth from Resource Use. *Procedia CIRP, 40,* 536-541.

Storn, R. (1996). On the usage of differential evolution for function optimization. In *Proceedings of the1996 Biennial Conference of the North American Fuzzy Information Processing Society* (pp. 519–523).

Storn, R., & Price, K. V. (1995). Differential Evolution - A simple and efficient adaptive scheme for global optimization over continuous spaces.

Storn, R. (2008). Differential evolution research—Trends and open questions BT - Advances in Differential Evolution. *Advances in Differential Evolution, 143*(1), 1–31. doi:10.1007/978-3-540-68830-3_1

Storn, R., & Price, K. (1996). Minimizing the real functions of the ICEC'96 contest by differential evolution. In *Proceedings of IEEE International Conference on Evolutionary Computation* (pp. 842–844). doi:10.1109/ICEC.1996.542711

Storn, R., & Price, K. (1997). Differential Evolution: A Simple and Efficient Heuristic for global Optimization over Continuous Spaces. *Journal of Global Optimization, 11*(4), 341–359. doi:10.1023/A:1008202821328

Süer, G. A., & Alhawari, O. (2011). Operator assignment decisions in a highly dynamic cellular environment. In *Operations Management Research and Cellular Manufacturing Systems: Innovative Methods and Approaches: Innovative Methods and Approaches* (p. 258).

Süer, G. A., & Dagli, C. (2005). Intra-cell manpower transfers and cell loading in labor-intensive manufacturing cells. *Computers & Industrial Engineering, 48*(3), 643–655. doi:10.1016/j.cie.2003.03.006

Süer, G. A., & Tummaluri, R. R. (2008). Multi-period operator assignment considering skills, learning and forgetting in labour-intensive cells. *International Journal of Production Research, 46*(2), 469–493. doi:10.1080/00207540601138551

Sundar, S., Suganthan, P. N., Jin, C. T., Xiang, C. T., & Soon, C. C. (2016). A hybrid artificial bee colony algorithm for the job-shop scheduling problem with no-wait constraint. *Soft Computing.*

Sung, C. S., & Jin, H. W. (2001). Dual-based approach for a hub network design problem under non-restrictive policy. *European Journal of Operational Research, 132*(1), 88–105. doi:10.1016/S0377-2217(00)00114-4

Sung, C., & Jin, H. (2000). A tabu-search-based heuristic for clustering. *Pattern Recognition, V*(33), 849–858. doi:10.1016/S0031-3203(99)00090-4

Suresh, N. C. (1991). Partitioning work centres for group technology: Insights from an analytical model. *Decision Sciences, 22*(4), 772–791. doi:10.1111/j.1540-5915.1991.tb00364.x

Susilawati, A., Tan, J., Bell, D., & Sarwar, M. (2015). Fuzzy logic based method to measure degree of lean activity in manufacturing industry. *Journal of Manufacturing Systems, 34,* 1–11. doi:10.1016/j.jmsy.2014.09.007

Tabassum, M., & Mathew, K. (2014). A genetic algorithm analysis towards optimization solutions. *International Journal of Digital Information and Wireless Communications, 4*(1), 124–142. doi:10.17781/P001091

Tahooned, A., & Ziarati, K. (2011, June 12-15). Using Artificial Bee Colony to Solve Stochastic Resource Constrained Project Scheduling Problem, Advances in Swarm Intelligence. In *Proceedings of the Second International Conference, ICSI 2011, Chongqing*, China (pp. 293-302).

Taj, S. (2008). LM performance in China: Assessment of 65 manufacturing plants. *Journal of Manufacturing Technology Management, 19*(2), 217–234. doi:10.1108/17410380810847927

Talbi, E.-G. (2009). *Metaheuristics: From Design to Implementation*. Lille: John Wiley & Sons, Inc. doi:10.1002/9780470496916

Tang, L., & Wang, X. (2013). A hybrid multiobjective evolutionary algorithm for multiobjective optimization problems. *IEEE Transactions on Evolutionary Computation, 17*(1), 20–45. doi:10.1109/TEVC.2012.2185702

Tang, O., & Musa, S. N. (2010). Designing fuzzy- genetic learner model based on multi-agent systems in supply chain management. *International Journal of Production Economics, 133*, 25–34. doi:10.1016/j.ijpe.2010.06.013

Tang, S. H., Nouri, H., & Motlagh, O. (2011). A bacteria foraging algorithm for solving multi-period cell formation and subcontracting production planning in a dynamic cellular manufacturing system. *South African Journal of Industrial Engineering, 22*(2), 80–99. doi:10.7166/22-2-17

Tan, K. C., Lee, T. H., & Khor, E. F. (2001). Evolutionary algorithms with dynamic population size and local exploration for multiobjective optimization. *IEEE Transactions on Evolutionary Computation, 5*(6), 565–588. doi:10.1109/4235.974840

Tasgetiren, M. F., Pan, Q. K., Suganthan, P. N., & Chen, A. H.-L. (2011). A discrete artificial bee colony algorithm for the total flowtime minimization in permutation flowshops. *Information Sciences, 181*(16), 3459–3475. doi:10.1016/j.ins.2011.04.018

Tasgetiren, M. F., Pan, Q. K., Suganthan, P. N., & Oner, A. (2013). A discrete artificial bee colony algorithm for the no-idle permutation flowshop scheduling problem with the total tardiness criterion. *Applied Mathematical Modelling, 37*(10-11), 6758–6779. doi:10.1016/j.apm.2013.02.011

Tavakkoli-Moghaddam, R., Gholipour-Kanani, Y., & Cheraghalizadeh, R. (2008). A genetic algorithm and memetic algorithm to sequencing and scheduling of cellular manufacturing systems. *International Journal of Management Science and Engineering Management, 3*(2), 119–130.

Tavakkoli-Moghaddam, R., Javadian, N., Khorrami, A., & Gholipour-Kanani, Y. (2010). Design of a scatter search method for a novel multi-criteria group scheduling problem in a cellular manufacturing system. *Expert Systems with Applications, 37*(3), 2661–2669. doi:10.1016/j.eswa.2009.08.012

Tavakkoli-Moghaddam, R., & Rahimi-Vahed, A. R. (2006). Multi-criteria sequencing problem for a mixed-model assembly line in a JIT production system. *Applied Mathematics and Computation, 181*(2), 1471–1481. doi:10.1016/j.amc.2006.02.033

Tavakkoli-Moghaddam, R., Safaei, N., & Gholipour, Y. (2006). A hybrid simulated annealing for capacitated vehicle routing problems with the independent route length. *Applied Mathematics and Computation, 17*(2), 445–454. doi:10.1016/j.amc.2005.09.040

Tawdross, P., & König, A. (2006). Particle swarm optimization for reconfigurable sensor electronics-case study: 3 Bit Flash ADC. *4th IEEE international workshop on intelligent solutions in embedded systems (WISES'06).*

Tawdross, P., & Koning, A. (2005). Investigation of Particle Swarm Optimization for dynamic reconfiguration of field programmable analog circuits. *Proc. of 5th Conference on hybrid intelligent systems*. doi:10.1109/ICHIS.2005.66

Tay, J. C., & Wibowo, D. (2004). An Effective Chromosome Representation for Evolving Flexible Job Shop Schedules. In K. Deb et al. (Eds.), GECCO 2004, LNCS (Vol. 3103, pp. 210-221). Springer-Verlag Berlin Heidelberg 2004.

Teodorovic, D., Lucic, P., Markovic, G., & Dell'Orco, M. (2006). Bee Colony Optimization: Principles and Applications. In *Proceedings of the 8th Seminar on Neural Network Applications in Electrical Engineering.*

Teoh, C. K., Wibowo, A., & Ngadiman, M. S. (2015). Review of state of the art for metaheuristic techniques in Academic Scheduling Problems. *Artificial Intelligence Review*, *44*(1), 1–21. doi:10.1007/s10462-013-9399-6

Tesfamariam, D., & Lindberg, B. (2005). Aggregate analysis of manufacturing systems using system dynamics and ANP. *Computers & Industrial Engineering*, *49*(1), 98–117. doi:10.1016/j.cie.2005.05.001

Thakker, R. A., Baghini, M. S., & Patil, M. B. (2009). Low-power low-voltage analog circuit design using hierarchical particle swarm optimization. *Proc. IEEE 22nd international conference on VLSI design*. doi:10.1109/VLSI.Design.2009.14

Thames, L., & Schaefer, D. (2016). Software-defined Cloud Manufacturing for Industry 4.0. *Procedia CIRP*, *52*, 12–17. doi:10.1016/j.procir.2016.07.041

Thevenin, S., Zufferey, N., & Potvin, J.-Y. (2017). Makespan minimization for a parallel machine scheduling problem with preemption and job incompatibility. *International Journal of Production Research*, *55*(6), 1588–1606. doi:10.1080/00207543.2016.1181285

Thevenin, S., Zufferey, N., & Widmer, M. (2015). Metaheuristics for a Scheduling Problem with Rejection and Tardiness Penalties. *Journal of Scheduling*, *18*(1), 89–105. doi:10.1007/s10951-014-0395-8

Thevenin, S., Zufferey, N., & Widmer, M. (2016b). Order acceptance and scheduling with earliness and tardiness penalties. *Journal of Heuristics*, *22*(6), 849–890. doi:10.1007/s10732-016-9321-x

Thomadsen, T., & Larsen, J. (2007). A hub location problem with fully interconnected backbone and access networks. *Computers & Operations Research*, *34*(8), 2520–2531. doi:10.1016/j.cor.2005.09.018

Thornhill, N. F., & Naim, M. M. (2006). An exploratory study to identify rogue seasonality in a steel companys supply network using spectral principal component analysis. *European Journal of Operational Research*, *172*(1), 146–162. doi:10.1016/j.ejor.2004.09.044

Thürer, M., Silva, C., Stevenson, M., & Land, M. (2014). Controlled order release: A performance assessment in job shops with sequence-dependent set-up times. *Production Planning and Control*, *25*(7), 603–615. doi:10.1080/09537287.2012.735801

Topal, A. O., & Altun, O. (2016). A novel meta-heuristic algorithm: Dynamic Virtual Bats Algorithm. *Information Sciences*, *354*, 222–235. doi:10.1016/j.ins.2016.03.025

Topoyan, M. (2011). An Integrated Measurement Model of Supply Chain Flexibility. *Ege Academic Review*, *11*(4).

Torabi, S. A., & Amiri, A. S. (2012). A possibilistic approach for designing hybrid cellular manufacturing systems. *International Journal of Production Research*, *50*(15), 4090–4104. doi:10.1080/00207543.2011.590827

Torralba, A., Chavez, J., & Franquelo, L. G. (1996). FASY: A fuzzy-logic based tool for analog synthesis. *IEEE Trans. on Computer-Aided Design of Integrated Circuits and Systems*, *15*(7), 705–715. doi:10.1109/43.503939

Tsai, L. H. (1995). Mixed-model sequencing to minimize utility work and the risk of conveyor stoppage. *Management Science*, *41*(3), 485–495. doi:10.1287/mnsc.41.3.485

Tulunay, G., & Balkir, S. (2008). A synthesis tool for CMOS RF low noise amplifiers. *IEEE Transactions on Computer-Aided Design of Integrated Circuits and Systems*, *27*(5), 977–982. doi:10.1109/TCAD.2008.917579

Turing Finance. (2016, 09 29). *Clustering using Ant Colony Optimization*. Retrieved 10 19, 2016, from http://www.turingfinance.com/ant-colony-optimization-finance/

UCI Machine Learning Repository. (2015). Iris datasets. Retrieved from https://archive.ics.uci.edu/ml/datasets/Iris

Uckun, S., Bagchi, S., Kawamura, K., & Miyabe, Y. (1993). Managing genetic search in Job-Shop scheduling. *IEEE Expert*, *8*(5), 15–24. doi:10.1109/64.236477

Upton, D. M. (1994). The management of manufacturing flexibility. *California Management Review*, *36*(2), 72–89. doi:10.2307/41165745

Ustundağ, A. (2016). *Endüstri 4.0 ve Akıllı Sistemler*. Retrieved from http://alpustundag.blogspot.com.tr/

Vahedi Nouri, B., Fattahi, P., & Ramezanian, R. (2013). Hybrid firefly-simulated annealing algorithm for the flow shop problem with learning effects and flexible maintenance activities. *International Journal of Production Research*, *51*(12), 3501–3515. doi:10.1080/00207543.2012.750771

Van Horenbeek, A., Pintelon, L., & Muchiri, P. (2010). Maintenance optimization models and criteria. *International Journal of System Assurance Engineering and Management*, *1*(3), 189–200. doi:10.1007/s13198-011-0045-x

Van Peteghem, V., & Vanhoucke, M. (2010). A Genetic Algorithm for the Preemptive and Non-Preemptive Multi-Mode Resource-Constrained Project Scheduling Problem. *European Journal of Operational Research*, *201*(2), 409–418. doi:10.1016/j.ejor.2009.03.034

Van Peteghem, V., & Vanhoucke, M. (2014). An Experimental Investigation of Metaheuristics for the Multi-Mode Resource-Constrained Project Scheduling Problem On New Dataset Instances. *European Journal of Operational Research*, *235*(1), 62–72. doi:10.1016/j.ejor.2013.10.012

van Steenbergen, M., Bos, R., Brinkkemper, S., van de Weerd, I., & Bekkers, W. (2010). The Design of Focus Area Maturity Models. In R. Winter (Ed.), *Global Perspectives on Design Science Research* (pp. 317–332). Berlin: Springer. doi:10.1007/978-3-642-13335-0_22

Varyani, A., Jalilvand-Nejad, A., & Fattahi, P. (2013). Determining the optimum production quantity in three-echelon production system with stochastic demand. *International Journal of Advanced Manufacturing Technology*, *72*(1-4), 119–133. doi:10.1007/s00170-014-5621-1

VDI/VDE. (2013). *Cyber-Physical Systems: Chancen und Nutzen aus Sicht der Automation*. Retrieved June 27, 2016 from https://www.vdi.de/uploads/media/Stellungnahme_Cyber-Physical_Systems.pdf

Venkataramanaiah, S. (2008). Scheduling in cellular manufacturing systems: An heuristic approach. *International Journal of Production Research*, *46*(2), 429–449. doi:10.1080/00207540601138577

Venkataramanaiah, S., & Krishnaiah, K. (2002). Hybrid heuristic for design of cellular manufacturing systems. *Production Planning and Control*, *13*(3), 274–283. doi:10.1080/09537280110073978

Viguier, F., & Pierreval, H. (2004). An approach to the design of a hybrid organization of workshops into functional layout and group technology cells. *International Journal of Computer Integrated Manufacturing*, *17*(2), 108–116. doi:10.1080/09511920310001593092

Vitasek, K. L., Manrodt, K. B., & Abbott, J. (2005). What makes a lean supply chain? *Supply chain management review*, *9*(7), 39-45.

Vogel-Heuser, B. (2014). Herausforderungen und Anforderungen aus Sicht der IT und der Automatisierungstechnik. In T. Bauernhansl, M. ten Hompel, & B. Vogel-Heuser (Eds.), *Industrie 4.0 in der Produktion, Automatisierung und Logistik* (pp. 37–48). Wiesbaden: Springer. doi:10.1007/978-3-658-04682-8_2

Vural, R., Bozkurt, U., & Yildirim, T. (2013). Analog active filter component selection with nature inspired metaheuristics. *Int. J. Electron. Commun.*, *67*(3), 197–205. doi:10.1016/j.aeue.2012.07.009

Vural, R., & Yildirim, T. (2012). Analog Circuit Sizing via Swarm Intelligence. *Int. J. Electron. Commun.*, *66*(9), 732–740. doi:10.1016/j.aeue.2012.01.003

Vural, R., Yildirim, T., Kadioglu, T., & Basargan, A. (2012). Performance evaluation of evolutionary algorithms for optimal filter design. *IEEE Transactions on Evolutionary Computation*, *16*(1), 135–147. doi:10.1109/TEVC.2011.2112664

Wang, G., T.N. Wong, C. Yu. (2011). A computational model for multi-agent E-commerce negotiations with adaptive negotiation behaviors, *J. Comput. Sci.* doi:10.1016/j.jocs.2011.10.003

Wang, C. H., & Lin, T. W. (2011). Improved particle swarm optimization to minimize periodic preventive maintenance cost for series-parallel systems. *Expert Systems with Applications*, *38*(7), 8963–8969. doi:10.1016/j.eswa.2011.01.113

Wang, C., Huang, R., & Wei, Q. (2015). Integrated pricing and lot-sizing decision in a two-echelon supply chain witha finite production rate. *International Journal of Production Economics*, *161*, 44–53. doi:10.1016/j.ijpe.2014.11.011

Wang, H. (2002). A survey of maintenance policies of deteriorating systems. *European Journal of Operational Research*, *139*(3), 469–489. doi:10.1016/S0377-2217(01)00197-7

Wang, H., & Pham, H. (2006). *Reliability and optimal maintenance.* Springer Science & Business Media.

Wang, J., Fan, X., Zhao, A., & Yang, M. (2015). A Hybrid Bat Algorithm for Process Planning Problem. *IFAC-PapersOnLine*, *48*(3), 1708–1713. doi:10.1016/j.ifacol.2015.06.332

Wang, K. (2003). *Intelligent Condition Monitoring and Diagnosis Systems – A computational Intelligence Approach.* Amsterdam: IOS Press.

Wang, L., Chu, J., & Mao, W. (2008). An optimum condition-based replacement and spare provisioning policy based on Markov chains. *Journal of Quality in Maintenance Engineering*, *14*(4), 387–401. doi:10.1108/13552510810909984

Wang, L., Chu, J., & Wu, J. (2007). Selection of optimum maintenance strategies based on a fuzzy analytical hierarchy process. *International Journal of Production Economics*, *107*(1), 151–163. doi:10.1016/j.ijpe.2006.08.005

Wang, S., & Liu, M. (2015). Multi-objective optimization of parallel machine scheduling integrated with multi-resources preventive maintenance planning. *Journal of Manufacturing Systems*, *37*, 182–192. doi:10.1016/j.jmsy.2015.07.002

Wang, W., Cha, K., Cheng, C., & And Qiu, L. (2009). A comparison of performance of several artificial intelligence methods for forecasting monthly discharge time series. *Journal of Hydrology (Amsterdam)*, *374*(3-4), 294–306. doi:10.1016/j.jhydrol.2009.06.019

Wang, X., Tang, J., & Yung, K. L. (2010). A scatter search approach with dispatching rules for a joint decision of cell formation and parts scheduling in batches. *International Journal of Production Research*, *48*(12), 3513–3534. doi:10.1080/00207540902922828

Wang, Y., Deng, C., Wu, J., Wang, Y., & Xiong, Y. (2014). A corrective maintenance scheme for engineering equipment. *Engineering Failure Analysis*, *36*, 269–283. doi:10.1016/j.engfailanal.2013.10.006

Watanabe, M., Ida, K., & Gen, M. (2005). A genetic algorithm with modified crossover operator and search area adaptation for the job-shop scheduling problem. *Computers & Industrial Engineering*, *48*(4), 743–752. doi:10.1016/j.cie.2004.12.008

Waters, D. (2003). *Logistics: an introduction to supply chain management*. Editorial Aardvark.

Webster, J., & Watson, R. T. (2002). Analyzing the past to prepare for the future: Writing a literature review. *Management Information Systems Quarterly*, *26*(2), xiii–xxiii.

Weck, M., & Brecher, C. (2009). *Werkzeugmaschinen 4 – Automatisierung von Maschinen und Anlagen*. Berlin: Springer.

Wee, H. M., & Widyadana, G. A. (2013). A production model for deteriorating items with stochastic preventive maintenance time and rework process with FIFO rule. *Omega*, *41*(6), 941–954. doi:10.1016/j.omega.2012.12.001

Wee, H., Lo, C., & Hsu, P. (2009). A multi-objective joint replenishment inventory model of deteriorated items in a fuzzy environment. *European Journal of Operational Research*, *197*(2), 620–631. doi:10.1016/j.ejor.2006.08.067

Weyer, S., Schmitt, M., Ohmer, M., & Gorecky, D. (2015). Towards Industry 4.0 - Standardization as the crucial challenge for highly modular, multi-vendor production systems. *IFAC-Papers OnLine*, *48*(3), 579-584

Wittenberg, C. (2016). Human-CPS Interaction - requirements and human-machine interaction methods for the Industry 4.0. *IFAC-Papers OnLine*, *49*(19), 420–425. doi:10.1016/j.ifacol.2016.10.602

Wolff, I., & Schulze, S. (2013). *Industrie 4.0 – Cyber Physical Systems in der Produktion. Clustermanagement IKT. NWR*. Wuppertal: Bergische Universität.

Womack, J. P., Jones, D. T., & Roos, D. (1990). *The Machine that Changed the World*. New York, NY: Simon and Schuster.

Wong, L. P., Low, M. Y. H., & Chong, C. S. (2008, May). A bee colony optimization algorithm for traveling salesman problem. In *Proceedings of the Second Asia International Conference on Modeling & Simulation AICMS '08* (pp. 818-823). IEEE. doi:10.1109/AMS.2008.27

Wong, L. P., Low, M. Y. H., & Chong, C. S. (2010). Bee colony optimization with local search for traveling salesman problem. *International Journal of Artificial Intelligence Tools*, *19*(03), 305–334. doi:10.1142/S0218213010000200

Wong, L., Puan, C. Y., Low, M. Y. H., & Chong, C. S. (2008). Bee Colony Optimization Algorithm with Big Valley Landscape Exploitation for Job Shop Scheduling Problems. In *Proceedings of the 2008 Winter Simulation Conference* (pp. 2050–2058). doi:10.1109/WSC.2008.4736301

Woolson, D., & Husar, M. (1997). Transforming a plant to lean in a large, traditional company: Delphi Saginaw Steering Systems, GM. In J. K. Liker (Ed.), *Becoming Lean: Inside Stories of U.S. Manufacturers* (pp. 120–158). Portland, OR: Productivity Press.

Xu, H., & Hu, W. (2008). Availability optimisation of repairable system with preventive maintenance policy. *International Journal of Systems Science*, *39*(6), 655–664. doi:10.1080/00207720701872057

Xu, Z., Ming, X. G., Zheng, M., Li, M., He, L., & Song, W. (2015). Cross-trained workers scheduling for field service using improved NSGA-II. *International Journal of Production Research*, *53*(4), 1255–1272. doi:10.1080/00207543.2014.955923

Yamada, T., & Nakano, R. (1992), A genetic algorithm applicable to a large scale Job-Shop problems. In R. Manner & B. Mandrick (Eds.), Proceedings of Parallel problem solving from nature (Vol. 2, pp. 281-290). Amsterdam: North Holland.

Yaman, H. (2008). Star p-hub median problem with modular arc capacities. *Computers & Operations Research*, *35*(9), 3009–3019. doi:10.1016/j.cor.2007.01.014

Yaman, H., Kara, B. Y., & Tansel, B. Ç. (2007). The latest arrival hub location problem for cargo delivery systems with stopovers. *Transportation Research Part B: Methodological, 41*(8), 906–919. doi:10.1016/j.trb.2007.03.003

Yang & Gan. (2007). Robust Calibration of an Improved Delta-Sigma Data Converter Using Convex Optimization. *IEEE Journal of Selected Topics in Signal Processing, 1*(4).

Yang, X. S. (2009, October). Firefly algorithms for multimodal optimization. In *Proceedings of the International symposium on stochastic algorithms* (pp. 169-178). Springer Berlin Heidelberg.

Yang, X. S. (2010). A new metaheuristic bat-inspired algorithm. In Proceedings of the Nature inspired cooperative strategies for optimization (NICSO 2010) (pp. 65-74). Springer Berlin Heidelberg. doi:10.1007/978-3-642-12538-6_6

Yang, X. S., & Deb, S. (2009, December). Cuckoo search via Lévy flights. In *Proceedings of the World Congress on Nature & Biologically Inspired Computing NaBIC '09* (pp. 210-214). IEEE.

Yang, X. S., Karamanoglu, M., & He, X. (2014). Flower pollination algorithm: A novel approach for multiobjective optimization. *Engineering Optimization, 46*(9), 1222–1237. doi:10.1080/0305215X.2013.832237

Yang, X.-S. (2012). *Flower pollination algorithm for global optimization. In Unconventional computation and natural computation* (pp. 240–249). Springer. doi:10.1007/978-3-642-32894-7_27

Yang, X.-S., & Deb, S. (2014). Cuckoo search: Recent advances and applications. *Neural Computing & Applications, 24*(1), 169–174. doi:10.1007/s00521-013-1367-1

Yanran, Z., Wang, Z., & Weimin, C. (1995). Numerical simulations for extrusion and ironing and die-angle optimization. *Journal of Materials Processing Technology, 55*(1), 48–52. doi:10.1016/0924-0136(95)01811-5

Yavuz, M., Inan, U. H., & Fığlalı, A. (2008). Fair referee assignments for professional football leagues. *Computers & Operations Research, 35*(9), 2937–2951. doi:10.1016/j.cor.2007.01.004

Yeh, W. C., & Hsieh, T. J. (2011). Solving reliability redundancy allocation problems using an artificial bee colony algorithm. *Computers & Operations Research, 38*(11), 1465–1473. doi:10.1016/j.cor.2010.10.028

Yıldırmaz, C., Karabat, S., & Sayın, S. (2009). Pricing and lot-sizing decisions in a two-echelon system with transportation costs. *OR-Spektrum, 31*(3), 629–650. doi:10.1007/s00291-008-0156-1

Yilmaz, O. F., & Durmusoglu, M.B. (in press). A Multi-objective scheduling of hybrid manufacturing systems with walking workers.

Yilmaz, O. F., Cevikcan, E., & Durmusoglu, M. B. (2016). Scheduling batches in multi hybrid cell manufacturing system considering worker resources: A case study from pipeline industry. *Advances in Production Engineering & Management, 11*(3), 192–206. doi:10.14743/apem2016.3.220

Yilmaz, O. F., & Erbiyik, H. (2016). *Effective Applications of Optimization Methods in the Manufacturing Environment in Turkey. Comparative Economics and Regional Development in Turkey*. Hershey, PA: Business Science Reference; doi:10.4018/978-1-4666-8729-5.ch015

Yılmaz, S., & Küçüksille, E. U. (2015). A new modification approach on bat algorithm for solving optimization problems. *Applied Soft Computing, 28*, 259–275. doi:10.1016/j.asoc.2014.11.029

Yu, Y., Tang, J., Sun, W., Yin, Y., & Kaku, I. (2013). Combining local search into non-dominated sorting for multi-objective line-cell conversion problem. *International Journal of Computer Integrated Manufacturing, 26*(4), 316–326. doi:10.1080/0951192X.2012.717717

Zandieh, M., Mahdavi, I., & Bagheri, A. (2008). Solving a flexible job-shop scheduling problems by a Genetic Algorithm. *Journal of Applied Sciences*, *8*(24), 4650–4655. doi:10.3923/jas.2008.4650.4655

Zezulka, F., Marcon, P., Vesely, I., & Sajdl, O. (2016). Industry 4.0 – An Introduction in the phenomenon. *IFAC-Papers OnLine*, *49*(25), 8–12. doi:10.1016/j.ifacol.2016.12.002

Zhang, H. P., & Gen, M. (2005). Multistage-based genetic algorithm for flexible job shop scheduling problem. *Journal of Complexity International*, *48*, 409–425.

Zhang, M., & Bell, P. C. (2007). The effect of market segmentation with demand leakage between market segments on a firms price and inventory decisions. *European Journal of Operational Research*, *182*(2), 738–754. doi:10.1016/j.ejor.2006.09.034

Zhang, M., Bell, P., Cai, G., & Chen, X. (2010). Optimal fences and joint price and inventory decisions in distinct markets with demand leakage. *European Journal of Operational Research*, *204*(3), 589–596. doi:10.1016/j.ejor.2009.11.032

Zhang, R. (2011). An Artificial Bee Colony Algorithm Based on Problem Data Properties for Scheduling Job Shops. *Procedia Engineering*, *23*, 131–136. doi:10.1016/j.proeng.2011.11.2478

Zhang, R., Song, S., & Wu, C. (2013). A Hybrid Artificial Bee Colony Algorithm for the Job Shop Scheduling Problem. *International Journal of Production Economics*, *141*(1), 167–178. doi:10.1016/j.ijpe.2012.03.035

Zhang, Y., & Jiao, J. (2007). An associative classification-based recommendation system for personalization in B2C e-commerce applications. *Expert Systems with Applications*, *33*(2), 357–367. doi:10.1016/j.eswa.2006.05.005

Zhao, S., Wang, L., & Zheng, Y. (2014). Integrating production planning and maintenance: An iterative method. *Industrial Management & Data Systems*, *114*(2), 162–182. doi:10.1108/IMDS-07-2013-0314

Zhou, A., Qu, B. Y., Li, H., Zhao, S. Z., Suganthan, P. N., & Zhang, Q. (2011). Multiobjective evolutionary algorithms: A survey of the state of the art. *Swarm and Evolutionary Computation*, *1*(1), 32–49. doi:10.1016/j.swevo.2011.03.001

Zhu, S. X. (2013). Dynamic replenishment, production, and pricing decisions, in the face of supply disruption and random price-sensitive demand. *International Journal of Production Economics*, *146*(2), 612–619. doi:10.1016/j.ijpe.2013.08.009

Ziarati, K., Akbari, R., & Zeighami, V. (2011). On the performance of Bee Algorithms for Resource Constrained Project Scheduling Problem. *Applied Soft Computing*, *11*(4), 3720–3733. doi:10.1016/j.asoc.2011.02.002

Zio, E., & Viadana, G. (2011). Optimization of the inspection intervals of a safety system in a nuclear power plant by Multi-Objective Differential Evolution (MODE). *Reliability Engineering & System Safety*, *96*(11), 1552–1563. doi:10.1016/j.ress.2011.06.010

Zitzler, E., Laumanns, M., & Thiele, L. (2002). SPEA2: Improving the strength Pareto evolutionary algorithm.

Zolfaghari, S., & Lopez Roa, E. V. (2006). Cellular manufacturing versus a hybrid system: A comparative study. *Journal of Manufacturing Technology Management*, *17*(7), 942–961. doi:10.1108/17410380610688250

Zufferey, N. (2015), October 22-24. Multiple Neighborhoods in Tabu Search: Successful Applications in Operations Management. In*Proceedings of the 4th M-Sphere Conference M-SPHERE 2015*, Dubrovnik, Croatia.

Zufferey, N., Cho, B. Y., & Glardon, R. (2016, February 23-25). Dynamic Multi-Trip Vehicle Routing with Unusual Time-Windows for the Pick-Up of Blood Samples and Delivery of Medical Material. In *Proceedings of the 5th International Conference on Operations Research and Enterprise Systems ICORES 2016*, Rome, Italy. doi:10.5220/0005733303660372

Zufferey, N. (2012). Metaheuristics: Some Principles for an Efficient Design. *Computer Technology and Applications, 3*(6), 446–462.

Zufferey, N. (2012b). Optimization by Ant Algorithms: Possible Roles for an Individual Ant. *Optimization Letters, 6*(5), 963–973. doi:10.1007/s11590-011-0327-x

Zufferey, N. (2016). Tabu Search Approaches for Two Car Sequencing Problems with Smoothing Constraints. In *Metaheuristics for Production Systems* (pp. 167–190). Springer International Publishing. doi:10.1007/978-3-319-23350-5_8

Zühlke, D., Schlick, J., & Stephan, P. (2012). Produktion 2020 – Auf dem Weg zur 4. industriellen Revolution. *IM Information Management und Consulting*, 26-33.

# About the Contributors

**Ömer Faruk Yılmaz** received his MS in Industrial Engineering from Istanbul University. He is a Research Assistant of Industrial Engineering Department at Yalova University. He obtained his Ph.D. in Industrial Engineering from the Istanbul Technical University. His research interests are heuristic and meta-heuristic applications to manufacturing systems. He is also the author several technical articles and chapters in these areas.

**Süleyman Tüfekçí** serves as associate professor of industrial and systems engineering at the University of Florida. His teaching and research focuses on manufacturing management, lean manufacturing, integrated product and process design, emergency management, reverse logistics and supply chain and operations planning in electronics global services industries. Through his Emulated Flexible Manufacturing Laboratory, Tufekçi focuses on developing methods, tools, and software for management systems in lean manufacturing control, supply chain models and emergency area evacuation systems.

* * *

**Debiprasad Priyabrata Acharya** received Ph.D. degree in Electronics & Communication Engineering from National Institute of Technology Rourkela India and is currently Associate Professor in Electronics & Communication Engineering at National Institute of Technology Rourkela India. He has served at Center for Development of Telematics (C-DOT), Bangalore, Wipro Technologies, Bangalore and Tata Elxsi Ltd, Bangalore under various capacities in the design and product development. His research interests include Integrated Circuits Design, Embedded System Development, Internet of Things (IoT) and Intelligent Signal Processing.

**Paulo Nocera Alves, Jr.** is currently a Ph.D. candidate at Production Engineering from University of São Paulo (USP) - School of Engineering of São Carlos (EESC). Master's at Production Engineering from São Paulo State University (UNESP) - College of Engineering of Bauru (FEB), 2014, and Bachelor's at Production Engineering from University of Franca (UNIFRAN), 2011. He has experience in Production Engineering, acting on the following subjects: information management, inventory control, operations research, optimal control theory, data envelopment analysis, multivariate analysis, and quantitative methods.

**Alperen Bal** is a Ph.D. candidate at Istanbul Technical University where he is currently a research assistant. His main areas of research interest are maintenance management, lean production and lean logistics.

**Freimut Bodendorf** holds the Chair of Information Systems (Services, Processes, Intelligence) and is Director of the Institute of Information Systems at the University Erlangen-Nuremberg. After completing his computer science studies in 1977 he earned his Ph.D. in the area of Information Systems. Afterwards he worked at various universities in Germany and Switzerland. Prior to his position at the University Erlangen-Nuremberg he was head of the Institute for Computer Science at the University of Fribourg/ Switzerland. Freimut Bodendorf is a member of many international scientific organizations and author of more than 250 refereed publications in journals, books, and conference proceedings.

**Elmer Pablo Tito Cari** received the Electrical Engineering degree (1999) from "Universidad Nacional de San Agustin de Arequipa", Perú, M.Sc (2005) and Ph.D. (2009) degrees from University of Sao Paulo (USP), Brazil. He is currently an Assistant Professor at Sao Paulo University. He also held Visiting Professor Fellow at the Duisburg-Essen University (Germany) in 2012 and 2015. His research interests include model identification and optimization in Electric Power Systems.

**Emre Çevikcan** is currently an associate professor of Industrial Engineering Department in Istanbul Technical University. He received the BS degree in Industrial Engineering from Yıldız Technical University, the MSc degree and PhD degree in Industrial Engineering from Istanbul Technical University. He studied the scheduling of production systems for his PhD dissertation. His research has so far focused on the design of production systems (assembly lines, production cells, etc.), lean production, scheduling. He has several research papers in International Journal of Production Research, Computers and Industrial Engineering, Assembly Automation, Expert Systems with Applications, International Journal of Information Technology & Decision Making. He is currently reviewer in OMEGA, European Journal of Operational Research, International Journal of Production Research, Applied Soft Computing, Journal of Intelligent Manufacturing and Journal of Intelligent and Fuzzy Systems.

**David Dal Molin** is a project associate of IPros SARL. He owns a Master's degree in Mechanical Engineering and a minor in MTE (Management, Technology and Entrepreneurship) from EPFL (Swiss Federal Institute of Technology at Lausanne). His interests are mainly focused on Operational and Supply Chain management. His work centers on developing active solutions for pharmaceutical, machine and watch-making industry with the help of simulation tools.

**Onur Doğan** has a master degree on Industrial Engineering at Istanbul Technical University. He has started doctorate in 2014 and worked as research assistant at the same university since 2011. He is closely interested in data mining, knowledge management, lean manufacturing, quality management tools. He provides training for process management, conflict management, lean six sigma, quality tools (QFD, FMEA, DOE) and TRIZ for companies. Also, Onur Dogan plays as a moderator and product developer in an innovative training method, occupational health and safety and lean transformation training with forum theatre.

**Mehmet Bülent Durmusoglu** is a Professor of Industrial Engineering at Istanbul Technical University. He obtained his Ph.D. in industrial engineering from the same university. His research interests are design and implementation of cellular/lean manufacturing systems. He is also the author numerous technical articles in these areas.

**Ahmet Erdoğan** works as an engineer in Republic of Turkey Ministry of National Defence. He graduated with a BS from Erciyes University in 2011, a MS from the Istanbul Technical University in 2015 and now continues PhD in Industrial Engineering Program in Yıldız Technical University. His research interest is focused on supply chain management.

**Fatemi Ghomi** was born in Ghom, Iran, in 1952. He received his B.S. degree in Industrial Engineering from Sharif University, Tehran, in 1973, and a Ph.D. degree in Industrial Engineering from Bradford University, England in 1980. He is currently Professor in the Department of Industrial Engineering at Amirkabir University of Technology in Tehran. His research and teaching interests include: Stochastic Activity Networks, Production Planning and Control, Scheduling, Queuing Systems, and Statistical Quality Control.

**Reza Ghasemy Yaghin** is currently Assistant Professor in Clothing Engineering and Management Group, Department of Textile Engineering at Amirkabir University of Technology, Tehran, Iran. His main research interests include production and inventory systems, pricing theory, fuzzy optimization and textile supply chains.

**Rémy Glardon** is Honorary Professor of the Swiss Federal Institute of Technology at Lausanne where he was the founder and director of the Laboratory for Production Management and Processes. He owns a Master in Mechanical Engineering and a PhD in Materials Science from the Swiss Federal Institute of Technology at Lausanne. After his graduate study at EPFL, he spent several years as a post-doc at UC-Berkeley where he published several research papers on the mechanical behavior and wear of materials. He then moved to industry and occupied management positions in Operations and Supply Chain Management. His main interests are related to the design, configuration and strategic, tactical & operational management of value adding networks. Rémy Glardon has initiated and completed several research and consulting projects with private companies in these areas, aiming at improving the cost/performance ratio in operations management. He is the co-founder and managing partner of the consulting company Ipros.

**Omer Faruk Gurcan** is a Research Assistant of Industrial Engineering at Istanbul Technical University (ITU). He graduated with a BS from Selcuk University in 2006, a MS from the ITU and now continues PhD in Industrial Engineering Program in ITU. His research interests are focused on knowledge and technology management, decision making techniques and some statistical applications.

**Mariem Gzara** obtained the Engineering Diploma on Production at the "ENSAIT" (French "Grande Ecole") in 1998 and the Master degree on automatic control and industrial computing in 1998 from the University of Lille 1, France. She received a PhD in automatic control and industrial computing from the University of Lille 1, France in 2001. She is currently an Associate Professor in Computer Science at the Higher School of Computer Science and Mathematics of Monastir, University of Monastir, Tunisia. She is a member of the Multimedia, Information systems and Advanced Computing Laboratory (Mir@cl), University of Sfax. Her research interests include data mining techniques, Optimization, sParallelization, Distributed Computation and Information Retrieval.

**Matthias Hafner** is a research assistant at the Chair of Information Systems III at the University of Erlangen-Nuremberg and an external Ph.D. student at the German company evosoft. His research focuses on maturity model building for SMEs, especially in the context of digitalization and Industrie 4.0. Matthias Hafner holds a master's degree in International Information Systems. During several work placements and apprenticeships, he was able to gain insights into the methods and the acquaintance of small and medium sized enterprises with the topics digitalization and Industrie 4.0, in order to strengthen his studies.

**Choon Sin Ho** received his degree in Mechanical Engineering from Universiti Tun Hussein Onn Malaysia, Johor, Malaysia in year 2016. He is currently a Ph.D candidate in Universiti Tun Hussein Onn Malaysia, Johor, Malaysia.

**Emek Köksoy Atiker** was born in Düsseldorf, Germany in 1988. She received the B.E. degree in industrial engineering from Yıldız Technical University, İstanbul, Turkey in 2011 and master degree in industrial engineering from İstanbul Technical University, İstanbul, Turkey in 2014. She started to work as industrial engineer at Ford Otosan in June, 2012 when she studied at master program. She has been in her current position as business analyst at Intertech Information Technology and Marketing since January 2014. She is responsible in treasury IT projects in core banking system.

**Matthias Lederer** is head of the Competence Center Processes at the Institute for Information Systems at the University of Erlangen-Nuremberg. His major field of research and studies is business process management and especially the strategy alignment as well as performance and compliance management of IT-enabled workflows. Mr. Lederer holds a PhD. information systems. As a BPM consultant at the German industrial manufacturing company REHAU as well as in several work placements and a trainee program, he earned practical experiences in the fields of workflow management and software engineering.

**Hadi Mosadegh** received his B.S. and M.S. degrees in Industrial Engineering from Yazd University and Amirkabir University of Technology (Iran), respectively. Currently, he is a PhD candidate of Industrial Engineering at Amirkabir University of Technology. His research interests are production and operations management, mixed-model assembly production planning as well as optimization and control techniques. He has published articles in, among others, International Journal of Production Research (IJPR) and Computers & Operations Research (COR).

**Fraj Naifar** Fraj NAIFAR is a senior researcher in Digital Reserch Center of Sfax, Tunisia. He obtained a Master in operation research from the University of Sfax, Tunisia. He is a PhD candidate in quantitative methods. His areas of research include combinatorial optimisation, flexible job-shop scheduling, packing, using heuristics and metaheuristics methods.

**Umakanta Nanda** received MTech and PhD degree in Electronics and Communication Engineering from National Institute of Technology, Rourkela, India in 2010 and 2017 respectively. Currently he is working as an Assistant Professor in Silicon Institute of Technology, Bhubaneswar, India. His research interests include design of analog and mixed signal integrated circuits, application specific processor design, and embedded systems design.

**Sina Niedermaier** received her master's degree in Industrial Engineering from the Friedrich-Alexander University of Erlangen-Nuremberg in 2016. During an internship in 2013 she worked as a purchaser for electro mechanical parts in the automotive industry at Robert Bosch LLC, Anderson, SC, United States. Afterwards she completed her bachelor thesis in the field of information systems for production planning and control at Robert Bosch GmbH, Bamberg. During her master studies she worked at the Fraunhofer Institute for Integrated Circuits, Nuremberg in the Group for Supply Chain Services. Furthermore, as a student employee she was in the engineering department for mainline transports at the Siemens AG, Erlangen. She is also member of the Siemens talent program YOLANTE.

**Mahir Öner** received his BS degree in ITU. Currently, he is working as a research assistant in Istanbul Technical University and he is a PHD candidate in the same university. His main research areas are real time tracking systems, RFID, Industry 4.0 applications.

**Sultan Ceren Öner** received her MS degree in Industrial Engineering Department from Çukurova University in 2011. In 2011, she was hired as a research assistant in Istanbul Technical University and is currently a PhD candidate in the same university. She writes and presents widely on issues of location based systems, data mining and fuzzy logic.

**Pauline Ong** received her Ph.D in applied mathematics from Universiti Sains Malaysia, Penang, Malaysia in year 2011, with specialiation in artificial neural networks. She is currently a senior lecturer with Universiti Tun Hussein Onn Malaysia, Johor, Malaysia. Her current research interests include artificial neural networks, mathematical modeling, and metaheuristic algorithm.

**Ercan Oztemel** is currently a Professor of Industrial Engineering in Marmara University. His research include artificial intelligence, simulation and modelling, knowledge based systems, strategic planning, quality engineering, decision support systems etc. He has already carried out various national and international projects and has many publications. Between 1993-2010, while he is teaching, he also worked for Turkish Science and Engineering Council doing AI and simulation related research. Meanwhile, he was scientific panel member of NATO SAS Panel and Steering Committee Member of WEAG (Western European Armament Group) Research Cell for CEPA 11 and CEPA 15. Between 2010-2015, he was the deputy president of Turkish Measurement, Selection and Placement Centre of Turkey. He has being taking part in the committees of various scientific journals (including Journal of Intelligent Manufacturing, Robotics and Autonomous Systems, Production Planning and Control, International Journal of Intelligent Systems and Applications in Engineering, etc.).

**Prakash Rout** received the Ph.D. degree from the department of Electronics and Communication Engineering, National Institute of Technology, Rourkela, India, and is currently working as an Associate Professor in the department of Applied Electronics and Instrumentation Engineering, Silicon Institute of Technology, Bhubaneswar, India. His research interests include analog and mixed signal integrated circuit design & optimization and computational intelligence.

**Mousumi Roy** is an Assistant Professor in Residence for the Management and Engineering for Manufacturing (MEM) program at the University of Connecticut. She currently teaches courses in Manufacturing Systems and Automation. Earlier, she had taught courses in Engineering, Business Statistics, Engineering Management and Mathematics courses at Johns Hopkins, Penn State at Worthington Scranton, and at Wilkes University.

**Peiman A. Sarvari** (PhD) is a researcher at Luxembourg Institute of Science and Technology (LIST). His major research area is logistics, big data analytics, and machine learning. He published numerous book sections and papers on different areas of Industrial Engineering.

**Sule Itir Satoglu** earned her master's and PhD degrees in Industrial Engineering, from Istanbul Technical University. She has been working as a faculty member of the Industrial Engineering Department in Istanbul Technical University, since 2009. Her interests include lean production, logistics and supply chain management, production and maintenance planning.

**Peter Schott** is a research assistant at the Chair of Information Systems II at the University of Erlangen-Nuremberg and consultant at the German industrial manufacturing company REHAU. His research emphasizes process and production optimization and complexity management, especially in the context of digitalization and the internet of things. Peter Schott holds a master's degree in industrial engineering and management. During several work placements, he gained insights into the process and production departments of globally active, large-scale enterprises and was able to underpin his theoretical knowledge with practical know-how.

**Ayse Aycim Selam** is a lecturer in the Industrial Engineering Department of Marmara University. She holds a BSc. degree in Mechanical Engineering, and a MSc. in Engineering Management from Istanbul Technical University, and a PhD. degree in Industrial Engineering from Marmara University. Her main research areas include bee colony optimization, project scheduling, sustainable project management, renewable energy, and sustainable energy.

**Christos Tsagkalidis** Managing Partner at IPros, Advisor at EPFL, IMD alumnus, World Economic Forum Global Shaper.

**Fatma Betül Yenı** is a PhD student at Istanbul Technical University. She holds a BSc. degree in Industrial Engineering from Karadeniz Technical University and also working as a research assistant at the university. Her major research areas include heuristics, simulation and fuzzy logic.

**Nicolas Zufferey** is a full professor of operations management at the University of Geneva in Switzerland. His research activities are focused on designing metaheuristics for difficult and large optimization problems, with applications mainly in transportation, scheduling, production, inventory management, network design, supply chain management and telecommunications.

# Index

Stay Current on the Latest Emerging Research Developments

# Become an IGI Global Reviewer for Authored Book Projects

## The overall success of an authored book project is dependent on quality and timely reviews.

In this competitive age of scholarly publishing, constructive and timely feedback significantly decreases the turnaround time of manuscripts from submission to acceptance, allowing the publication and discovery of progressive research at a much more expeditious rate. Several IGI Global authored book projects are currently seeking highly qualified experts in the field to fill vacancies on their respective editorial review boards:

**Applications may be sent to:**
development@igi-global.com

Applicants must have a doctorate (or an equivalent degree) as well as publishing and reviewing experience. Reviewers are asked to write reviews in a timely, collegial, and constructive manner. All reviewers will begin their role on an ad-hoc basis for a period of one year, and upon successful completion of this term can be considered for full editorial review board status, with the potential for a subsequent promotion to Associate Editor.

If you have a colleague that may be interested in this opportunity, we encourage you to share this information with them.

www.ingramcontent.com/pod-product-compliance
Lightning Source LLC
LaVergne TN
LVHW062107300926
849161LV00010B/2388
*9781522529446*